AF606301

APPLIED VIROLOGY RESEARCH

Volume 2

Virus Variability, Epidemiology, and Control

APPLIED VIROLOGY RESEARCH

Volume 1 NEW VACCINES AND CHEMOTHERAPY
Edited by Edouard Kurstak, R. G. Marusyk, F. A. Murphy, and M. H. V. Van Regenmortel

Volume 2 VIRUS VARIABILITY, EPIDEMIOLOGY, AND CONTROL
Edited by Edouard Kurstak, R. G. Marusyk, F. A. Murphy, and M. H. V. Van Regenmortel

---

APPLIED VIROLOGY RESEARCH

Volume 2

Virus Variability, Epidemiology, and Control

Edited by

Edouard Kurstak

University of Montreal
Montreal, Quebec, Canada

R. G. Marusyk

University of Alberta
Edmonton, Alberta, Canada

F. A. Murphy

Centers for Disease Control
Atlanta, Georgia

and

M. H. V. Van Regenmortel

Institute of Molecular and Cellular Biology
Strasbourg, France

PLENUM MEDICAL BOOK COMPANY
NEW YORK AND LONDON

Library of Congress Cataloging-in-Publication Data

Virus variability, epidemiology, and control / edited by Edouard
Kurstak ... [et al.].
p. cm. -- (Applied virology research ; v. 2)
Includes bibliographical references.
Includes index.
ISBN 0-306-43359-1
1. Viruses--Variation. 2. Viral genetics. 3. Virus diseases-
-Epidemiology. I. Kurstak, Edouard. II. Series.
[DNLM: 1. Virus Diseases--epidemiology. 2. Virus Diseases-
-prevention & control. 3. Viruses--genetics. W1 AP516R v. 2 / WC
500 V8225]
QR392.V57 1990
616'.0194--dc20
DNLM/DLC
for Library of Congress 90-7440
CIP

233 Spring Street, New York, N.Y. 10013

Plenum Medical Book Company is an imprint of Plenum Publishing Corporation

Printed in the United States of America

Contributors

D. J. Alexander • Poultry Department, Central Veterinary Laboratory, Weybridge, New Haw, Surrey KT15 3NB, England

A. Al-Tuwaijri • Department of Microbiology and Immunology, Faculty of Medicine, University of Montreal, Montreal, Quebec, H3C 3J7 Canada

M. E. Andrew • CSIRO Australian Animal Health Laboratory, Geelong, Victoria 3220, Australia

A. R. Bellamy • Department of Cellular and Molecular Biology, University of Auckland, Auckland, New Zealand

C. C. Bergmann • Department of Cellular and Molecular Biology, University of Auckland, Auckland, New Zealand

G. W. Both • CSIRO Division of Biotechnology, Laboratory for Molecular Biology, North Ryde, NSW 2113, Australia

D. B. Boyle • CSIRO Australian Animal Health Laboratory, Geelong, Victoria 3220, Australia

Lynn G. Bruce • Chemical Defence Establishment, Porton Down, Salisbury SP4 0JQ, England

Susan Carpenter • Laboratory of Persistent Viral Diseases, Rocky Mountain Laboratories, Hamilton, Montana 59840; *present address*: Department of Veterinary Microbiology and Preventive Medicine, Iowa State University, Ames, Iowa 50011.

Mary E. Chamberland • Division of HIV/AIDS, Center for Infectious Diseases, Centers for Disease Control, Public Health Service, U.S. Department of Health and Human Services, Atlanta, Georgia 30333

Bruce Chesebro • Laboratory of Persistent Viral Diseases, Rocky Mountain Laboratories, Hamilton, Montana 59840

May C. Chu • Division of Vector-Borne Viral Diseases, Center for Infectious Diseases, Centers for Disease Control, Public Health Service, U.S. Department of Health and Human Services, Fort Collins, Colorado 80522

J. M. Coffin • Tufts University School of Medicine, Department of Molecular Biology and Microbiology, Boston, Massachusetts 02111

Nancy J. Cox • Influenza Branch, Division of Viral Diseases, Center for Infectious Diseases, Centers for Disease Control, Atlanta, Georgia 30333

James W. Curran • Division of HIV/AIDS, Center for Infectious Diseases, Centers for Disease Control, Public Health Service, U.S. Department of Health and Human Services, Atlanta, Georgia 30333

Julian I. Delic • Chemical Defence Establishment, Porton Down, Salisbury SP4 0JQ, England

Esteban Domingo • Centro de Biología Molecular, Universidad Autónoma de Madrid, Canto Blanco 28049, Madrid, Spain

Joaquín Dopazo • Laboratorio de Genética, Universidad de Valencia, 46100 Burjassot, Valencia, Spain

Stephen M. Eley • Chemical Defence Establishment, Porton Down, Salisbury SP4 0JQ, England

Leonard H. Evans • Laboratory of Persistent Viral Diseases, Rocky Mountain Laboratories, Hamilton, Montana 59840

George E. Fox • University of Houston, Department of Biochemical and Biophysical Sciences, Houston, Texas 77004

Matthew A. Gonda • Laboratory of Cell and Molecular Structure, Program Resources, Inc., NCI-Frederick Cancer Research Facility, Frederick, Maryland 21701

Maurice W. Harmon • Influenza Branch, Division of Viral Diseases, Center for Infectious Diseases, Centers for Disease Control, Atlanta, Georgia 30333

Robert M. Henstridge • Chemical Defence Establishment, Porton Down, Salisbury SP4 0JQ, England

A. Hossain • Department of Microbiology and Immunology, Faculty of Medicine, University of Montreal, Montreal, Quebec, H3C 3J7 Canada

David L. Huso • Division of Comparative Medicine, The Johns Hopkins University School of Medicine, Baltimore, Maryland 21205

Alan P. Kendal • Influenza Branch, Division of Viral Diseases, Center for Infectious Diseases, Centers for Disease Control, Atlanta, Georgia 30333

Olen M. Kew • Division of Viral Diseases, Center for Infectious Diseases, Centers for Disease Control, Atlanta, Georgia 30333

Srisakul C. Kliks • University of California, School of Public Health, Department of Biomedical and Environmental Health Sciences, Berkeley, California 94705

C. Kurstak • Department of Microbiology and Immunology, Faculty of Medicine, University of Montreal, Montreal, Quebec, H3C 3J7 Canada

E. Kurstak • Department of Microbiology and Immunology, Faculty of Medicine, University of Montreal, Montreal, Quebec, H3C 3J7 Canada

Charles L. Manske • University of Houston, Department of Biochemical and Biophysical Sciences, Houston, Texas 77004

Miguel A. Martínez • Departamento de Sanidad Animal, Instituto Nacional de Investigaciones Agrarias, 28012 Madrid, Spain

Mauricio G. Mateu • Centro de Biología Molecular, Universidad Autónoma de Madrid, Canto Blanco 28049, Madrid, Spain

Thomas P. Monath • SGRD-UIV, Virology Division, USAMRIID, Fort Derrick, Frederick, Maryland 21701–5011.

Norman F. Moore • Chemical Defence Establishment, Porton Down, Salisbury SP4 0JQ, England

Andrés Moya • Laboratorio de Genética, Universidad de Valencia, 46100 Burjassot, Valencia, Spain

Brian Murphy • Laboratory of Infectious Diseases, National Institutes of Health, Bethesda, Maryland 20892

Opendra Narayan • Division of Comparative Medicine and Department of Neurology, The Johns Hopkins University School of Medicine, Baltimore, Maryland 21205

Baldev K. Nottay • Division of Viral Diseases, Center for Infectious Diseases, Centers for Disease Control, Atlanta, Georgia 30333

Peter Palese • Department of Microbiology, Mount Sinai School of Medicine, New York, New York 10029

Mark A. Pallansch • Division of Viral Diseases, Center for Infectious Diseases, Centers for Disease Control, Atlanta, Georgia 30333

Rebecca Rico-Hesse • Division of Viral Diseases, Center for Infectious Diseases, Centers for Disease Control, Atlanta, Georgia 30333; *present address*: Department of Epidemiology and Public Health, Yale University School of Medicine, New Haven, Connecticut 06510

P. H. Russell • Department of Veterinary Pathology, The Royal Veterinary College, London NW1 0TU, England

A. C. R. Samson • Department of Biochemistry and Genetics, University of Newcastle upon Tyne, Newcastle upon Tyne NE2 4HH, England

M. Sevoian • Department of Veterinary and Animal Sciences, University of Massachusetts, Amherst, Massachusetts 01002

Francisco Sobrino • Departamento de Sanidad Animal, Instituto Nacional de Investigaciones Agrarias, 28012 Madrid, Spain

S. C. Stirzaker • CSIRO Division of Biotechnology, Laboratory for Molecular Biology, North Ryde, NSW 2113, Australia

John Treanor • Department of Medicine, Infectious Disease Unit, University of Rochester, Rochester, New York 14642

Dennis W. Trent • Division of Vector-Borne Infectious Diseases, Center for Infectious Diseases, Centers for Disease Control, Public Health Service, U.S. Department of Health and Human Services, Fort Collins, Colorado 80522

David Arthur John Tyrrell • Medical Research Council Common Cold Unit, Harvard Hospital, Salisbury, Wilts SP2 8BW, England

Guido van der Groen • Institute of Tropical Medicine, 2000 Antwerp, Belgium

Kathleen van Wyke Coelingh • Laboratory of Infectious Diseases, NIAID, National Institutes of Health, Bethesda, Maryland 20892; *present address:* Protein Design Labs, Inc., Palo Alto, California 94304

Makoto Yamashita • Bioscience Research Laboratories, Sankyo Co. Ltd., Shinagawa-Ku, Tokyo, Japan

Preface to the Series

Viral diseases contribute significantly to human morbidity and mortality and cause severe economic losses by affecting livestock and crops in all countries. Even with the preventive measures taken in the United States, losses caused by viral diseases annually exceed billions of dollars. Five million people worldwide die every year from acute gastroenteritis, mainly of rotavirus origin, and more than one million children die annually from measles. In addition, rabies and viral hepatitis continue to be diseases of major public health concern in many countries of the Third World, where more than 200 million people are chronically infected with hepatitis B virus. The recent discovery of acquired immunodeficiency syndrome (AIDS), which is caused by a retrovirus, mobilized health services and enormous resources. This virus infection and its epidemic development clearly demonstrate the importance of applied virology research and the limits of our understanding of molecular mechanisms of viral pathogenicity and immunogenicity.

The limitations of our knowledge and understanding of viral diseases extend to the production of safe and reliable vaccines, particularly for genetically unstable viruses, and to antiviral chemotherapy. The number of antiviral drugs currently available is still rather limited, despite extensive research efforts. The main problem is finding compounds that selectively inhibit virus replication without producing toxic effects on cells. Indeed, the experimental efficacy of several drugs, for example, new nucleoside derivatives, some of which are analogues of acyclovir, makes it clear that antiviral chemotherapy must come of age because many new compounds show promise as antiviral agents. In the field of antiviral vaccine production, molecular biologists are using a wide variety of new techniques and tools, such as genetic engineering technology, to refine our understanding of molecular pathogenicity of viruses and of genetic sequences responsible for virulence. Identification of genes that induce virulence is vital to the construction of improved antiviral vaccines.

Novel types of vaccines are presently receiving particular attention. For example, the protein that carries the protective epitopes of hepatitis B virus, which is produced by expressing the appropriate viral gene in yeast or in mammalian cell systems, is now available.

Another group of new vaccines are produced by using vaccinia virus as a vector for the expression of genes of several viruses. Vaccines for rabies, influenza, respiratory syncytial disease, hepatitis B, herpes infection, and AIDS, which are based on greatly enhanced expression of the viral genes in vaccinia virus, are being tested. Also of interest is baculovirus, an insect cell vector system now used in the development of recombinant DNA vaccines for a variety of important human and animal virus diseases. This system yields very large quantities of properly processed and folded proteins from the rabies, hepatitis B, AIDS, and Epstein-Barr viruses.

The synthetic peptides, which act as specific immunogens, have also received attention as new antiviral vaccines. The recent experimental performance of new synthetic peptides of foot-and-mouth disease virus, as well as peptide-based vaccines for poliovirus, rotavirus, hepatitis B, and Venezuelan equine encephalitis virus, gives strong support for this group of specific immunogens. However, testing of these synthetic peptide vaccines is in the early stages and future research will have to answer several questions about their safety, efficacy, and immune responses.

Current attempts at developing synthetic vaccines are based either on recombinant DNA technology or on chemical-peptide synthesis. Several virus proteins have been produced in bacterial, yeast, or animal cells through the use of recombinant DNA technology, while live vaccines have been produced by introducing relevant genes into the genome of vaccinia virus. By using solid-phase peptide synthesis, it has been possible to obtain peptides that mimic the antigenic determinants of viral proteins, that elicit a protective immunity against several viruses. Both the chemical and the recombinant approaches have led to the development of experimental vaccines. It should become clear within a few years which approach will lead to vaccines superior to the ones in use today.

The recent development of monoclonal antibody production techniques and enzyme immunoassays permits their application in virology research, diagnosis of viral diseases, and vaccine assessment and standardization. These techniques are useful at different stages in the development of vaccines, mainly in the antigenic characterization of infectious agents with monoclonal antibodies, in assessment schemes in research and clinical assays, and in production.

This new series, entitled *Applied Virology Research,* is intended to promote the publication of overviews on new virology research data, which will include within their scope such subjects as vaccine production, antiviral chemotherapy, diagnosis kits, reagent production, and instrumentation for automation interfaced with computers for rapid and accurate data processing.

We sincerely hope that *Applied Virology Research* will serve a large audience of virologists, immunologists, geneticists, biochemists, chemists, and molecular biologists, as well as specialists of vaccine production and experts of health services involved in the control and treatment of viral diseases of plants, animals, and man. This series will also be of interest to all diagnostic laboratories, specialists, and physicians dealing with infectious diseases.

Edouard Kurstak

Montreal, Canada

Preface to Volume 2

The limitations of our knowledge of viral diseases extend to the production of safe and reliable vaccines, particularly for genetically unstable viruses like influenza viruses or retroviruses. Despite new techniques and tools, such as genetic engineering, the variability of several viruses remains the major obstacle in the production of efficient vaccines. Identification of the genes that induce virulence and our understanding of their variability and role in the pathogenicity of viruses are vital to the construction of improved antiviral vaccines.

This second volume in the series *Applied Virology Research* is devoted to virus variability, epidemiology, and control of viral diseases, and emphasizes the central role played by the intrinsic variability of viral genomes in determining the biological properties of viruses. Sequencing studies of picornaviruses, influenza viruses, and certain bacteriophages have clearly established the quasispecies nature of their RNA genomes. That is, each individual viral clone, derived from a single virion, always contains a master genome sequence together with a large number of mutants appearing at a rate of about 10^{-3} per nucleotide per replication cycle. This mixed population of variant genomes, including the master genome, constitutes the viral quasispecies and is the collective target that is subjected to natural selection. Viral adaptation corresponds to a shift in the quasispecies equilibrium, resulting from interaction with an altered environment or a new ecological niche.

This volume describes our current knowledge of the intrinsic variability present in the major families of vertebrate viruses and illustrates the relevance of genome and antigen variability for understanding viral epidemiology and controlling viral diseases. A similar treatise has not been previously available.

The second volume of *Applied Virology Research* comprises 18 chapters summarizing the latest research data related to the genetic organization, mutations, and recombinations of several viruses of importance to medical and veterinary practice. The mutation, recombination, and selection events are discussed in depth in relation to pathogenicity and to the understanding of virus evolution and the need for new vaccines based on the molecular epidemiology of viruses. Special attention is given to the variation of human immunodeficiency virus in the context of genetic evolutionary changes observed for several other retroviruses. Detailed information is presented on how viruses escape from immune surveillance, with specific examples given.

The contributors to this volume are leading experts in the genetics and molecular epidemiology of viruses, as well as in vaccine production. The editors would like to express their sincere gratitude to them for the effort and care with which they prepared their well-documented chapters.

This volume will be a useful tool for all those concerned with the epidemiology and control of viral diseases, as well as with virus pathogenicity, genetics, immune response, and the strategy of virus vaccine application.

The editors' thanks are also addressed to the staff of Plenum Publishing Corporation for their efforts in the production of this series.

E. Kurstak
Montreal, Quebec, Canada
R. G. Marusyk
Edmonton, Alberta, Canada
F. A. Murphy
Atlanta, Georgia
M. H. V. Van Regenmortel
Strasbourg, France

Contents

IV. VIRUS HEMORRHAGIC FEVERS: DIVERSITY AND EPIDEMIOLOGY

Introduction

Virus Variability and Impact on Epidemiology and Control of Diseases

E. Kurstak and A. Hossain

I. INTRODUCTION

An important number of virus infections and their epidemic developments demonstrate that ineffectiveness of prevention measures is often due to the mutation rate and variability of viruses (Kurstak *et al.*, 1984, 1987). The new human immunodeficiency retroviruses and old influenza viruses are only one among several examples of virus variation that prevent, or make very difficult, the production of reliable vaccines. It could be stated that the most important factor limiting the effectiveness of vaccines against virus infections is apparently virus variation. Not much is, however, known about the factors influencing and responsible for the dramatically diverse patterns of virus variability.

II. MUTATION RATE AND VARIABILITY OF HUMAN AND ANIMAL VIRUSES

Mutation is undoubtedly the primary source of variation, and several reports in the literature suggest that extreme variability of some viruses may be a consequence of an unusually high mutation rate (Holland *et al.*, 1982; Domingo *et al.*, 1985; Smith and Inglis, 1987). The mutation rate of a virus is defined as the probability that during a single replication of the virus genome a particular nucleotide position is altered through substitution, deletion, insertion, or recombination. Different techniques have been utilized to measure virus mutation rates, and these have been noted to vary in their accuracy and in the extent of application to different viruses.

A. Direct Methods

One of the most direct techniques is to measure the error rate *in vitro* of the polymerase responsible for the replication of the viral genome. Error rate between $10^{-2.4}$ and $10^{2.8}$ has been estimated for the reverse transcriptase encoded by retroviruses based on the frequency of nucleotide incorporation during transcription *in vitro* of homopolymers of ribonucleotides or deox-

E. Kurstak and A. Hossain • Department of Microbiology and Immunology, Faculty of Medicine, University of Montreal, Montreal, Quebec, Canada H3C 3J7.

yribonucleotides (Loeb and Kunkel, 1982), with polymerase error rates and nucleotide substitution frequencies recalculated from the published data so that they represent the frequency at which a given nucleotide at a particular position in a virus genome is replaced by any of the other three nucleotides. Using a different technique, the error rate of the vesicular stomatis virus (VSV) RNA polymerase has been estimated to be $10^{3.15}$ (Steinhauer and Holland, 1986). It would be of interest to see these techniques applied to other viruses. Virus mutation rates can also be estimated from the frequency of nucleotide substitution in the genomes of a population of virus derived from a single purified clone. Such estimates may not accurately reflect virus mutation rates since some mutant genomes will be inviable and since the number of replicative events involved in the production of a virus population may be uncertain. Nevertheless, it is clear that the analysis of nucleotide substitution frequencies in virus genomes will provide useful information about the relative variability of different viruses.

Nucleotide substitution frequencies can also be measured by direct nucleotide sequence analysis of numerous virus clones each derived from the same plaque: A frequency of $10^{-4.1}$ has been obtained for the influenza virus nonstructural (NS) segment, and a significantly lower frequency of less than $10^{-5.0}$ for the poliovirus type 1 VP1 gene (Parvin *et al.*, 1986). Although quantitation is difficult, both the influenza virus NS segment and the poliovirus VP1 gene are capable of variation, and thus this different substitution frequency may reflect a difference in mutation rates of the two viruses.

B. Indirect Methods

Substitution frequencies can also be estimated from the frequency of variants displaying a particular phenotype in a cloned virus stock. Phenotypes that can be influenced by many different mutations in several genes, such as temperature-sensitive (ts) growth, plaque morphology, host range, and pathogenicity, will be of little use for this type of analysis. Whichever method is used to measure virus mutation rates, it should be noted that different regions of a virus genome may have different mutation rates because of effects of secondary structure on the fidelity of replication (Weddell *et al.*, 1986). The mutation rate may also depend on whether a transition or transversion mutation is required for the variant phenotype. Different isolates of the same virus may even have different mutation rates if they encode altered forms of the virus polymerase (Hall *et al.*, 1984). Furthermore, sources of mutation other than nucleotide misincorporation, such as deletion, insertion, duplication, and inversion, will each occur at a particular frequency (Buonagurio *et al.*, 1984). Both deletion and insertion are involved in the generation of antigenic diversity, as evidenced by nucleotide sequence analysis of influenza virus (Buonagurio *et al.*, 1958) and human immunodeficiency virus (HIV), responsible for acquired immunodeficiency syndrome (AIDS) (Hahn *et al.*, 1986).

Given the many difficulties involved in the measurement of virus mutation rates, it would be of interest to see whether any trends emerge from the data available in the literature. Different methods of measuring virus mutation rates can sometimes give very different results, and since several sources of inaccuracy could be associated with indirect measurements of substitution frequencies (Smith and Inglis, 1987), ideally comparison between viruses would be based on direct measurements of the mutation rate. At present, this is limited to four estimates, each for a different virus and made using three different methods (Loeb and Kunkel, 1982; Parvin *et al.*, 1986; Steinhauer and Holland, 1986).

III. FACTORS OF VIRUS VARIABILITY AS WELL AS EPIDEMIOLOGIC OUTBREAKS

All viruses have relatively high mutation rates and the rate of field variation of a virus may actually alter the intensity of selection by other factors. For example, while some viruses are accessible to immune attack throughout infection, other viruses, such as herpes simples virus (HSV), replicate in a site that is protected from immune attack and so only encounter antigenic selection in infecting a previously infected individual (Smith and Inglis, 1987). The intensity of selection will also be relatively low for viruses uncommon in the population since reinfection of individuals would be rare. On the other hand, a potential source of intense selection would be "founder effect," which could occur should infection be initiated by an extremely small number of virus particles (Buonagurio *et al.*, 1985). The rate of virus variation may also be dependent on the inherent properties of viral antigens (targets of immune attack) and may be affected by the extent of existence as cocirculating strains. Collectively, these factors may determine the rate of viral variation by either limiting or promoting the spread of variants in the population. Although the relatively high mutation rate of eukaryotic viruses is obviously important in generating a pool of mutant genomes from which new mutants can emerge, it is possible that differences between viruses in their rate of variation reflect differences in selective forces rather than differences in mutation rates (Smith and Inglis, 1987).

The oligonucleotide fingerprints of the genome RNA has offered a useful tool for the understanding of virus variability of yellow fever. Examination by oligonucleotide fingerprinting of the 40S genome RNA of isolates of yellow fever virus from Africa (Senegambia, Central African Republic, Ivory Coast, Burkina Faso) and from South America (Panama, Ecuador, Trinidad) revealed that geographically isolated and epidemiologically unrelated viruses were very distinct (Deube *et al.*, 1986). Likewise, by utilization of oligonucleotide fingerprinting, the genetic relatedness within Venezuelan equine encephalitis strains in Colombia has been shown to be a function of geography rather than epidemiology (Rico-Hesse *et al.*, 1988). The temporal incidence of respiratory viruses identified in children admitted to an Edinburgh hospital over a period of 14 years failed to find any evidence of interference between respiratory syncytial virus (RSV) and influenza virus and indicated that parainfluenza viruses were the least predictable in their epidemiologic behavior (Winter and Inglis, 1987). In a recent study (Hossain *et al.*, 1988), the cross-sectional seroepidemiology of respiratory infections caused by RSV, influenza A, B viruses, and parainfluenza types 1, 2, 3 viruses in a developing country with subtropical climatic conditions was assessed. The pattern of age-specific antibody prevalence rate demonstrated a unique situation with regard to infections with adenoviruses in this population with a rapid increase in the level of antibodies detected in infants 7–12 months of age and thereafter no progressive increase in the level of antibodies. Furthermore, it was also interesting to note that in the same population studies with other viruses such as Epstein-Barr virus (Hossain, 1987), HSV-1 (Hossain, 1989), and measles (Bakir *et al.*, 1988) indicated antibody levels comparable with those found in all three types of parainfluenza viruses in infants of 6 months and under. A study of Norwalk-like virus gastroenteritis (epidemic) indicated person-to-person transmission from the epidemic curve (Leers *et al.*, 1987).

IV. GENE REASSORTMENT, GENETIC RECOMBINATION WITH EXAMPLES OF VIRUSES

Genetic recombination with reassortment of genes in segmented RNA viruses has been well documented in the case of reoviruses (Hossain and Graham, 1978) and notably in influenza A viruses (Palese and Young, 1982). This variability of influenza viruses is in marked contrast to the antigenic properties of other viral agents, such as poliovirus or measles virus, which appear to remain essentially unchanged, and also differs from that of herpes viruses or rhinoviruses, which coexist with a number of variants in the population but do not undergo the rapid changes observed with influenza viruses.

Influenza viruses are capable of constantly changing the genes coding for their surface proteins as well as for their nonsurface proteins. The occurrence of antigenic changes has been attributed to the impression that these antigenic changes could give rise to pandemics of influenza at 10- to 15-

year intervals and to minor changes that cause epidemics at 1- to 2-year intervals (Kilbourne, 1975). The major changes that occur in the hemagglutinin (HA) or neuraminidase molecule are called antigenic shift and probably result from a recombination event between animal and human influenza viruses. The minor changes, called antigenic drift, involve gradual changes in the surface glycoproteins of influenza viruses and are thought to result from the selection of spontaneously occurring mutant visions by an immune population. In contrast, many other viruses, such as Sendai virus and VSV, appear to be antigenically stable with no antigenic variation being so far reported. Re-emergence of strains that had previously been in circulation could also lead to different subtypes of influenza viruses.

Events leading to variation within subtypes of influenza viruses could include (1) point mutations occurring in genes coding for surface as well as nonsurface proteins, (2) short deletions and insertions in genes coding for nonsurface proteins, and (3) reassortment leading to exchange of genes coding for nonsurface proteins. Antigenic shift occurs when HA is replaced by a new viral strain with an antigenically novel HA. Antigenic drift has been observed to occur with both influenza A and B but shift only in influenza A (Krystal *et al.*, 1983). Antigenic drift in influenza A and B occurs mainly through a series of single amino acid changes in HA and along a common evolutionary lineage. Furthermore, the rate of antigenic drift among influenza B appears to be lower than among influenza A viruses (Krystal *et al.*, 1983). Examination of genetic variation in the NS gene of influenza C viruses allowed the comparison of genetic stability over time of influenza A and C viruses. Extensive studies reported on the evolution of influenza A viruses isolated over long periods of time through comparative sequence analysis of NS genes (Chanock *et al.*, 1958) reveal that influenza A viruses undergo rapid evolutionary change with time and in contrast the C virus NS genes possibly evolve more slowly and that genes belonging to the same lineage may not exhibit mutations over a period of only one or two decades. Although the epidemiology of influenza C viruses is different from that of the A (and possibly B) strains, it may be similar to that of most other RNA and DNA viruses.

Human type 3 parainfluenza virus (PIV3), recognized since 1957 as a significant cause of acute lower respiratory diseases in humans (Portner *et al.*, 1980) and second only to RSV as a cause of bronchiolitis and pneumonia in infants, appears to be endemic worldwide with no strict seasonal pattern. Because of its significant role in childhood illness, there is considerable interest in development of an effective vaccine. From plaque-purified virus preparations antigenic variants of PIV3 with an average frequency of $10^{-6.1}$ have been reported, and these are similar to that obtained for other RNA viruses such as influenza A, Sendai, and VSV (Nishikawa *et al.*, 1983). The variation in PIV3 appears to represent genetic heterogeneity among strains and is in contrast to the progressive accumulation of antigenic changes with time characteristic of influenza A virus. Furthermore, the antigenic variation of PIV3 appears to correlate in some instances with the geographic origin of the strain in similarity to antigenic variation of field isolates of Newcastle disease virus (van Wyke Coelingh *et al.*, 1986). Analysis of the hemagglutinin-neuraminidase (HN) gene sequences of antigenic variants selected *in vitro* in the presence of cross-reactive hemagglution inhibition-monoclonal antibodies has indicated the overall degree of homology between the bovine and human PIV3 HN proteins to be higher than that between the human PVI3 HN and Sendai virus HN proteins and reflects their evolutionary relationships and single-point mutations identified in the HN gene (van Wyke Coelingh *et al.*, 1986).

Unlike antigenic variation in influenza with mutations responsible for widespread disease outbreaks and such antigenic variation driven by antibody selection being epidemiologically significant (Yewdell and Gerhard, 1981), it is unclear at this time whether antigenic variation of HSV can lead to changes in the immunobiology of natural infections. It is also not known whether rapid antigenic variation does indeed occur during the cause of infection with HSV. However, antigenic variation could contribute to outbreaks of recurrent infection and transmission of HSV to new hosts (Holland *et al.*, 1983).

The rates of fixation of mutations during the evolution of the foot-and-mouth disease virus

(FMDV) in nature have been estimated by hybridization of viral RNA to cloned cDNAs representing defined FMDV genome segments and compared by T_1 RNase oligonucleotide fingerprinting. The proportion of mutation between two viral RNAs did not increase significantly with the time elapsed between the two isolations, suggesting a cocirculation of multiple, related, nonidentical FMDVs (evolving quasispecies) as the mode of evolution of this agent. This emphasizes the heterogeneous nature of an FMDV population at any given location and time. The extensive genetic heterogeneity of field isolates suggests that in FMDV, antigenic variability most likely is a consequence of a general genetic variability affecting many RNA sites (Sobrino *et al.*, 1986).

Rabies viruses isolated from different animal species in various parts of the world were in the past considered to be antigenically closely related. Only when antibodies produced in animals immunized with whole virions or viral components were assayed by the plaque reduction method were some minor differences detected in the antigenic composition of the various rabies strains. The rabies virus was demonstrated to have a similar potential as influenza to undergo antigenic variation under suitable experimental conditions *in vitro* and essentially involve single-point mutations in the glycoprotein (Kurstak and Marusyk, 1984).

V. CONTROL OF VIRAL DISEASES BY VIRUSES WITH HIGH MUTATION RATES AND VACCINES AGAINST SUCH VIRUSES

The present era has seen the successful eradication of smallpox virus. Although poliomyelitis and measles eradication seems to be a distinct possibility, the prevention of epidemic and pandemic influenza undoubtedly underscores our capabilities at the present time. A number of factors may contribute to the problem of influenza control, but the major factor that has limited our ability to control the disease is the capacity of influenza viruses in nature to vary rapidly and undergo changes in antigenic structure (Kilbourne, 1975) that circumvent the protective effects of a patient's immune response. Variation in influenza viruses could well be a multifaceted phenomenon involving many different mechanisms. The influenza viruses show a uniquely high mutation frequency. Factors other than mutation rates may influence the selection of influenza virus variants in nature, thereby resulting in what appears to be a higher mutation rate. For example, in the case of influenza the immune response in some patients may lead to subneutralization conditions favoring the selection of variants through several cycles of virus replication (Palese and Young, 1982). Direct sequence analysis enabled the study of rate of mutation in tissue culture for the NS gene of influenza virus and VP1 gene of poliovirus type 1, with mutation rates of 1.5×10^{-5} and less than 2.1×10^{-6} mutations per nucleotide per infectious cycle being obtained for influenza and polioviruses, respectively. The higher rate of mutation rate of influenza virus than poliovirus possibly correlates well with the rapidity of evolution of this virus in nature and lack of vaccination. The probability of successful vaccination against a particular virus is usually determined by the mutation rate of that virus, and knowledge of this parameter is generally believed to influence the strategy of vaccine design. Thus, specifically vaccine strains used against influenza A viruses have to be changed frequently (at least every 2–3 years) to protect against an evolving virus population, in contrast to vaccines against poliovirus and other viruses based on strains used for the last several decades without loss in efficacy.

It has also been demonstrated that the rabies virus has a similar potential as influenza virus to undergo antigenic variation (Wiktor and Koprowski, 1980). However, unlike influenza, the existence of antigenic variants of rabies in nature only became fully apparent when monoclonal antibodies were applied to the analysis of field strains isolated in different parts of the world. In view of this, the selection of vaccine strains and the methods used to evaluate the potency of the rabies vaccines need to be looked into.

The ability of a virus to evade or compromise the immune response of an infected individual can greatly influence its pathogenesis and transmissibility. Recent evidence suggests that human immunodeficiency viruses may be able to do both. The surface glycoproteins of all retroviruses are quite similar in organization, although highly divergent in sequence, and appear to resemble the receptor binding proteins of a number of other enveloped viruses such as the HA protein of the influenza virus. The overall variation from one human immunodeficiency virus isolate to another does not seem particularly striking and is not at variance from other retroviruses. Variation is noticeable in variation concentrated within several "hypervariable" regions within *env*, particularly in the sequence coding for the larger glycoprotein. It has been postulated that the variability of these regions is responsible for significant antigenic variation of the virus (Coffin, 1986). The cost of variation observed with *env* gene of human immunodeficiency viruses perhaps suggests that there might be difficulties in the development of effective vaccines against the AIDS retroviruses.

The use of monoclonal antibodies to select neutralization-resistant mutants successfully identified the important antigenic sites of a number of viruses, including influenza, rabies, parainfluenza, and poliovirus. Utilization of this technique identified three regions of the rotavirus glycoprotein involved in serotype-specific neutralization. A practical application would be in the design and construction of expression vectors containing cloned glycoprotein genes and bacterial strains capable of expressing serotype-specific antigens for use as oral vaccines could be developed (Dyall-Smith *et al.*, 1986). This indicates that for adequate protection of infants, a human rotavirus vaccine should include all known human serotypes.

REFERENCES

Bakir, T. M. F., Hossain A., Ramia, S. and Sinha, N. P. (1988). *J. Trop. Paediatr.* **34,** 254–256.

Buonagurio, D. A., Krystal, M., Palese, O., De-Borde, D. C. and Massab, H. F. (1984). *J. Virol.* **49,** 418–425.

Buonagurio, D. A., Nakada, S., Desselberger, U., Krystal, M. and Palese, P. (1985). *Virology* **146,** 221–232.

Chanock, R. M., Parrott, R. H., Cook, K., Andrews, B. E., Bell, J. A., Reichelderfer, T., Kapikian, A. Z., Mastrola, F. M., and Heubner, R. J. (1958). *N. Engl. J. Med.* **258,** 207–213.

Coffin, J. M. (1986). *Cell* **46,** 1–4.

Deube, I. V., Digoutte, J. P., Monath, T. P., and Girard, M. (1986). *J. Gen. Virol.* **67,** 209–213.

Domingo, E., Martinez-Salas, E., Sobrino, F., DeLaTorre, C. D., Pontela, J., Ortin, J., Lopez-Galindez, C., Pentz-Bolna, P., Villanueva, N., Najera, R., Van De Pol, S., Steinhaueur, D., De Polo, N., and Holland, J. (1985). *Gene* **40,** 1–8.

Dyall-Smith, M. L., Lazdins, I., Tregear, G. W., and Holmes, I. H. (1986). *Proc. Natl. Acad. Sci. USA* **83,** 3465–3468.

Hahn, B. H., Shaw, G. M., Taylor, M. E., Redfield, R. R., Markhum, P. D., Salahuddin, S. Z., Wong-Staal, F., Gallo, R. C., Parks, E. S., and Parks, W. P. (1986). *Science* **232,** 1548–1553.

Hall, J. D., Coen, D. M., Fisher, B. L., Weisslitz, M., Randall, S., Almy, R. E., Gelez, P. T., and Schaffer, P. A. (1984). *Virology* **132,** 26–37.

Holland, J., Spindler, K., Horodyski, F., Grabau, E., Nichol, S. and van de Pol, S. (1982). *Science* **232,** 1548–1553.

Holland, T. C., Marlin, S. D., Levine, M., and Glorioso, J. (1983). *J. Virol.* **45,** 672–682.

Hossain, A. (1987). *J. Trop. Paediatr.* **33,** 257–260.

Hossain, A. (1989). *J. Trop. Paediatr.* **35,** 19–22.

Hossain, A., and Graham, A. F. (1978). In *Viruses and Environment* (E. Kurstak and K. Maramorosch, eds.), pp. 663–668, Academic Press, New York.

Hossain, A., Bakir, T. F., Zakzouk, S. M., and Sengupta, D. K. (1988). *Ann. Trop. Paediatr.* **8,** 108–111.

Kilbourne, E. D. (1975). *The Influenza Viruses and Influenza* (E. D. Kilbourne, ed.), pp. 483–538, Academic Press, New York.

Krystal, M., Buonagurio, D., Young, J. F., and Palese, P. (1983). *Proc. Natl. Acad. Sci. USA* **80,** 4527–4531.

Kurstak, E., and Marusyk, R. G. (1984). *Control of Virus Diseases,* 584 pages, Dekker, New York.

Kurstak, E., Al-Nakib, W., and Kurstak, C. (1984). *Applied Virology,* 518 pages, Academic Press, New York.

Kurstak, E., Marusyk, R. G., Murphy, F. A., and Van Regenmortel, M. H. V. (1987). *Applied Virology Research, Vol. 1, New Vaccines and Chemotherapy,* 306 pages, Plenum Press, New York.
Leers, W. D., Kasupski, G., Fralick, R., Wartman, S., Garcia, J., and Gany, W. (1987). *Publ. Hlth.* **77,** 291–295.
Loeb, L. A., and Kunkel, T. A. (1982). *Annu. Rev. Biochem.* **51,** 429–457.
Nishikawa, K., Isomura, S., and Suzuki, S. (1983). *Virology* **130,** 318–330.
Palese, P., and Young, J. F. (1982). *Science* **215,** 1468–1474.
Parvin, J. D., Moscona, A., Pan, W. T., Leider, J. M., and Palese, P. (1986). *J. Virol.* **59,** 377–383.
Portner, A., Webster, R. G., and Bean, W. J. (1980). *Virology* **104,** 235–238.
Rico-Hesse, R., Roehrig, J. T., Trent, D. W., and Dikerman, R. W. (1988). *Am. J. Trop Med. Hyg.* **38,** 195–204.
Smith, D. B., and Inglis, S. C. (1987). *J. Gen. Virol.* **68,** 2729–2740.
Sobrino, F., Palma, E. L., and Beck, E. (1986). *Gene* **50,** 149–159.
Steinhauer, D. A., and Holland, J. J. (1986). *J. Virol.* **57,** 219–228.
van Wyke Coelingh, K. J., Winter, C. C., Murphy, B. R., John, M. R., and Kimball, P. C. (1986). *J. Virol.* **60,** 90–96.
Weddell, G. N., Yansura, D. G., and Dowbencko, D. J. (1986). *Proc. Natl. Acad. Sci. USA* **82,** 2618–2622.
Wiktor, T. J., and Koprowski, H. (1980). *J. Exp. Med.* **152,** 99–112.
Winter, G. F., and Inglis, J. M. (1987). *J. Infect.* **15,** 103–107.
Yewdell, J. W., and Gerhard, W. (1981). *Annu. Rev. Microbiol.* **35,** 185–206.

I

Genome and Antigenic Variability of Retroviruses

1

Genetic Variation in Retroviruses

John M. Coffin

I. INTRODUCTION

Retroviruses are a highly specialized group of viruses whose members are closely related in genetic organization, virion structure, and mode of replication, yet within this common framework they display an unparalleled diversity of biologic effects on their host. On a large scale, retroviruses have evolved to occupy a wide variety of distinct lifestyles—ranging from a completely benign transposable elementlike association with host germline that can span millennia to horizontal infections that lead to the death of the infected host with lytic, immunologic, or neoplastic disease. On a microscale, retroviruses can often undergo dramatic changes in genome structure during infection of a single host. Examples of this include the reproducible acquisition of oncogenes by slowly oncogenic viruses, the elaborately orchestrated series of recombinations and mutational events that create pathogenic out of benign viruses during the lifetime of certain mice, and the rapid variation in nucleotide sequence seen in the envelope gene of human immunodeficiency virus (HIV) and other lentiviruses.

My approach to this short review will be first to summarize relevant features of the retrovirus life cycle; second, to establish some general mechanistic principles of mutation, recombination, and selection; third, to describe some specific examples of retrovirus variation in the context of the principles; and finally, to discuss the implications of these to our understanding of retrovirus evolution. My goal will be to establish the following thesis: that mutation and variation are not at all the same. Retrovirus genomes are extraordinarily plastic and have impressively high rates of mutation and recombination. However, an equally important and often overlooked consideration in understanding retrovirus variation is the selective forces that act on the genome during replication and spread, and it is these that mold the retrovirus genome into the multitude of forms we find today.

II. GENERAL PRINCIPLES

A. The Retrovirus Life Cycle

The root of most variation in retroviruses is aberrations that occur during the course of replication. To understand this better, it is worthwhile to quickly review important aspects of the virus life cycle. For more detailed information, several excellent recent reviews should be consulted

John M. Coffin • Tufts University School of Medicine, Department of Molecular Biology and Microbiology, Boston, Massachusetts 02111.

(Varmus, 1987, 1988; Varmus and Swanstrom 1982, 1985; Temin, 1985; Goff, 1984). Retroviruses are unique in that their replication is accomplished by three distinct enzyme systems, each responsible for a different phase of the life cycle. The first phase consists of synthesis of viral DNA using systems that enter the cell within the virion. The second phase is replication of the provirus integrated into a host chromosome during cell division, using the usual cellular DNA polymerases. The third phase is the synthesis of progeny genomes using cellular transcription machinery.

1. Viral DNA Synthesis

The replication cycle is summarized in Fig. 1. Following entry into the cell, the virion nucleocapsid (or core) comprising the diploid genome, primer tRNA, reverse transcriptase and integrase, and at least some capsid proteins carries out the synthesis of double-stranded DNA. The product DNA is molecule colinear with the genome except for the duplication at each end of sequences uniquely present near one end or the other of the genome. This reduplication to form the long terminal repeat (LTR) is a key feature of reverse transcriptase. It is accomplished by a set of

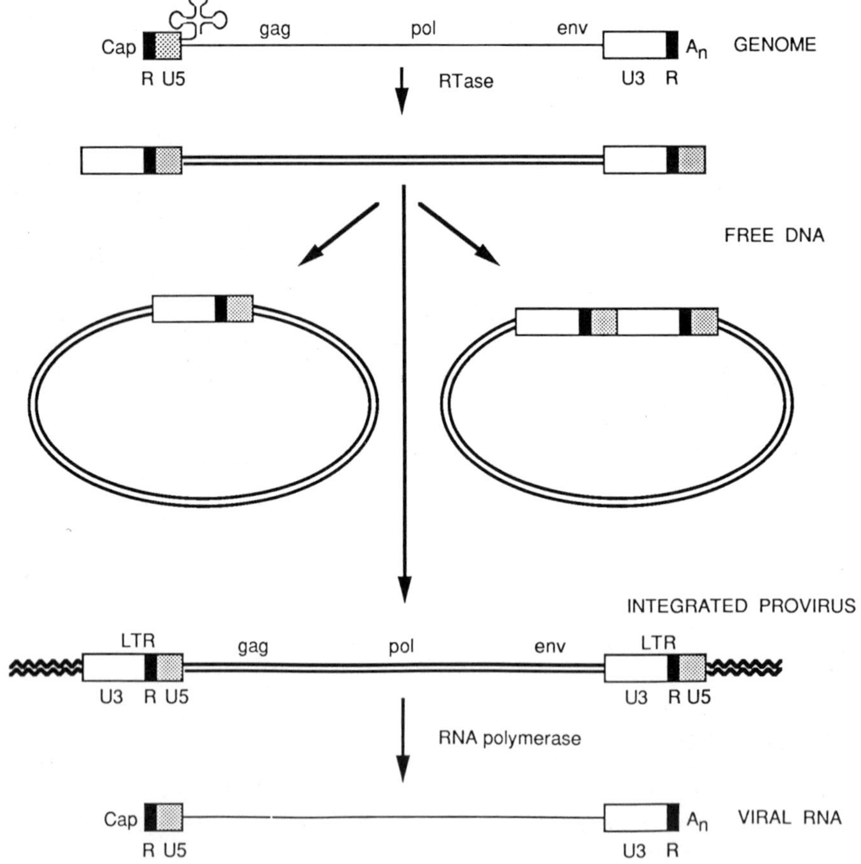

Figure 1. Retrovirus replication cycle. The flow of information from the RNA genome through the unintegrated viral DNA, the integrated provirus, and back to RNA is shown. The circular DNA forms shown are probably side products and do not participate in replication. (Courtesy of S. A. Herman.)

"jumps" in which the nascent DNA molecule, upon being extended up to a block to further elongation (such as the 5′ end of the genome), is freed of template by the action of the associated RNase H (which removes RNA from an RNA–DNA hybrid) and used to form a new template–primer pair by base-pairing with a complementary sequence at the other end. This proclivity of the enzyme system for jumping from template to template is a unique characteristic of reverse transcriptase. It is most likely also a characteristic that figures strongly in the introduction of genetic variation, contributing to high frequencies of both homologous and illegitimate recombination as well as deletions, reduplications, and other gross genomic rearrangements. It is important to keep in mind that the accuracy of the jumps (as well as other key early events) seems to be critically dependent on the integrity of the core structures and precise jumping has not been possible to duplicate in *in vitro* systems using purified enzyme and template.

Reverse transcriptase is a DNA polymerase capable of using either RNA or DNA more or less equally as a template or primer and containing an exonucleolytic RNase H activity necessary for removal of the template as synthesis of the minus strand DNA proceeds (Varmus and Swanstrom, 1982, 1985). It lacks an exonucleolytic (proofreading) activity and seems to have a corresponding high frequency of base misincorporation *in vitro*. Error frequencies ranging over a factor of 100 from 3×10^{-3} to 3×10^{-5} have been reported for various viruses (Gopinathan *et al.*, 1979; Kunkel *et al.*, 1983; Loeb *et al.*, 1986; Roberts *et al.*, 1989; Mizutani and Temin, 1976; Smith and Inglis, 1987). Since the extremely high rates are based on misincorporation with homopolymeric templates and probably do not closely resemble the natural case, a more reliable estimate would lie within a tenfold range centered on about one error in 10^{-4} bases incorporated, a value in rough agreement with measured mutation rates. The reverse transcriptase of HIV seems to be much more error-prone when compared head to head with that of avian myeloblastosis virus (AMV) or murine leukemia virus (MuLV) (Roberts *et al.*, 1988; Preston *et al.*, 1988). The significance of this suggestive finding remains to be explored. The rate of misincorporation is not likely to be uniform across the whole genome. Rather, it seems to be strongly dependent on the specific context, with some types of errors much more likely than others (Roberts *et al.*, 1988, 1989).

2. Integration

Once synthesized, viral DNA, still associated with the remains of the core, is transported to the nucleus and integrated into the cell DNA to give rise to the provirus. Like correct DNA synthesis, integration *in vitro* seems to require the integrity of the core and is mediated primarily (or entirely) by proteins that entered the cell in the virion, most notably the virus-specified integrase (Brown *et al.*, 1987; Fujiwara and Mizuuchi, 1988; Hagino-Yamagishi *et al.*, 1987; Quinn, 1988). The majority of integrations seem to be into randomly distributed sites [with a possible affinity toward relatively "open" chromatin (Rohdewohld *et al.*, 1987; Vijaya *et al.*, 1986)], although a fraction of very strongly preferred sites used by a minority of integrations has also been reported (Shih *et al.*, 1988).

The integration event can be of considerable mutagenic importance to the host, leading to activation of protooncogenes neighboring the integration site (for review, see Bishop and Varmus, 1985) or (more commonly) to inactivation of genes into which the insertion has occurred. Indeed, in animals such as mice whose germline has suffered numerous assaults of retroviral integration leading to large numbers of inherited or "endogenous" proviruses, this may be an important source of genetic diversity among present-day strains (Stoye and Coffin, 1985; Stoye *et al.*, 1988). With the possible exception of its indirect participation in oncogene "capture" (see Section IV.B) the integration process does not seem to be an important source of genetic variation of the viral genome. Although gross errors in the process—in particular autointegration of a DNA molecule into itself—are readily detectable both *in vivo* (Shoemaker *et al.*, 1980) and *in vitro* (Y. M. Lee and J. M. Coffin, submitted), these are so disruptive to the genome structure that they only create dead ends incapable of further replication. On the other hand, minor errors in integration, affecting only a few bases at the ends of the provirus, are lost during replication since the ends of the genome are derived from sequences within the LTRs.

3. Replication of the Provirus

Once integrated, a provirus can be considered a permanent fixture of the infected cell in that there is no mechanism coded by the virus for either excision or direct transposition of the provirus, although random loss by deletion of large regions of chromosome or recombination across the LTRs can occur as an infrequent event. Furthermore, proviruses seem to contain no signals to specify their independent replication. If not integrated, they are lost from the infected cell. Provirus replication then occurs only as a regular part of chromosome replication and must have a similar error rate, which is likely to be so low as to be negligible for the purposes of this discussion. For example, analysis of a 200-base region of five endogenous proviruses inserted independently into the mouse germline prior to inbreeding and therefore separated by at least 100 animal generations from one another revealed no difference in sequence—an error frequency of less than 10^{-5} per animal generation. Indeed, an endogenous provirus known to have been inserted into the primate germline more than 5 million years ago still displays considerable sequence similarity to exogenous viruses of mice (Repaske *et al.*, 1985; Steele *et al.*, 1986).

4. Genome RNA Synthesis

Synthesis of all retroviral genomes is carried out by the action of host RNA polymerase, probably unmodified by virus-specified components. Accordingly, all retroviruses contain within the LTR complicated combinations of enhancers and other transcription factor binding sites that can specify expression and therefore replication in a cell-specific manner. Indeed, small changes in LTR sequences can have major effects on tissue specificity of replication and pathogenesis (see Section IV.A.2).

Processing of reverse transcripts, including splicing, polyadenylation, and transport, is also conducted by host systems responding to signals within the transcripts. While some retroviruses, such as HIV, encode proteins that affect these processes in poorly understood ways, this is not a general feature of retrovirus biology. As with integrations, aberrations of RNA processing do not seem to be important sources of variation, again, with the possible exception of oncogene capture (see Section IV.B).

The error rate of RNA polymerase II is not known (and could be very difficult to measure), but is unlikely to be much different from that of other RNA synthesizing enzymes (or of reverse transcriptase), i.e., about 10^{-4} plus or minus a factor of 3. If so, then it is worth noting that errors in genomic RNA synthesis may be as important as those in reverse transcription in generating sequence variation. Also noteworthy is that there would be no point for the virus to encode a highly accurate reverse transcriptase when half the replication cycle is carried out by relatively inaccurate pol II.

B. Mechanism of Variation

1. Theory

The major point I wish to make in this chapter is as follows. Variation is not the same as mutation. The rate of variation at a given position in a genome retrovirus is a function of three factors: the mutation rate per replication at that position, the relative growth rate (selective advantage or disadvantage) of the mutant relative to its parent, and the number of replication cycles per unit time.

The number of mutant genomes (M) and wild-type genomes (W) at cycle n can be estimated as

$$M_n = skM_{n-1} + k\mu_f W_{n-1} - k\mu_r M_{n-1}$$

and

$$W_n = kW_{n-1} + k\mu_r M_{n-1} - k\mu_f W_{n-1}$$

where K is the growth constant for wild-type virus, s the relative growth rate for mutant virus, μ_f the forward mutation rate, and μ_r the reverse mutation rate. The proportion of mutant genomes P_n is then equal to

$$\frac{M_n}{M_n + W_n}$$

or

$$\frac{skM_{n-1} + k\mu_f W_{n-1} - k\mu_r M_{n-1}}{skM_{n-1} + k\mu_f W_{n-1} - k\mu_r M_{n-1} + kW_{n-1} + k\mu_r M_{n-1} - k\mu_f W_{n-1}}$$

This rearranges to

$$\frac{(s - \mu_r)M_{n-1} + \mu_f W_{n-i}}{sM_{n-1} + W_{n-1}}$$

or in terms of P_{n-1},

$$\frac{(s - \mu_r) + \mu_f(1/P_{n-1} - 1)}{s + (1/P_{n-1} - 1)}$$

This is a bit complicated, and a solution, relating the proportion of mutant genomes to the number of replication cycles (with some limiting assumptions), was derived by Batschelet *et al.* (1976). Despite the complexity of this formula, it is instructive to consider a specific example. Imagine starting an infection with a cloned virus for which the mutation rate per cycle is 10^{-4} misincorporations per base per replication cycle (which would correspond to about one mutation per genome per replication cycle). Initially, this will also be the rate of accumulation of mutations, and shortly the resulting virus population will diverge into a "quasi-species" with a modal genome sequence, but in which no two members are exactly alike. In the absence of selective effects (i. e., $s = 1$), the population will diverge at a rate set by the relationship of the forward and reverse mutation rates until the proportion of mutants at any one position reaches an equilibrium value also determined by the ratio of the two rates. It is, however, highly unlikely that any change in the genome is selectively neutral, given the large number of functions (both coding and noncoding) embedded in it. Imagine, for example, that a change in a given position results in a 1% growth disadvantage. This is a very small difference. It would correspond to a decrease in the rate of virus production by the cell from 100 virions/hr to 99 virions/hr, or an increase in the probability of neutralization by antibody from 99% to 99.01%. Under these conditions, the population will reach an equilibrium at about 1% mutant genomes at which time the rate of appearance of new mutants will exactly balance the rate of overgrowth of mutant by wild-type virus. If all bases in the genome behave as in this example, then, the population will not drift aimlessly apart by unchecked mutational random walking, but will quite gradually diverge to a steady state in which each member differs from the modal sequence at about 1% of all positions (or about 100 bases over a complete genome). In the opposite case, if the mutation is selectively advantageous, i.e., confers a 1% growth advantage, then when the mutant genomes constitute 1% of the population, the contribution of mutants arising *de novo* and those "old" mutants selected by overgrowth becomes approximately equal and, as time goes on, selective advantage becomes more and more important in determining the rate of appearance of the mutant.

2. A Direct Experiment

To visualize these effects at work, we performed an *in vitro* "evolution" experiment in which a cloned avian retrovirus (Rous sarcoma virus) was passaged repeatedly (every 3 or 4 days) through fresh host cells (Coffin *et al.*, 1980). At intervals, the genomes were sampled using a sensitive fingerprinting technique that displays the genome as oligonucleotides comprising about 500 bases (or about 5% of the genome) and can detect and quantitate a single change in any one of these bases. Since the virus was no longer required to transform cells, but only to replicate rapidly, this passaging regime had the effect of subjecting the virus to a selective environment rather different from the one it had been maintained in previously. Several types of gross changes were observed during the course of the experiment (Fig. 2), including (1) deletion in two steps of the oncogene (src); and (2) appearance of virus containing a large deletion (ld virus) extending from the NP region of *gag* to the TM region of *env*. The *src* deletion was the result of a homologous recombination; the other due to an illegitimate recombination between unrelated sequences (Voynow and Coffin, 1985a,b). The ld virus clearly overgrew the parental nondefective virus in mixed infection; we do not know the basis for this selection.

In addition to rearrangements, point mutations were also observed. The appearance of one, in a conserved region near the NH_2 terminus of the *env* protein, is plotted in Fig. 3. Comparison of the

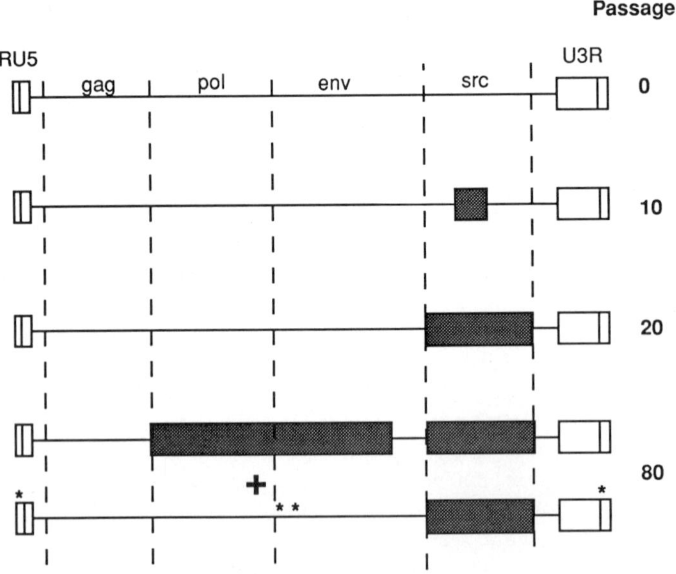

Figure 2. Variants arising during retrovirus passages. A cloned stock of Rous sarcoma virus was passaged repeatedly at 3- or 4-day intervals through susceptible host cells. At the times shown the genome structure was analyzed (see Coffin *et al.*, 1980 for details); the shaded boxes indicate deletions; and asterisks show the location of some selected point mutations.

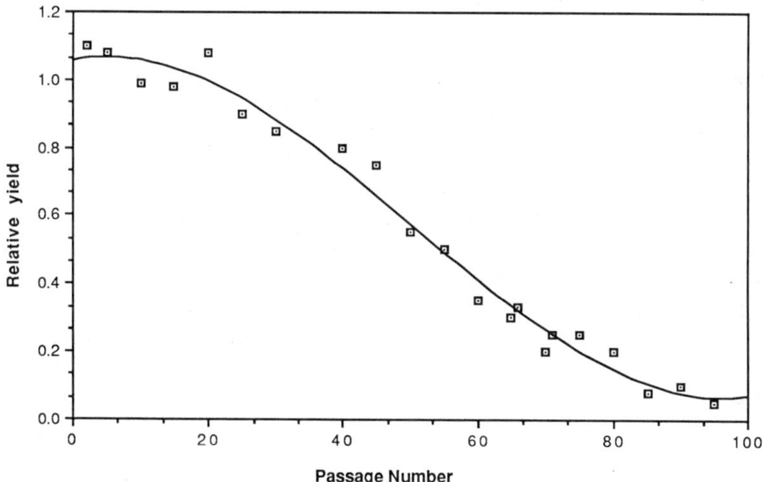

Figure 3. Selection of a point mutation on passaging. A specific point mutation in the NH_2-terminal portion of *env*, detectable as loss of an oligonucleotide from the genome fingerprint, was detected during the course of the passaging experiment described in Fig. 2. The proportion of the wild-type sequence relative to complete genomes is plotted here as a function of passage number. The line approximates a theoretical curve derived from the formula of (Batschelet *et al.*, 1976) assuming a mutation rate of about 3×10^{-4} per passage and a relative growth rate of mutant to wild type of 1.05.

rate of appearance of this specific mutation with theoretical curves derived from the approach of Batschelet *et al.* (1976) imply a relative growth rate for this mutant of about 1.05 that of wild type and a mutation rate on the order of 3×10^{-4} or so per passage, corresponding very roughly to about 10^{-4} mutations per base per replication cycle. It is noteworthy that the selective advantage could be estimated quite precisely; the mutation rate only very roughly, consistent with the idea that selective forces can be much more important than mutation rates in determining variation.

The final—and perhaps most striking—point to emerge from this experiment came from a study of viruses recloned from the mixture after 75 passages (perhaps 200 replication cycles). Analysis of these genomes showed virtually no additional alterations in sequence that were not also detectable in the mass population. About $200 \times 500 \times 10^{-4} = 10$ detectable mutations per genome would have been expected in the absence of selection. Thus, we conclude that the virus population is not subject to random mutational drift, but that all positions analyzed were most likely under some selective pressure either for a specific change or to remain constant.

3. Conclusions

A final point concerning mechanisms of variation is that the observed rate of variation with time is also a function of the rate of replication. This is particularly crucial with retroviruses, since replication of the integrated provirus with the host cell DNA has a much lower error rate and is subject to very different selective forces than is a cycle of reverse transcription and transcription. Thus, the rate of variation of a retrovirus carried in culture by passage of infected cells will be much lower than that of one passaged by repeated infection cycles (as in the example described in Section II.B.2). Similarly, viruses whose pathogenic mechanism involves killing of infected cells and reinfection of new ones (such as HIV) will have a much larger number of replication cycles (and display a correspondingly higher extent of variation) than ones (like HTLV-1) which do not kill infected cells but rather establish a long-term infection and persist as infected cells.

It is important to keep in mind that not only is the mutation rate of the virus replication cycle very different from that of the cell, but the selective forces are also quite different. Rapid virus replication can be inimical to cell growth and survival, so that viruses which are spread and persist as proviruses, such as endogenous viruses, tend to have relatively low replication rates (Coffin, 1982; Stoye and Coffin, 1985). Selection in such cases can thus be for lack of deleterious effect on the infected cell, or even (as in the case of oncogene capture) for promotion of increased cell growth. In some cases (as has been proposed for the case of HIV infection) progression from an acute to latent and back to acute infection may involve changing selective pressures and thus lead to a reproducible "evolution" of virus genomes, during the course of infection of a single individual.

From the arguments just presented, it should be clear that as long as the selection advantage (or disadvantage) of a mutation is significantly greater than the mutation rate, the rate of variation of the population will be much more strongly a function of selective processes than mutation rates. Four obvious warnings should be (but often have not been) heeded by those interested in following genetic variation of HIV or other retroviruses:

1. In general, mutation rates are only indirectly related to observed rates of variation.

2. Because the selective effect of a mutation at one position will be quite different from that at another, the rates of variation at one position give no guide for the rate at another position. *It is always incorrect to extrapolate evolutionary history from modern-day variation* (as some evolutionary biologists in the HIV field have been inclined to do) (Smith *et al.*, 1988).

3. The process of variation with time as a population diverges from a few founding members is highly nonlinear, proceeding instead as a sigmoidal curve approaching an asymptote determined by relative rates of forward and reverse mutation and selection. It is thus meaningless to determine a "rate of variation" by simple measurement of the amount of variants at different times.

4. Real insight into virus variation requires understanding of selective forces involved as much as (or more than) knowledge of the mechanism and role of mutation.

III. MECHANISMS OF VARIATION

As discussed in Section II.A, much of the variation observed in retrovirus genomes is attributable to errors in the reverse transcription process. In this section, I discuss a few specific examples in somewhat more detail.

A. Point Mutations

Although much work (cited in Section II.A) has been done on the error rate of reverse transcriptases *in vitro*, relatively few studies have addressed the mutation rate (as opposed to the rate of variation) *in vivo*. The passaging experiment shown above permitted a rough estimate of around 10^{-4} per generation (plus or minus a factor of 3) for Rous sarcoma virus (RSV). A similar estimate (1.4×10^{-4}) was obtained from an experiment where cloned virus was subjected to rapid recloning and the progeny analyzed for point changes by a hybridization technique (Leider *et al.*, 1988). Using a rather different approach—the reversion of a specific point mutation in a dominant selectable gene inserted into a spleen necrosis virus-derived vector—Dougherty and Temin (1986) estimated a lower rate of 2×10^{-5} per generation. These results, although obtained with different methods, are reasonably consistent with our prior estimate. Considering that there is likely to be considerable variation in rate from one base to the next, the first two approaches are likely to be biased toward the more frequent events. All are within the range of error rates estimated for reverse transcriptase *in vitro*. It will be interesting to see whether the mutation rate for HIV *in vivo* is higher, as might be expected from the *in vitro* studies.

These mutation rates are in the range of rates estimated for RNA viruses in general. Similar figures have been obtained for bacteriophage Qβ (Batschelet *et al.*, 1976), and vesicular stomatitis virus (Steinhauer and Holland, 1982), but up to tenfold lower rates have been estimated for poliovirus and influenza virus (Parvin *et al.*, 1986). Thus, while mutation rates of retroviruses may be higher than those of most viruses, they are not truly exceptional.

B. Rearrangements

Retrovirus genomes are subject to a high (but undetermined) rate of intragenomic rearrangements—deletions, duplications, inversions, or combinations of these. Indeed, a survey of the literature on cloning of viral DNA from infected cells indicates that perhaps half of the DNAs so obtained have suffered some sort of rearrangement (for example, see Shimotohno and Temin, 1982). In the majority of cases, the rearrangements are so severe as to render the genome inactive, but there are some important exceptions. For example, rearrangements in the LTR of MuLV are important determinants of differential pathogenicity (Short *et al.*, 1987; Golemis *et al.*, 1988); genetic variation within the *env* gene of different HIV isolates seems to include reduplication of short sequences (Coffin, 1986); and specific types of defective mutants such as spleen focus-forming virus or some feline leukemia viruses (see Coffin, 1985; Overbaugh *et al.*, 1988) can be important pathogens in their own right.

Although the formation of such variants has not been analyzed in detail, it is almost certainly a consequence of specific aberrations in reverse transcription. [While rearrangements involving integrated viral DNA can sometimes be detected if the selection is strong enough (for example, see Levantis *et al.*, 1986), such events seem to occur at a very low rate (a few in a million) and cannot account for the vast majority of such events.] Lesions in the process of reverse transcription include mispriming, premature termination of one end or the other, incorrect end-to-end strand transfer, and foldback copying of a newly completed strand to form inversions. The most common type of rearrangement, however, is reduplication or deletion of sequence, most likely due to aberrant "jumping" of reverse transcriptase from one RNA template to another or within a template. These mutants can then be considered to be the consequence of an incorrect recombination event (see next section). It is noteworthy that, while deletion and reduplication often involve the use of homologous sequences (Omer *et al.*, 1983; Hughes and Kosick, 1984), they do not always do so, and there are instances of such rearrangements in the observance of detectable homology or other sequence features (Voynow and Coffin, 1985a,b).

C. Recombination

One of the most remarkable features of retrovirus genetics is the extraordinarily high rate of recombination. Probably no other biologic system displays the capacity for exchange of genetic information to the same extent as retroviruses, and all retrovirus systems tested, including HIV (M. Martin, personal communication), undergo recombination at high rates. Although most commonly observed between infecting exogenous viruses, recombination can also readily be observed between exogenous and endogenous viruses and between virus and unrelated host cell information (as in oncogene capture). Although the frequency of recombination has not been precisely measured, it is so high that in a usual experiment (i.e., co-infecting a cell culture with a mixture of viruses differing in two selectable markers, e.g., transforming ability and host range, and then selecting a recombinant between them), markers as close a 1 kb are found to segregate independently, as if unlinked (Coffin, 1979). This free exchange implies that a population of virus containing sequence variations and allowed to interact (by co-infection of the same cells) should be considered to be homogenized across the genome such that all possible combinations of variants are present at any one time. This has important implications for the variation of pathogens such as HIV and the potential of populations of viruses to generate new combinations of variants which might have very different properties, for example, antigenic variants.

Although two plausible models for retrovirus recombination have been proposed (Coffin, 1979; Junghans *et al.*, 1982), neither has been subjected to a critical test. It is clear that recombination only occurs following infection with heterozygous virions, i.e., those produced by cells co-infected with two parental viruses (Linial and Blair, 1982). If a cell is simply doubly infected, recombinants are not observed. From this requirement, it follows that recombination must be an early event, occurring prior to integration, presumably related to the process of viral DNA synthesis. [It is important to note that cotransfection of naked retroviral DNA (as with all other DNA's) into cells leads to high rates of recombination (Bandyopadhyay *et al.*, 1984), but this process—although useful for mapping—has nothing to do with the natural case.]

The two models proposed are illustrated in Fig. 4. The first (Junghans *et al.*, 1982) proposes that exchange of a newly made fragmentary + strand copied from one genome onto a − strand copy of the other could lead to heteroduplex proviruses which would segregate wild-type and recombinant virus after integration and cell division. While concordant with electron microscope observation of the products of *in vitro* reverse transcriptase reactions, this scheme requires that both members of the diploid genome serve as templates simultaneously and it does not readily explain recombination involving large regions of nonhomology, such as around oncogenes.

The second model (Coffin, 1979) (which I favor, for obvious reasons) is based on two observations: (1) reverse transcriptase efficiently switches templates when it encounters an end while still elongating a growing chain, and (2) retrovirus genomes can have considerable numbers of breaks and yet remain associated in an intact complex. Thus, when reverse transcriptase encounters such a break, it can behave exactly as it would at the end of a genome, i.e., switch to the homologous sequence on the other genome and continue synthesis. The product of this reaction would be a − strand containing a mosaic of information from both genomes (depending on the number of preexisting breaks). This mechanism could not only promote high-frequency exchange of information, but would have the important additional benefit of repairing preexisting RNA breakage and relieving the virus of using extraordinary measures to shield its genome from the ravages of the extracellular environment. Since reverse transcriptase can make strand transfers in the absence of extended homology (indicating that affinity of enzyme for template may be a stronger driving force for jumping than base pairing), a similar mechanism can be readily invoked to explain other types of genomic rearrangements, as well as capture of cellular sequences.

IV. SOME EXAMPLES OF RETROVIRUS VARIATION

In the examples that follow, I draw from the experience of the virologist to discuss a number of specific examples of retrovirus variation, the underlying mechanisms involved, and the biological consequences. Since genetic variation of HIV is covered extensively in other chapters of this volume, it will be discussed in passing, with discussion centered largely on issues arising with other retroviruses.

A. Strain Variation

Within a group of retroviruses, different isolates can show considerable variation in host range, pathogenicity, and lifestyle. For example, among closely related avian leukosis viruses, seven distinct host range variants (designated A–G) can be identified by their utilization of distinct receptors on chicken cells (Weiss, 1982). Several variants in pathogenicity [ranging from no disease induction to high levels of lymphoma or osteopetrosis (Teich *et al.*, 1985)] and two distinct lifestyles [as an exogenous infectious agent transmitted both horizontally and vertically and as an endogenous

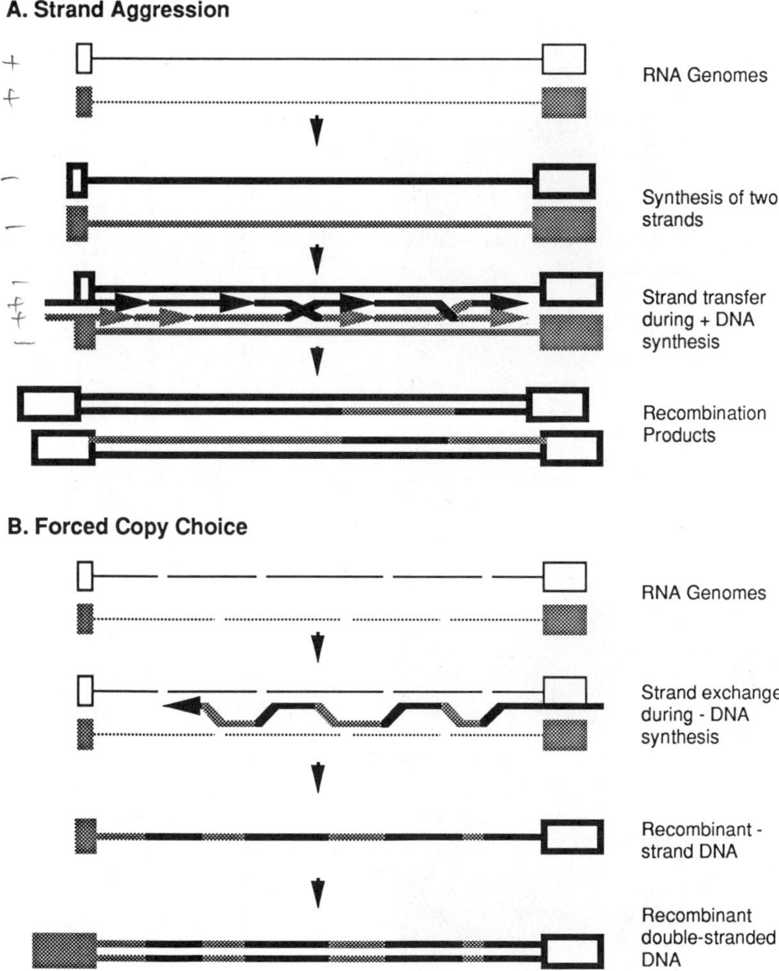

Figure 4. Models of retroviral recombination. RNA genomes are shown as thin lines, DNA as thick. A, In the strand aggression model (Junghans *et al.*, 1982), − strand DNA molecules copied from the two genomes simultaneously serve as templates for + strand DNA synthesis in short fragments. These are postulated to displace the corresponding strand of the other molecule by crossing over during synthesis, and they give rise to the products shown. B, In "forced copy" choice (Coffin, 1979), preexisting nicks in the RNA cause the reverse transcriptase to jump back and forth, leading to synthesis of a single recombinant − strand. Note the clear difference in structure of the double-strand molecules predicted from the two schemes.

element transmitted as a germline provirus (Coffin, 1982; Stoye and Coffin, 1985)] can be distinguished. Similar ranges of variation are observable within other virus groups as well, but I will limit the following discussion largely to the avian and murine retroviruses.

1. The env Gene and Host Range

Comparison of the nucleotide sequence of the *env* gene of a number of different ALV host range subgroups reveals a striking pattern (Fig. 5), consisting of blocks of highly variable sequence interspersed with strongly conserved sequence (Dorner *et al.*, 1985; *et al.*, 1986; Bova *et al.*, 1988).

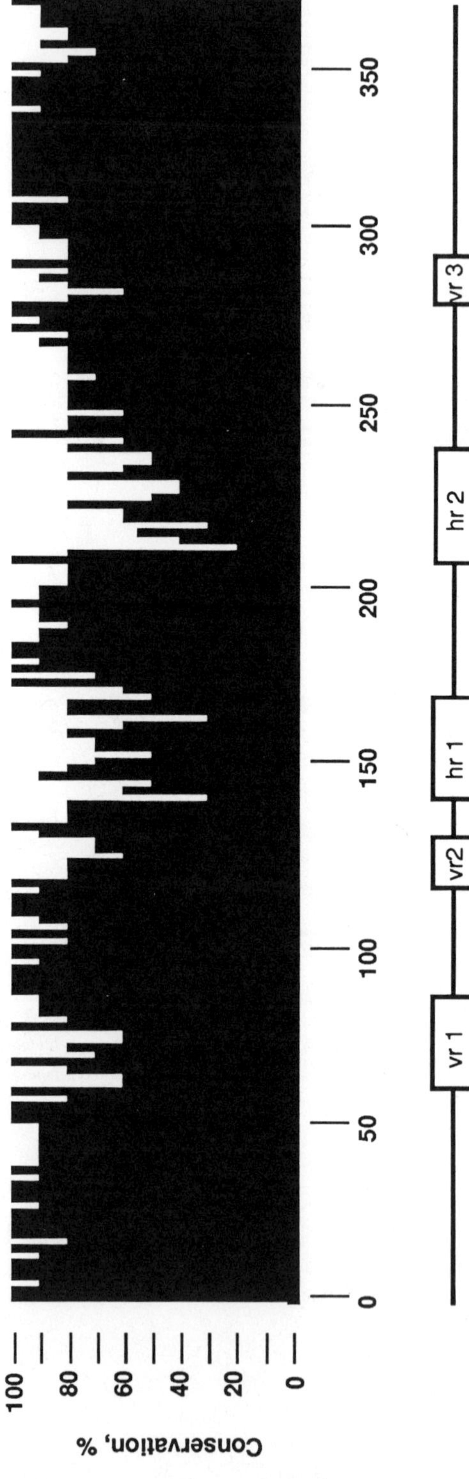

Figure 5. Genetic variability in the ALV SU protein. The amino acid sequences of gp85 from six subgroups (A–F) of ALV are compared (Dorner and Coffin, 1986; Dorner *et al.*, 1985; A. J. Dorner, J. P. Stoye, and J. M. Coffin, in preparation). The vertical bar at each position indicates the number of genomes with the consensus amino acid at that position. Regions hr1, hr2, and vr3 have been shown to encode the receptor binding specificity of the protein.

In the central portion of the major surface (SU) protein (also known as gp85, analogous to HIV gp120) are two long blocks of variable sequence (hr1 and hr2) flanking a conserved region. Analysis of recombinant virus genomes shows that these two regions together comprise the principal determinant of host range variation. Furthermore, most recombinants that mix these two regions from different subgroups are noninfectious, yet at least one such recombinant encodes a receptor specificity that is a combination of the two parents (Dorner and Coffin, 1986). This result is consistent with these two regions encoding two separated segments of protein which together form the receptor binding site.

Clearly, neighboring regions of the same protein have evolved at very different rates (Fig. 6), and it is instructive to consider what evolutionary forces might have been involved, even if it might be difficult to design experiments to test our ideas. First, it is important to point out that most chickens carry in their genomes a few copies of an endogenous provirus, closely related in sequence to exogenous avian leukosis viruses. These endogenous proviruses are always of a specific host range subgroup (subgroup E), not found among exogenous viruses (Weiss, 1982). Furthermore, it is worth noting that most modern-day chickens, but not birds of other species, lack the receptor recognized by subgroup E viruses. Since endogenous proviruses have most likely been present in chickens for a long time and survive with little change through long tenure in the germline, it is reasonable to suppose that they are fossil remnants of ancient viruses, and that they therefore represent an ancestral type. Furthermore, it is a reasonable speculation that the absence in most (although not all) chickens of subgroup E virus receptors is a response of the host to selective pressure imposed by the virus itself, that the divergence of the virus into recognition of other receptors is a response by the virus, and that this evolutionary back-and-forth process continues today. The selective pressure forcing divergence of *env* genes and receptor binding is clear. The exceptional conservation of portions of the gene outside those encoding host range could well be a consequence of the presence of the endogenous provirus itself, by an indirect immunologic mecha-

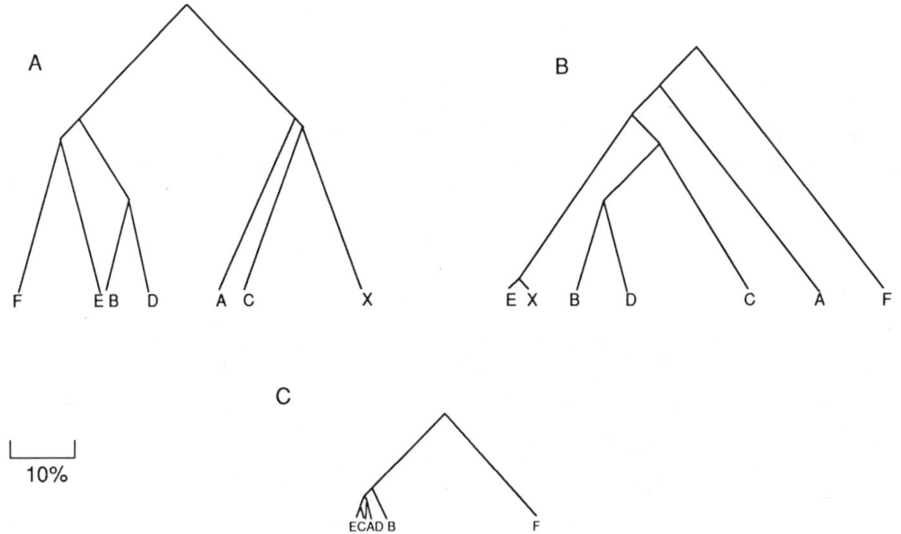

Figure 6. Evolution of different parts of the ALV SU protein. Relationship trees for the indicated regions of gp85 were drawn from amino acid sequence information (Dorner and Coffin, 1986; Dorner *et al.*, 1985; A. J. Dorner, J. P. Stoye, and J. M. Coffin, in preparation). Note that host range subgroups A–E are all found in viruses of chickens whereas subgroup F is an endogenous virus of pheasants. X is an apparently novel recombinant detected in the course of nucleotide sequencing of a subgroup C virus (Hunter *et al.*, 1983).

nism. Expression (even at a low level) of the endogenous *env* protein would be expected to render the animal tolerant to this protein, and therefore infecting viruses whose *env* proteins resembled those of the endogenous viruses as closely as possible would have an advantage over those with greater antigenic variation. That such selection for antigen similarly indeed occurs is suggested by the observation that anti-*env* antibodies raised in chickens are invariably subgroup specific, suggesting that they react only with the variable host range coding regions (Weiss, 1982). In addition to antigenic enforcement of strong similarity to endogenous proteins, the progeny of endogenous proviruses can recombine with infecting viruses to provide a uniform source of such sequence. The observation that endogenous viruses of pheasants (subgroup F) exhibit relatively greater divergence from the others in the conserved than in the variable regions is consistent with this picture.

It is worth contrasting this situation with that of HIV, where there is a complete lack of divergence in receptor utilization among isolates—even Simian immunodeficiency virus (SIV) isolates use the same CD_4 receptors—and probably strong selection for antigenic variation to allow the virus to escape neutralization during long residence in the same host. The pattern of genetic variation observed is thus quite different—different isolates display variability in many small blocks of "hypervariable" sequence scattered along the gp120 sequence, with relatively strong conservation of sequence within the candidate receptor binding region (Lasky *et al.*, 1987; Modrow *et al.*, 1987; Starcich *et al.*, 1986) (Fig. 7). It is hard to resist speculating that the hypervariable regions reflect surface features of the protein whose sequence can vary considerably without affecting the function of the protein (Coffin, 1986). Such regions would be highly visible to the immune system, but could respond by rapid variation as necessary to evade it. Note from the argument given in Section II.B that even a relatively modest difference in sensitivity to neutralization (one that might be virtually impossible to measure by standard techniques) could give rise to extremely rapid evolution of variants in a region that was under no selective constraint for conservation. It is imperative to keep in mind that such regions are themselves highly specialized structures and must not be considered typical of the whole genome or even the rest of the protein. Indeed, the rest of the HIV genome does not display a significantly greater variability from one isolate to another than does that of other retroviruses, such as avian leukosis virus (ALV).

2. *The LTR and Pathogenicity*

Among closely related isolates of many retroviruses, the greatest sequence divergence is found within the LTR, particularly within that portion of U_3 which contains the transcriptional enhancers.

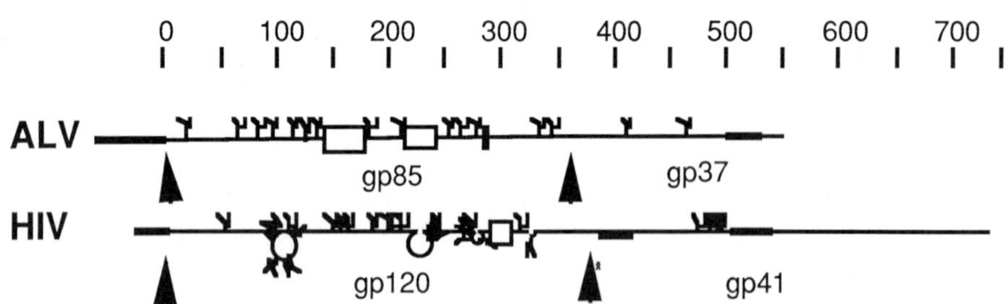

Figure 7. Comparison of the structure of ALV and HIV *env* proteins. The thick lines indicate hydrophobic regions (signal peptides, anchor sequences, etc.); the crooked symbols show the location of *N*-linked glycosylation sites; vertical arrows the position of cleavage sites; and open boxes the location of candidate receptor binding domains. The loops in the HIV structure indicate the location of "hypervariable" regions (Coffin, 1986; Willey *et al.*, 1986).

This plasticity permits the virus ready access to a variety of niches by selection of variants whose level and cell specificity of expression are well matched to the particular lifestyle.

For example, endogenous avian leukosis viruses have an approximately tenfold lower replication rate than their exogenous cousins and a virtually complete lack of pathogenic effect (Coffin, 1982; Stoye and Coffin, 1985). These differences are probably crucial to the endogenous lifestyle: An endogenous virus that replicated rapidly and was strongly pathogenic would be counterselected owing to early death of its host. Both these functions can be attributed to extensive sequence differences in the U_3 region of the LTR (Brown and Robinson, 1988; Tsichlis and Coffin, 1980); indeed, most of the enhancer activity detectable in exogenous virus LTRs is absent from endogenous viruses (Cullen *et al.*, 1983).

More subtle changes in the LTR can also have pronounced effects on the biology of the virus. The enhancer (i.e., transcription-factor binding) portion of the MuLV LTR is fairly well conserved among related viruses, yet a small number of rearrangements within this region are apparently sufficient to convert a benign endogenous virus of mice into a virus capable of inducing T-cell lymphoma (Boral *et al.*, 1988; Short *et al.*, 1987). An even smaller difference in the same region is apparently largely responsible for another type of pathogenic variation. Detailed mapping studies have shown that a difference in a few bases is the important factor distinguishing a virus that causes T-cell lymphoma from one that causes erythroleukemia (Golemis *et al.*, 1988; Li *et al.*, 1987). This difference affects a binding site of one transcription factor (Speck and Baltimore, 1987) and, it is presumed, confers target cell specificity in expression and therefore replication and oncogenesis.

B. Oncogene Capture

1. General Principles

Some of the greatest gifts that have been presented to cancer biologists are the oncogenes found in many retrovirus isolates. Analysis of the structure and function of these and the role of their cellular homologues in normal and malignant cell processes has permitted a deeper insight into the molecular basis of cancer than any other system (for review, see Bishop and Varmus, 1985). Oncogenes have also been important in virology; many types of retrovirus first came to light because of the presence of an oncogene that made them much more visible by conferring increased pathogenic potential, ability to visibly transform cells in culture, and greater attractiveness to the curious scientist.

Retrovirus oncogenes can be defined as nonviral sequences that have been inserted into one of a variety of locations in the virus genome in place of or in addition to normal viral genes. They are derived from. variety of normal cellular genes (protooncogenes) which themselves seem to be important in regulating the growth of the normal cell. When incorporated into a virus genome, they are altered in important ways so that their expression leads—usually directly—to transformation of the host cell. Important alterations include increased expression under the influence of the viral LTR; loss of regulation of expression; and point or deletion mutations in the oncogene itself that alter the regulation of the oncogene product at a functional level. Since similar changes to some of the same protooncogenes can also be observed in naturally occurring cancers in the absence of viral involvement, study of these genes and their acquisition has been central to developing a molecular theory of cancer.

2. Mechanisms of Capture

Retroviruses are the only group of animal viruses that acquire cellular genes in this way, so the mechanism of their acquisition must be connected with their special features. An attractive model has been developed from the original observation of Goldfarb and Weinberg (1981), who studied a deletion mutant of Harvey murine sarcoma virus (Ha-MSV) that contained the oncogene but lacked any homology to MLV at its 3′ end. They found that virus heterozygous for this truncated genome and a wild-type MuLV genome would yield transforming virus with good frequency, indicating that the 3′ end of the genome could be readily repaired by illegitimate recombination. Since such partial genomes closely resemble those that might be derived from aberrancies surrounding the integration event, the following model was proposed (Swanstrom *et al.*, 1983) (Fig. 8, left).

1. The provirus integrates upstream from a protooncogene. Integrations of this sort can also lead directly to oncogene "activation" and cell transformation under certain circumstances (Hayward *et al.*, 1981).

2. A deletion at the DNA level joins the protooncogene to a portion of the provirus.

3. The joint viral–oncogene transcript is processed (by splicing of the oncogene portion, for example) and incorporated into heterozygous virions along with a wild-type genome (from another provirus in the same cell).

4. An illegitimate recombination event during reverse transcription restores the 3′ end of the viral genome.

Note that the final structure of the oncogene containing virus genome may also be determined by additional rearrangements, as well as point mutations within the oncogene itself, so that the structure observed in viruses need not exactly reflect the initial events. Recently, we have suggested a variation of this model, based on the observation that correct formation of viral 3′ ends (by cleavage and polyadenylation) is an inefficient process. About 15% of the ALV genomes are the result of polyadenylation of primary transcripts in sequences derived from cell DNA downstream of the provirus and are thus joint virus–cell products (Herman and Coffin, 1986). This model is shown in Fig. 8, right (Herman and Coffin, 1987) and differs from the other as follows. After integration adjacent to an oncogene as before, a joint readthrough transcript (perhaps processed by splicing in the oncogene portion) is incorporated into heterozygous virus, along with a wild-type virus genome, and then a pair of illegitimate recombination events during reverse transcription incorporates the oncogene into a viral genome.

The latter scheme requires two illegitimate recombinants rather than one, but these are likely to be much more common (and active in a much larger population of genomes) than a specific deletion of cell DNA. In any case, it should be possible to test these ideas using transformation techniques and appropriate constructs with selectable markers.

3. *Selective Forces*

No matter the mechanism, the capture of an oncogene must be a rare event. Nevertheless, about two dozen different oncogenes have been found in hundreds of virus isolates, and in some specific systems oncogene-containing virus can be reproducibly isolated from animals infected with the ancestral wild-type viruses (Miles and Robinson, 1985; Stewart *et al.*, 1986). Oncogene-containing viruses are almost certainly not passed from animal to animal as infectious agents in the wild, since they are poorly transmitted and rapidly lethal. Rather, they seem always to have arisen in the animal from which they are isolated. The apparent ease of isolating the products of rare events is clearly the result of powerful selection. In the infected animal there must be at least 10^{10} replication events. If any one of these gave rise to a rapidly oncogenic virus, it would be readily detectable as an animal with a tumor. Since the most productive oncogene sources have been in chickens at slaughterhouses and among cats presented to veterinarians with malignant disease, in screens that include tens of thousands of animals, events as rare as 10^{-14} or less can be detected in this way. If not taken out of the animal in which it arose and given a good home in the laboratory, the virus would surely have died with the animal in which it arose.

Selective forces can also be inferred to operate at individual steps in this process. For example, insertion of a provirus in an appropriate location relative to a specific protooncogene must be a very rare event, on a per-cell basis, although statistically certain in a whole animal. If such cells acquire a growth advantage relative to their neighbors, then a much larger population of targets for subsequent events will become available.

C. In vivo Evolution

One of the most complicated systems in virology was created when some strains of mice were inbred to select for high incidence of spontaneous leukemia (Coffin, 1982; Stoye and Coffin, 1985; Teich *et al.*, 1982). Mice of strains such as AKR and C58 typically die at about a year of age with a T-cell lymphoma induced by a retrovirus derived from proviruses endogenous to the particular strains. While the ultimate molecular event in oncogenesis involves insertional activation of one or more oncogenes by new proviruses (Villemur *et al.*, 1987; Selten *et al.*, 1984), and possibly other events as well (Davis *et al.*, 1987), a complex set of rearrangements is involved in the generation of the ultimate oncogenic viruses themselves (Fig. 9).

All laboratory mice contain 60 or so endogenous proviruses related to MuLV. These can be divided into four host range classes (E, P, M, X). Each of these proviruses is individually benign and expressed at a low level. Three of them, however, conspire in a complex way to generate the ultimate oncogenic agent.

The specific steps inferred in this process are as follows (J. P. Stoye, C. Moroni, and J. M. Coffin, in preparation):

1. Expression of an E-class provirus and its widespread replication in the animal around the time of birth leading to viremia.
2. Recombination (presumably by copy choice) with an X-class provirus to provide a new LTR.
3. A specific rearrangement in the LTR sequence to duplicate the enhancer region.
4. Recombination with a P-class provirus to alter the envelope SU (gp70) protein.
5. Integration adjacent to a protooncogene and transformation (by a poorly understood set of events) of the T-cell target.

These specific events occur synchronously (as can be detected using specific DNA probes, for example) in virtually every mouse of these strains. There is no imaginable basis by which the specific recombinations and rearrangements can be directed to occur; rather, the products seen most are those selected out of a very large pool of randomly occurring events.

The selective forces involved are not well understood. The LTR alterations presumably act to enhance the ability of the virus to replicate in the target tissue. The env alteration provides the recombinant virus [known as mink cell focus-forming (MCF) virus] with the ability to use a different receptor (Rein, 1982), but the role of this difference is less than obvious. In all of this genetic contortion it is striking that the only genomic variation seen is that selected for. Comparison of nucleotide sequence of recombinant and parental viruses reveals very little random sequence variation (Buller *et al.*, 1987; Holland *et al.*, 1983; Laigret *et al.*, 1988). Thus, this example points up the critical role of selective forces in shaping retroviral genomes.

V. CONCLUSIONS AND IMPLICATIONS FOR RETROVIRAL EVOLUTION

In this chapter, I have summarized what I consider to be important aspects of retrovirus genetic variation and some specific examples from my own experience. The plasticity of retrovirus genomes and their ability to evolve under selective pressure should by now be apparent. Genomic plasticity clearly results from specialized aspects of viral replication, the tendency of reverse transcriptase to make certain kinds of errors, and the uniquely intimate association the virus has with its host. Selective pressures favoring such evolution can be imposed at many different levels, from the scientist taking home chicken tumors to the acquired immunodeficiency syndrome patient with

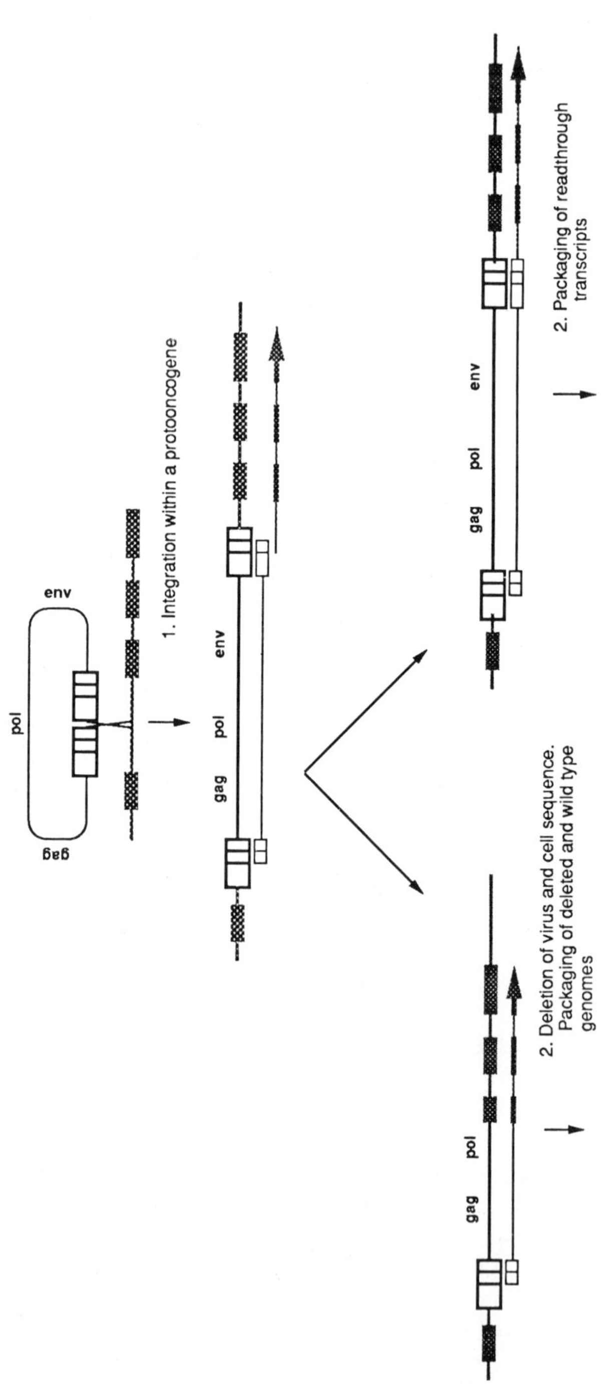
gag
pol
env
1. Integration within a protooncogene
gag
pol
env
gag
pol
env
2. Packaging of readthrough transcripts
gag
pol
2. Deletion of virus and cell sequence.
Packaging of deleted and wild type genomes

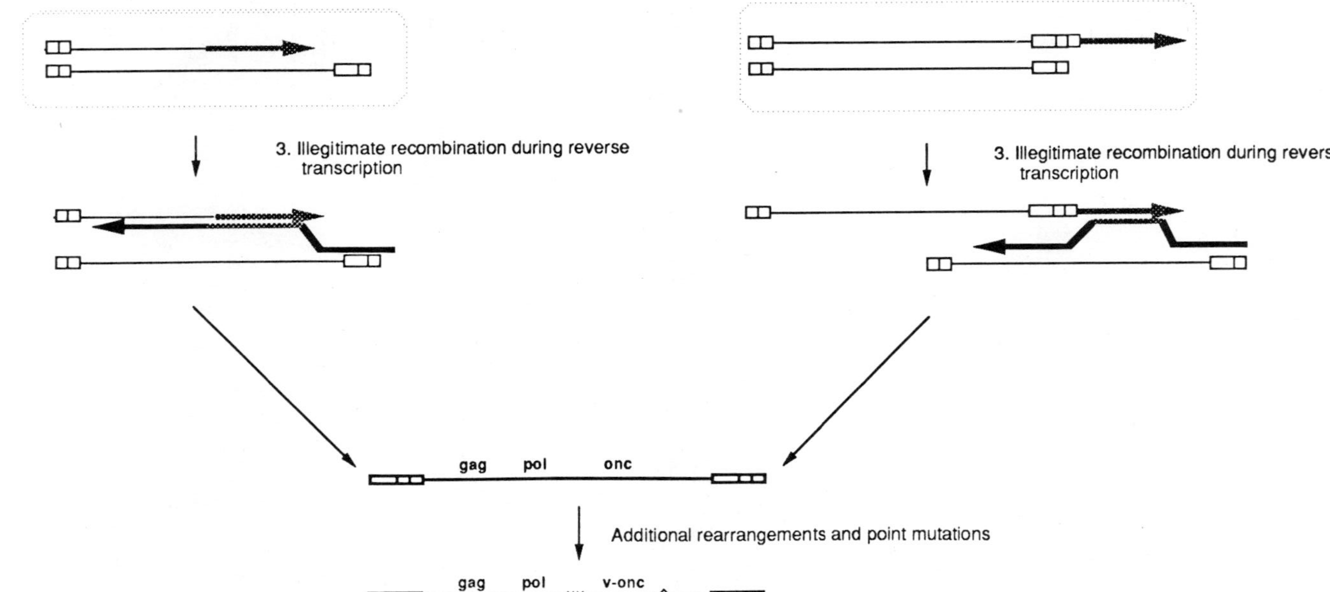

Figure 8. Two mechanisms of oncogene capture. The initial step in the two schemes is integration of a provirus within a protooncogene, in this example between the first two exons (thick lines). In one scheme (left), this is followed by a specific deletion that fuses viral and oncogene sequences (Swanstrom *et al.*, 1983). A joint transcript is then copackaged with a wild-type genome and illegitimate recombination generates the final product. In the alternative scheme (right), a readthrough transcript containing complete viral and (possibly spliced) protooncogene sequences is incorporated into virus, where it can serve as a substitute for illegitimate recombination during reverse transcription (Herman and Coffin, 1987).

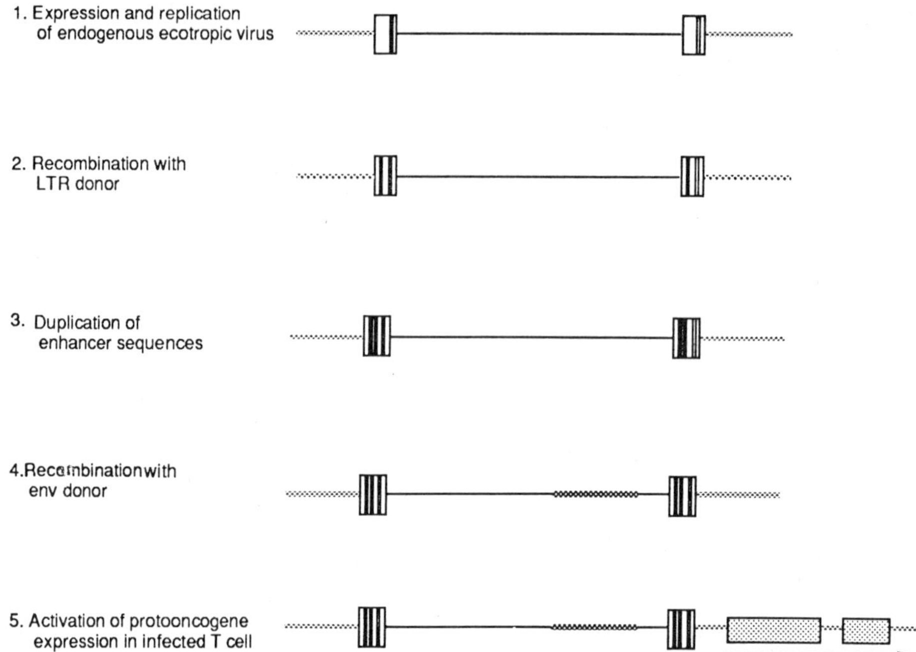

Figure 9. Virological event during leukemogenesis in some mice. The genomes of the various variant viruses detectable during aging of virtually all AKR mice are shown along with the events that presumably give rise to them (J. P. Stoye, C. Moroni, and J. M. Coffin, in preparation).

circulating antibodies. Given that retroviruses can be found to do almost anything, accurate description of selective effects is thus more important to understanding mechanisms of variation than is understanding of the mechanisms involved.

In view of the probable importance of understanding variation and evolution in HIV and the amount of nucleotide sequence information available, it has become recently popular to draw phylogenetic trees relating different HIV (and lentivirus) isolates (Myers *et al.*, 1988; Smith *et al.*, 1988) and, on a grander scale, all retrovirus groups (Doolittle *et al.*, 1989; McClure *et al.*, 1988) to one another. This can be a useful exercise for understanding the divergence of retroviruses. Such charts cannot be used to infer times of divergence or rates of past evolution since to do so requires the untenable assumption that rates of variation are uniform for all parts of the genome (or that the author can identify changes that represent truly "neutral" mutations). Making such assumptions leads to ridiculous conclusions such as that all retroviruses evolved from a common ancestor within the past 10,000 years or that HIV-1 and HIV-2 [which differ by about 35% of amino acid sequence in the most strongly conserved region of *pol* (Myers *et al.*, 1988)] diverged 40 years ago (Smith *et al.*, 1988)!

Estimates like these are probably in error by 4 or more orders of magnitude. Retroviruses are extremely widespread among vertebrates and probably other animals as well. They are also the only group to have left a "fossil record"—the endogenous proviruses found in vertebrate genomes (Coffin, 1982; Stoye and Coffin, 1985). These clearly reveal a history of repeated waves of virus infection of the germline, progressing back through evolutionary history. For instance, the endogenous proviruses found in the same location in the human and chimpanzee genome (Repaske *et al.*, 1985; Steele *et al.*, 1986) clearly belong to the group that includes endogenous and exogenous mammalian retroviruses, such as MuLV, indicating that the group has existed in modern form for much longer than 5 million years. Proviruses that are still much older but obviously related to modern-day groups can also be found (Dunwiddie *et al.*, 1986).

In conclusion, although retroviruses are subject to rapid variation and adaptation to new niches, they are not necessarily subject to rapid evolution. There is every reason to believe that retroviruses closely resembling those we see today have been in existence throughout vertebrate history. Considering their present importance in disease and their past importance in modifying the germline, it behooves us to understand the evolutionary relationships of these special agents with their hosts to the greatest extent possible.

REFERENCES

Bandyopadhyay, P. K., Watanabe, S., and Temin, H. M. (1984). *Proc. Natl. Acad. Sci. USA* **81,** 3476–3480.

Batschelet, E., Domingo, E., and Weissmann, C. (1976). *Gene* **1,** 27–32.

Bishop, J. M., and Varmus, H. E. (1985). In *RNA Tumor Viruses* 2/supplements and appendixes (R. Weiss, N. Teich, H. Varmus, and J. Coffin, eds.), pp. 249–356, Cold Spring Harbor Laboratory, Cold Spring Harbor, NY.

Boral, A. L., Okenquist, S. A., and Lenz, J. (1988). *J. Virol.* **63,** 76–84.

Bova, C. A., Manfredi, J. P., and Swanstrom, R. (1986). *Virology* **152,** 343–354.

Bova, C. A., Olsen, J. C., and Swanstrom, R. (1988). *J. Virol.* **62,** 75–83.

Brown, D. W., and Robinson, H. L. (1988). *Virology* **162,** 239–242.

Brown, P. O., Bowerman, B., Varmus, H. E., and Bishop, J. M. (1987). *Cell* **49,** 347–356.

Buller, R. S., Ahmed, A., and Portis, J. L. (1987). *J. Virol.* **61,** 29–34.

Coffin, J. M. (1979). *J. Gen. Virol.* **42,** 1–26.

Coffin, J. M. (1982). In *RNA Tumor Viruses* 1/text (R. Weiss, N. Teich, H. Varmus, and J. Coffin, eds.), pp. 1109–1204, Cold Spring Harbor Laboratory, Cold Spring Harbor, NY.

Coffin, J. M. (1985). In *RNA Tumor Viruses* 2/supplements and appendixes (R. Weiss, N. Teich, H. Varmus, and J. Coffin, eds.), pp. 17–74, Cold Spring Harbor Laboratory, Cold Spring Harbor, NY.

Coffin, J. M. (1986). *Cell* **46,** 1–4.

Coffin, J. M., Tsichlis, P. N., Barker, C. S., and Voynow, S. (1980). *Ann. NY Acad. Sci.* **354,** 410–425.

Cullen, B. R., Skalka, A. M., and Ju, G. (1983). *Proc. Natl. Acad. Sci. USA* **80,** 2946–2950.

Davis, B. R., Brightman, B. K., Chandy, K. G., and Fan, H. (1987). *Proc. Natl. Acad. Sci. USA* **84,** 4875–4879.

Doolittle, R. F., Feng, D. F., Johnson, M. S., and McClure, M. A. (1989). *Q. Rev. Biol.* **64,** 1–30.

Dorner, A. J., and Coffin, J. M. (1986). *Cell* **45,** 365–374.

Dorner, A. J., Stoye, J. P., and Coffin, J. M. (1985). *J. Virol.* **53,** 32–39.

Dougherty, J. P., and Temin, H. M. (1986). *Mol. Cell. Biol.* **6,** 4387–4395.

Dunwiddie, C. T., Resnick, R., Boyce-Jacino, M., Alegre, J. N., and Faras, A. J. (1986). *J. Virol.* **59,** 669–675.

Fujiwara, T., and Mizuuchi, K. (1988). *Cell* **54,** 497–504.

Goff, S. P. (1984). *Curr. Topics Microbiol. Immunol.* **112,** 45–69.

Goldfarb, M. P., and Weinberg, R. A. (1981). *J. Virol.* **38,** 136–150.

Golemis, E., Li, Y., Fredrickson, T. N., Hartley, J. W., and Hopkins, N. (1988). *J. Virol.* **63,** 328–337.

Gopinathan, K. P., Weymouth, L. A., Kunkel, T. A., and Loeb, L. A. (1979). *Nature* **278,** 857–859.

Hagino-Yamagishi, K., Donehower, L. A., and Varmus, H. E. (1987). *J. Virol.* **61,** 1964–1971.

Hayward, W. S., Neel, B. G., and Astrin, S. M. (1981). *Nature* **290,** 475–480.

Herman, S. A., and Coffin, J. M. (1986). *J. Virol.* **60,** 497–505.

Herman, S. A., and Coffin, J. M. (1987). *Science* **236,** 845–848.

Holland, C. A., Wozney, J., and Hopkins, N. (1983). *J. Virol.* **47,** 413–420.

Hughes, S. H., and Kosick, E. (1984). *Virology* **136,** 89–99.

Hunter, E., Hill, E., Hardwick, M., Bhown, A., Schwartz, D. E., and Tizard, R. (1983). *J. Virol.* **46,** 920–936.

Junghans, R. P., Boone, L. R., and Skalka, A. M. (1982). *Cell* **30,** 53–62.

Kunkel, T. A., Schaaper, R. M., and Loeb, L. A. (1983). *Biochemistry* **22,** 2378–2384.

Laigret, F., Repaske, R., Boulukos, K., Rabson, A. B., and Khan, A. S. (1988). *J. Virol.* **62,** 376–386.
Lasky, L. A., Nakamura, G., Smith, D. H., Fennie, C., Shimasaki, C., Patzer, E., Berman, P., Gregory, T., and Capon, D. J. (1987). *Cell* **50,** 975–985.
Leider, J. M., Palese, P., and Smith, F. I. (1988). *J. Virol.* **62,** 3084–3091.
Levantis, P., Gillespie, D. A. F., Hart, K., Bissell, M. J., and Wyke, J. A. (1986). *J. Virol.* **57,** 907–916.
Li, Y., Golemis, E., Hartley, J. W., and Hopkins, N. (1987). *J. Virol.* **61,** 693–700.
Linial, M., and Blair, D. (1982). In *RNA Tumor Viruses* 1/text (R. Weiss, N. Teich, H. Varmus, and J. Coffin, eds.), pp. 649–783, Cold Spring Harbor Laboratory, Cold Spring Harbor, NY.
Loeb, D. D., Padgett, R. W., Hardies, S. C., Shehee, W. R., Comer, M. B., Edgell, M. H., and Hutchinson, C. A. H., III. (1986). *Mol. Cell. Biol.* **6,** 168–182.
McClure, M. A., Johnson, M. S., Feng, D., and Doolittle, a. R. F. (1988). *Proc. Natl. Acad. Sci. USA* **85,** 2469–2473.
Miles, B. D., and Robinson, H. L. (1985). *J. Virol.* **54,** 295–303.
Mizutani, S., and Temin, H. M. (1976). *Biochemistry* **15,** 1510–1516.
Modrow, S., Hahn, B. H., Shaw, G. M., Gallo, R. C., Wong-Staal, F., and Wolf, H. (1987). *J. Virol.* **61,** 570–578.
Myers, G., Rabson, A. B., Josephs, S. F., and Wong-Staal, F., eds. (1988). *Human Retroviruses and AIDS 1988,* Los Alamos National Laboratory, Los Alamos, NM.
Omer, C. A., Pogue-Geile, K., Guntaka, R., Staskis, K. A., and Faras, A. J. (1983). *J. Virol.* **47,** 380–382.
Overbaugh, J., Donahue, P. R., Quackenbush, S. L., Hoover, E. A., and Mullins, J. I. (1988). *Science* **239,** 906–910.
Parvin, J. D., Moscona, A., Pan, W. T., Leider, J. M., and Palese, P. (1986). *J. Virol.* **59,** 377–383.
Preston, B. D., Poiesz, B. J., and Loeb, L. A. (1988). *Science* **242,** 1168–1171.
Quinn, T. P. and Grandgenett, D. P. (1988). *J. Virol.* **62,** 2307–2312.
Rein, A. (1982). *Virology* **120,** 251–257.
Repaske, R., Steele, P. E., O'Neill, R. R., Rabson, A. B., and Martin, M. A. (1985). *J. Virol.* **54,** 764–772.
Roberts, J. D., Bebenek, K., and Kunkel, T. A. (1988). *Science* **242,** 1171–1173.
Roberts, J. D., Preston, B. D., Johnston, L. A., Soni, A., Loeb, L. A., and Kunkel, T. (1989). *Mol. Cell. Biol.* **9,** 469–476.
Rohdewohld, H., Weiher, H., Reik, W., Jaenisch, R., and Breindl, M. (1987). *J. Virol.* **61,** 336–343.
Selten, G., Cuypers, H. T., Zijlstra, M., Melief, C., and Berns, A. (1984). *EMBO J.* **3,** 3215–3222.
Shih, C., Stoye, J. P., and Coffin, J. M. (1988). *Cell* **53,** 531–537.
Shimotohno, K., and Temin, H. M. (1982). *J. Virol.* **41,** 163–171.
Shoemaker, C. S., Goff, S., Gilboa, E., Paskind, M., Mitra, S. W., and Baltimore, D. (1980). *Proc. Natl. Acad. Sci. USA* **77,** 3932–3936.
Short, M. K., Okenquist, S. A., and Lenz, J. (1987). *J. Virol.* **61,** 1067–1072.
Smith, D. B., and Inglis, S. C. (1987). *J. Gen. Virol.* **68,** 2729–2740.
Smith, T. F., Srinivasan, A., Schochetman, G., Marcus, M., and Myers, G. (1988). *Nature* **333,** 573–575.
Speck, N. A., and Baltimore, D. (1987). *Mol. Cell. Biol.* **7,** 1101–1110.
Starcich, B. R., Hahn, B. H., Shaw, G. M., McNeely, P. D., Modrow, S., Wolf, H., Parks, E. S., Parks, W. P., Josephs, S. F., Gallo, R. C., and Wong-Staal, F. (1986). *Cell* **45,** 637–648.
Steele, P. E., Martin, M. A., Rabson, A. B., Bryan, T., and O'Brien, S. J. (1986). *J. Virol.* **59,** 545–550.
Steinhauer, D. A., and Holland, J. J. (1982). *J. Virol.* **57,** 219–228.
Stewart, M. A., Forrest, D., McFarlane, R., Onions, D., Wilkie, N., and Neil, J. C. (1986). *Virology* **154,** 121–134.
Stoye, J. P., and Coffin J. M. (1985). In *RNA Tumor Viruses* 2/supplements and appendixes (R. Weiss, N. Teich, H. Varmus, and J. Coffin, eds.), pp. 357–404, Cold Spring Harbor Laboratory, Cold Spring Harbor, NY.
Stoye, J. P., Fenner, S., Greenoak, G. E., Moran, C., and Coffin, J. M. (1988). *Cell* **54,** 383–391.
Swanstrom, R., Parker, R. C., Varmus, H. E., and Bishop, J. M. (1983). *Proc. Natl. Acad. Sci. USA* **80,** 2519–2523.
Teich, N., Wyke, J., Mak, T., Bernstein, A., and Hardy, W. (1982). In *RNA Tumor Viruses* 1/text (R. Weiss, N. Teich, H. Varmus, and J. Coffin, eds.), pp. 785–998, Cold Spring Harbor Laboratory, Cold Spring Harbor, NY.
Teich, N., Wyke, J., and Kaplan, P. (1985). In *RNA Tumor Viruses* 2/supplements and appendixes (R. Weiss,

N. Teich, H. Varmus, and J. Coffin, eds.), pp. 187–248, Cold Spring Harbor Laboratory, Cold Spring Harbor, NY.

Temin, H. M. (1985). *Mol. Biol. Evol.* **6,** 455–468.

Tsichlis, P. N., and Coffin, J. M. (1980). *J. Virol.* **33,** 238–249.

Varmus, H. E. (1987). *Sci. Am.* **257,** 56–66.

Varmus, H. (1988). *Science* **240,** 1427–1435.

Varmus, H. E. and Swanstrom, R. (1982). In *RNA Tumor Viruses* 1/text (R. Weiss, N. Teich, H. Varmus, and J. Coffin, eds.), pp. 369–512, Cold Spring Harbor Laboratory, Cold Spring Harbor, NY.

Varmus, H. E., and Swanstrom, R. (1985). In *RNA Tumor Viruses* 2/supplements and appendixes (R. Weiss, N. Teich, H. Varmus, and J. Coffin, eds.), pp. 74–134, Cold Spring Harbor Laboratory, Cold Spring Harbor, NY.

Vijaya, S., Steffen, D. L., and Robinson, H. L. (1986). *J. Virol.* **60,** 683–692.

Villemur, Monczak, Y., Rassart, E., Kozak, C., and Jolicoeur, P. (1987). *Mol. Cell. Biol.* **7,** 512–522.

Voynow, S. L., and Coffin, J. M. (1985a). *J. Virol.* **55,** 67–78.

Voynow, S. L., and Coffin, J. M. (1985b). *J. Virol.* **55,** 79–85.

Weiss, R. A. (1982). In *RNA Tumor Viruses* 1/text (R. Weiss, N. Teich, H. Varmus, and J. Coffin, eds.), pp. 209–260, Cold Spring Harbor Laboratory, Cold Spring Harbor, NY.

Willey, R. L., Rutledge, R. A., Dias, S., Folks, T., Theodore, T., Buckler, C. E., and Martin, M. A. (1986). *Proc. Natl. Acad. Sci. USA* **83,** 5038–5042.

2

Human Immunodeficiency Virus Variation and Epidemiology of Acquired Immunodeficiency Syndrome and Human Immunodeficiency Virus Infection

Mary E. Chamberland and James W. Curran

I. HUMAN IMMUNODEFICIENCY VIRUSES

Substantial progress has been achieved during the last few years in elucidating a clearer understanding of the molecular biology of human immunodeficiency virus-1 (HIV-1) and its mechanisms of infectivity and pathogenesis. HIV-1 selectively infects cells that have a surface antigen known as CD4 (Klatzmann *et al.*, 1984; Maddon *et al.*, 1986). The T-helper lymphocyte is particularly susceptible; infection with HIV-1 can destroy T-helper cells and interfere with T-helper cell function (Barre-Sinoussi *et al.*, 1983; Gallo, *et al.*, 1984; Fauci, 1988). CD4-positive monocytes, macrophages, B lymphocytes, and certain cells in the brain and gastrointestinal tract can also be infected with HIV-1 (Montagnier *et al.*, 1984; Gartner *et al.*, 1986; Wiley *et al.*, 1986; Nelson *et al.*, 1988). The collective result is profound immunosuppression manifested by opportunistic infections and malignancies, as well as other clinical features including hematologic abnormalities, arthritis, pneumonitis, nephritis, certain diarrheal and wasting syndromes, and neurologic dysfunction, all of which may be direct viral effects rather than secondary effects of immune suppression (Report of the Workgroup on Clinical Manifestations and Pathogenesis, 1989).

Different strains of HIV-1 vary in their individual genotypes and in biologic properties, such as host cell range, infection kinetics, replication, and cytopathogenicity (Sakai *et al.*, 1988). Some studies have indicated that a person may be infected with multiple HIV-1 strains, which may account in part for differences in the clinical expression of HIV-1 infection, the rate of disease progression, and differences in infectivity and susceptibility, but the data are not conclusive (Hahn *et al.*, 1986; Sakai *et al.*, 1986).

A second retrovirus, HIV-2, has been isolated and associated with the development of serious

Mary E. Chamberland and James W. Curran • Division of HIV/AIDS, Center for Infectious Diseases, Centers for Disease Control, Public Health Service, U.S. Department of Health and Human Services, Atlanta, Georgia 30333.

opportunistic diseases clinically indistinguishable from acquired immunodeficiency syndrome (AIDS) caused by HIV-1. HIV-1 and HIV-2 are genetically and immunologically distinct. HIV-2 has only approximately 40% DNA homology with HIV-1 (Guyader *et al.*, 1987). Most cases of HIV-2 infection have been reported from countries in west Africa and have been associated with the same modes of transmission as HIV-1 (Brun-Vezinet *et al.*, 1987; Clavel *et al.*, 1987; CDC, 1988a; Horsburgh and Holmberg, 1988). Although the reported spectrum of disease is similarly broad, the relative pathogenicity of the two viruses is unknown. Simultaneous HIV-1 and HIV-2 infection in the same person has been reported (Evans *et al.*, 1988; Rayfield *et al.*, 1988).

II. EPIDEMIOLOGY OF AIDS IN THE UNITED STATES

A. Incidence of AIDS and Associated Mortality

AIDS cases in the United States are reported to the Centers for Disease Control (CDC) by all 50 states, the District of Columbia, and U.S. possessions and territories, using a uniform surveillance case definition. The original definition has evolved from one which included a limited number of specific opportunistic diseases diagnosed by microscopy and histologic or cytologic techniques in patients with no other known causes of immunodeficiency to one which incorporates a wider spectrum of HIV-1-associated clinical and immunologic manifestations. The most recent revision (September 1987) incorporated (1) a broader range of AIDS-indicative diseases, most notably encephalopathy and wasting syndrome; (2) a limited number of indicator diseases that are diagnosed presumptively, without confirmatory laboratory evidence of the opportunistic disease; and (3) HIV-1 diagnostic tests to further improve the sensitivity and specificity of the definition (Table 1) (CDC, 1987a).

As of December 18, 1988, 79,698 cases of AIDS in adults and adolescents older than 13 years were reported in the United States; an additional 1298 cases were reported in children under 13 years of age. The number of cases has increased rapidly over time. Between June 1981 and June 1982 less than 500 cases were reported. By comparison, in the most recent 12-month period, over 32,000 cases were reported. In 1992 alone, 80,000 AIDS cases are projected to be diagnosed, and the cumulative number of diagnosed cases is estimated to total between 205,000 and 440,000 by the end of 1992 (Fig. 1) (Report of the Workgroup on Epidemiology and Surveillance, 1989).

Fifty-six percent of all adult AIDS patients are reported to have died. Reported fatalities increase with time from diagnosis of AIDS: 25% of patients are reported to have died within 1 year of diagnosis; 48% and 69% are reported to have died 2 and 3 years, respectively, after diagnosis. A similar proportion (56%) of children with AIDS are reported to have died. However, reported fatality rates in children vary significantly by age: 67% of infants younger than 1 year at diagnosis are reported to have died compared with 47% of children older than 1 year. Reporting of deaths is known to be incomplete and fatality rates for both children and adults are much higher (Long-Term Survivor Collaborative Study Group and Hardy, 1987).

One measure of the impact of premature mortality is the years of potential life lost before age 65 (YPLL). This measure emphasizes deaths of children and young adults whereas crude mortality statistics represent all deaths, which largely occur among the elderly. The leading causes of YPLL have changed only minimally from 1979 to 1987, except for AIDS and HIV infection. Fewer than five AIDS deaths were retrospectively identified in 1979, but by 1987 AIDS and HIV infection had become the seventh leading cause of YPLL in the United States (CDC, 1989). The impact of AIDS on patterns of premature death is particularly important in areas such as New York City and San Francisco, where AIDS has been the leading cause of YPLL for single men aged 25–44 since 1984 (Curran *et al.*, 1988). AIDS is the leading cause of death for men aged 30–39 years and for women aged 25–29 years in New York City (Kristal, 1986; Chiasson *et al.*, 1987).

Table 1. Revised Surveillance Case Definition for AIDS, Centers for Disease Control, September 1987

I. Indicator diseases diagnosed definitively in absence of other causes of immunodeficiency and laboratory tests for HIV
- Candidiasis of the esophagus, trachea, bronchi, or lungs
- Cryptococcus, extrapulmonary
- Cryptosporidiosis with diarrhea >1 month
- Cytomegalovirus disease exclusive of liver, spleen, or lymph nodes in patients >1 month of age.
- Herpes simplex virus infection, causing a mucocutaneous ulcer >1 month; or bronchitis, pneumonitis, or esophagitis in patients >1 month of age
- Kaposi's sarcoma in patients <60 years of age
- Lymphoma of the brain (primary) in patients <60 years of age
- Lymphoid interstitial pneumonia and/or pulmonary lymphoid hyperplasia in patients <13 years of age
- *Myocobacterium avium* complex or *M. kansasii* disease, disseminated
- *Pneumocystis carinii* pneumonia
- Progressive multifocal leukoencephalopathy
- Toxoplasmosis of the brain in patients >1 month of age

II. Indicator diseases diagnosed definitively regardless of other causes of immunodeficiency and with laboratory evidence of HIV
- All indicator diseases listed in Section I
- Bacterial infections, recurrent or multiple in patients <13 years of age caused by *Haemophilus, Streptococcus,* or other pyogenic bacteria
- Coccidioidomycosis, disseminated
- HIV encephalopathy
- Histoplasmosis, disseminated
- Isosporiasis with diarrhea >1 month
- Kaposi's sarcoma at any age
- Lymphoma of the brain (primary), at any age
- Non-Hodgkin's lymphoma of B-cell or unknown immunologic phenotype including small noncleaved lymphoma or immunoblastic sarcoma
- Mycobacterial disease exclusive of *M. tuberculosis,* disseminated
- *M. tuberculosis,* extrapulmonary
- *Salmonella* septicemia, recurrent
- HIV wasting syndrome

III. Indicator diseases diagnosed presumptively with laboratory evidence of HIV infection
- Candidiasis, esophageal
- Cytomegalovirus retinitis with loss of vision
- Kaposi's sarcoma
- Lymphoid interstitial pneumonia and/or pulmonary lymphoid hyperplasia in patients <13 years of age
- Mycobacterial disease, disseminated
- *Pneumocystis carinii* pneumonia
- Toxoplasmosis, brain, in patients >1 month of age

IV. Indicator diseases diagnosed definitively in absence of other causes of immunodeficiency and laboratory tests for HIV are negative
- *Pneumocystis carinii* pneumonia
- Other indicator diseases listed in Section I and a T-helper/inducer (CD4) lymphocyte count <400/mm^3

B. Patient Exposure Groups

For surveillance purposes, AIDS cases are counted only once in a hierarchy of exposure categories (Table 2). Persons with more than one reported mode of exposure to HIV-1 are classified in the exposure category listed first in the hierarchy, except for men with a history of both homosexual/bisexual contact and intravenous (IV)-drug use, who make up a separate category.

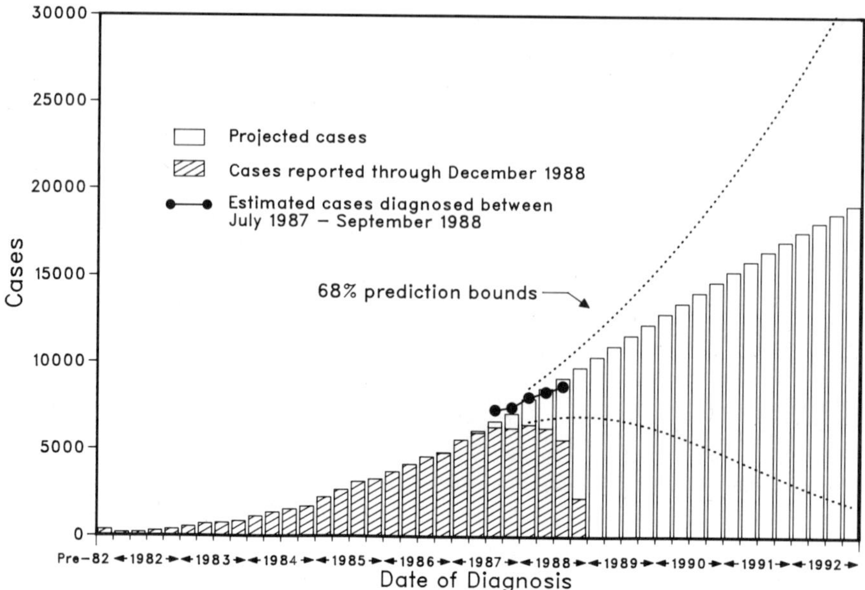

Figure 1. Reported (striped bars) and projected (open bars) AIDS cases in the United States. Projected cases were derived by extrapolating trends in cases diagnosed through June 1987 and reported through March 1988. For comparison with projections, estimated cases diagnosed between July 1987 and September 1988 and reported through December 1988 are shown (solid line with bullets). Both projected and estimated cases have been adjusted for reporting delays.

Table 2. Demographic Characteristics of Adults and Adolescents >13 Years with AIDS Reported in the United States through December 19, 1988

Exposure group	No. cases	Age at diagnosis (yr)		Sex (%)		Race/ethnicity (%)			
		Mean	Range	Male	Female	White	Black	Hispanic	Other
Homosexual/bisexual man	49,294	37.0	13–86	100	NA	73	16	10	1
IV-drug user	15,782	35.4	15–72	78	22	20	50	29	1
Homosexual man and IV-drug user	5,760	34.0	15–75	100	NA	60	25	14	1
Hemophilia/coagulation disorder	764	35.5	13–87	97	3	84	6	8	2
Heterosexual transmission									
Heterosexual contact	2,263	35.4	15–78	24	76	30	47	23	1
Born in country with high incidence of heterosexual transmission	1,230	33.5	16–73	76	24	<1	99	1	0
Transfusion recipient	2,003	54.1	13–90	63	37	74	16	8	2
Undetermined	2,602	40.0	14–84	80	20	38	38	22	2
Total	79,698	36.9	13–90	91	9	58	26	15	1
U.S. population				49	52	80	12	6	2

Since the beginning of the epidemic, cases in homosexual and bisexual men have consistently represented the largest number, as well as the largest proportion, of AIDS cases. While the number of AIDS cases reported each year continued to increase in all patient exposure groups, the percent increase from 1987 through mid-December 1988 has been smaller (25%) for homosexual or bisexual adult men than for other patient groups (88%). The patient increase over this period for heterosexual IV-drug users was 99%. The proportion of cases in homosexual and bisexual men remained relatively stable at about 65% until 1988, when it decreased to 57% of cases reported that year. There has been a corresponding increase from 19% to 24% in the proportion of cases attributable to IV-drug use by heterosexuals.

These changes in the proportional distribution of AIDS cases reflect in part the larger percent increase in cases among nonhomosexual or bisexual men, particularly among heterosexual IV-drug users. Some of these changes can be attributed to different rates by homosexual/bisexual men and heterosexual IV-drug users of HIV-1 antibody testing and presumptive diagnostic methods incorporated in the 1987 revised case definition. In the first 12 months after implementation of the September 1987 definition, cases reported in heterosexual IV-drug users were significantly more likely (45%) than those in exclusively homosexual or bisexual men (21%) to fulfill only the new criteria in the 1987 case definition.

The proportion of cases in adults with hemophilia or other coagulation disorders requiring administration of clotting factors has remained stable at 1%. Recipients of transfused blood or blood components account for 3% of all AIDS cases. While there has been a small, but steady increase in the proportion of transfusion-associated cases from 1% of adult AIDS cases reported in 1982, to 3% of cases reported in 1988, all but a few of these cases reflect transmission from blood or blood products received before screening for HIV-1 antibody was available (Ward *et al.*, 1988a). Because of the long period from the date of transfusion to the development of AIDS, additional transfusion-associated cases will be diagnosed and reported.

Of the 3493 adults in the heterosexual transmission category, 2263 (65%) reported contact with a person with or at increased risk for HIV-1 infection, and 1230 (35%) were born in areas, such as Haiti or central Africa, where heterosexual transmission is a major route of HIV-1 infection.

Excluding AIDS patients of Haitian or central African origin, persons with heterosexually acquired AIDS had partners who were IV-drug users (71%), bisexual men (11%), born in Haiti or central Africa (2%), or received blood or blood components (3%); for 13% the risk of the partner was unreported or under investigation.

Although the overall proportion of heterosexually acquired AIDS cases has remained relatively stable at about 4% since 1983, the composition of this group has changed significantly over time. Of the 107 heterosexual transmission cases reported in 1983, 84 (79%) were in persons of Haitian or central African origin. In contrast, only 360 (25%) of the 1453 heterosexual transmission cases reported in 1988 were in persons from Haiti or central Africa. The decline in the proportion of cases in Haitians may reflect a concomitant decline in immigration to the United States following a large influx from 1978 to 1981 (Hardy *et al.*, 1985). In addition, the percent increase from 1987 through mid-December 1988 has been larger for persons who reported heterosexual contact with a specific person at risk of or known to have HIV-1 infection than for persons of Haitian or central African origin (81% versus 35%).

Heterosexual contact patients, excluding persons born in Haiti or central Africa, increased from 1.1% of all cases reported in 1983 to 3.6% reported in 1988. The proportional increase in this group of heterosexually acquired cases was most striking for women: the proportion of women with AIDS in the heterosexual contact category (excluding women born in Haiti or central Africa) increased from 15% of cases reported in 1983 to 25% in 1988. The proportion of men with AIDS in this category increased from 0.1% to 1.1% over the same period.

Overall, 2602 (3%) of all adult AIDS cases have no risk factor reported and are classified as undetermined. The undetermined exposure category comprises persons who could not be reclassified after follow-up investigation (13%); persons who have died, refused to be interviewed, or were lost to follow-up (18%); and persons still under investigation (69%). When follow-up information is available, risk factors can be identified for over 70% of these patients, and they are reclassified into the appropriate exposure categories (Castro *et al.*, 1988a). The race/ethnicity and gender of AIDS patients with an undetermined risk are demographically distinctive within the U.S. population (Table 2), which suggests that they probably represent a mixture of persons who either have undetected risk factors they are unwilling to report or are sex partners of persons in risk groups (Chamberland *et al.*, 1984; Lekatsas *et al.*, 1986; Castro *et al.*, 1988a). The possibility of infection through heterosexual contact can be inferred from follow-up investigations of patients with undetermined risk: nearly 40% report a history of sexually transmitted diseases and one-third of interviewed men give a history of heterosexual contact with a prostitute.

Children born to HIV-infected mothers account for over three-quarters of reported pediatric AIDS cases (Table 3). Seventy-three percent of these mothers were IV-drug users or sex partners of IV-drug users. Risk factors for the remaining children include transfusion (13%) and hemophilia (6%); 3% had an undetermined risk. Like adults, when follow-up information is available, most children who are initially reported with an undetermined risk are reclassified. In one study, half of those reclassified had mothers who used IV-drugs or were partners of men who used IV-drugs (Lifson *et al.*, 1987).

C. Demographic Characteristics

Overall, 88% of adults with AIDS are 20–49 years of age; only 10% are older than 49 years. However, the age distribution of persons reported with AIDS varies by exposure group (Table 2).

Overall, 58% of reported adult AIDS cases are in white persons, 26% in blacks, and 15% in Hispanics (Table 2). However, the overall cumulative incidence of AIDS (number of cases/million population of individual racial/ethnic group) for black and Hispanic adults is approximately 3 times the rate for whites (Selik *et al.*, 1988). Among women, the disproportionate incidence rates of AIDS in blacks and Hispanics compared with whites are most striking for those who report IV-drug use (18 and 11 times greater than whites, respectively) or heterosexual contact with a male IV-drug user (17 and 19 times greater). Similarly, among heterosexual men with AIDS who report IV-drug use, the cumulative incidence for blacks and Hispanics is 18 times greater than for whites.

Men account for 91% of all adult AIDS cases. However, there has been a significant increase in the proportion of adult AIDS patients who are women: from 7% of cases reported in 1982 to 10% in 1988. Heterosexual contact cases (excluding persons born in Haiti or central Africa) are the only transmission group with a predominance of women (76%); 77% of all other heterosexuals with AIDS are men. The smaller number of men with heterosexually acquired AIDS probably reflects the

Table 3. Demographic Characteristics of Children <13 Years with AIDS Reported in the United States through December 19, 1988

Exposure group	No. cases	Age at diagnosis (mo)		Sex (%)		Race/ethnicity (%)			
		Mean	Range	Male	Female	White	Black	Hispanic	Other
Hemophilia/coagulation disorder	82	102.9	21–144	98	2	72	12	13	3
Parent with/at risk of HIV	1,006	20.7	<1–144	50	50	15	61	24	<1
Transfusion recipient	165	51.5	4–144	62	38	56	22	21	1
Undetermined	45	21.3	<1–144	51	49	18	62	20	0
Total	1,298	29.5	<1–144	55	45	24	53	23	<1

much smaller reservoir of infection in U.S. women and the possible lower efficiency of female-to-male transmission (Peterman and Curran, 1986; Guinan and Hardy, 1987).

Most reported pediatric cases of AIDS are in children under 5 years of age at the time of initial diagnosis (82%), with 24% under 6 months of age. Children with hemophilia have the highest mean and median ages (8.6 and 10.0 years, respectively) (Table 3).

The racial/ethnic distribution of children with AIDS is very similar to that of heterosexual adult patients with AIDS: 53% are black, 22% are Hispanic, and 23% are white (Table 3). The cumulative incidence rates in black and Hispanic children are 11.2 and 7.6 times the rate for white children, with the highest rates of AIDS reported for transmission categories associated with IV-drug use (Selik *et al.*, 1988). For children whose mothers use IV-drugs, the rates for blacks and Hispanics are 24.7 and 17.9 times greater, respectively, than that for whites, and for children whose mothers have sex partners who use IV-drugs, 15.9 and 16.3 times greater, respectively.

There is an equal distribution by sex for perinatal cases and cases with an undetermined risk (Table 3). Boys are overrepresented in hemophilic children (98%) and transfusion recipients (62%). The overrepresentation of boys among transfusion-associated cases most likely reflects the higher incidence of transfusions in male infants (Friedman *et al.*, 1980).

D. Geographic Distribution

Five states—New York, California, Florida, Texas, and New Jersey—have reported the majority (66%) of cases. However, the proportion of cases from these states has decreased steadily from 86% of cases reported in 1982 to 62% in 1988. This trend is primarily due to the more rapid increase in the number of cases reported from the remaining 45 states. In general, the cumulative reported incidence of AIDS by state roughly parallels the prevalence of HIV-1 infection in military recruits, with the sex-adjusted prevalence of HIV-1 in military recruit applicants through September 1988 three to ten times higher than the cumulative incidence of reported AIDS (Fig. 2).

The distribution of cases by state of residence varies among the exposure groups: nearly half the AIDS cases in homosexual or bisexual men have been reported from California and New York, while 63% of cases in heterosexual IV-drug users have been reported from New Jersey and New York. Most patients who acquired AIDS through heterosexual contact with an IV-drug user also have been from New York and New Jersey. AIDS patients who received blood or blood product transfusions or clotting factor concentrates are more uniformly distributed throughout the United States. The geographic distribution of perinatally acquired cases in children is similar to that for women with AIDS in that 67% were reported from New York, New Jersey, Florida, and California, compared with 48% of transfusion-acquired cases, and 31% of hemophilia-associated cases in children.

E. Associated Opportunistic Diseases

The two most frequently reported opportunistic diseases among adult with AIDS are *Pneumocystis carinii* pneumonia (61%) and Kaposi's sarcoma (16%).

P. carinii pneumonia is the most commonly reported opportunistic infection for every exposure group. When analyzed by year of diagnosis, the percentage of AIDS patients initially presenting with *P. carinii* pneumonia has increased from 37% in 1982 to 68% in 1988. The reason for this proportional increase is unknown but does not appear to be related to changes in the distribution of cases by transmission group, race/ethnicity, geographic location, or use of presumptive diagnostic measures (Selik *et al.*, 1987).

Kaposi's sarcoma is nearly eight times more frequently reported in homosexual and bisexual men than in heterosexual IV-drug users. When analyzed by year of diagnosis, the percentage of AIDS patients initially presenting with Kaposi's sarcoma has decreased from 28% in 1982 to 10% in

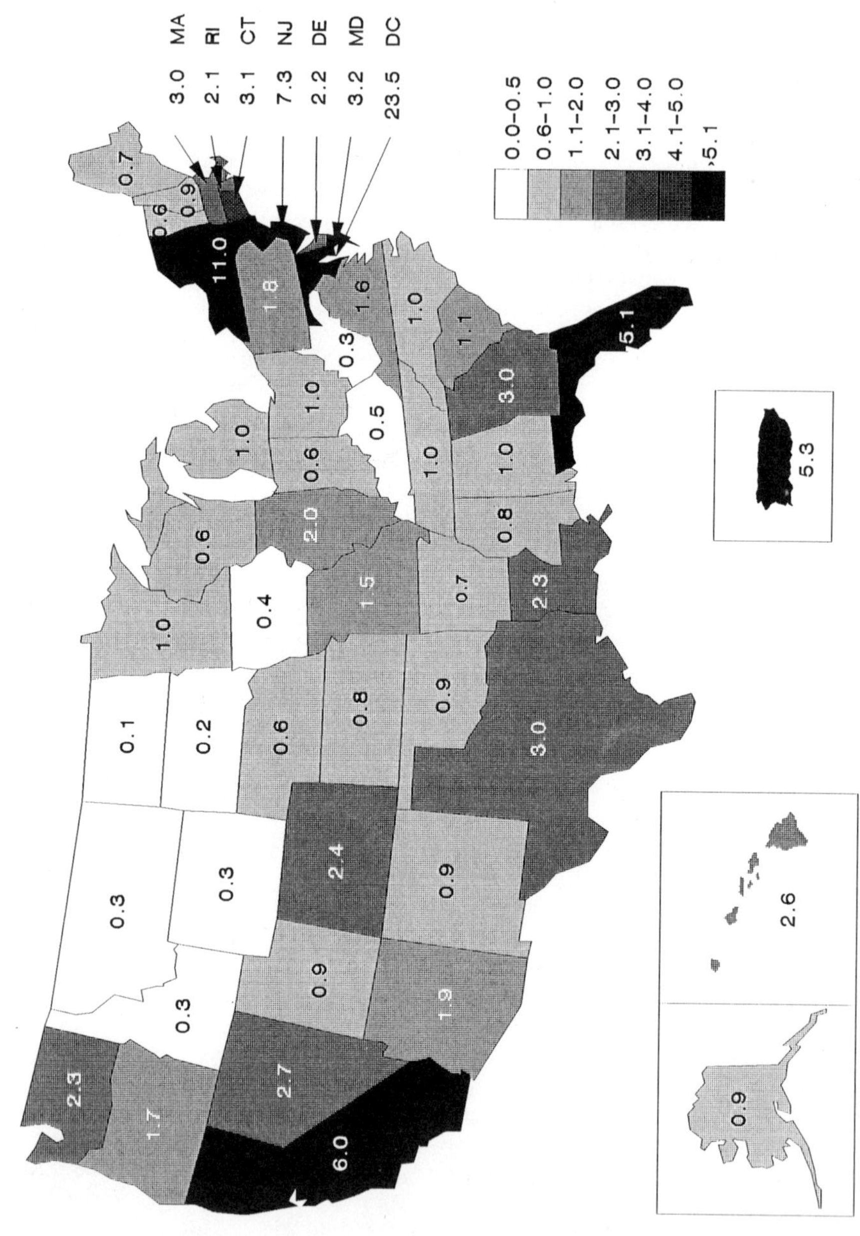
A
3.0 MA
2.1 RI
3.1 CT
7.3 NJ
2.2 DE
3.2 MD
23.5 DC
0.0–0.5
0.6–1.0
1.1–2.0
2.1–3.0
3.1–4.0
4.1–5.0
›5.1

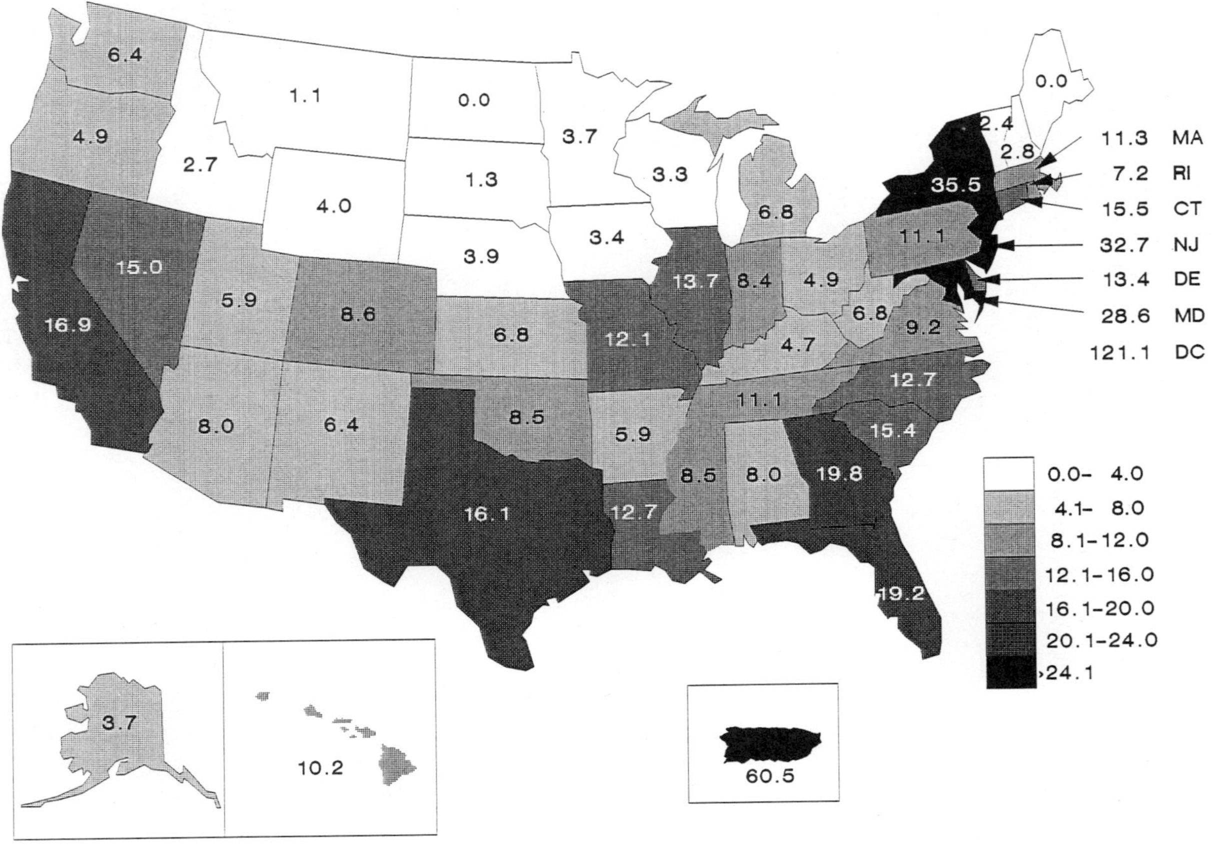

Figure 2. (A) Reported cumulative incidence rates of AIDS, as of December 1988 and (B) HIV seroprevalence rates in U.S. military recruit applicants, by state, from October 1985 through September 1988. Rates are per 10,000 population; military recruit applicant data are sex-adjusted.

1988. Recent studies suggest that this proportional decline is not secondary to diagnostic bias or selective underreporting of Kaposi's sarcoma (Lifson *et al.*, 1988a), but may reflect either a concomitant decrease in exposure to an unknown cofactor necessary for the development of Kaposi's sarcoma or the reciprocal increase in *P. carinii* pneumonia.

Other opportunistic infections or cancers that have been reported in ≥1% of all AIDS patients include esophageal or bronchial candidiasis (16%), wasting syndrome (8%), extrapulmonary cryptococcosis (7%), disseminated *Mycobacterium avium* complex (5%), cytomegalovirus (CMV) disease (5%), herpes simplex (4%), toxoplasmosis of the brain (4%), chronic cryptosporidiosis (3%), encephalopathy (3%), immunoblastic sarcoma (2%), extrapulmonary tuberculosis (2%), CMV retinitis (1%), and other disseminated mycobacterial diseases (1%). The reported frequency of each of these opportunistic diseases represents a minimum estimate of the actual incidence among AIDS patients, because there is significant underreporting of opportunistic diseases diagnosed after the initial case report to CDC.

III. MODES OF TRANSMISSION

A. Sexual Transmission

Sexual transmission is the major mode of HIV-1 transmission throughout the world. Multiple epidemiologic studies have identified specific risk factors associated with HIV-1 infection in homosexual men. These include an increased number of sex partners, receptive anal intercourse, and other practices associated with rectal trauma, such as "fisting" and douching (Marmor *et al.*, 1982; Jaffe *et al.*, 1983; Stevens *et al.*, 1986; Winkelstein *et al.*, 1987; Kingsley *et al.*, 1987; Darrow *et al.*, 1987; Moss *et al.*, 1987).

Heterosexual transmission of HIV-1 has been documented from men to women and, less frequently, from women to men. While transmission of HIV-1 has been reported after a single heterosexual contact (Peterman *et al.*, 1988), other studies have reported no evidence of heterosexual transmission, even after hundreds of contacts (Kreiss *et al.*, 1985; Jason *et al.*, 1986; Fischl *et al.*, 1987; Padian *et al.*, 1987; Smiley *et al.*, 1987; Peterman *et al.*, 1988). The highest reported rates of heterosexual transmission of HIV-1 have been reported for the partners of infected IV-drug users and persons born in Haiti or central Africa (Curran *et al.*, 1988). The differences in these rates may reflect characteristics of the source and recipient partner, as yet poorly differentiated, which may in turn modulate infectivity and susceptibility.

At least one study suggests that infectivity varies over time. In a prospective study of infected hemophilic men, transmission of HIV-1 to their female sex partners was correlated with a significant reduction in the absolute number of T-helper lymphocytes in the men (Goedert *et al.* 1987). Laboratory data indicate that the ability to isolate HIV-1 *in vitro* significantly increases as the number of T-helper cells declines and the patient's clinical course worsens (Redfield *et al.*, 1987). However, other epidemiologic studies have not demonstrated an association between transmission and stage of clinical illness (Padian *et al.*, 1987; Peterman *et al.*, 1988).

Other factors related to heterosexual transmission of HIV-1 include concurrent sexually transmitted diseases (STDs), large numbers of different heterosexual partners, and specific sexual practices. Certain STDs may promote HIV-1 transmission by disrupting genital epithelium. Genital ulcer disease, particularly chancroid, has been associated with HIV-1 infection in both heterosexual men and female prostitutes in Africa (Kreiss *et al.*, 1986; Greenblatt *et al.*, 1987). Similarly, studies of homosexual men show an association between syphilis and herpes simplex virus type 2 and HIV-1 infection independent of the number of partners and specific sex practices (Handsfield *et al.*, 1987; Holmberg *et al.*, 1988). Most heterosexual transmission of HIV-1 occurs through penile–vaginal contact, but in two studies, receptive anal intercourse was associated with an increased risk of transmission between heterosexual partners (Padian *et al.*, 1987; Steigbigel *et al.*, 1987), while in other studies, no such association was demonstrated.

B. Exposure to Blood and Blood Products

AIDS in persons with hemophilia and recipients of transfusions has clearly documented contaminated blood and blood products as a potential vehicle of HIV-1 transmission (Curran *et al.*, 1984; Evatt *et al.*, 1984). The likelihood of becoming infected with HIV-1 after receiving a single-donor blood transfusion documented to be HIV-1 seropositive approaches 100% (Ward *et al.*, 1987). Transmission of HIV-1 has also been linked to recipients of organs, tissues, and semen from donors at increased risk for, or documented to have, HIV-1 infection (L'age-Stehr *et al.*, 1985; Stewart *et al.*, 1985; CDC, 1988b). Serologic screening of blood and blood products, organs, and tissues, heat treatment of clotting factors, and donor deferral procedures have vastly reduced the risk of transmission of HIV-1 by infected donors (Rabkin *et al.*, 1987; CDC, 1987b; Zuck, 1987). The rare exception is recently infected donors who have not yet developed detectable HIV-1 antibody. The rate of such transmission by HIV-1-seronegative blood has been estimated at 26 per 1 million transfusions in the United States (Ward *et al.*, 1988a).

Among IV-drug users, HIV-1 is transmitted by parenteral exposure to HIV-1-contaminated needles and other equipment used for injection. Specific risk factors that have been associated with HIV-1 seropositivity include duration of IV-drug use since 1978, needle sharing, number of persons with whom needles are shared, number of injections, median number of injections in "shooting galleries," and residence in an area with a high prevalence of HIV-1 infection (Schoenbaum *et al.*, 1986; Weiss *et al.*, 1986; Chaisson *et al.*, 1987).

Parenteral, nonintact skin, and mucous membrane exposures to blood have infrequently resulted in occupationally acquired HIV-1 infections in health-care workers (CDC, 1988c). Data from several prospective surveillance projects of health-care workers indicate that the risk of seroconversion following a needlestick exposure to blood from an HIV-1-infected patient is approximately 0.5% (Henderson *et al.*, 1986; Gerberding *et al.*, 1987a; Marcus *et al.*, 1988). Although HIV infection following exposure of nonintact skin or mucous membranes to HIV-infected blood or concentrated virus has been reported in a few instances, the level of risk associated with this type of exposure is likely far lower than a needlestick exposure (CDC, 1987c, 1988c.)

C. Perinatal Transmission

Most perinatal transmission of HIV-1 probably occurs during pregnancy and/or at delivery through exposure to maternal blood or other infected fluids. This is consistent with the laboratory isolation of HIV-1 from both fetal tissues (Jovaisas *et al.*, 1985; LaPointe *et al.*, 1985) and cord blood (Nzilambi *et al.*, 1987). The isolation of HIV-1 from breast milk (Thiry *et al.*, 1985), as well as several case reports of breast-feeding mothers who became infected after delivery and subsequently transmitted HIV-1 to their infants, indicate that breast feeding can transmit HIV-1 (Ziegler *et al.*, 1988; Colebunders *et al.*, 1988; Oxtoby, 1988; Weinbreck *et al.*, 1988). In prospective studies of infants born to infected mothers, the reported rate of perinatally acquired HIV-1 infection is 23–45% (Ryder *et al.*, 1988; European Collaborative Study, 1988; Blanche *et al.*, 1988; Scott *et al.*, 1988; Mendez *et al.*, 1988). Preliminary data suggest that the risk of perinatal transmission increases with the progression of HIV-1-associated immunosuppression and disease in the mother (Nzilambi *et al.*, 1987; Mok *et al.*, 1987).

D. Other Modes of Transmission

Even though HIV-1 has been isolated from a number of body secretions and excretions, only blood, semen, vaginal secretions, and breast milk have been documented as vehicles of transmission (Friedland and Klein, 1987; Lifson, 1988). Considerable attention has focused on the possibility of HIV-1 transmission via saliva following human bites and following mucous membrane and cutaneous contact in the health care setting. At least 15 persons bitten by HIV-infected persons did not become infected with HIV-1 (Lifson, 1988; Rogers *et al.*, 1986; Tsoukas *et al.*, 1986). Prospective studies of more than 100 health-care workers who had parenteral, mucous membrane, or nonintact skin exposure to the saliva of HIV-1-infected persons have not documented transmission of HIV-1 (Henderson *et al.*, 1988; Gerberding *et al.*, 1987a; Marcus *et al.*, 1988). In other studies, over 1800 dental professionals have been evaluated (Flynn *et al.*, 1987; Gerberding *et al.*, 1987b; Klein *et al.*, 1988), and only one dentist, who reported no behavioral risk factors for AIDS but who had frequent cutaneous contact with blood and saliva, as well as several needlesticks, was HIV-1-seropositive (Klein *et al.*, 1988). In one case report of two siblings infected with HIV-1, the authors suggested that a bite from the index sibling was the route of transmission for the previously uninfected sibling (Wahn *et al.*, 1986). Because the bite did not break the skin or result in bleeding and because the timing of the infection in the second sibling in relation to the bite is uncertain, the precise mode of transmission remains unclear. Laboratory data indicate the virus can occasionally be isolated from saliva, but saliva and saliva filtrates contain components that inactivate HIV-1 (Fultz, 1986). Thus, laboratory and epidemiologic evidence are consistent with a very low risk, if any, of HIV-1 infection following exposure to saliva.

To examine the risk of HIV-1 transmission through casual contact, more than 700 household members have been evaluated after extended, nonsexual contact with adults and children with HIV-1 infection (Redfield *et al.*, 1985; Jason *et al.*, 1986; Fischl *et al.*, 1987; Friedland *et al.*, 1987; Peterman *et al.*, 1988; Lifson, 1988). No transmission of HIV-1 has been documented except for sex partners, children born to infected mothers, and persons who themselves had risk factors for AIDS.

Laboratory and epidemiologic studies have failed to demonstrate either replication of HIV-1 within insects or transmission of HIV-1 through biting or blood-sucking insects (Miike, 1987; Srinivasan *et al.*, 1987). The possibility of insect-mediated HIV-1 infection has been studied in Belle Glade, Florida, as well as in Haiti and Africa. In these studies, HIV-1 infection was not associated with either epidemiologic or laboratory measures of exposure to mosquitoes, including the presence of antibodies to various arboviruses and malaria (Pape *et al.*, 1987; Castro *et al.*, 1988b; Greenberg *et al.*, 1988). Furthermore, in Belle Glade, HIV-1 infection was not detected in any adults older than 60 years or in children aged 2–10 years (Castro *et al.*, 1988b).

E. Transmission Patterns outside the United States

While similar modes of transmission exist throughout the world, their geographic distribution and relative importance vary (Piot *et al.*, 1988). Homosexual or bisexual men and heterosexual IV-drug users are the major infected population groups in North America, Australia, New Zealand, and parts of South America. In Africa and the Caribbean high rates of HIV-1 infection in heterosexual men and women have been documented. Between 3% and 11% of pregnant women in some urban areas of central and east Africa are seropositive, which has led to significant levels of perinatally acquired infection (Mann *et al.*, 1986; Francis *et al.*, 1987; Nzilambi *et al.*, 1987; Ryder *et al.*, 1988; Braddick *et al.*, 1988).

IV. MANIFESTATIONS OF HIV-1 INFECTION OTHER THAN AIDS

A. Spectrum and Progression of HIV-1 Infection

Patients infected with HIV-1 may have a spectrum of clinical manifestations ranging from asymptomatic infection to severe immunodeficiency associated with serious secondary infections, neoplasms, and other conditions (Stoneburner *et al.*, 1988a). Classification systems that incorporate serologic, clinical, and immunologic data have been developed to categorize the manifestations of HIV-1 infection in children and adults (Haverkos *et al.*, 1985; CDC, 1986, 1987d; Redfield *et al.*, 1986).

An acute, mononucleosis-like illness has been described in association with initial infection with the virus (Cooper *et al.*, 1985; Ho *et al.*, 1985; Tindall *et al.*, 1988). The rate of symptomatic primary HIV-1 infection is difficult to determine because symptoms tend to be nonspecific, may be fairly benign and not discerned by the patient, and are often assessed retrospectively. In patients with well-characterized exposures to HIV-1 and compatible symptoms, the interval between exposure and symptomatic illness was 6–58 days (C. R. Horsburgh, Jr, unpublished data, 1988).

After initial infection with HIV-1, antigen can be detected within 1–4 weeks, followed by the development of HIV-1 anticore (p24) and antienvelope antibodies (gp41) usually within 3–12 weeks of the infection (Cooper *et al.*, 1985; Ho *et al.*, 1985; Allain *et al.*, 1986; Gaines *et al.*, 1987; Simmonds *et al.*, 1988; Ward *et al.*, 1989). Virus has been recovered from persons infected for 6 years or more (Feorino *et al.*, 1985; Jaffe *et al.*, 1985), suggesting that infection with HIV-1 is lifelong. Loss of serologically detectable antibodies to HIV-1 is rare in clinically asymptomatic persons. In a longitudinal study of 1000 HIV-1-seropositive homosexual men without AIDS, serologic tests for HIV-1 antibody for four men (0.4%) apparently reverted from positive to negative, although polymerase chain reaction (gene amplification) assays demonstrated HIV-1 provirus in all four (Farzadegan *et al.*, 1988). Long-term follow-up will be required to determine whether serologic reexpression of HIV-1 occurs and clinical evidence of infection develops.

The period between initial infection with HIV-1 and the development of AIDS is variable but often quite long. Many infected persons remain asymptomatic for years. Of 6700 homosexual and bisexual men enrolled in 1978–1980 for studies of hepatitis B in San Francisco, 70% are estimated to be infected with HIV-1 (Hessol *et al.*, 1987). For a sample of 179 men from this study who had known dates of seroconversion, the cumulative proportion who have developed AIDS after 10 years is 48% (Lifson *et al.*, 1988b). Fewer than 5% were diagnosed with AIDS in the first 2 years after infection, however. The rate of disease progression to AIDS in hemophilic persons is similar to that reported for homosexual men (Eyster and Goedert, 1988; Lifson *et al.*, 1988b; Jason *et al.*, 1989). Blood transfusion recipients may have a slightly more rapid rate of progression to AIDS (Lifson *et al.*, 1988b; Ward *et al.*, 1988b). Because HIV-1 causes latent infection, it is likely that longer mean latency periods will be observed with additional follow-up time (Medley *et al.*, 1987; Lui *et al.*, 1988).

B. Prevalence and Incidence of HIV-1 Infection

The Public Health Service estimated in 1988 that approximately 1–1.5 million persons were infected with HIV-1 in the United States (CDC, 1987e). To help monitor the levels and trends of HIV-1 infection, the U.S. Public Health Service, in collaboration with state and local health departments and other agencies, has initiated a program of HIV-1 seroprevalence surveys in more than 30 major metropolitan areas. To date, a consistent pattern of HIV-1 infection has emerged—a pattern similar to that seen for AIDS cases. In general, males have higher prevalence rates than females, blacks and Hispanics have higher prevalence rates than whites, and persons 20–45 years of age have higher prevalence rates than persons in other age groups (CDC, 1988d).

Reported HIV-1 seroprevalence rates in homosexual and bisexual men range from 10% to as high as 70%, with rates in most major metropolitan areas 20–50% (CDC, 1987e). HIV-1 infection rates in IV-drug users attending drug treatment facilities vary geographically. The highest rates (usually >50%) have been reported from New York, New Jersey, and Puerto Rico, while in most other areas of the country, infection rates are generally lower than 5% (CDC, 1987e). In sexually active heterosexuals who attended sexually transmitted disease clinics and who did not acknowledge homosexual contact, IV-drug use, or sexual contact with persons at high risk, the level of HIV-1 infection ranged from 3% of men and 2% of women in Baltimore (Quinn *et al.*, 1988), to 1% of men and 0.4% of women in New York City (Stoneburner *et al.*, 1988b), and 0.2% of men and 0 women in Denver (Judson *et al.*, 1988).

Several population groups have been studied to monitor the spread of HIV-1 infection in the general population. The two selected populations that have been followed the longest are blood donors and military recruit applicants. Because these two groups are prescreened to exclude persons likely to be infected, the observed prevalence of HIV-1 infection in these groups is likely to be much lower than the actual prevalence in the general population. Among blood donors, 0.020% of 12.6 million American Red Cross blood donations collected from April 1985 through May 1987 were seropositive for HIV-1 (CDC, 1987e). The overall level decreased from 0.035% in mid-1985 to 0.012% in mid-1987, primarily as a result of eliminating previously identified seropositive persons from the donor pool. Among first-time donors, who provide a measure of the prevalence of infection in the geographic area from which they are drawn, the level of HIV-1 seropositivity averaged 0.043% during 1985–1987 (CDC, 1987e).

The overall prevalence of HIV-1 among 1.5 million civilian applicants for military service from October 1985 through March 1988 decreased from 1.5 to 1.2 per 1000 applicants (CDC, 1988e). This decline was observed in men only and suggests that men are increasingly refraining from enlistment. In contrast, during this period, the prevalence of HIV-1 antibody remained stable in female applicants at 0.7/1000. Among active-duty U.S. army personnel who have been tested for HIV-1 more than once, 0.77/1000 per year have been detected to have HIV since their initial negative test (CDC, 1987e).

Population-based estimates of HIV infection have been derived from blinded surveys of blood samples collected on filter paper from newborns for routine metabolic assays, such as phenylketonuria. HIV seroprevalence rates measured by this approach have ranged from 0.09% in suburban/rural Massachusetts to 1.6% in New York City (CDC, 1988d; Hoff *et al.*, 1988).

V. CONCLUSIONS

Epidemiologic and laboratory data have demonstrated that HIV-1 is transmitted through intimate sexual contact, contact with contaminated blood, and perinatally, from an infected mother to her infant. Infection with HIV-1 results in a spectrum of clinical illness, ranging from an asymptomatic carrier state to the development of fatal infections, malignancies, and other conditions. At this time, there is no vaccine to prevent HIV infection and only a limited number of effective antiviral agents are available to modulate disease progression once infection has been established. Control efforts must focus on strategies that prevent sexual, bloodborne, and perinatal transmission.

Significant reductions in high-risk sexual behavior, such as having sex with nonsteady partners and practicing anal intercourse, have been reported among homosexual and bisexual men and appear to be correlated with decreased rates of rectal and pharyngeal gonorrhea, syphilis, and incident HIV-1 infection (CDC, 1984; Winkelstein *et al.*, 1988). Prevention of HIV-1 infection through contaminated blood, blood products, and donated tissues and organs has largely been achieved by implementation of donor deferral programs, serologic screening of donated blood and plasma, and heat treatment of clotting factor concentrates. Interruption of transmission through IV-drug use and perinatally acquired HIV-1 infection remains the biggest challenge. Routine HIV-1 counseling and testing in family planning and obstetrical settings should be considered a standard of care for women residing in areas of high prevalence (CDC, 1987f; Landesman *et al.*, 1987). HIV-1-infected women should be advised to avoid pregnancy because of the high probability of infecting their unborn children. Effective strategies to reduce HIV-1 infection associated with IV-drug use hinge on educational and therapeutic approaches to prevent and treat IV-drug use (DesJarlais and Friedman, 1987; Report of the Presidential Commission on the Human Immunodeficiency Virus Epidemic, 1988). To control this epidemic most effectively it will be necessary to undertake all of these measures, which will require substantial resources.

REFERENCES

Allain, J-P., Laurian, Y., Paul, D. A., Senn, D., and Members of the AIDS-Haemophilia French Study Group. (1986). *Lancet* **2,** 1233–1236.

Barre-Sinoussi, F., Chermann, J. C., Rey, F., Nugeyre, M. T., Chamaret, S., Gruest, J., Dauguet, C., Axler-Blin, C., Vezinet-Brun, F., Rouzioux, C., Rozenbaum, W., and Montagnier, L. (1983). *Science* **220,** 868–871.

Blanche, S., Rouzioux, C., Tricoire, J., Baechler, E., Mayaux, M. J., and Griscelli, C. (1988). In *Proceedings of the Fourth International Conference on AIDS, Book 1, June 12–16, 1988,* p. 434, Swedish Ministry of Health and Social Affairs, Stockholm.

Braddick, M., Kreiss, J., Embree, J., Datta, P., Ndinya-Achola, J., Vercauteren, G., Piot, P., Coombs, R., Holmes, K., Quinn, T., and Plummer, F. A. (1988). In *Proceedings of the Fourth International Conference on AIDS, Book 1, June 12–16, 1988,* p. 346, Swedish Ministry of Health and Social Affairs, Stockholm.

Brun-Vezinet, F., Rey, M. A., Katlama, C., Girard, P. M., Roulot, D., Yeni, P., Lenoble, L., Clavel, F., Alizon, M., Gadelle, S., Madjar, J. J., and Harzic, M. (1987). *Lancet* **1,** 128–132.

Castro, K. G., Lifson, A. R., White, C. R., Bush, T. J., Chamberland, M. E., Lekatsas, A. M., and Jaffe, H. W. (1988a). *JAMA* **259,** 1338–1342.

Castro, K. G., Lieb, S., Jaffe, H. W., Narkunas, J. P., Calisher, C. H., Bush, T. J., Witte, J. J., and The Belle Glade Field-Study Group. (1988b). *Science* **239,** 193–197.

Centers for Disease Control. (1984). *MMWR* **33,** 295–297.

Centers for Disease Control. (1986). *MMWR* **35,** 334–339.

Centers for Disease Control. (1987a). *MMWR* **36** (Suppl. 1), 1–15.

Centers for Disease Control. (1987b). *MMWR* **36,** 121–124.

Centers for Disease Control. (1987c). *MMWR* **36,** 285–289.

Centers for Disease Control. (1987d). *MMWR* **36,**225–236.

Centers for Disease Control. (1987e). *MMWR* **36,** (Suppl. 6), 1–48.

Centers for Disease Control. (1987f). *MMWR* **36,** 509–515.

Centers for Disease Control. (1988a). *MMWR* **37,** 33–35.

Centers for Disease Control. (1988b). *MMWR* **37,** 597–599.

Centers for Disease Control. (1988c). *MMWR* **37,** 229–234, 239.

Centers for Disease Control. (1988d). *MMWR* **37,** 223–226.

Centers for Disease Control. (1988e). *MMWR* **37,** 677–679.

Centers for Disease Control. (1989). *MMWR* **38,** 27–29.

Chaisson, R. E., Moss, A. R., Onishi, R., Osmond, D., and Carlson, J. R. (1987). *Am. J. Public Health* **77,** 169–172.

Chamberland, M. E., Castro, K. G., Haverkos, H. W., Miller, B. I., Thomas, P. A., Reiss, R., Walker, J., Spira, T. J., Jaffe, H. W., and Curran, J. W. (1984). *Ann. Intern. Med.* **101,** 617–623.

Chiasson, M. A., Fleischer, E., Petrus, D., and Miller, B. (1987). In *Proceedings of the Third International Conference on AIDS, June 1–5, 1987,* p. 174, U.S. Department of Health and Human Services and the World Health Organization, Washington, DC.

Clavel, F., Mansinho, K., Chamaret, S., Guetard, D., Favier, V., Nina, J., Santos-Ferreira, M-D., Champalimaud, J-L., and Montagnier, L. (1987). *N. Engl. J. Med.* **316,** 1180–1185.

Colebunders, R., Kapita, B., Nekwei, W., Bahwe, Y., Lebughe, I., Oxtoby, M., and Ryder, R. (1988). *Lancet* **2,** 1487.

Cooper, D. A., Gold, J., Maclean, P., Donovan, B., Finlayson, R., Barnes, T. G., Michelmore, H. M., Brooke, P., and Penny, R. for the Sydney AIDS Study Group. (1985). *Lancet* **1,** 537–540.

Curran, J. W., Lawrence, D. N., Jaffe, H., Kaplan, J. E., Zyla, L. D., Chamberland, M., Weinstein, R., Lui, K-L., Schonberger, L. B., Spira, T. J., Alexander, W. J., Swinger, G., Ammann, A., Solomon, S., Auerbach, D., Mildvan, D., Stoneburner, R., Jason, J. M., Haverkos, H. W., and Evatt, B. L. (1984). *N. Engl. J. Med.* **310,** 69–75.

Curran, J. W., Jaffe, H. W., Hardy, A. M., Morgan, W. M., Selik, R. M., and Dondero, T. J. (1988). *Science* **239,** 610–616.

Darrow, W. W., Echenberg, D. F., Jaffe, H. W., O'Malley, P. M., Byers, R. H., Getchell, J. P., and Curran, J. W. (1987). *Am. J. Public Health* **77,** 479–483.

DesJarlais, D. C., and Friedman, S. R. (1987). *AIDS* **1,** 67–76.

European Collaborative Study. (1988). *Lancet* **2,** 1039–1042.

Evans, L. A., Moreau, J., Odehouri, K., Seto, D., Thomson-Honnebier, G., Legg, H., Barboza, A., Cheng-Mayer, C., and Levy, J. A. (1988). *Lancet* **2,** 1389–1391.

Evatt, B. L., Ramsey, R. B., Lawrence, D. N., Zyla, L. D., and Curran, J. W. (1984). *Ann. Intern. Med.* **100,** 499–504.

Eyster, M. E., and Goedert, J. J. (1988). In *Proceedings of the Fourth International Conference on AIDS, Book 2, June 12–16, 1988,* p. 365, Swedish Ministry of Health and Social Affairs, Stockholm.

Farzadegan, H., Polis, M. A., Wolinsky, S. M., Rinaldo, C. R., Sninsky, J. J., Kwok, S., Griffith, R. L., Kaslow, R. A., Phair, J. P., Polk, B. F., and Saah, A. J. (1988). *Ann. Intern. Med.* **108,** 785–790.

Fauci, A. S. (1988). *Science* **239,** 617–622.

Feorino, P. M., Jaffe, H. W., Palmer, E., Peterman, T. A., Francis, D. P., Kalyanaraman, V. S., Weinstein, R. A., Stoneburner, R. L., Alexander, W. J., Raevsky, C., Getchell, J. P., Warfield, D., Haverkos, H. W., Kilbourne, B. W., Nicholson, J. K. A., and Curran, J. W. (1985). *N. Engl. J. Med.* **312,** 1293–1296.

Fischl, M. A., Dickinson, G. M., Scott, G. B., Kilmas, N., Fletcher, M. A., and Parks, W. (1987). *JAMA* **257,** 640–644.

Flynn, N. M., Pollet, S. M., Van Horne, J. R., Elvebakk, R., Harper, S. D., and Carlson, J. R. (1987). *West. J. Med.* **146,** 439–442.

Francis, H., Lubaki, N., Duma, M. P., Ryder, R. W., Mann, J., and Quinn, T. C. (1987). In *Proceedings of the Third International Conference on AIDS, June 1–5, 1987,* p. 214, U.S. Department of Health and Human Services and the World Health Organization, Washington, DC.

Friedland, G. H., and Klein, R. S. (1987). *N. Engl. J. Med.* **317,** 1125–1135.

Friedland, G. H., Kahl, P., Feiner, C., Rogers, M., Mayers, M., and Klein, R. S. (1987). In *Proceedings of the Third International Conference on AIDS, June 1–5, 1987,* p. 196, U.S. Department of Health and Human Services and the World Health Organization, Washington, DC.

Friedman, B. A., Burns, T. L., and Schork, M. A. (1980). A Study of National Trends in Transfusion Practice, pp. 1–283, The University of Michigan Medical School and School of Public Health, Ann Arbor.

Fultz, P. N. (1986). *Lancet* **2,** 1215.

Gaines, H., von Sydow, M., Sonnerborg, A., Albert, J., Czajkowski, J., Pehrson, P. O., Chiodi, F., Moberg, L., Fenyo, E. M., Asjo, B., and Forsgren, M. (1987). *Lancet* **1,** 1249–1253.

Gallo, R. C., Salahuddin, S. Z., Popovic, M., Shearer, G. M., Kaplan, M., Haynes, B. F., Palker, T. J., Redfield, R., Oleske, J., Safai, B., White, G., Foster, P., and Markham, P. D. (1984). *Science* **224,** 500–503.

Gartner, S., Markovits, P., Markovitz, D. M., Kaplan, M. H., Gallo, R. C., and Popovic, M. (1986). *Science* **233,** 215–219.

Gerberding, J. L., Bryant-LeBlanc, C. E., Nelson, K., Moss, A. R., Osmond, D., Chambers, H. F., Carlson, J. R., Drew, W. L., Levy, J. A., and Sande, M. A. (1987a). *J. Infect. Dis.* **156,** 1–8.

Gerberding, J. L., Nelson, K., Greenspan, D., Greenspan, J., Greene, J., and Sande, M. S. (1987b). In *Proceedings and Abstracts of the 27th Interscience Conference on Antimicrobial Agents and Chemotherapy, Oct. 4–7, 1987,* p. 219, American Society for Microbiology, New York.

Goedert, J. J., Eyster, M. E., and Biggar, R. J. (1987). In *Proceedings of the Third International Conference on AIDS, June 1–5, 1987,* p. 106, U.S. Department of Health and Human Services and the World Health Organization, Washington, DC.

Greenberg, A. E., Nguyen-Dinh, P., Mann, J. M., Kabote, N., Colebunders, R. L., Francis, H., Quinn, T. C., Baudoux, P., Lyamba, B., Davachi, F., Roberts, J. M., Kabeya, N., Curran, J. W., and Campbell, C. C. (1988). *JAMA* **259,** 545–549.

Greenblatt, R. M., Lukehart, S. L., Plummer, F. A., Quinn, T. C., Critchlow, C. W., and D'Costa, L. J. (1987). In *Proceedings of the Third International Conference on AIDS, June 1–5, 1987,* p. 174, U.S. Department of Health and Human Services and the World Health Organization, Washington, DC.

Guinan, M. E., and Hardy, A. (1987). *JAMA* **257,** 2039–2042.

Guyader, M., Emerman, M., Sonigo, P., Clavel, F., Montagnier, L., and Alizon, M. (1987). *Nature* **326,** 662–669.

Hahn, B. H., Shaw, G. M., Taylor, M. E., Redfield, R. R., Markham, P. D., Salahuddin, S. Z., Wong-Staal, F., Gallo, R. C., Parks, E. S., and Parks, W. P. (1986). *Science* **232,** 1548–1553.

Handsfield, H. H., Ashley, R. L., Rompalo, A. M., Stamm, W. E., Wood, R. W., and Corey, L. (1987). In *Proceedings of the Third International Conference on AIDS, June 1–5, 1987,* p. 206, U.S. Department of Health and Human Services and the World Health Organization, Washington, DC.

Hardy, A. M., Allen, J. R., Morgan, W. M., and Curran, J. W. (1985). *JAMA* **253,** 215–220.
Haverkos, H. W., Gotlieb, M. S., Killen, J. Y., and Edelman, R. (1985). *J. Infect. Dis.* **152,** 1095.
Henderson, D. K., Saah, A. J., Zak, B. J., Kaslow, R. A., Lane, H. C., Folks, T., Blackwelder, W. C., Schmitt, J., LaCamera, D. J., Masur, H., and Fauci, A. S. (1986). *Ann. Intern. Med.* **104,** 644–647.
Henderson, D. K., Fahey, B. J., and Willy, M. E. (1988). In *Proceedings of the Fourth International Conference on AIDS, Book 1, June 12–16, 1988,* p. 480, Swedish Ministry of Health and Social Affairs, Stockholm.
Hessol, N. A., Rutherford, G. W., O'Malley, P. M., Doll, L. S., Darrow, W. W., and Jaffe, H. W. (1987). In *Proceedings of the Third International Conference on AIDS, June 1–5, 1987,* p. 1, U.S. Department of Health and Human Services and the World Health Organization, Washington, DC.
Ho, D. D., Sarngadharan, M. G., Resnick, L., Dimarzo-Veronese, F., Rota, T. R., and Hirsch, M. S. (1985). *Ann. Intern. Med.* **103,** 880–883.
Hoff, R., Berardi, V. P., Weiblen, B. J., Mahoney-Trout, L., Mitchell, M. L., and Grady, G. F. (1988). *N. Engl. J. Med.* **318,** 525–530.
Holmberg, S. D., Stewart, J. A., Gerber, A. R., Byers, R. H., Lee, F. K., O'Malley, P. M. and Nahmias, A. J. (1988). *JAMA* **259,** 1048–1050.
Horsburgh, C. R., Jr., and Holmberg, S. D. (1988). *Transfusion* **28,** 192–195.
Jaffe, H. W., Choi, K., Thomas, P. A., Haverkos, H. W., Auerbach, D. M., Guinan, M. E., Rogers, M. F., Spira, T. J., Darrow, W. W., Kramer, M. A., Friedman, S. M., Monroe, J. M., Friedman-Kien, A. E., Laubenstein, L. J., Marmor, M., Safai, B., Dritz, S. K., Crispi, S. J., Fannin, S. L., Orkwis, J. P., Kelter, A., Rushing, W. R., Thacker, S. B., and Curran, J. W. (1983). *Ann. Intern. Med.* **99,** 145–151.
Jaffe, H. W., Feorino, P. M., Darrow, W. W., O'Malley, P. M., Getchell, J. P., Warfield, D. T., Jones, B. M., Echenberg, D. F., Francis, D. P., and Curran, J. W. (1985). *Ann. Intern. Med.* **102,** 627–628.
Jason, J. M., McDougal, J. S., Dixon, G., Lawrence, D. N., Kennedy, M. S., Hilgartner, M., Aledort, L., and Evatt, B. L. (1986). *JAMA* **255,** 212–215.
Jason, J., Lui, K-J., Ragni, M. V., Hessol, N. A., and Darrow, W. W. (1989). *JAMA* **261,** 725–727.
Jovaisas, E., Koch, M. A., Shafer, A., Stauber, M., and Lowenthal, D. (1985). *Lancet,* **2,** 1129.
Judson, F. N., Douglas, J., and Cohn, D. (1988). In *Proceedings of the Fourth International Conference on AIDS, Book 1, June 12–16, 1988,* p. 263, Swedish Ministry of Health and Social Affairs, Stockholm.
Kingsley, L. A., Detels, R., Kaslow, R., Polk, B. F., Rinaldo, C. R., Jr., Chmiel, J., Detre, K., Kelsey, S. F., Odaka, N., Ostrow, D., VanRaden, M., and Visscher, B. (1987). *Lancet* **1,** 345–348.
Klatzmann, D., Barre-Sinoussi, F., Nugeyre, M. T., Dauguet, C., Vilmer, E., Griscelli, C., Brun-Vezinet, F., Rouzioux, C., Gluckman, J. C., Chermann, J-C., and Montagnier, L. (1984). *Science* **225,** 59–63.
Klein, R. S., Phelan, J. A., Freeman, K., Schable, C., Friedland, G. H., Trieger, N., and Steigbigel, N. H. (1988). *N. Engl. J. Med.* **318,** 86–90.
Kreiss, J. K., Kitchen, L. W., Prince, H. E., Kasper, C. K., and Essex, M. (1985). *Ann. Intern. Med.* **102,** 623–626.
Kreiss, J. K., Koech, D., Plummer, F. A., Holmes, K. K., Lightfoote, M., Piot, P., Ronald, A. R., Ndinya-Achola, J. O., D'Costa, L. J., Roberts, P., Ngugi, E. N., and Quinn, T. C. (1986). *N. Engl. J. Med.* **314,** 414–418.
Kristal, A. R. (1986). *JAMA* **255,** 2306–2310.
L'age-Stehr, J., Schwarz, A., Offermann, G., Langmaack, H., Bennhold, I., Niedrig, M., and Koch, M. A. (1985). *Lancet* **2,** 1361–1362.
Landesman, S., Minkoff, H., Holman, S., McCalla, S., and Sijin, O. (1987). *JAMA* **258,** 2701–2703.
LaPointe, N., Michaud, J., Pekovic, D., Chausseau, J. P. and Dupey, J. M. (1985). *N. Engl. J. Med.* **312,** 1325–1326.
Lekatsas, A. M., Walker, J., O'Donnell, R., Garcia, N., Thomas, P., and Stoneburner, R. (1986). In *Proceedings of the International Conference on AIDS, June 23–25, 1986,* p. 151, L'Association pour la Recherche sur les Deficits Immunitaires Viro-Induits (ARDIVI), Paris.
Lifson, A. R. (1988). *JAMA* **259,** 1353–1356.
Lifson, A. R., Rogers, M. F., White, C., Thomas, P., O'Donnell, R., and Scott, G. (1987). *Pediatr. Infect. Dis. J.* **6,** 292–293.
Lifson, A. R., Darrow, W. W., O'Malley, P. M., Bodecker, T. W., Barnhart, J. L., Hessol, N. A., Jaffe, H. W., and Rutherford, G. W. (1988a). In *Proceedings of the Fourth International Conference on AIDS, Book 1, June 12–16, 1988,* p. 290, Swedish Ministry of Health and Social Affairs, Stockholm.
Lifson, A. R., Rutherford, G. W., and Jaffe, H. W. (1988b). *J. Infect. Dis.* **158,** 1360–1367.

Long-Term Survivor Collaborative Study Group and Hardy, A. M. (1987). In *Proceedings and Abstracts of the 27th Interscience Conference on Antimicrobial Agents and Chemotherapy, Oct. 4–7, 1987,* p. 98, American Society for Microbiology, New York.

Lui, K-J., Peterman, T. A., Lawrence, D. N., and Allen, J. R. (1988). *Stat. Med.* **7,** 395–401.

Maddon, P. J., Dalgleish, A. G., McDougal, J. S., Clapham, P. R., Weiss, R. A., and Axel, R. (1986). *Cell* **47,** 333–348.

Mann, J. M., Francis, H., Davachi, F., Baudoux, P., Quinn, T. C., Nzilambi, N., Bosenge, N., Colebunders, R. L., Piot, P., Kabote, N., Asila, P. K., Malonga, M., and Curran, J. W. (1986). *Lancet* **2,** 654–657.

Marcus, R. and the CDC Collaborative Needlestick Surveillance Group. (1988). *N. Engl. J. Med.* **319,** 1118–1123.

Marmor, M., Friedman-Kien, A. E., Laubenstein, L., Byrum, R. D., William, D. C., D'Onofrio, S., and Dubin, N. (1982). *Lancet* **1,** 1083–1085.

Medley, G. F., Anderson, R. M., Cox, D. R., and Billard, L. (1987). *Nature* **328,** 719–721.

Mendez, H., Willoughby, A., Hittelman, J., Minkoff, H., Berthaud, M., Moroso, G., Sunderland, A., Holman, S., Bihari, B., Goedert, J., and Landesman, S. (1988). In *Proceedings of the Fourth International Conference on AIDS, Book 2, June 12–16, 1988,* p. 295, Swedish Ministry of Health and Social Affairs, Stockholm.

Miike, L. (1987). Do Insects Transmit AIDS? Health Program, Office of Technology Assessment, U.S. Congress, pp. 1–43.

Mok, J. Q., Giaquinto, C., DeRossi, A., Grosch-Worner, I., Ades, A. E., and Peckham, C. S. (1987). *Lancet* **1,** 1164–1168.

Montagnier, L., Gruest, J., Chamaret, S., Dauguet, C., Axler, C., Guetard, D., Nugeyre, M. T., Barre-Sinoussi, F., Chermann, J-C., Brunet, J. B., Klatzmann, D., and Gluckman, J. C. (1984). *Science* **225,** 63–66.

Moss, A. R., Osmond, D., Bacchetti, P., Chermann, J-C., Barre-Sinoussi, F., and Carlson, J. (1987). *Am. J. Epidemiol.* **125,** 1035–1047.

Nelson, J. A., Wiley, C. A., Reynolds-Kohler, C., Reese, C. E., Margaretten, W., and Levy, J. A. (1988). *Lancet* **1,** 259–262.

Nzilambi, N., Ryder, R. W., Behets, F., Francis, H., Bayende, E., Nelson, A., and Mann, J. M. (1987). In *Proceedings of the Third International Conference on AIDS, June 1–5, 1987,* p. 158, U.S. Department of Health and Human Services and the World Health Organization, Washington, DC.

Oxtoby, M. J. (1988). *Pediatr. Infect Dis. J.* **7,** 825–835.

Padian, N., Marquis, L., Francis, D. P., Anderson, R. E., Rutherford, G. W., O'Malley, P. M., and Winkelstein, W. (1987). *JAMA* **258,** 788–790.

Pape, J. W., Stanback, M. E., Pamphile, M., Verdier, R., Deschamps, M-M., and Johnson, W. D., Jr. (1987). In *Proceedings of the Third International Conference on AIDS, June 1–5, 1987,* p. 6, U.S. Department of Health and Human Services and the World Health Organization, Washington, DC.

Peterman, T. A., and Curran, J. W. (1986). *JAMA* **256,** 2222–2226.

Peterman, T. A., Stoneburner, R. L., Allen, J. R., Jaffe, H. W., and Curran, J. W. (1988). *JAMA* **259,** 55–58.

Piot, P., Plummer, F. A., Mhalu, F. S., Lamboray, J-L., Chin, J., and Mann, J. M. (1988). *Science* **239,** 573–579.

Quinn, T. C., Glasser, D., Cannon, R. O., Matuszak, D. L., Dunning, R. W., Kline, R. L., Campbell, C. H., Israel, E., Fauci, A. S., and Hook, E. W., III. (1988). *N. Engl. J. Med.* **318,** 197–203.

Rabkin, C. S., Van Devanter, N., Ewing, W. E., and Pindyck, J. (1987). In *Proceedings of the Third International Conference on AIDS, June 1–5, 1987,* p. 103, U.S. Department of Health and Human Services and the World Health Organization, Washington, DC.

Rayfield, M., De Cock, K., Heyward, W., Goldstein, L., Krebs, J., Kwok, S., Lee, S., McCormick, J., Moreau, J. M., Odehouri, K., Schochetman, G., Sninsky, J., and Ou, C-Y. (1988). *J. Infect. Dis.* **158,** 1170–1176.

Redfield, R. R., Markham, P. D., Salahuddin, S. Z., Samgadharan, M. G., Bodner, A. J., Folks, T. M., Ballou, W. R., Wright, C., and Gallo, R. C. (1985). *JAMA* **253,** 1571–1573.

Redfield, R. R., Wright, D. C., and Tramont, E. C. (1986). *N. Engl. J. Med.* **314,** 131–132.

Redfield, R. R., Wright, D. C., Khan, N. C., and Burke, D. S. (1987). In *Proceedings of the Third International Conference on AIDS, June 1–5, 1987,* p. 180, U.S. Department of Health and Human Services and the World Health Organization, Washington, DC.

Report of the Presidential Commission on the Human Immunodeficiency Virus Epidemic (1988). Submitted to the President of the United States, June 24, 1988, pp. 1–201, U.S. Government Printing Office, Washington, DC.

Report of the Workgroup on Clinical Manifestations and Pathogenesis. (1989). *Public Health Rep.* **103,** 28–40.

Report of the Workgroup on Epidemiology and Surveillance. (1989). *Public Health Rep.* **103,** 10–18.

Rogers, M. F., White, C. R. and Sanders, R. (1986). In *Proceedings and Abstracts of the 26th Interscience Conference on Antimicrobial Agents and Chemotherapy, Sept. 28–Oct. 1, 1986,* p. 284, American Society for Microbiology, New Orleans.

Ryder, R. W., Rayfield, M., Quinn, T., Kashamuka, M., Francis, H., Vercauteren, G., and Piot, P. (1988). In *Proceedings of the Fourth International Conference on AIDS, Book 1, June 12–16, 1988,* p. 345, Swedish Ministry of Health and Social Affairs, Stockholm.

Sakai, K., Casareale, D., Sonnabend, J., and Volsky, D. J. (1986). *Ann Inst. Pasteur Virol.* **137E,** 311–315.

Sakai, K., Dewhurst, S., Ma, X., and Volsky, D. J. (1988). *J. Virol.* **62,** 4078–4085.

Schoenbaum, E. E., Selwyn, P. A., Klein, R. S., Rogers, M. F., Freeman, K., and Friedland, G. H. (1986). In *Proceedings of the International Conference on AIDS, June 23–25, 1986,* p. 111, L'Association pour la Recherche sur les Deficits Immunitaires Viro-Induits (ARDIVI), Paris.

Scott, G., Hutto, C., Mastrucci, T., Mitchell, C., and Parks, W. (1988). In *Proceedings of the Fourth International Conference on AIDS, Book 2, June 12–16, 1988,* p. 292, Swedish Ministry of Health and Social Affairs, Stockholm.

Selik, R. M., Starcher, E. T., and Curran, J. W. (1987). *AIDS* **1,** 175–182.

Selik, R. M., Castro, K. G., and Pappaioanou, M. (1988). *Am. J. Public Health* **78,** 1539–1545.

Simmonds, P., Lainson, F. A. L., Cuthbert, R., Steel, C. M., Peutherer, J. F., and Ludlam, C. A. (1988). *Br. Med. J.* **296,** 593–598.

Smiley, L., White, G. C., II, Macik, G., Becherer, P., Weinhold, K. J., and Matthews, T. J. (1987). In *Proceedings of the Third International Conference on AIDS, June 1–5, 1987,* p. 23, U.S. Department of Health and Human Services and the World Health Organization, Washington, DC.

Srinivasan, A., York, D., and Bohan, D. (1987). *Lancet* **1,** 1094–1095.

Steigbigel, N. H., Maude, D. W., Feiner, C. J., Harris, C. A., Saltzman, B. R., and Klein, R. S. (1987). In *Proceedings of the Third International Conference on AIDS, June 1–5, 1987,* p. 106, U.S. Department of Health and Human Services and the World Health Organization, Washington, DC.

Stevens, C. E., Taylor, P. E., Zang, E. A., Morrison, J. M., Harley, E. J., Rodriguez de Cordoba, S., Bacino, C., Ting, R. C. Y., Bodner, A. J., Sarngadharan, M. G., Gallo, R. C., and Rubinstein, P. (1986). *JAMA* **255,** 2167–2172.

Stewart, G. J., Tyler, J. P. P., Cunningham, A. L., Barr, J. A., Driscoll, G. L., Gold, J., and Lamont, B. J. (1985). *Lancet* **2,** 581–585.

Stoneburner, R. L., DesJarlais, D. C., Benezra, D., Gorelkin, L., Sotheran, J. L., Friedman, S. R., Schultz, S., Marmor, M., Mildvan, D., and Maslansky, R. (1988a). *Science* **242,** 916–919.

Stoneburner, R. L., Chiasson, M. A., Lifson, A. R., Hildebrandt, D., Schultz, S., and Jaffe, H. W. (1988b). In *Proceedings of the Fourth International Conference on AIDS, Book 1, June 12–16, 1988,* p. 377, Swedish Ministry of Health and Social Affairs, Stockholm.

Thiry, L., Sprecher-Goldberger, S., Jonckheer, T., Levy, J., Van de Perre, P., Henrivaux, P., Cogniaux-LeCleve, J., and Clumeck, N. (1985). *Lancet* **2,** 891–892.

Tindall, B., Barker, S., Donovan, B., Barnes, T., Roberts, J., Kronenberg, C., Gold, J., Penny, R., Cooper, D., and the Sydney AIDS Study Group. (1988). *Arch. Intern. Med.* **148,** 945–949.

Tsoukas, C., Hadjis, T., Theberge, L., Gold, P., O'Shaughnessy, M., and Feorino, P. (1986). In *Proceedings of the International Conference on AIDS, June 23–25, 1986,* p. 125, L'Association pour la Recherche sur les Deficits Immunitaires Viro-Induits (ARDIVI), Paris.

Wahn, V., Kramer, H. H., Voit, T., Bruster, H. T., Scrampical, B., and Scheid, A. (1986). *Lancet* **2,** 694.

Ward, J. W., Deppe, D. A., Samson, S., Perkins, H., Holland, P., Fernando, L., Feorino, P. M., Thompson, P., Kleinman, S., and Allen, J. R. (1987). *Ann. Intern. Med.* **106,** 61–62.

Ward, J. W., Holmberg, S. D., Allen, J. R., Cohn, D. L., Critchley, S. E., Kleinman, S. H., Lenes, B. A., Ravenholt, O., Davis, J. R., Quinn, M. G., and Jaffe, H. W. (1988a). *N. Engl J. Med.* **318,** 473–478.

Ward, J., Perkins, H., Pepkowitz, S., Lieb, L., and Holmberg, S. (1988b). In *Proceedings of the Fourth International Conference on AIDS, Book 2, June 12–16, 1988,* p. 352, Swedish Ministry of Health and Social Affairs, Stockholm.

Ward, J. W., Schable, C., Dickinson, G. M., Makowka, L., Yanaga, K., Caruana, R., Chan, H., Salazar, F., Schochetman, G., and Holmberg, S. (1989). *Transplantation* **47,** 722–724.

Weinbreck, P., Loustaud, V., Denis, F., Vidal, B., Mounier, M., and deLumley, L., (1988). *Lancet* **1,** 482.

Weiss, S. H., Ginzburg, H. M., Altman, R., Taylor, F., Durako, S., and Blattner, W. A. (1986). In *Proceedings of the International Conference on AIDS, June 23–25, 1986,* p. 124, L'Association pour la Recherche sur les Deficits Immunitaires Viro-Induits (ARDIVI), Paris.

Wiley, C., Schrier, R. D., Nelson, J. A., Lampert, P. W., and Oldstone, M. B. A. (1986). *Proc. Natl. Acad. Sci. USA* **83,** 7089–7093.

Winkelstein, W., Jr., Lyman, D. M., Padian, N., Grant, R., Samuel, M., Wiley, J. A., Anderson, R. E., Lang, W., Riggs, J., and Levy, J. A. (1987). *JAMA* **257,** 321–325.

Winkelstein, W., Jr., Wiley, J. A., Padian, N. S., Samuel, M., Shiboski, S., Ascher, M. S., and Levy, J. A. (1988). *Am. J. Public Health* **78,** 1472–1474.

Ziegler, J. B., Stewart, G. J., Penny, R., Stuckey, M., and Good, S. (1988). In *Proceedings of the Fourth International Conference on AIDS, Book 1, June 12–16, 1988,* p. 339, Swedish Ministry of Health and Social Affairs, Stockholm.

Zuck, T. F. (1987). *Transfusion* **27,** 447–448.

3

Acquired Immunodeficiency Syndrome

Molecular Virology, Management, Control, and New Therapeutic Approaches

A. Hossain, A. Al-Tuwaijri, C. Kurstak, and E. Kurstak

I. INTRODUCTION

Since 1982 when the acquired immunodeficiency syndrome (AIDS) was first reported, at least 215,144 diagnosed cases of AIDS are presently known, with 120,069 cases reported from the North American continent and with 29,725, 40,519, 511, and 1,596 cases respectively from Europe, Africa, Asia and Australia. Countrywise, the United States tops the list with 117,781 reported cases, followed by Brazil, France, Uganda, Kenya, Tanzania, Italy, and Zaire with 9,555, 8,025, 7,375, 6,004, 5,627, 5,307, and 4,636 cases, respectively (*WHO Wkly. Epidemiol. Rep. 2.02,* 1990). AIDS is now known to occur worldwide (Brunet and Ancelle, 1985); the agent responsible has been isolated and determined to be a human retrovirus. The antigenic and molecular characteristics of the virus are indicative of its uniqueness among the human and animal retroviruses (Wong-Staal and Gallo, 1985). Its transmission is exclusively by sexual contact, prenatal transmission, or inoculation of infected blood/blood products, rather than casual contact (Friedland and Klein, 1987).

In 1983, Montagnier and co-workers at Pasteur Institute, in France, isolated a virus from a patient with AIDS-related lymphadenopathy. Despite its characterization as a retrovirus, it differed widely from the two human retroviruses (human T-cell leukemia viruses HTLV-I and II) previously characterized by Gallo and co-workers at the National Institute of Health, in Bethesda, Maryland. The initial viral isolate was termed lymphadenopathy-associated virus (LAV) (Barre-Sinoussi *et al.*, 1983), and the subsequent isolates from AIDS patients at the Pasteur Institute, with certain characteristics different from LAV, were labeled immunodeficiency-associated viruses (IDAV) (Montagnier, 1985). In the United States Gallo and co-workers also isolated and reported a new virus from several individuals with AIDS and AIDS-related complex, (ARC) and named the virus human T-lymphotropic virus type III (HTLV-III) (Gallo *et al.*, 1984; Gallo and Wong-Staal, 1985). Genetic and molecular structure studies of these isolates indicate them to be closely related to the same virus group and they have since been referred as HTLV-III/LAV or "AIDS virus." Recently, it has been proposed that all AIDS-related viruses be designated by one name, human immunodeficiency virus (HIV) (Coffin *et al.*, 1986); this is in conformity with the current system of retrovirus nomenclature based on the species of the organisms infected and the type of disease caused. Furthermore, future

---

A. Hossain, A. Al-Tuwaijri, C. Kurstak, and E. Kurstak • Department of Microbiology and Immunology, Faculty of Medicine, University of Montreal, Montreal, Quebec, H3C 3J7 Canada.

human retroviral isolates should preferably be compared by nucleic acid homology to HIV reference strains in order to determine whether they are related or unrelated viruses (Janda, 1987).

II. MOLECULAR VIROLOGY

The molecular structure of HIV has been described and is similar to other retroviruses. Essentially, they are enveloped viruses containing a single-stranded RNA genome and magnesium-dependent reverse transcriptase involved in transcribing a DNA copy of the viral RNA in the infected cell (Wong-Staal and Gallo, 1985). The viral DNA then becomes integrated into the host cell genome and is also present in multiple copies in an unintegrated state (Folks *et al,*. 1986). Like other human retroviruses, HIV is T-cell tropic and specifically infects T4 helper lymphocytes; studies with monoclonal antibodies against the T4 cell surface antigen of lymphocytes suggest that the cell receptor of HIV may, in fact, be an epitope of the T4 molecule itself (McDougal *et al.*, 1986).

The genetic "map" of HIV resembles other that of retroviruses but also includes unique nucleic acid sequences. In keeping with other retroviruses, the RNA genome of HIV contains gag, pol, and env genes, which code for viral core proteins, reverse transcriptase, and virus-specific envelope proteins, respectively (Essex *et al.*, 1985; di Marzo Veronese *et al.*, 1986; Kurstak *et al.*, 1986). The specific gene products of some of these genes have been isolated and characterized, and antibodies to various viral antigens detected in sera of individuals infected with HIV using the Western immunoblot technique.

Flanking these genes are sequences of nucleotides called long terminal repeats (LTRs) which contain regions (promoters) for the specific binding of polymerase for viral replication and enhancer elements that allow certain genes to be transcribed (and subsequently translated) more frequently than others. This promoter–polymerase interaction leads to the production of larger amounts of specific gene products. The expression of viral-encoded regulatory proteins (like the *tat*-III gene product), may also be activated by host cell transcriptional control elements (Jones, 1986). Several other unique gene sequences in HIV have been characterized, and at present efforts are underway to identify their specific gene products and their role in the pathogenecity of the virus (Allan, 1985; Sodroski *et al.*, 1985; Janda, 1987). These genes and their location are shown in Fig. 1. Included are the *tat*-III, 3′-*orf* (open reading frame), and *sor* (short open reading frame) genes. Restriction endonuclease enzymes and bacterial plasmids have been utilized as cloning vectors and HIV genomes with specific gene deletions constructed and transfected into lymphocyte cell lines in an effort to assess the effect of specific gene deletions on the ability of the virus to replicate and produce cytopathic effects (Sodroski, 1986). The *tat*-III gene, for example, codes for an enhancer product that stimulates rates of viral replication. Specific deletion of this gene and, thus, its specific gene product results in less efficient viral replication (Sodroski *et al.*, 1985). The 3′-*orf* and *sor* genes code for 27,000- and 23,000-Da molecular weight proteins, respectively (Allan, 1985; Kan, 1986; Lee, 1986; Sodroski, 1986). Neither of these viral products has, however, been demonstrated to be necessary for either HIV replication or production of cytopathic effects. Furthermore, viral regulatory proteins may also affect the expression of host cell genes (Gallo and Wong-Staal, 1985; Wong-Staal and Gallo, 1985). Certain host-derived molecules may also have a role in viral gene expression and virulence. Antibodies against thymosin alpha 1, a thymic hormone, have been shown to inhibit HIV replication in cultured lymphocytes by combining with gag gene products compressing part of the viral core proteins (Sarin, 1986). Research along such lines may eventually be able to determine which gene products and/or virus/host cell interactions indeed determine virulence of HIV and induction of immunodeficiency.

Following the discovery of HIV and its association with AIDS, other related viruses have been detected in both animals and humans. A primate virus designated T-lymphotropic virus type III (STLV-III) was isolated both from macaques sick with an AIDS-like disease and from healthy with African green monkeys. There is a similarity in the structural and functional virologic characteristics

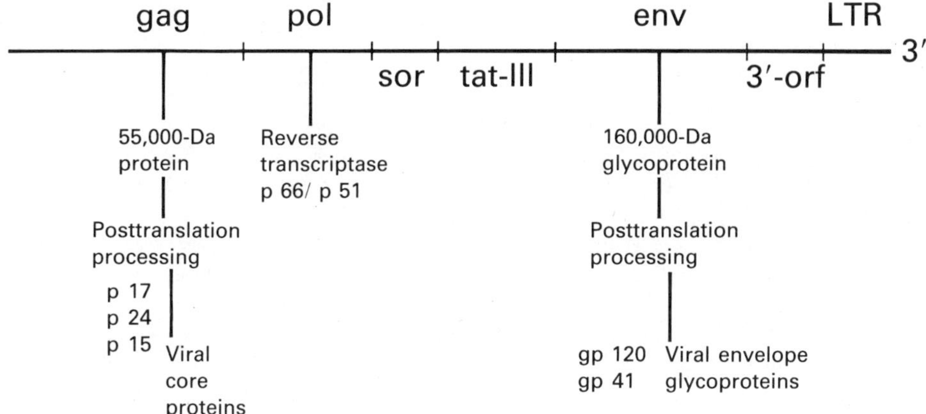

Figure 1. Schematic diagram of HIV (HTLV-III/LAV) RNA genome and gene products.

with HIV, and the STLV-III proteins are similar to and cross-react immunogically with those of HIV. Recent epidemiologic investigation of AIDS in Africa (van der Groen and Piot, 1989) have indicated that sera from healthy individuals from West Africa reacted more strongly with STLV-III antigens than with HIV antigens. Culture of peripheral blood lymphocytes from these healthy STLV-III-antibody-positive individuals resulted in the isolation of yet another virus designated as the human T-lymphotropic virus type IV (HTLV-IV). While this virus showed strong immunologic relatedness with all major antigens of STLV-III, only certain antigenic determinants were shared with HIV. Western blot analysis of immunologic reactivity showed that HTLV-IV-antibody-positive sera reacted with HIV viral core antigens (i.e., *gag*-gene-coded proteins) but not with HIV envelope antigens (i.e., *env*-gene-coded proteins). Simultaneously another virus was isolated from two West African patients with AIDS (Clavel, 1986). This virus, designated LAV type II, demonstrated related core antigen epitopes with HIV (of LAV type I) but possessed substantially different envelope glycoprotein antigens. Serum samples from these two West African AIDS patients failed to react in the HIV ELISA test but did react in a serologic test using STLV-III virus lysates as the antigen; serum from STLV-III-infected macaques reacted in serologic tests with lysates of viral isolates from the two patients. These studies indicate the LAV type II is probably more closely related to STLV-III than to the prototype AIDS-related retrovirus presently known as HIV. Thus, there appears to be little doubt that STLV-III, HTLV-IV, LAV type II, and HIV are related and probably share a common evolutionary origin. The discovery of these new retroviruses and investigations of their relatedness to other simian and human retroviruses may shed light on the origins and virulence mechanisms of HIV as well as provide information on shared antigenic epitopes among the viruses. Information of this type may be useful for the development of HIV vaccines. Furthermore, observations on the induction of an AIDS-like illness by "STLV-III" in macaques, the absence of disease among similarly infected African green monkeys, and the absence of illness in individuals with HTLV-IV infection may provide information on the role of the immune system in disease development. The STLV-III/macaque relationship may also serve as a useful animal model for testing and evaluation of antiretroviral drugs and vaccines.

III. MANAGEMENT AND CONTROL

Each patient, irrespective of whether the patient evaluation reveals a healthy patient with AIDS anxiety or an infected patient, requires education about HIV disease and its transmission. Education could be accomplished by emphasizing that HIV could very well affect men, women, gays, straights, children; that it is not casually transmitted but is transmitted through sexual intercourse or drug use; that those infected can be infectious even though currently asymptomatic; and that women infected with HIV can pass it to their children during pregnancy or birth. Above all, potential exposure to the HIV virus can be controlled through simple precautions, limiting the number of sexual relationships, practicing "safe sex," and eliminating drug abuse. The needs of the various risk/infection groups vary considerably and thus counseling should be preferably tailored to the clinical findings.

Patients at low risk of developing AIDS but anxious because of a perceived exposure to the HIV virus may need counseling to overcome AIDS anxiety. These "worried well" generally show symptoms of anxiety, depression, sleep disorders, and sudden abnormal sexual activity. For some of these patients, antibody-negative test results will not eliminate anxiety.

The asymptomatic patient with a positive HIV antibody test result should be assured that not all asymptomatic, infected patients with early symptoms progress to AIDS. Reports in literature suggest that only 5–20% of infected patients progress to AIDS (Brettman, 1986). They should practice safe sex to protect themselves and others, but this does necessitate social isolation. The physician's psychologic support can help alleviate the patient's preoccupation with developing AIDS.

The apparent fatality associated with AIDS poses a challenge to the medical profession. A behavioral approach to symptom management could include a recommendation to avoid reading or listening to media about AIDS and to minimize contacts with other anxious or depressed patients and instead focus on diet, exercise, and activities or hobbies to break the obsession. Some (30%) AIDS patients develop psychiatric disorders caused by undiagnosed brain disease (Christ and Wiener, 1985; Loewenstein and Rubinow, 1987). These symptoms generally occur after other symptoms have developed and may be a late manifestation of the disease. Early symptoms include headaches, sadness, inability to concentrate, sleep disturbances, and decreased libido (Black, 1985). Referral to a neurologist would be most appropriate.

Prevention is of extreme importance in controlling HIV infection due to the unavailability yet of vaccines and effective therapy. Control measures could include:

1. The sexual route of transmission of AIDS should be controlled by effective public educational programs.
2. Noninfected or low-risk group individuals should avoid sexual contacts with high-risk persons, such as drug addicts, bisexuals, sexually active people.
3. Infected individuals should definitely limit sexual contacts.
4. All pregnant women at high risk should be screened to effectively limit prenatal transmission.
5. Barriers such as condoms should be used during sexual contact.
6. Blood and blood products should be screened to prevent parenteral transmission.

IV. NEW THERAPEUTIC APPROACHES

Several therapeutic agents are available for treatment of opportunistic infections and carcinomas in AIDS patients, but no therapy is currently available for reversal of underlying immunodeficiency. Traditional therapy rarely is successful when Kaposi's sarcoma strikes. Antitumor regiments using standard chemotherapeutic agents such as vinblastine, Adriamycin, and bleomycin have not been promising. Vincristine is known to occasionally produce partial reduction in tumor size. Initial trials using immune-modulating agents as alpha-interferon, gamma-interferon, isoprenosine, and interleukin-2 found little positive effect on immune system functions (Lotze, 1985; Skeen, 1985). Trimethoprim/sulfamethoxazole or pentamidine and ventilatory support have been reported to be useful treatments for *Pneumocystis carinii* pneumonia infection, which occurs in more than half of AIDS patients and carries a high mortality rate of 53% (Skeen, 1985).

Antiviral chemotherapy would be another method to reduce the effects of HIV infection. Suramin, commonly used to treat parasitic infections in the human, has been shown to contain the infectivity and cytopathic effect of HIV (Fishinger and Bolognesi, 1985); however, the toxic effect of suramin on tissue may limit its potential use. Zidovudine (3′-azido-3′-deoxythymidine-AZT), a reverse transcriptase inhibitor, has recently been marketed. This drug can alter nucleoside metabolism, leading to a reduction of the level of its natural, thymidine triphosphate, for viral and cellular DNA polymerases. The effect of such alteration on the antiretrovirus activity and cellular toxicity is unknown. Zidovudine crosses the blood–brain barrier and may improve the neurologic involvement observed in HIV patients. Reversible bone marrow suppression and anemia have been seen with zidovudine and may require intermittent treatment schedules (Jeffries, 1989).

A drug intended to stop the spread of the AIDS virus in the body by mimicking a part of the blood cells it usually infects has proved to be highly effective in experimental use on monkeys. The animal research provides the first clear evidence outside the test tube that this strategy has a chance of slowing and perhaps arresting the disease in humans. Even if the treatment works as well in people as it does in monkeys, it will not cure AIDS but it might possibly make life better for persons with the virus. Recently, testing has been started of the drug CD4 on people with AIDS. CD4 is designed to be a decoy target for HIV, the AIDS virus, deflecting it from healthy blood cells that it takes over and kills. CD4 is a natural molecule to which the AIDS virus is attracted on the surface of some blood cells. The treatment obviously cannot wipe out the virus when hidden inside the body's cells, but loose viruses in the bloodstream could latch onto the molecule instead of new cells. The latest research conducted at the Harvard–New England Regional Primate Research Center showed that the drug dramatically reduced levels of a closely related virus in rhesus monkeys. The monkeys were infected with simian immunodeficiency virus of macaques (SIV). The virus is related to HIV. Monkeys that catch SIV grow sick and die of an illness that closely resembles human AIDS. This is the first evidence *in vivo* in a living animal that the drug has antiviral effects with chronic administration in a relevant animal model. Should the drug be well tolerated in humans, patients could hopefully live significantly longer and in a better way.

Investigators are also experimenting with combinations of antiviral and immunomodulating drugs to fight the infection and strengthen the immune system (Marwick, 1986).

Many viral diseases have been controlled through a vaccine containing avirulent virus, inactivated virus, or viral subunits. Several animal retroviruses are being used in large-scale immunization toward developing a vaccine for AIDS. The problem limiting AIDS vaccine is that HIV virus appears in many antigenic forms. Furthermore, the virus is capable of constantly altering its antigenic coating. This may necessitate several vaccines, as with influenza, owing to the variability of HIV. Currently, several approaches to develop an AIDS vaccine are under investigation (Fauci, 1989). Four principal candidate vaccines are being tested in humans: a whole killed virus vaccine, a gp160–vaccinia virus recombinant, autologous cells infected with a gp160–vaccinia virus recombinant, and a baculovirus-expressed gp160 recombinant.

REFERENCES

Allan, J. A. (1985). *Science* **230,** 810–813.

Barre-Sinoussi, F., Chermann, J. C., Rey, R., Nugeyre, M. F., Chamaret, J., Gruest, J., Daugnet, C., Axier-Blin, C., Vezinet, Brun, F., Rouzioux, C., Rozenbaum, P. H., and Montagnier, L. (1983). *Science* **220,** 868–871.

Black, P. H. (1985). *N. Engl. J. Med.* **313,** 1538–1540.

Brettman, L. R. (1986). *Infections Medi.* **3,** 18–33.

Brunet, J. B., and Ancelle, R. A. (1985). *Ann. Intern. Med.* **103,**670–674.

Christ, G. H., and Wiener, L. S. (1985). In *AIDS: Etiology, Diagnosis, Treatment and Prevention* (J. B. De Vita, Jr. ed.), pp. 275–297, Lippincott, Philadelphia.
Clavel, F. (1986). *Science* **233,** 343–346.
Coffin, J., Haase, A., and Levy, J. A. (1986). *Science* **232,** 687.
di Marzo Veronese, F., Copeland, T. D., and De Vico, A. L. (1986). *Science* **231,** 1289–1291.
Essex, M., Allan, J., and Kanki, P. (1985). *Ann. Intern. Med.* **103,** 700–703.
Fauci, S. A. (1989). *Ann. Intern. Med.* **110,** 373–385.
Fishinger, P. J., and Bolognesi, D. P. (1985). In *AIDS: Etiology Diagnosis, Treatment, and Prevention* (J. B. De Vita, Jr, ed.), pp. 55–88, Lippincott, Philadelphia.
Folks, T., Powell, D. M., Lightfoote, M. M., Benn, S., Martin, M. A., and Fauci, A. S. (1986). *Science* **231,** 600–602.
Friedland, G. H., and Klein, R. S. (1987). *N. Engl. J. Med.* **317,** 1125–1135.
Gallo, R. C., and Wong-Staal, F. (1985). *Ann. Intern. Med.* **103,** 679–689.
Gallo, R. C., Salahuddin, S. Z., Popovic, M., Shearer, G. M., Kaplan, M., Haynes, B. F., Palker, T. J., Redfield, R., Oleske, J., Safai, B., White, G., Foster, P., and Markham, P. D. (1984). *Science* **224,** 500–503.
Janda, W. M. (1987). *Clin. Microbiol. Newslett.* **9**(2), 9–15.
Jeffries, D. J. (1989). *J. Infection* **18**(1), 5–13.
Jones, K. A. (1986). *Science* **232,** 751–759.
Kan, N. C. (1986). *Science* **231,** 1553–1555.
Kurstak, E., Tijssen, P., Kurstak, C., and Morisset, R. (1986). In *Advances in Sexually Transmitted Diseases* (R. Morisset and E. Kurstak, eds.), pp. 209–216, VNU Science Press, The Netherlands.
Lee, T. H. (1986). *Science* **231,** 1546–1549.
Loewenstein, R. J., and Rubinow, D. R. (1987). In *Viruses, Immunity, and Mental Disorders* (E. Kurstak, Z. J. Lipowski, and P. V. Morozov, ed.), pp. 95–107, Plenum Press, New York.
Lotze, M. T. (1985). In *AIDS: Etiology, Diagnosis, Treatment and Prevention* (J. B. DeVita, Jr., ed.), pp. 235–263, Lippincott, Philadelphia.
Marwick, C. (1986). *JAMA* **255,** 1233–1242.
McDougal, J. S., Kennedy, M. S., and Sligh, J. M. (1986). *Science* **231,** 382–385.
Montagnier, L. (1985). *Ann. Intern. Med.* **103,** 689–693.
Sarin, P. S. (1986). *Science* **232,** 1135–1137.
Skeen, W. F. (1985). *Ann. Emerg. Med.* **14,** 267–273.
Sodroski, J. (1986). *Science* **231,** 1549–1553.
Sodroski, J., Rosen, C., and Wong-Staal, F. (1985). *Science* **228,** 171–173.
Van der Groen, G., and Piot, P. (1989). *Ann. Saudi Med.* **9**(1), 105–106.
WHO Weekly Epidemiological Report. (1989). 1.05.
Wong-Staal, F., and Gallo, R. C. (1985). *Blood* **65,** 253–263.

4

Escape of Lentiviruses from Immune Surveillance

David L. Huso and Opendra Narayan

I. INTRODUCTION

Lentiviruses are nononcogenic retroviruses that cause chronic disease in persistently infected individuals. The viruses are transmitted exogenously and replicate productively in terminally differentiated cells of the immune system. Lentiviruses are transmitted in body fluids and infection often remains inapparent for months or years. Disease begins insidiously and infected individuals eventually develop a chronic multisystemic disease manifest as wasting, immune dysfunction, dyspnea, or neurologic impairment. The viruses are species-specific and members of this group infect a variety of hosts causing unique disease syndromes in each host species. However, the members of the lentivirus group share a remarkable number of biologic, morphologic, and structural features which define this subfamily of viruses (Narayan and Clements, 1987; Haase, 1986).

Lentiviruses cause various diseases in different host species. Equine infectious anemia virus (EIV) causes anemia and cyclical febrile illness of horses (Cheevers and McGuire, 1985). Visna-maedi virus (VMV) of sheep and caprine arthritis–encephalitis virus (CAEV) of goats are two closely related viruses that cause wasting, pneumonia, arthritis, and encephalitis in their respective hosts (Narayan and Cork, 1985). Human Immunodeficiency Virus (HIV) is the lentivirus of man and causes wasting, encephalitis and immunodeficiency with opportunistic infections (Fauci, 1988). Simian immunodeficiency virus (SIV) infects nonhuman primates and similarly causes immunodeficiency with opportunistic infections and encephalitis in macaques (Daniel *et al.*, 1985). Bovine immunodeficiency virus (BIV) causes encephalitis, lymphocytosis, and emaciation of infected cattle (Braun *et al.*, 1988). Finally, cats infected with the feline immunodeficiency virus (FIV) manifest leukopenia and immunodeficiency with opportunistic infections (Pedersen *et al.*, 1987).

One of the hallmarks of lentiviruses is that they cause persistent infections in their respective hosts. These agents effectively elude host defense mechanisms during infection, allowing them to survive indefinitely within the infected host. They not only persist, but replicate continuously, often in the face of competent immune responses which are directed toward their elimination. There is no single, unifying mechanism by which these agents avoid the host immune response. Rather, these viruses have a number of unique properties that function in concert during various stages of infection and ensure persistence of the virus in the infected host. In this chapter we summarize the features of

David L. Huso • Division of Comparative Medicine, The Johns Hopkins University School of Medicine, Baltimore, Maryland 21205. *Opendra Narayan* • Division of Comparative Medicine and Department of Neurology, The Johns Hopkins University School of Medicine, Baltimore, Maryland 21205.

lentiviruses that allow them to maintain a persistent infection in their host while escaping immune surveillance.

II. STRUCTURE AND REPLICATION

The lentiviruses have an RNA genome that is approximately 10 kb in length and contains three genes that encode the major viral structural proteins (Molineaux and Clements, 1983). These are (5′-3′), the *gag, pol,* and *env* genes, common to all retroviruses. The genes are encoded in one of three reading frames. The *gag* and *pol* genes are encoded in different reading frames and overlap slightly. The large *env* gene is separated from the *pol* gene, and there are multiple short open reading frames located between and overlapping the *pol* and *env* genes. These short open reading frames encode small regulatory proteins that control viral gene expression and influence virus infectivity. This combination of structural and regulatory genes and the larger overall size of the genome set the lentivirus genome apart from other retroviruses (Ratner *et al.*, 1985; Sanchez-Pescador *et al.*, 1985; Sonigo *et al.*, 1985; Chakrabarti *et al.*, 1987).

The *gag* gene encodes a large polyprotein which is cleaved to form the structural core proteins of the virus. These proteins are immunogenic in the infected individual and evoke a strong antibody response. Serologic cross-reactivity among members of the lentivirus group is often due to the immunologic relatedness of the *gag* proteins (Pyper *et al.*, 1984). Seroreactivity to the core antigen forms the basis of many diagnostic tests used to screen for the presence of infection.

The *pol* gene sequence is highly conserved among lentiviruses. It is expressed by suppression of the *gag* translation termination codon and subsequent fusing of the *gag* to the *pol* reading frame. The fusion of these two gene products during translation results in synthesis of a *gag/pol* precursor polyprotein. The polyprotein is then cleaved at specific sites and forms the three unique protein products of the *pol* gene. These three products have four catalytic activities. One of the products is a viral protease that is responsible for cleaving the viral precursor polyproteins. The second product is the viral reverse transcriptase with an associated ribonuclease H activity and functions in viral replication. Finally, an integrase/endonuclease protein is produced and it functions in the integration of proviral DNA into the host cell genome (Farmerie *et al.*, 1987).

The *env* gene product is synthesized as a large precursor polyprotein and is subsequently cleaved intracellularly at a specific endoproteolytic cleavage site. Cleavage yields a heterodimer composed of a large hydrophilic outer membrane protein and a smaller, hydrophobic, transmembrane protein anchor (McCune *et al.*, 1988). The envelope glycoproteins of lentiviruses characteristically are highly glycosylated. The *env* gene sequence contains about 25–30 potential *N*-linked glycosylation sites (Sanchez-Pescador *et al.*, 1985; Sonigo *et al.*, 1985). More than two-thirds of these potential sites are glycosylated in the native state (Geyer *et al.*, 1988). *O*-linked carbohydrates are a structural component of many viruses and also are present in lentiviruses (Huso *et al.*, 1988; Abel *et al.*, 1987). Carbohydrates therefore comprise a significant portion of the membrane glycoprotein of these viruses. The envelope glycoprotein mediates virus-induced cell-to-cell fusion; it is responsible for binding of virus to cell surface receptors; it contains the epitopes that induce neutralizing antibodies as well as epitopes of cell-mediated immunity; and it forms a protective coat around the viral genome during extracellular phases of the viral life cycle.

Lentiviruses bind to cell surface receptors as the first step in the viral replication cycle. Receptor specificity determines, in part, the host and cellular tropism of the virus. Visna and CAE viruses infect both macrophages and fibroblasts (Gendelman *et al.*, 1985; Narayan *et al.*, 1980, 1982), while HIV has a tropism for macrophages and T4 lymphocytes (Gartner *et al.*, 1986; Klatzman *et al.*, 1984). Following receptor binding, the lentiviruses may enter target cells by direct fusion to the cell plasma membrane or through receptor-mediated endocytosis. Viral particles are uncoated and the viral nucleoprotein is released into the cytoplasm. The viral reverse transcriptase makes a proviral DNA copy of the viral genome and the proviral DNA is transported into the nucleus. A portion of the proviral DNA becomes circularized and integrates into the host cell genome, catalyzed by the viral integrase enzyme. Multiple copies of both linear and circularized proviral DNA also remain unintegrated (Clements *et al.*, 1979). Transcription from the proviral DNA yields full-length genomic RNA copies and shorter, spliced messenger RNA (Davis *et al.*, 1987).

During the intricate priming steps necessary to initiate synthesis of proviral DNA, sequences unique to each end of the RNA genome(U_3,U_5) are duplicated and added to repetitious sequences (R) present at the ends of the proviral DNA. These combinations of sequences (U_3,R,U_5) are present at both ends of the linear proviral DNA and are called long terminal repeats (LTR). The LTRs function in the circularization and integration of the proviral DNA into the host genome. The LTR contains elements that initiate gene expression of the viral genome (Hess *et al*,. 1985, 1986). It contains the viral promoter, the 5′ cap site, which is the starting site for synthesis of viral messenger and genomic RNA, and enhancer sequences, which modulate the rate of virus gene expression from the viral promoter. Virus-encoded proteins also affect gene expression. For example, the viral *tat* gene acts to up-regulate viral gene expression while the *nef* gene down-regulates gene expression (Arya *et al.*, 1985; Haseltine and Wong-Staal, 1988). In the infected individual, complex interactions between these factors and the host cell transcriptional control proteins may be responsible for different rates of viral replication in different populations of cells and may play an important role in the persistence of lentiviruses.

Following translation of viral proteins, the lentivirus envelope glycoprotein is glycosylated in the endoplasmic reticulum and Golgi by the host cell glycosylation enzymes. From there it is transported to the cell surface and is anchored in the plasma membrane. Other viral structural proteins and viral genomic RNA localize beneath the viral glycoproteins and virus particles assemble as the virus buds from the cell membrane (Fig. 1). The enveloped lentivirus particles are pleiomorphic and measure 80–120 nm. They have a characteristic dense cylindrical core which tapers at one end (Gonda *et al.*, 1985). The capsid proteins line the inner surface of the viral envelope and encircle the core. The large, highly glycosylated envelope glycoprotein forms spikes which stud the outer surface of the lentivirus particles (Gelderblom *et al.*, 1987).

III. HUMORAL RESPONSE

Coincident with their ability to persist indefinitely and replicate continuously in their respective hosts, lentiviruses are poor inducers of neutralizing antibodies. Low levels of virus neutralizing activity are present in the serum of persons infected with HIV (Weiss *et al.*, 1985; Robert-Guroff *et al.*, 1985). Similarly, visna virus infection results in induction of low levels of neutralizing antibodies while CAEV almost never induces detectable neutralizing antibodies (Narayan *et al.*, 1984).

The neutralizing antibodies induced by the CAEV and visna are relatively inefficient (Kennedy-Stoskopf and Narayan, 1986). The antibodies neutralize virus slowly, requiring long preincubations of the virus–serum mixtures in order to be effective. The slow kinetics of neutralization reflects a low affinity of antibodies for neutralizing epitopes on the virus. In contrast, the affinity of the virus for its receptor is higher. Since the kinetics of virus–receptor binding is more rapid than the kinetics of virus–antibody interaction, this may enhance the viruses' chances of spreading cell to cell before being neutralized.

Most of the antiviral antibodies produced by infected or immunized hosts do not neutralize infectious virions. The antibodies bind to immobilized viral antigens or will precipitate soluble antigen in antigen–antibody complexes, but do not prevent virus from infecting cells *in vitro*. These antibodies are useful diagnostically, but their role in the development of lentiviral disease is not known.

The lack of neutralizing antibodies consistently observed in goats infected or immunized with CAEV could not be overcome by repeated immunizations with purified virus, purified envelope

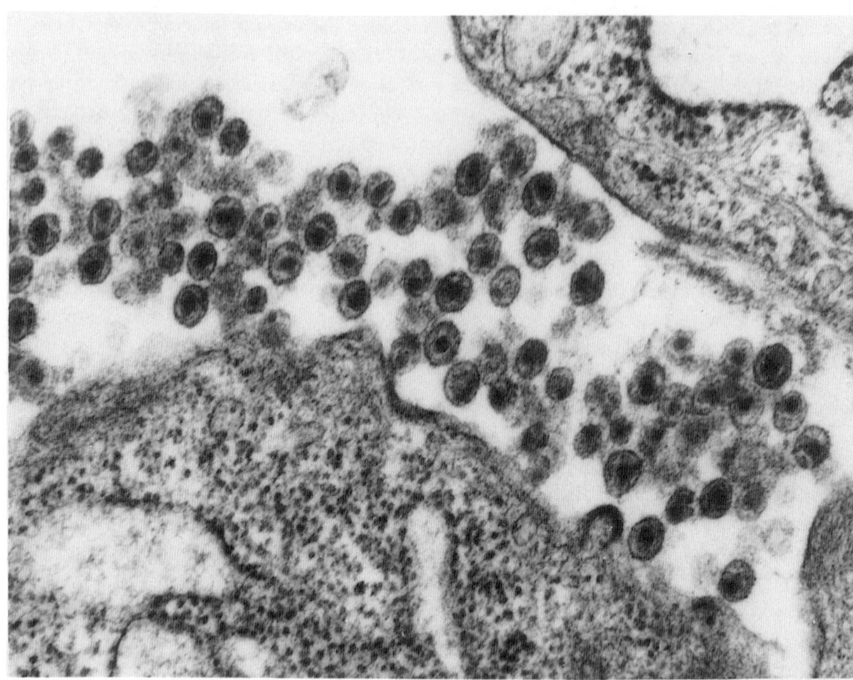

Figure 1. Visna virus particles are initially agglutinated on the surface of an infected cell by neutralizing antibodies. Infected cells, cultured in the presence of neutralizing antibodies from infected sheep, produce antigenically variant viruses that can escape neutralization by the original serum and infect other cells in the culture. (The micrograph at a magnification of 30,000 was made by Edwards and Huso.)

glycoprotein, inactivated virus, virus disrupted with detergent, or cells infected with virus. Initially, the failure of CAEV to induce neutralizing antibodies was thought to be due to a lack of neutralizing epitopes on the virus particles. However, the virus particles do have neutralizing epitopes. Low titers of neutralizing antibodies were induced when virus was administered with *Mycobacterium tuberculosis*. These neutralizing antibodies still only neutralized virus at low dilutions and with slow kinetics (Narayan *et al.*, 1984).

Neutralization determinants are not only poor at inducing antibodies, but these antibodies usually do not protect individuals against infection after challenge with infectious virus. Humans develop neutralizing antibodies to human immunodeficiency virus (HIV) following immunization with the appropriate antigen (Zagury *et al.*, 1988), but there are no reports of protective antibodies. The development of viral-specific antibodies during HIV infection does not correlate with any improvement of the infected individual (Groopman *et al.*, 1987). Similarly, neutralizing antibodies failed to alter the course of SIV infection in immunized and challenged macaque monkeys (Letvin *et al.*, 1987). In visna virus infection, neutralizing antibodies and virus coexist in the animal. Therefore, neutralizing antibodies that do develop are inefficient and may be of little beneficial value to the host (Narayan *et al.*, 1987). Recently, immunization of macaque monkeys with SIV immunogens has suggested that protective immunity can be induced (Desrosiers *et al.*, 1989; Murphy-Corb *et al.*, 1989). Therefore, further investigation of lentiviral immune responses is needed.

IV. VIRAL ENTRY INTO MACROPHAGES

Lentiviruses infect cells of the monocyte–macrophage lineage. Infection in these cells not only provides a reservoir of virus, but these cells also carry the virus throughout the body. Macrophages are specialized cells with receptors that function in the uptake of substances that have been opsonized by serum proteins. Even though immune serum may block lentiviral entry into some target cells, serum proteins may actually enhance viral entry into macrophages. Therefore, immune sera may provide alternate means by which lentivirus particles can enter macrophages in the face of an antilentiviral immune response.

Immunoglobulins from immune serum enhance the uptake of lentivirus particles by macrophages (Jolly *et al.*, 1989a). Antibodies bind to lentiviruses and these immune complexes bind to Fc receptors (FcR) on macrophages. Macrophages then take up these complexes by phagocytosis. In studies of visna virus, preincubated mixtures of virus and antibodies added to fibroblasts and macrophages resulted in much greater uptake of the virus by the macrophages. This effect was not obtained when $F(ab)^2$ fragments of the antibodies were used. Neutralizing $F(ab)^2$ fragments still protected fibroblasts from infection, but they lacked the ability to enhance binding, penetration, and uncoating of the virus in macrophages. These effects must have been caused by interaction of the Fc portion of the immunoglobulin molecule with the FcR of the macrophage. FcR-mediated entry of visna virus into macrophages did not promote enhancement of replication of the virus. However, inoculation of macrophages with HIV preincubated with subneutralizing concentrations of immune sera resulted in enhancement of replication of the virus. Heat-aggregated immunoglobulins blocked the enhancement presumably by competing for Fc binding sites on the macrophages. Moreover, the enhancement occurred only if the antiviral antibodies had functional Fc domains. This suggested the enhancement was mediated by uptake of virus–antibody complexes by Fc receptors on the macrophages (Jolly *et al.*, 1989b; Takeda *et al.*, 1988).

In addition to antibodies, a heat-labile component of serum (complement) is capable of enhancing HIV infection of macrophages (Robinson *et al.*, 1988). This heat-labile factor was present in normal serum, but was absent from serum that was deficient in the third component of complement (C3). The enhancing activity of the serum could not be removed by protein A chromatography, but was blocked by the anticomplementary activity of cobra venom, further suggesting the enhancing factor was a complement component. Macrophages express complement receptors at their surface. Therefore, following fixation of complement by lentivirus particles, it appears that enhancement of lentiviral entry into target macrophages may also occur via the complement receptor.

Finally, mannose sugars found on the lentiviral envelope may mediate entry into macrophages by one of two routes. The mannose receptor on the surface of macrophages mediates endocytosis following binding of a mannose ligand. Since the envelopes of the lentiviruses contain mannose, virions may be a ligand, binding to the mannose receptor of macrophages with resulting stimulation of endocytosis and uptake of virus particles (Robinson *et al.*, 1987). In addition, soluble mannose binding proteins present in serum are capable of binding to lentiviruses and blocking infectivity in nonmacrophage cell types. Whether the soluble mannose binding proteins may also opsonize the mannose-containing lentivirus particles for uptake by macrophages is not known (Ezekowitz *et al.*, 1989). Therefore, specialized receptors on the macrophage act independently or in concert with the viral receptor to enhance infection of macrophages.

Fusion plays an important role in the entry of lentiviruses into target cells. Fusion of the viral envelope to the cellular membrane appears to be a prerequisite for entry of the viral genome into the cell following binding of virus to cell receptors (Stein *et al.*, 1987). In addition, the envelope glycoprotein, which is expressed on the surface of infected cells, mediates cell-to-cell fusion in infected cultures (Liffson *et al.*, 1986; Narayan *et al.*, 1980).

Pure cultures of macrophages infected with lentiviruses restrict viral replication and are resistant to virus-induced fusion. However, infected macrophages readily fuse to uninfected, nonmacrophage cells. Primary adherent macrophages were grown in culture and infected with the goat lentivirus, CAEV. By 3 days after infection, infectious virions were being synthesized, but at minimal levels in the macrophages. However, when uninfected fibroblast cells were added to the infected macrophages, the macrophages rapidly began to fuse with the uninfected fibroblast cells. Viral messenger RNA was transmitted from the infected macrophage to the fused, previously uninfected cells. The appearance of the lentiviral glycoprotein on the surface of infected macrophages mediated fusion and transfer of viral RNA from the infected macrophage to contiguous, uninfected cells. This suggests that lentivirus infectivity can be transferred from cell to cell by fusion without participation of infectious particles.

In summary, immune serum that would usually block virus-receptor-mediated entry into target cells may be ineffective at blocking lentivirus entry into macrophages. The soluble mannose binding protein, virus-specific antibodies, and complement in the serum of infected individuals may direct opsonized virus to enter macrophages via alternate pathways such as by Fc receptors or C3 receptors. Lentiviruses may enter macrophages following binding to the mannose receptor, an endocytic receptor found on the surface of macrophages. Finally, infectivity can be transferred directly from infected macrophages to uninfected cells by virus-mediated cell-to-cell fusion without the requirement for extracellular virus particles.

V. RESTRICTED VIRAL REPLICATION IN MACROPHAGES

Lentivirus replication in macrophages is highly regulated so that the virus replicates at restricted levels within the host. Immature monocytes can be infected by lentiviruses both *in vitro* and *in vivo* but produce few viral transcripts. As the monocytes mature into macrophages, these cells become more permissive for viral replication. Therefore restriction of virus replication is influenced by maturation factors from macrophages (Gendelman *et al.*, 1985, 1986; Narayan *et al.*, 1983; Folks *et al.*, 1988).

During infection of macrophages with the ungulate lentiviruses visna virus and CAEV, interferon is produced following interaction of infected macrophages and lymphocytes. Supernatant fluids containing the interferon activity have a number of biologic activities related to restricted replication of lentiviruses. It has a direct inhibitory effect on viral gene expression in mature macrophages at the level of transcription. It also inhibits virus-mediated cell-to-cell fusion perhaps by its effect on membrane fluidity. In addition, interferon inhibits maturation of monocytes to macrophages, indirectly causing a down-regulation of virus replication (Zink and Narayan, 1989; Narayan *et al.*, 1985).

Tissue-specific factors also play a role in restriction of the virus life cycle. For example, CAEV viral RNA was found abundantly in synovial, splenic, alveolar, and brain macrophages in infected animals. However, other macrophage populations such as Kupffer cells contained only minimal viral RNA. Transgenic mice that had part of the HIV genome inserted into their cells showed high levels of expression in Langerhans cells of the skin but not in other macrophage populations (Leonard *et al.*, 1988).

Finally, the genetic makeup of the host may influence the restriction of viral gene expression. Icelandic sheep develop brain lesions of encephalitis following infection with visna virus. However, North American sheep appear to be resistant to the development of encephalitis and have very little viral RNA in the brain following infection. These two populations of sheep are genetically distinct from each other. Thus, host genetic factors may influence the development of encephalitis following infection with visna virus.

VI. ANTIGENIC VARIATION

During replication, lentiviruses are prone to frequent mutations of their genomic sequence. Unlike cellular replication enzymes, the reverse transcriptase of lentiviruses has no proofreading capabilities. Therefore, mistakes made during the synthesis of proviral DNA are not corrected. These mutations are passed on to messenger RNA, viral proteins, viral genomic RNA, and progeny virus particles. The mutations in the nucleotide sequence of a viral gene may change the amino acid sequence and the properties of the protein that the gene encodes. While some mutations render the protein nonfunctional, some changes are slight so that the altered protein is still at least partially functional. This results in progeny virus which, due to mutation, has a slightly different genome and protein structure than the parental virus from which it was derived. This altered protein structure may result in altered antigenic reactivity compared to the parental virus.

Certain regions of the lentiviral genome are more likely to contain mutations than other regions. While the *gag* and *pol* genes of lentiviruses are quite conserved, the *env* gene is a highly variable region among lentiviral isolates (Pyper *et al.*, 1986; Saag *et al.*, 1988). Mutations in the *env* gene may alter not only the amino acid sequence of the glycoprotein, but also its glycosylation pattern. Mutations that occur within the glycosylation site on the protein backbone may directly alter the final glycosylation pattern of the molecule. However, mutations may also alter glycosylation indirectly by altering the folding of glycoproteins. The cellular transferase and glycosidase enzymes that mediate glycosylation must recognize their specific glycosylation target site before they can act. Any mutation that alters protein conformation could potentially alter glycosylation indirectly by making the glycosylation site inaccessible to the host cell glycosylation enzymes. Also, mutations may result in addition of glycosylation sites such that carbohydrates may cover regions of the glycoprotein that are functionally important. Therefore, it would be expected that different virus isolates that have undergone mutation would vary in biologic properties mediated by the envelope glycoprotein. These functions include receptor binding, neutralization, and ability to cause fusion. In fact, genetically variant lentiviruses frequently arise due to the high mutation rates seen among lentiviruses.

Genetic variation may affect virus interference, a conserved property of retroviruses. Once a cell is infected with one type of virus, the infection prevents superinfection of the cell by a virus of the same subtype. The virus glycoprotein apparently binds to and saturates receptor sites, making them unavailable to other viruses that recognize the same ligand. However, variant viruses may recognize different epitopes on their receptor or completely different receptors such that the viruses do not interfere with each other (Steck and Rubin, 1966). When genetic variants of visna and CAEV were tested for homologous and heterologous interference, some variants interfered with each other while other variants did not interfere. Surprisingly superinfection with a noninterfering virus resulted in enhancement of infection by the second virus. Thus, some genetic variants appear to differ in the receptor epitope that they recognize, and the potential for dual infection of cells exists (Jolly and Narayan, 1989).

In addition to differences in antigenicity and receptor recognition, the variant viruses may also differ in their replication potential in certain cell types. For example, one strain of visna virus replicates to high levels in cells derived from the sheep choroid plexus cells of the brain, whereas another genetically variant strain does not replicate in these cells but only replicates in macrophages or in cells derived from synovial membranes of the joint. Similarly, some isolates of HIV replicate well in T cells, but replicate poorly in macrophages, whereas other HIV isolates are highly macrophage-tropic viruses (Koyanagi *et al.*, 1987).

Variant viruses also may vary in their ability to cause fusion. When high concentrations of visna virus are inoculated onto cells, the cells begin to fuse within 2–4 hr. This rapid fusion occurs before any progeny virus has been synthesized and is called "fusion from without." Viruses obtained from animals vary in their ability to cause cell fusion. This is particularly evident among CAEV strains. Whereas some cause almost no fusion from without, others fuse with the efficiency of visna virus. These differences in fusion potential of the lentiviruses appear to be due to variations in the envelope protein which mediates fusion.

During infection with visna virus, sheep develop neutralizing antibodies to the virus slowly. The viral envelope contains the determinants for virus neutralization. The envelope gene is highly variable and closely related virus isolates can often be distinguished antigenically by differences in neutralization with immune serum. These changes in reactivity result from mutations described earlier.

In studies of sheep inoculated with biologically cloned visna virus, virus was readily isolated from peripheral blood leukocytes (PBLs) of the infected animals throughout infection. Over a period of months, the infected animals developed neutralizing antibodies to the parental virus. These "early neutralizing" antibodies neutralized virus recovered from the animals early in the infection but failed to neutralize virus obtained 1 year after injection. Serum that neutralized the parental virus (1514) did not neutralize a new virus isolate (LV1-1), indicating that LV1-1 was antigenically different than the parental virus. This suggested that the original parental virus which had been inoculated into the animals had undergone mutations during replication in the sheep. The mutations had resulted in the appearance of a new virus that no longer was neutralized by antibodies directed against the neutralizing epitopes on the parental virus (Narayan *et al.*, 1977). Over time, neutralizing antibodies developed to the mutated virus such that the animal's serum could neutralize both parental and mutated LV1-1 virus and this serum was called "late neutralizing serum."

Further analysis showed that numerous antigenically variant viruses were present simultaneously in the infected sheep. All of these isolates were antigenically stable during amplification of the individual isolates in cell culture provided that neutralizing antibodies were not incorporated in the medium. When tested for replication, all isolates shared a similar replication potential in the cells tested, suggesting that differences in replication in the presence of neutralizing serum was due to antigenic differences between the viruses. Each virus maintained its antigenic phenotype in culture and was defined by specific neutralization patterns when tested against a panel of immune sera (Narayan *et al.*, 1978).

Antigenic variants of visna virus were induced readily in cell culture when infected cultures were maintained in medium containing "early neutralizing antibodies" to the parental virus. The presence of the "early neutralizing serum" prevented the spread of the parental virus in the culture. However, after approximately 2–3 weeks of culture, cytopathic effects began to appear in the monolayers. Titration of the cell supernatant revealed the presence of an antigenic variant that no longer could be neutralized by the "early neutralizing serum." However, the "late neutralizing serum" was quite effective at neutralizing this variant virus. The late serum also prevented the development of antigenically variant viruses even when infected cultures were held for long periods of time in the presence of the late neutralizing serum. The mutants selected by the early serum were similar genetically and serologically to the mutants obtained from sheep 2 years after inoculation. Therefore, these antibodies may have been important in selecting virulent, nonneutralizable mutant viruses that appeared both in cell culture and in the infected animal (Narayan *et al.*, 1981). Similar results have been obtained with HIV-inoculated cell cultures treated with neutralizing antibodies.

Antigenic variant viruses had point mutations in the 3′ end of the genome where the *env* gene is located (Clements *et al.*, 1980). The point mutations accumulated such that later isolates had more mutations than early isolates (Clements *et al.*, 1982). The point mutations in the viral envelope correlated with antigenic changes that were detected by a panel of monoclonal antibodies (Stanley *et al.*, 1987). Therefore, relatively minor sequence changes in the envelope gene resulted in conformational rearrangements in the envelope protein.

EIAV, the lentivirus of horses, causes cyclical episodes of hemolytic crises and fever in horses and each episode is associated with emergence of a new antigenic variant of the virus followed by the development of neutralizing antibody to the new virus (Cheevers and McGuire, 1985). Comparison of various EIAV isolates showed that, like other lentiviruses, it had an envelope gene composed of conserved and variable regions. Within the variable region was a hypervariable region which had the highest rate of amino acid divergence between isolates. The variable region of the protein contained several potential *N*-linked glycosylation sites(Asn-X-Ser/Thr). The number and placement of glycosylation sites varied between each antigenic variant virus isolated. Among the isolated variant viruses, approximately 40% of the amino acid substitutions involved part of the *N*-linked glycosylation site (asparagine residues). The number of glycosylation sites in the envelope gene varied by up to almost 50% (Payne *et al.*, 1987a,b). Two distinct antigenic variants of visna virus had no differences in *N*-linked glycosylation sites. Antigenic variation in this case may have involved only protein epitopes, or it is possible that mutations affecting sites of *O*-linked glycosylation may also be important in antigenic variation. Variability in glycosylation sites in the *env* gene is a frequent observation in other lentiviruses and has led to the suggestion that carbohydrates on the viral envelope may affect the immunogenicity or immune reactivity of the lentivirus envelope.

VII. GLYCOSYLATION

The carbohydrates on the lentiviral envelope have important biologic functions. Site-directed mutagenesis of glycosylation sites on the HIV envelope gene identified a site that is essential for an early step in the viral replication cycle of the virus although it was not essential for receptor binding. Mutation of this glycosylation site resulted in production of noninfectious particles (Willey *et al.*, 1988).

Metabolic inhibitors of glycosylation interfere with lentivirus infectivity and virus-mediated fusion. When cells infected with CAEV were cultured in the presence of inhibitors of *N*-linked glycosylation and oligosaccharide processing, there was reduced cell-to-cell fusion (*N*-methyl-1-deoxynojirimycin) and certain inhibitors (tunicamycin) resulted in the release of noninfectious virus into the cultures. Culture of HIV-infected cells in the presence of chemicals that block trimming of *N*-linked oligosaccharides resulted in reduced fusion and viral replication in the infected cultures (Gruters *et al.*, 1987; Walker *et al.*, 1987).

Studies of influenza virus, a well-known model of antigenic variation, showed that addition of carbohydrate side chains to the viral envelope glycoprotein (HA) not only altered the intracellular transport of the molecule, but also altered its antigenic reactivity. A mutation in the influenza HA gene resulting in the addition of a new glycosylation site near a neutralization epitope on the viral envelope resulted in lack of binding of a neutralizing monoclonal antibody to the epitope and allowed this mutated variant virus to escape neutralization by the monoclonal antibody (Skehel *et al.*, 1984). In other studies, enzymatic removal of *N*-linked carbohydrates from the influenza glycoprotein destroyed epitopes that were recognized by antiserum against the virus. Together, these studies suggest that virus epitopes can either be dependent on the presence of carbohydrates or, in some instances, be masked by the presence of adjacent carbohydrate side chains (Alexander and Elder, 1984). In addition to its effect on antigenicity, glycosylation of lentivirus particles affords protection of the virus against proteolytic enzymes. This may be important in the natural history of these viruses.

CAEV, the naturally occurring lentivirus pathogen of goats, is the cause of an epidemic of milk-borne infection among dairy goats in the United States. The management practice of pooling milk from all animals and feeding the offspring from this common stock has resulted in spreading the virus extensively among the dairy goat population (Adams *et al.*, 1983). Macrophages in the mammary gland are host cells for virus replication in the infected female (Kennedy-Stoskopf *et al.*, 1985). Since colostrum and milk contain numerous macrophages, these body fluids are highly infectious and constitute a major means of virus dissemination. The virus is passed from mother to offspring soon after birth when the newborn animal ingests colostrum or milk. The virus then enters the new host by one of two mechanisms. First, infected macrophages present in the milk may directly cross the intestinal villi and act as a vector in carrying the infection to the new host. Alternatively, cell-free virus present in the milk or derived by proteolytic digestion of virus-laden macrophages may infect intestinal lining cells or be taken up by dome epithelial cells and go on to infect macrophages in the new host.

This mode of transmission, as well as the poor immunogenicity of CAEV particles discussed earlier, suggested that surface components of the infectious virus particle must provide an effective barrier to recognition of infectious particles by adverse environmental factors. CAEV particles survive the proteases of the digestive tract, avoid neutralization by antibodies, and successfully persist in the goat population. The requirements for recognition of viral proteins by proteolytic enzymes in some ways are similar to those for recognition of viral antigens by antibodies. Since carbohydrates have been shown to perform protective functions on mammalian glycoproteins and since lentiviruses are highly glycosylated on their envelope, we investigated further the role of carbohydrates in recognition events mediated by the envelope of CAEV.

Our studies of CAEV particles showed that they were highly resistant to digestion by proteinase K. This resistance to proteinase K was mediated in part by the highly sialylated CAEV envelope. Sialic acid is a highly charged, terminal carbohydrate and was found on the virus linked to both *O*-linked and *N*-linked oligosaccharides. The exoglycosidase neuraminidase removed the viral sialic acids without affecting viral infectivity directly. However, desialylation of viral particles made the infectious particles more sensitive to digestion by proteinase K. (Huso *et al.*, 1988).

Sialic acids also played a role in determining the affinity of antibodies for viral epitopes. Noneutralizing serum was obtained from an animal infected with CAEV. This serum, which failed to neutralize native virus particles, neutralized desialylated virus. The kinetics of neutralization was improved in another serum by desialylation of virus particles. Desialylated CAEV was neutralized faster and to a greater extent than control virus. This indicated that the removal of sialic acids improved the affinity of the antibodies within the serum for neutralization epitopes on the virus particle. In addition, removal of sialic acids from the surface of infected macrophages enhanced the lysis of these cells by antibody-complement-mediated lysis.

Carbohydrates on the lentiviral envelope therefore play a central role in determining viral persistence in macrophages. They mediate the binding of virus to the soluble mannose binding protein or the mannose receptor on macrophages discussed earlier. Carbohydrates appear to stabilize envelope glycoprotein conformations that are important in binding of the virus to its cellular receptor and in virus-induced cell-to-cell fusion. Carbohydrates also play a role in the poor affinity of antibodies for viral epitopes both on the virus particle and on virus antigens on the surface of infected cells. Finally, the envelope glycosylation sites are targets for mutations that occur during antigenic variation of lentiviruses.

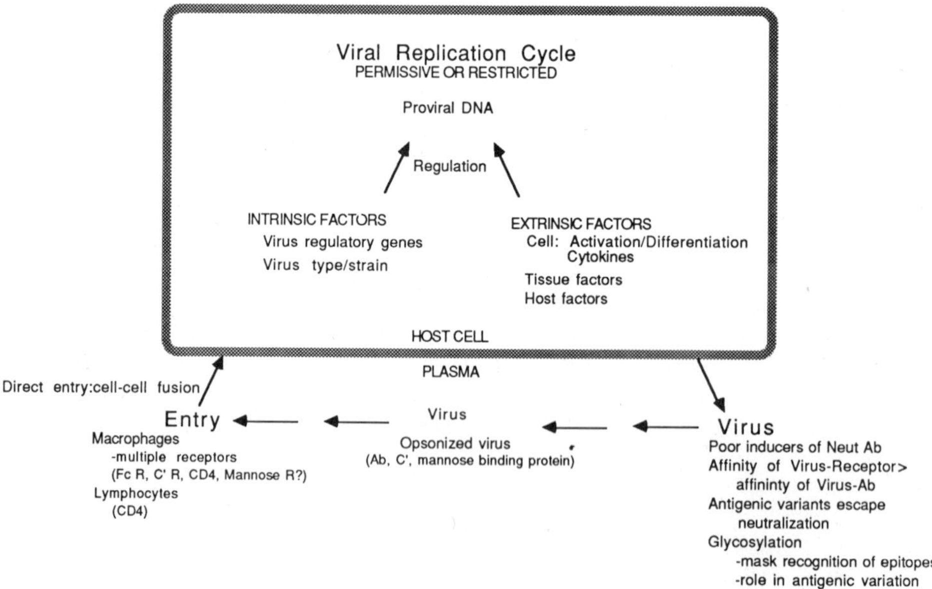

Figure 2. Features of lentiviruses that mediate escape from immune surveillance.

In summary, the escape of lentiviruses from immune surveillance is mediated by a combination of factors (Fig. 2). Lentivirus proviral DNA integrates into the genome of host cells and may therefore be spread directly to progeny cells during cell division. Gene expression from the proviral DNA is regulated by a number of factors, including maturation or activation of host cells, factors that are tissue specific, viral proteins that act in *trans,* or cytokines such as interferon. Regulation of viral replication allows the virus to persist in a latent state within the host for long periods of time. Lentivirus genetic variation produces variant viruses with different cell tropisms and virulence as well as antigenic variants that can escape neutralization by antibodies. Lentiviruses are poor inducers of neutralizing antibodies, and the neutralizing antibodies that are produced have a low affinity for the highly glycosylated virions and are not protective. The Fc, complement, or mannose receptors on macrophages that normally function in clearance may provide alternate routes for viral entry into macrophages in the presence of immune serum (antibodies, complement, mannose binding protein). In addition, fusion of infected macrophages to uninfected cells provides a means of spread of the viral genome without exposure to the effector arms of the immune system.

ACKNOWLEDGMENTS

This work is supported by Public Health Service Grants AI 25774, NS 12127, NS 21916, and RR00130 from the National Institutes of Health.

REFERENCES

Abel, C. A., Nobel, E. L., Raymond, W. W., Mielke, C. H., and Klock, J. C. (1987). *Fed. Proc.* **46,** 1318.

Adams, D. S., Klevjer-Anderson, P., Carlson, J. L., McGuire, T. C., and Gorham, J. R. (1983). *Am. J. Vet. Res.* **44,** 1670–1675.

Alexander, S., and Elder, J. H. (1984). *Science* **226,** 1328–1330.

Arya, S. K., Guo, C., Josephs, S. F., and Wong-Staal, F. (1985). *Science* **229,** 69–73.

Braun, M. J., Lahn, S., Boyd, A. L., Kost, T. A., Nagashima, K., and Gonda, M. A. (1988). *Virology* **167,** 515–523.

Chakrabarti, L., Guyader, M., Alizon, M., Daniel, M. D., Desrosiers, R. C., Tiollais, P., and Sonigo, P. (1987). *Nature* **328,** 543–547.

Cheevers, W. P., and McGuire, T. C. (1985). *Rev. Infect. Dis.* **7,** 83–88.

Clements, J. E., Narayan, O., Griffin, D. E., and Johnson, R. T. (1979). *Virology* **93,** 377–386.

Clements, J. E., Pedersen, F. S., Narayan, O., and Haseltine, W. A. (1980). *Proc. Natl. Acad. Sci.* **77,** 4454–4458.

Clements, J. E., D'Antonio, N., and Narayan, O. (1982). *J. Mol. Biol.* **158,** 415–434.

Daniel, M. D., Letvin, N. L., King, N. W., Kannagi, M., Sehgal, P. K., Hunt, R. D., Kanki, P. J., Essex, M., and Desrosiers, R. C. (1985). *Science* **228,** 1201–1204.

Davis, J. L., Molineaux, S., and Clements, J. E. (1987). *J. Virol.* **61,** 1325–1331.

Desrosiers, R. C., Wyand, M. S., Kodawa, T., Rivgler, D. J., Arthur, L. O., Sehgal, P. K., Letvin, N. L., King, N. W., and Daniel, M. D. (1989). *Proc. Natl. Acad. Sci. USA* **86,** 6353–6357.

Ezekowitz, R. A. B., Kuhlman, M., Groopman, J. E., and Byrn, R. A. (1989). *J. Exp. Med.* **169,** 185–196.

Farmerie, W. G., Loeb, D. D., Casavant, N. C., Hutchinson, C. A., Edgell, M. H., and Swanstrom, R. (1987). *Science* **236,** 305–308.

Fauci, A. S. (1988). *Science* **239,** 617–622.

Folks, T. M., Kessler, S. W., Orenstein, J. M., Justement, J. S., Jaffe, E. S., and Fauci, A. S. (1988). *Science* **242,** 919–922.

Gartner, S. P., Markovitz, D. M., Markovitz, M. H., Kaplan, M. H., Gallo, R. C., and Popovic, M. (1986). *Science* **223,** 215–219.

Gelderblom, H. R., Hausman, E. H. S., Oyel, M., Pauli, G., and Kolb, M. A. (1987). *Virology* **156,** 171–176.

Gendelman, H. E., Narayan, O., Molineaux, S., Clements, J. E., and Ghotbi, Z. (1985). *Proc. Natl. Acad. Sci. USA* **82,** 7086–7090.

Gendelman, H. E., Narayan, O., Kennedy-Stoskopf, S., Kennedy, P. G., Ghotbi, Z., Clements, J. E., Stanley, J., and Pezeshkpour, G. (1986). *J. Virol.* **58,** 67–74.

Geyer, H., Holschbach, C., Hunsmann, G., and Schneider, J. (1988). *J. Biol. Chem.* **263,** 11760–11767.

Gonda, M. A., Wong-Staal, F., Gallo, R. C., Clements, J. E., Narayan, O., and Gilden, R. (1985). *Science* **227,** 173–177.

Groopman, J. E., Benz, P. M., Ferriani, R., Mayer, K., Allan, J. D., and Weymouth, L. A. (1987). *AIDS Res. Hum. Retro.* **3**(1), 71–85.

Gruters, R. A., Meejfes, J. J., Tersmetta, M., deGoede, R. E. Y., Tulp, A., Huisman, H. G., Meidema, F., and Ploegh, H. L. (1987). *Nature* **330,** 74–77.

Haase, A. T. (1986). *Nature* **322,** 130–136.

Haseltine, W. A., and Wong-Staal, F. (1988). *Sci. Am.* **252,** 59–69.

Hess, J. L., Clements, J. E., and Narayan, O. (1985). *Science* **229,** 482–485.

Hess, J. L., Pyper, J. M., and Clements, J. E. (1986). *J. Virol.* **60,** 385–393.

Huso, D. L., Narayan, O., and Hart, G. W. (1988). *J. Virol.* **62,** 1974–1980.

Jolly, P. E., and Narayan, O. (1989). *J. Virol.* **63,** 4682–4688.

Jolly, P. E., Huso, D. L., Sheffer, D., and Narayan, O. (1989a). *J. Virol.* **63,** 1811–1813.

Jolly, P. E., Huso, D. L., Hart, G. W., and Narayan, O. (1989b). *J. Gen. Virol.* **70,** 2221–2226.

Kennedy-Stoskopf, S., and Narayan, O. (1986). *J. Virol.* **59,** 37–44.

Kennedy-Stoskopf, S., Narayan, O., and Strandberg, J. D. (1985). *J. Comp. Pathol.* **95,** 609–617.

Klatzman, D., Barre-Sinoussi, F., Nugeyre, M. T., Dauguet, C., Vilmer, G., Grisielli, C., Brun-Vezinet, F., Rouzioux, C., Gluckman, J. C., Charman, J., and Montagnier, C. (1984). *Science* **225,** 59–63.

Koyanagi, Y., Miles, S., Mitsuyasu, R., Merrill, J., Vinters, H., and Chen, I. (1987). *Science* **236,** 819–822.

Leonard, J. M., Abramczuk, J. W., Pezen, D. S., Rutledge, R., Belcher, J. H., Hakim, F., Shearer, G., Lamperth, L., Travis, W., Fredrickson, T., Notkins, A. L., and Martin, M. A. (1988). *Science* **242,** 1665–1670.

Letvin, N. L., Daniel, M. D., King, N. W., Kiotaki, M., Kannagi, M., Chalifoux, L. V., Sehgal, P. K., Desrosiers, R. C., Arthur, L. O., and Allison, A. C. (1987). *Vaccines* **87,** 209–213.

Liffson, J. A., Feinberg, M. B., Reyes, G. R., Robin, L., Banapour, B., Chakrabarti, S., Moss, B., Wong-Staal, F., Steimer, K. S., and Engleman, E. G. (1986). *Nature* **323,** 725–728.

McCune, J. M., Rabin, L. B., Feinberg, M. N., Lieberman, M., Kosek, J. C., Reyes, G. R., and Weissman, I. L. (1988). *Cell* **53,** 55–67.

Molineaux, S., and Clements, J. E. (1983). *Gene* **23,** 37–148.

Murphy-Corb, M., Martin, L. N., Davison-Fairburn, B., Montelaro, R. C., Miller, M., West, M., Ohkawa, S., Baskin, G. B., Zhavy, J., Putney, S. D., Allison, A. C., and Eppstein, D. A. (1989). *Science* **246,**1293–1297.

Narayan, O., and Clements, J. E. (1987). In *Retrovirus Biology: An Emerging Role in Human Disease* (F. Wong-Staal, and R. C. Gallo, eds.), Marcel Dekker, New York.

Narayan, O., and Cork, L. C. (1985). *Rev. Infect. Dis.* **7,** 85–98.

Narayan, O., Griffin, D. E., and Chase, J. (1977). *Science* **197,** 376–378.

Narayan, O., Griffin, D. E., and Clements, J. E. (1978). *J. Gen. Virol.* **41,** 343–352.

Narayan, O., Clements, J. E., Strandberg, J. D., Cork, L. C., and Griffin, D. E. (1980). *J. Gen. Virol.* **50,** 69–79.

Narayan, O., Clements, J., Griffin, D. E., and Wolinsky, J. S. (1981). *Infect. Immun.* **32,** 1045–1050.

Narayan, O., Wolinsky, J. S., Clements, J. E., Strandberg, J. D., Griffin, D. E., and Cork, L. C. (1982). *J. Gen. Virol.* **59,** 345–356.

Narayan, O., Kennedy-Stoskopf, S., Sheffer, D., Griffin, D. E., and Clements, J. E. (1983). *Infect. Immun.* **41,** 67–73.

Narayan, O., Sheffer, D., Griffin, D. E., Clements, J. E., and Hess, J. (1984). *J. Gen. Virol.* **59,**349–355.

Narayan, O., Sheffer, D., Clements, J. E., and Tennekoon, G. (1985). *J. Exp. Med.* **162,** 1954–1969.

Narayan, O., Clements, J. E., Kennedy-Stopkopf, S., and Royal, W. (1987). In *Antigenic Variation: Molecular and Genetic Mechanisms of Relapsing Disease.* Contributions to Microbiology and Immunology (J. M. Cruze, ed.), pp. 60–76, Karger, Basel.

Payne, S. L., Salinovich, O., Nauman, S. M., Issel, C. J., and Montelaro, R. C. (1987a). *J. Virol.* **61,** 1266–1270.

Payne, S. L., Fang, F-D., Liu, C-P., Dhruva, B. R., Rwambo, P., Issel, C. J., and Montelaro, R. C. (1987b). *Virology* 161, 321–331.

Pedersen, N. C., Ho, E. W., Brown, M. L., and Yamanaty, J. K. (1987). *Science* **235,** 790–793.

Pyper, J. M., Clements, J. E., Molineaux, S. M., and Narayan, O. (1984). *J. Virol.* **51,** 713–721.

Pyper, J. M., Clements, J. E., Gonda, M. A., and Narayan, O. (1986). *J. Virol.* **58,** 665–670.

Ratner, L., Haseltine, W., Pararca, R., Livak, K. J., Storich, B., Josephs, S. F., Doran, E. R., Rafalski, J. A., Whitehorn, E. A., Baumeister, K., Ivaloff, L., Petteway, S. E., Pearson, M. L., Loutenberger, J. A, Papas, T. S., Ghroyet, J., Chang, N. T., Gallo, R. C., and Wong-Staal, F. (1985). *Nature* **313,** 277–284.

Robert-Guroff, M., Brown, M., and Gallo, R. C. (1985). *Nature* **316,** 72–74.

Robinson, W. E., Montefiori, D. C., and Mitchell, W. M. (1987). *AIDS Res. Hum. Retro.* **3,** 265–282.

Robinson, W. E., Montefiori, D. C., and Mitchell, W. M. (1988). *Lancet* **1,** 790–794.

Saag, M. S., Hahn, B. H., Gibson, J., Li, Y., Parks, E. S., Parks, W. P., and Shaw, G. M. (1988). *Nature* **334,** 440–444.

Sanchez-Pescador, R., Power, M. D., Barr, P. J., Steimer, K. S., Stempien, M. M., Brown-Shiver, S. L., Gee, W. W., Revard, A., Randolf, A., Levy, J. A., Dina, D., Luciw, P. A. (1985). *Science* **227,** 484–492.

Skehel, J. J., Stevens, D. J., Daniels, R. S., Douglas, A. R., Krossow, M., Wilson, I. A., and Wiley, D. C. (1984). *Proc. Natl. Acad. Sci. USA* **81,** 1779–1783.

Sonigo, A., Alizon, M., Stoskus, K., Klatzmann, D., Cole, D., Danos, O., Retyel, E., Tiollais, P., Haase, A., and Wain-Hobson, S. (1985). *Cell* **42,** 369–382.

Stanley, J., Bhaduri, L. M., Narayan, O., and Clements, J. E. (1987). *J. Virol.* **61,** 1019–1028.

Steck, F. T., and Rubin, H. (1966). *Virology* **29,** 628–653.

Stein, B. S., Gonda, S. D., Lifson, J. D., Penhallow, R. C., Bensch, K. G., and Engleman, E. G. (1987). *Cell* **49,** 659–668.

Takeda, A., Tuazon, C. U., and Ennis, F. A. (1988). *Science* **242,** 580–583.

Walker, B. D., Kowalski, M., Goh, W. C., Kozarsky, K., Krieger, M., Rosen, C., Rohrschneider, L., Haseltine, W. A., and Sodroski, J. (1987). *Proc. Natl. Acad. Sci. USA* **84,** 8120–8124.

Weiss, R. A., Clapman, P. R., Cheingsong-Papou, R., Dalgleish, A. G., Carne, C. A., Weller, I. V. D., and Tedder, R. S. (1985). *Nature* **316,** 72–74.

Willey, R. L., Smith, D. H., Laskey, L. A., Theodore, T. I., Earl, P. L., Moss, B., Capon, D. J., and Martin, M. A. (1988). *J. Virol.* **62,** 139–147.

Zagury, D., Bernard, J., Cheynier, R., Desportes, I., Leonard, R., Fouchard, M., Reveil, B., Ittele, D., Lurhuma, Z., Mbayo, K., Wane, J., Salaun, J-J., Goussard, B., Dechazal, L., Burny, A., Nara, P., and Gallo, R. C. (1988). *Nature* **332,** 728–731.

Zink, M. C., and Narayan, O. (1989). *J. Virol.* **63,** 2578–2584.

5

Visna Virus Genome

Variability and Relationship to Other Lentiviruses

Matthew A. Gonda

I. INTRODUCTION

Visna virus was originally recognized in the 1930s as the etiologic agent of an unusual and sporadic outbreak of a transmissible neurologic disease that simultaneously appeared with several chronic pneumonic diseases affecting sheep in Iceland (Sigurdsson, 1954, 1957; Sigurdsson and Palsson, 1958). Visna means "wasting" in Icelandic and became the name of the disease because of the progressive emaciation that accompanied the neural pathology and paralytic consequences of virus infection in sheep. Visna virus belongs to a broader group of genetically related sheep and goat viruses with variable neurologic and pneumonic disease potential; this group of viruses includes maedi (Icelandic for "shortness of breath") virus, ovine progressive pneumonia virus (OPPV), and caprine arthritis encephalitis virus (CAEV) (Thormar and Palsson, 1967; Kennedy *et al.*, 1968; Cutlip and Laird, 1976; Crawford *et al.*, 1980; Haase, 1975, 1986; Gonda, 1988; Gonda *et al.*, 1989).

These viruses share certain aspects of disease induction. The pathologic consequences of infection initially occur without apparent signs and are usually slow in onset but chronic and progressive. Often multisystemic, the disease progression is variable and may be accompanied by opportunistic infections; individual hosts may display a mosaic of pathology related to the virus infection. The clinical course generally leads to serious disease and death.

How the virus persists in the face of a strong immune response by the host has been the subject of much controversy. Several facets of the life history of lentiviruses and of the host–virus interaction appear to contribute to virus persistence and pathogenicity, including virus latency, the virus's (cell's?) ability to down-regulate or restrict viral replication, virus-induced cytopathogenicity, antigenic variation in the viral envelope, and the inability of the host to synthesize strong neutralizing sera or, alternatively, the ability of the virus to mask neutralizing epitopes. The molecular basis of pathogenesis and persistence of ovine and caprine lentiviruses has only recently become amenable to study with the molecular cloning (Molineaux and Clements, 1983; Harris *et al.*, 1984; Yaniv *et al.*, 1985; Pyper *et al.*, 1986) and sequencing (Sonigo *et al.*, 1985; Braun *et al.*, 1987) of individual isolates.

While lentiviruses have received attention for some time as putative models of human disease, their study has taken on new importance with the recent discovery that the human immunodeficien-

Matthew A. Gonda • Laboratory of Cell and Molecular Structure, Program Resources, Inc., NCI–Frederick Cancer Research Facility, Frederick, Maryland 21701.

cy virus (HIV), the cause of the acquired immunodeficiency syndrome (AIDS) is a member of the lentivirus subfamily or retroviruses (Gonda *et al.*, 1985, 1986, 1987; Sonigo *et al.*, 1985; Chiu *et al.*, 1985; Stephens *et al.*, 1986). This presentation on the visna virus genome and its variability takes this evolutionary relationship into consideration. The biology and molecular genetics of animal lentiviruses are predictive of the challenges ahead in developing a vaccine against lentiviruses, including those of human origin.

II. VIROLOGIC AND EVOLUTIONARY PERSPECTIVES

A. Retrovirus Properties of Visna Virus

Visna virus has a high-molecular-weight RNA genome that is in the form of + −stranded mRNA. The viral RNA is encapsulated in a proteinaceous matrix that can be subdivided into two compartments, the outer envelope and the inner core. The envelope contains the specific cell recognition receptors and surrounds the inner core, or nucleocapsid, which contains the viral RNA. Visna virus is a retrovirus because it has these features and, more important, because it has RNA-dependent DNA polymerase or reverse transcriptase (RT) activity (Stone *et al.*, 1971; see also Haase, 1986; Gonda, 1988; Gonda *et al.*, 1989, for detailed reviews of lentivirus biology, evolutionary relationships, molecular genetics, and virus structure).

B. Lentivirus Subfamily of Retroviruses

The retrovirus family can be divided into three subfamilies: Oncovirinae, Lentivirinae, and Spumavirinae. The lentivirus subfamily of retroviruses is a group of exogenous, nononcogenic viruses that cause a spectrum of chronic, persistent, debilitating diseases (Table 1) and are readily distinguishable from their oncogenic cousins. The spumaviruses have not been etiologically associated with any diseases (Weiss, 1988), although *in vitro* they display many of the cytopathic features

Table 1. Clinical Manifestations of Lentivirus Infections in Natural Hosts[a]

Lentivirus	Disease description
Ovine visna, maedi, and progressive pneumonia virus	Generalized wasting, chronic encephalomyelitis, progressive lethal interstitial pneumonia, mastitis, spasticity, paralysis, lymphadenopathy, opportunistic infections
Caprine arthritis encephalitis virus	Generalized wasting, chronic leukoencephalomyelitis, progressive arthritis, osteoporosis, paralysis
Equine infectious anemia virus	Fever, intermittent anemia, general proliferation of lymphoid cells in reticuloendothelial system, glomerulonephritis, central nervous system lesions
Bovine immunodeficiency-like virus	Persistent lymphocytosis, lymphadenopathy, central nervous system lesions, emaciation
Feline immunodeficiency virus	Immunodeficiency-like syndrome, generalized lymphadenopathy, leukopenia, fever, anemia, emaciation
Simian immunodeficiency virus	Immunodeficiency, neuropathologic changes, wasting, opportunistic infections
Human immunodeficiency virus	Immunodeficiency accompanied by opportunistic infections, lymphadenopathy, encephalopathy, Kaposi's sarcoma, interstitial pneumonia in infant AIDS

[a]Adapted from Gonda (1988), with permission.

of lentiviruses, for example, rapid and extensive syncytia induction that leads to the death of the multinucleated cell.

Classically, visna virus is considered prototypic of the lentiviruses. Nevertheless, visna was not the first lentivirus disease described. Decades earlier, equine infectious anemia had been attributed to a filterable agent by Vallée and Carré (1904). Later, this filterable agent, equine infectious anemia virus (EIAV), which was responsible for the severe anemia seen in horses, was also determined to be a retrovirus of the lentivirus subfamily (Charman *et al.*, 1976; Gonda *et al.*, 1978). Since these initial discoveries, lentiviruses have been isolated or recently characterized from goats (Crawford *et al.*, 1980), cats (Pedersen *et al.*, 1987), cattle (Van Der Maaten *et al.*, 1972; Gonda *et al.*, 1987; Braun *et al.*, 1988), monkeys (Daniel *et al.*, 1985), and humans (Barré Sinoussi *et al.*, 1983; Gallo *et al.*, 1984; Clavel *et al.*, 1986).

C. Evolutionary Relationship of Visna Virus to Other Lentiviruses

In addition to features of its biology, the morphogenesis and fine structure of visna virus and related ovine lentiviruses is most like that of other lentiviruses (Fig. 1) (Gonda *et al.*, 1978, 1985, 1989; Gonda, 1988). Thus, to account for the similarity in structure, it was anticipated that the lentiviruses would be genetically related; this was later demonstrated (Pyper *et al.*, 1984, 1986; Gonda *et al.*, 1985, 1986, 1987; Sonigo *et al.*, 1985; Chiu *et al.*, 1985; Stephens *et al.*, 1986).

The overall homology between the genomes of lentiviruses of different species is limited ($\leq$ 40%), except among the ovine and caprine lentiviruses and the simian lentiviruses. This finding is not unexpected considering the molecular genetics and pathogenicity of this retrovirus group that inevitably leads to the great variability seen in their genomes, which will be discussed in Section IV. Nevertheless, there are regions of the polymerase (*pol*) gene that are highly conserved and that have provided a basis for determining phylogenetic relationships for retroviruses (Chiu *et al.*, 1984). Using amino acid residues deduced from DNA sequences at the amino terminus of the RT domain of the *pol* gene, which has been the most conserved region in the evolution of retroviruses, one can derive a phylogenetic tree of retroviral relationships. The amino acids of the RT domain of the *pol* gene (deduced from DNA sequences available in GenBank) have been examined in a panel of retroviruses, including most of the known lentiviruses, and various investigators have reached similar conclusions (Sonigo *et al.*, 1985, 1986; Gonda *et al.*, 1987, 1989; Mclure *et al.*, 1988; Yokoyama *et al.*, 1988; Gonda, 1988). Pairwise comparisons of these sequences (Fig. 2) show that the strongest homology is between the lentiviruses, which have 50% or greater matching residues and no gaps, while all other retroviruses share only 40% or fewer matched residues with the lentiviruses, with gaps.

A Fitch–Margoliash phylogenetic tree deduced for the residues in Fig. 2 is shown in Fig. 3. The tree was rooted with Moloney murine leukemia virus (Mo-MuLV) as the outgroup taxon because it consistently had the lowest alignment scores with other retroviruses and because its RT preferentially uses Mn^{2+}. There are two major branches in the tree topology: One leads to the viruses whose RT preferentially uses Mn^{2+} and the other to the viruses whose RT preferentially uses Mg^{2+} cations. Among the retroviruses whose RTs use Mg^{2+}, three groups are apparent: the bovine leukosis virus (BLV)/human T-cell leukemia virus (HTLV) group, the Rous sarcoma virus (RSV)/simian Type D retrovirus (SRV-1) group, and the lentiviruses. Within the lentivirus group, there are several more closely related pairs: visna virus and CAEV, simian (macaque) immunodeficiency virus (SIV_{mac}) and HIV-2, and HIV-1 and simian (African green monkey) immunodeficiency virus (SIV_{agm}). While this analysis has SIV_{agm} on the evolutionary branch that gives rise to HIV-1, SIV_{agm}, HIV-1 and the SIV_{mac}/HIV-2 pair are about equidistant from each other. The large animal lentiviruses, CAEV, EIAV, visna virus, and the bovine immunodeficiency-like virus (BIV), are all more distantly related to HIV-1 than are lentiviruses of more recent primate ancestry. Complete sequences of the feline immunodeficiency virus (FIV) have not yet been published, but early indications are that it is about as distantly related to HIV as the large animal lentiviruses.

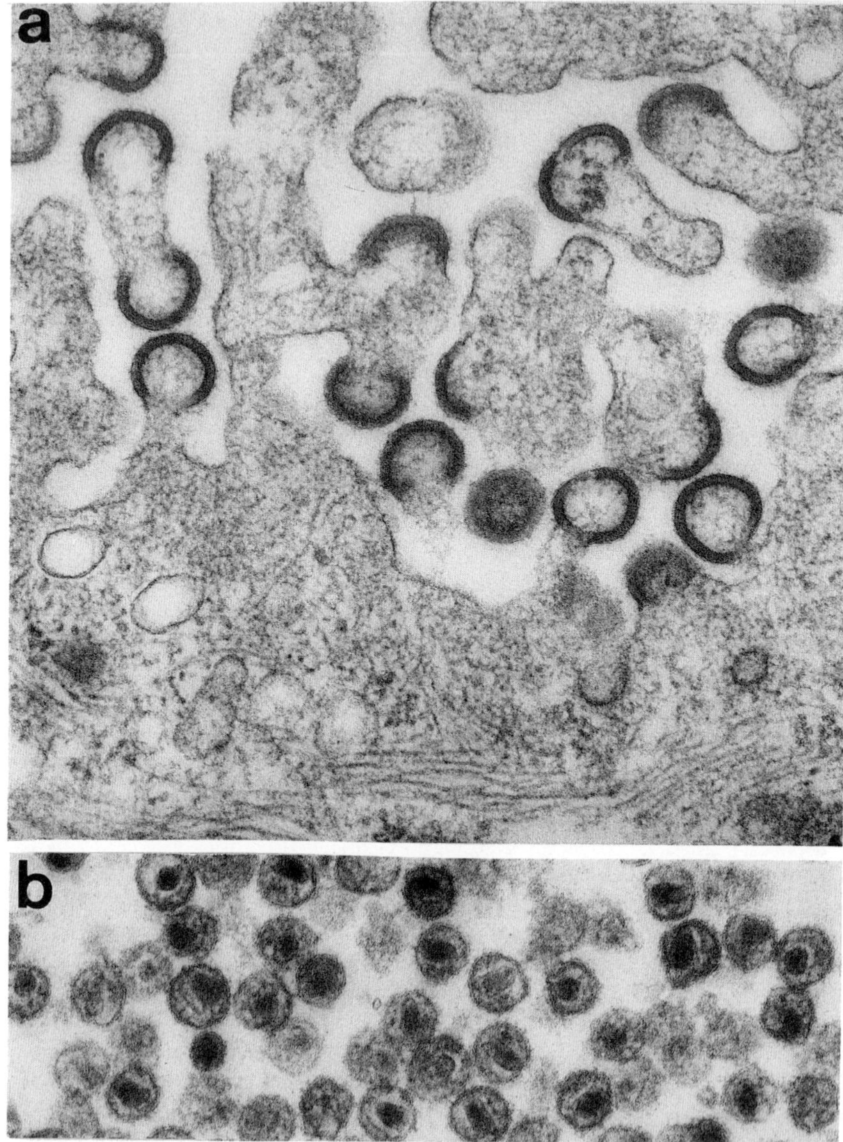

Figure 1. Electron micrograph of an ovine lentivirus-infected sheep choroid plexus cell. (a) Budding and immature extracellular virus particles. 60,000×. (b) Mature extracellular virus particles. 60,000×.

Clearly, the lentiviruses form a natural group distinct from all other viruses on the tree and visna virus clusters more closely with the caprine lentivirus, CAEV (Fig. 3). While the derivation of evolutionary trees may seem trivial at first glance, the phylogenetic relationship of lentiviruses suggests that the similarities between animal lentivirus diseases and AIDS have a genetic basis. Thus, information gained from research on lentivirus models may prove useful in understanding the pathogenesis of HIV.

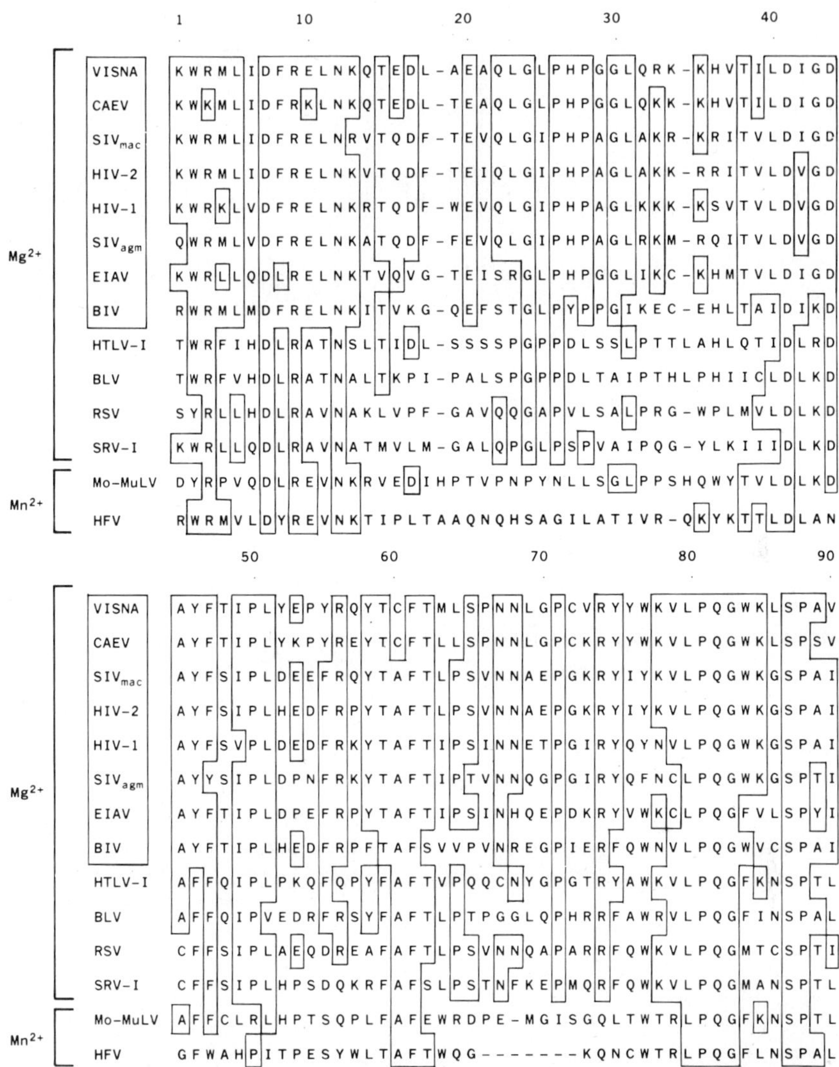

Figure 2. Alignment of amino acid residues of amino terminus of RT domain of a panel of retroviruses. The viruses are HIV-1, HIV-2, visna virus, CAEV, SIV_{agm}, SIV_{mac}, EIAV, BIV, HTLV-1, BLV, RSV, SRV-1, Mo-MuLV, and the putative human foamy virus (HFV) (Maurer and Flügel, 1988). The sequences used in this analysis were obtained from GenBank; amino acids were deduced from the DNA sequence. The alignment is generally that optimal for visna virus; slight improvements in other pairwise alignments can be made by minor shifts in the placement of gaps. Boxes are drawn around identical residues when five or more lentiviruses share that residue. A box is drawn around the abbreviated names for the lentiviruses to highlight their location. The retroviruses are further divided into those whose RTs preferentially use Mg^{2+} and Mn^{2+} as cation. (From Gonda *et al.*, 1989, with permission.)

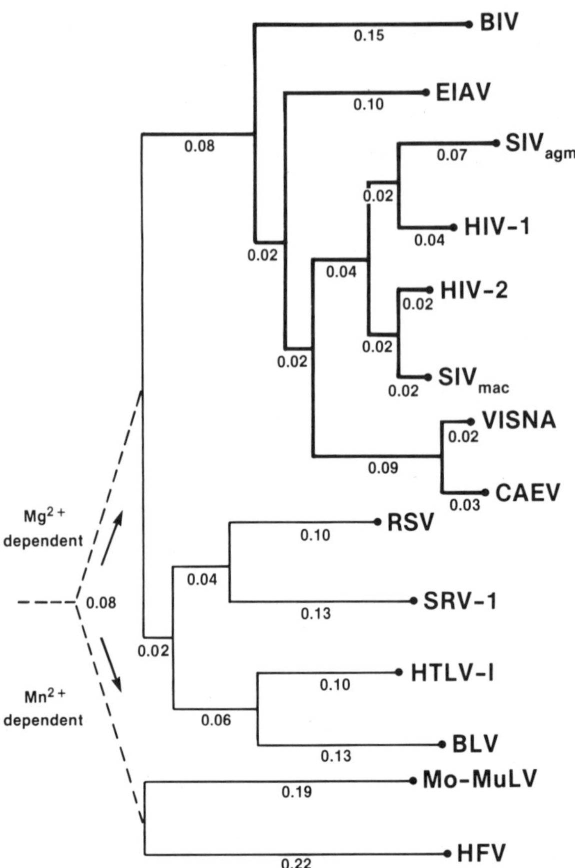

Figure 3. Fitch–Margoliash phylogenetic tree of retroviral relationships based on the *pol* segments shown in Fig. 2. Branch lengths are drawn to scale and are in units of $-\log(M)$ where M is the frequency of matching amino acid residues. The tree was rooted with Mo-MuLV and HFV as the outgroup taxa because they consistently had the lowest alignment scores with the other retroviruses and because their RTs preferentially use Mn^{2+}. The average percent standard deviation was 5.4. (From Gonda *et al.*, 1989, with permission.)

III. MOLECULAR GENETICS OF VISNA VIRUS

A. Organization of Replication-Competent Retrovirus Genome

The general organization of a replication-competent retrovirus genome is typified by that of the oncogenic type C virus, Mo-MuLV (Fig. 4a). The sequence predicts three major open reading frames (ORFs), for the *gag* (coreproteins), *pol* (RT, integrase/endonuclease), and *env* (envelope) genes. The protease gene (*pro*) can be found in various locations in the genome, usually in or overlapping the *gag* or *pol* genes. In addition, the retrovirus genome contains regulatory sequences, called long terminal repeats (LTRs), that initiate, enhance, and terminate transcription. The LTRs are derived from sequences unique to the 5′ end (U_5) or the 3′ end (U_3) or repeated at both ends (R)

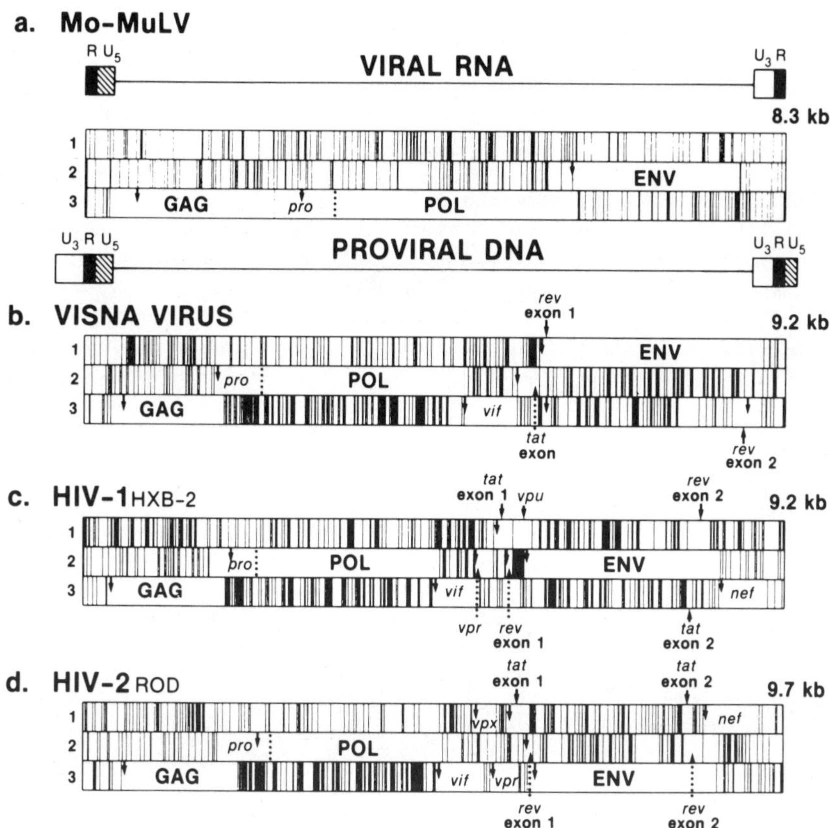

Figure 4. Molecular organization of retrovirus genomes. The retroviruses are Mo-MuLV, visna virus, HIV-1, and HIV-2. All genomes are given in the form of the viral RNA; the lengths in kilobases (kb) are given at the 3′ end of each genome. The ORF analysis of sequences is shown in all three reading frames and was obtained using a program written by G. Smythers. The vertical bars represent stop codons and the solid arrows represent the first AUG initiation codon encountered for each ORF of functional significance. The major structural *(gag, pol,* and *env)* and nonstructural *(vif, tat, rev, vpx, vpr, vpu,* and *nef)* genes are identified. The relative position of the protease gene *(pro)* for each virus is indicated. (a) Mo-MuLV. (b) Visna virus. (c) HIV-1. (d) HIV-2. In (a) the gross organization and location of LTRs in the viral RNA and proviral genomes of Mo-MuLV are shown above (viral RNA) and below (proviral DNA) the ORF analysis. (Adapted with modifications from Gonda *et al.*, 1989, with permission.)

of the viral RNA. In the proviral DNA, the structural genes of retroviruses are flanked on the 5′ and 3′ ends by the LTRs (Varmus, 1988; Gonda, 1988; Gonda *et al.*, 1989).

B. Topography of the Visna Virus Genome

Two visna virus genomes derived from a neurotropic Icelandic strain (their origins are discussed in Section III D) have been sequenced and these investigations have yielded similar conclusions regarding the architecture of the genome, including LTRs and assignment of major ORFs [(Sonigo *et al.*, 1985; Braun *et al.*, 1987); the details of the sequence analysis can be found in these two references. Sonigo *et al.* (1985) estimates molecular weights of major ORFs. Salient features of the genomes are summarized below.

The visna virus genome (parental strain 1514) in the form of the viral RNA (Fig. 4b) is 9203 nucleotides long (Braun *et al.*, 1987). The genome contains *gag, pol,* and *env* structural genes. The *gag* and *pol* genes overlap and are in different reading frames. The *pol* ORF contains the protease (*pro*) coding sequences at the 5′ end, which is followed by the RT and integrase/endonuclease domains. There is a gap between the *pol* and *env* genes called the "central region," which contains several small ORFs. The *env* ORF of the prototype strain 1514 sequence is open throughout. The complete visna virus LTR, including U_5, R, and U_3 sequences, is 415 nucleotides (Braun *et al.*, 1987). Thus, the overall genetic organization of the visna virus proviral genome is 5′ LTR-*gag-pol*-central region-*env*-LTR 3′. The genomes of HIV-1 and HIV-2 are shown for comparison in Fig. 4, c and d. There are notable similarities between the overall genetic organization of the human lentivirus genomes and that of visna virus. The most striking exception, which will be discussed, is the absence of the analogous ORF for *nef,* which is 3′ of the *env* ORF of HIV, in the visna virus genome.

C. Nature of Nonstructural–Regulatory Exons of Visna Virus Genome

The central region is one of the hallmarks of lentiviruses. Its gene content can vary between isolates of different species. It is present in the genomes of HIV-1 and HIV-2 (Fig. 4, c and d, respectively) and in other lentiviruses sequenced to date, but it is probably most striking in complexity in the primate lentiviruses. The central region contains exons of nonstructural genes involved in regulating virus replication and infection and that may be involved in the persistence and pathogenesis of lentiviruses. Because of the tremendous research activity surrounding the molecular mechanisms of HIV replication and pathogenesis, these nonstructural/regulatory genes have been extensively examined in the AIDS virus. Thus, more is known about the molecular biology and functioning of the central region ORFs of HIV and related primate lentiviruses.

On the basis of functional studies and assumptions, various terminologies have been used to describe these genes. Gallo *et al.* (1988) and Laurence (1988) have proposed a standard nomenclature for the nonstructural/regulatory genes of HIV that we have adapted for use with related areas of the genome of visna virus, where functional or structural analogies exist. Since more is known about the replication and pathogenesis of HIV, the central region and 3′ ORFs of HIVs (Fig. 4, c and d) and other primate lentiviruses will be described before the comparable regions of visna virus are reviewed.

The regulatory genes of HIV are *vif,* for viral infectivity factor; *tat* and *rev,* for transactivator and regulator of expression of viral proteins, respectively; and *nef,* for negative factor. *nef* is believed to play a role in down-regulating virus expression (Luciw *et al.*, 1987). The regulatory network and interactions of these genes are quite complex and have recently been reviewed by Haseltine (1988). Considering the genomic complexity and pattern of gene usage of lentiviruses, it is not surprising that a number of the mature mRNAs of the central region and 3′ ORF arise from multiple-splicing events. Each of the mRNAs for the structural and nonstructural genes begins at the site of transcription initiation in the 5′ LTR and terminates in a poly A tract derived from 3′ LTR sequences. Various regions of the genome are incorporated into each functional message depending on the location, number, and pattern of splicing events. Only the coding exons of the mature message are presented in the discussion here. The *vif* coding exon is derived from sequences located in the central region, which overlaps but is in a different reading frame than that of the *pol* coding exon. Both *tat* and *rev* have two coding exons; the first exon of each is located in the central region (*rev* exon 1 overlaps the amino terminus of *env*) and the second is located in the transmembrane domain of the *env* ORF, but in reading frames other than *env*. The *nef* coding exon is found after or overlapping the end of the *env* ORF and the LTR.

The functions of three additional proteins encoded by the central region of primate lentiviruses are either still speculative or unknown. Nevertheless, these proteins are immunoreactive with sera from infected individuals, suggesting a role in the virus life cycle. HIV-1, HIV-2, and SIV_{mac} each have an ORF that encodes a protein called *vpr* (Wong-Staal *et al.*, 1987). HIV-2, SIV_{agm}, and SIV_{mac} have an ORF that encodes a protein called *vpx* (Fukasawa *et al.*, 1988; Henderson *et al.*, 1988; Yu *et al.*, 1988); *vpx* is not found in HIV-1. Moreover, a new ORF encoding a protein called *vpu* was recently discovered (Cohen *et al.*, 1988; Matsuda *et al.*, 1988; Strebel *et al.*, 1988); *vpu* is not found in HIV-2, SIV_{mac}, or SIV_{agm}. Although the functions of these proteins are unknown, the proteins may serve as diagnostic tools to immunologically discriminate between different isolates.

The study of the central region of visna virus (Fig. 4b) has slowly begun to yield significant information about its contents. Visna virus contains an ORF with a high arginine amino acid content that is structurally analogous to *vif* in the HIVs and all other lentiviruses sequenced to date. Its topographic location in the genome is also similar, overlapping *pol* but in a different reading frame. Further studies are needed to determine the function of the *vif* ORF in visna virus. Visna virus also transactivates its LTR (Hess *et al.*, 1985). There is an ORF in the central region analogous in location to the *tat* gene of HIVs and SIVs. There is a small, but significant degree of amino acid homology in the conserved "cysteine-rich" region between the putative *tat* genes of visna virus and HIV-1 (Sonigo *et al.*, 1985). It is not known whether visna virus *tat* has a second coding exon like the primate lentiviruses, although the second exon of *tat* does not appear to be required for *tat* function in HIVs.

Recently, Mazarin *et al.* (1988) described a second gene in the visna virus genome with transactivating potential that was identified by cDNA cloning and sequencing. It is comprised of four small exons and is derived by a multiple-splicing event. The initial two exons are noncoding and are derived from the 5′ half of the genome, and the remaining coding exons are derived from the 3′ half of the genome. The first coding exon is in frame with the *env* ORF and begins at the first methionine in the *env* sequence. The second coding exon is in a different reading frame but overlaps the *env* gene. Because this visna virus gene transactivates its LTR, does not incorporate *tat* gene sequences, is in an analogous position to the *rev* gene in HIV-1, and shares some constitutive basic amino acids present in nucleic acid binding proteins and the *rev* gene of HIV-1, it is suggested that this may be the *rev* gene of visna virus (Mazarin *et al.*, 1988). At present, there has been no description of *vpx, vpr,* or *vpu* genes in visna virus, nor does visna virus appear to code for a *nef* gene.

D. Origins and Functionality of Visna Virus Molecular Clones

The visna virus (Molineaux and Clements, 1983) sequence reported by Braun *et al.* (1987) was determined by sequencing a clone of the prototypic neurotropic Icelandic strain 1514. The first published sequence of visna virus (Sonigo *et al.*, 1985) was originally reported to be that of the prototypic visna virus strain 1514. However, comparisons by Gdovin and Clements (unpublished data) of the sequences of an antigenic variant of 1514, termed LV1-1, and the sequences reported by Sonigo *et al.* (1985), in conjunction with immunologic and biologic analysis of the original virus stocks (Braun *et al.*, 1987) used to obtain the molecular clone (Harris *et al.*, 1984) sequenced by Sonigo *et al.* (1985), revealed that the clone previously thought to be 1514 (Harris *et al.*, 1984) is actually that of the antigenic variant, LV1-1, and thus, Sonigo *et al.*'s (1985) sequence is that of the variant. Therefore, the sequence of Braun *et al.* (1987) is the only *bona fide* sequence of visna virus strain 1514, which is the progenitor of LV1-1 and other antigenic variants in wide use (Narayan *et al.*, 1978).

Although the structural organization of the two visna virus genomes (Sonigo *et al.*, 1985; Braun *et al.*, 1987) is the same, an important difference between the molecular clones exists. The LV1-1 clone has a premature stop codon in the *env* domain, whereas the 1514 strain is open throughout. Nevertheless, attempts to demonstrate biologic activity of the 1514 clone by transfection and microinjection of cloned DNA into cells that support visna virus replication have not been successful, even though the cloned LTR responds to transciptional activators *in vitro* (Hess *et al.*, 1985) and all reading frames of the major genes are open (Braun *et al.*, 1987).

IV. VARIABILITY OF THE VISNA VIRUS GENOME

A. Sequence Comparison of Visna Virus Strains 1514 and LV1-1

Genetic variation is a remarkable feature of lentivirus biology that can greatly influence viral pathogenesis, transmissibility, and survival. Variation in the glycosylated outer membrane glycoprotein (OMP) of the *env* gene, in particular, presents one of the single most perplexing difficulties in preparing a vaccine to prevent infection. Genetic variation is known to occur in visna virus, EIAV, and HIV (Narayan *et al.*, 1977, 1987; Payne *et al.*, 1984; Alizon *et al.*, 1986; Hahn *et al.*, 1986; Starcich *et al.*, 1986; Salinovich *et al.*, 1986; McGuire *et al.*, 1987; Modrow *et al.*, 1987) and has been shown to correlate with changes in antigenicity in EIAV and visna virus. The exact molecular mechanism for this phenomenon is not clearly understood. However, the availability of DNA sequences for visna virus strain 1514 and LV1-1 has allowed a direct molecular comparison of the genomic changes occurring in a lentivirus antigenic variant following *in vivo* passage of the biologically cloned parental stock (Narayan *et al.*, 1978).

The sequence of strain 1514 (Braun *et al.*, 1987) differs from that of LV1-1 in 28 of 9203 nucleotides (0.3% sequence divergence), as summarized in Table 2. Of the 28 nucleotide substitutions, 27 represent true sequence divergence and one is a sequencing artifact. Of the 27 substitutions, 26 are in potential coding regions and, of these, 16 resulted in amino acid substitutions. Eleven of these are relatively conservative; five, however, result in charge changes. Of the substitutions that occur in coding exons, five are in the first base, 12 are in the second base, and ten are in the third base. This distribution does not differ from the random expectation of an equal number of substitutions at each position (X^2 test). The bias toward third-base substitutions usually caused by natural selection has yet to become apparent at this early stage of sequence divergence.

B. Hypervariable Site in the env Gene

The distribution of substitutions appears to be random throughout the genome except for a cluster of mutations in the OMP of the *env* gene (Fig. 5). This cluster of mutations occurs in the carboxy-terminus of the glycosylated OMP over a 27-nucleotide stretch (nucleotide positions 7864–7890). There are one synonymous (nucleotide position 7864) and three nonsynonymous (positions 7875, 7878, and 7890) substitutions, with two of the nonsynonymous substitutions resulting in a charge change (Table 2). To eliminate the probability that this could have occurred by chance, a series of Monte Carlo simulations were performed and statistically analyzed by Braun *et al.* (1987), who found it unlikely that the cluster in the glycosylated OMP of the envelope was random and that it may have resulted from an accumulation of mutations in this area.

When genomic variation is measured at the nucleotide level in HIV, numerous isolates of which have been sequenced, the pattern of substitutions is not consistent over the entire genome (reviewed in Coffin, 1986). The *gag* and *pol* genes, which encode internal proteins, are considerably more conserved. In contrast, nucleotide differences in the *env* gene are more frequent and also more frequently code for new amino acids, thus potentially changing the topography of external immunoreactive domains. The glycosylated OMP is an important protein in the survival strategy of lentiviruses, undergoing rapid antigenic variation in response to pressure from the immune system. However, the extent of the variation within *env* is not equally distributed. There is a pattern of interdispersed variable and conserved regions. This is not unexpected, since not only is it the external site of the virus, involved in interactions with the host's immune system, but it also serves

Table 2. Nucleotide Sequence Differences between Visna Virus Strains 1514 and LV1-1[a]

Functional unit	Nucleotide position	Base in 1514	Base in LV1-1	Amino acid in 1514	Amino acid in LV1-1
U5	104	C	None[b]	Noncoding	Noncoding
gag	551	T	C	Val	Ala
	838	A	G	Arg	Gly[c]
	1421	T	C	Met	Thr
	1629	A	C	Gln	His[b]
	1662	G	A	Gly	Gly
pol	2399	G	T	Pro	Pro
	2910	C	T	Leu	Leu
	3019	A	G	Lys	Arg
	3232	T	C	Ile	Thr
	3399	G	C	Val	Leu
	4430	A	G	Glu	Glu
	4679	C	T	Asn	Asn
sor	5076	G	A	Gln	Gln
tat	5701	C	T	Ala	Val
	5767	G	A	Gly	Asp[c]
env	5991	T	C	Met	Thr
	6309	A	G	Asn	Ser
	6804	G	A	Arg	Lys
	7459	C	T	Asn	Asn
	7756	C	T	Ser	Ser
	7864	A	G	Ser	Ser
	7875	C	A	Ala	Glu[c]
	7878	A	G	Gln	Arg[c]
	7890	A	G	Lys	Arg
	8074	A	G	Leu	Leu
	8162	C	T	Arg	Stop
U3	8966	G	A	Noncoding	Noncoding

[a]From Braun *et al.* (1987), with permission of ASM.
[b]Possibly a sequencing error. See text.
[c]This substitution results in a change in charge at this position.

as a receptor molecule early in the infection process. The envelopes of lentiviruses are very specific for their cell receptor, with a restricted host and cell type range. Any changes in the receptor regions of the envelope would severely handicap the virus's ability to interact with its target cell. Although this discussion has focused on point mutations, they are not the only source of variability; duplications and deletions also play a significant role (Modrow *et al.*, 1987).

Too few visna virus isolates have been sequenced to date to correlate the distribution of variable and conserved regions over the entire genome. However, Braun *et al.* (1987) described a cluster of mutations in the carboxy terminus of the glycosylated OMP of the *env* gene of LV1-1 (Table 2 and Fig. 5). A logical explanation for this distribution of mutations in the envelope, in which 15% of all LV1-1 substitutions occur in 0.3% of the genome, is that this site represents the (or one of the) major epitope(s) that make the virus antigenically and biologically different from the prototype strain 1514 in terms of neutralizing capacity. Newly arisen *env* variants have a temporal growth advantage over their progenitor, which initially allows them to escape the immune response. This hypervariable site in the OMP of visna virus may also be a hypermutable site in the sense that more mutations occur at

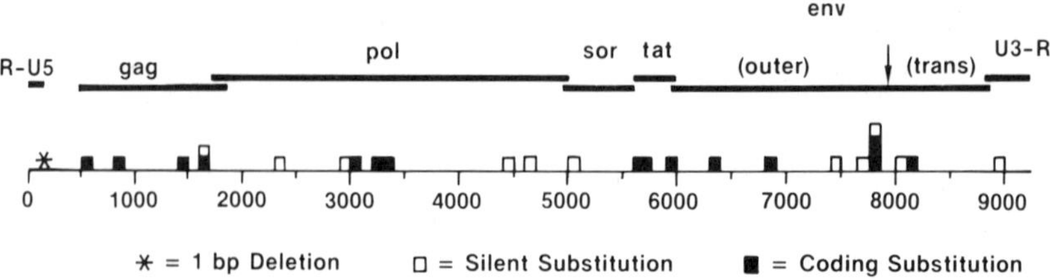

Figure 5. Distribution of nucleotide substitutions between visna virus strains 1514 and LV1-1 (Sonigo *et al.*, 1985; Braun *et al.*, 1987). The position of the major ORFs is shown by solid bars above the graph. The arrow indicates the OMP/TMP cleavage site in the *env* gene. Each square represents a single nucleotide substitution. (From Braun *et al.*, 1987, with permission of ASM.)

this site, whether or not they are selectively advantageous to the virus. The cloning and sequencing of additional antigenic variants of visna virus may provide information in regard to this.

C. Computer-Assisted Predictions of Functional Envelope Structure

The envelope proteins of lentiviruses contain the structural and functional features of all retroviruses and are grossly represented by a glycosylated OMP and transmembrane protein (TMP). In addition, they have two unique distinguishing features. They are both the largest and the most heavily glycosylated envelope proteins found in the retroviruses. To better analyze the *env* gene product of visna virus 1514, we have used Chou and Fausman computer algorithms to predict *env*-polypeptide secondary structure and calculated values for hydrophobicity, flexibility, and surface probability (Devereaux *et al.*, 1984; Jameson and Wolf, 1988) to predict probable antigenic determinants. These determinants are located in regions predicted to have a high probability of appearing on the surface of the polypeptide. Glycosylation sites and other local features of interest for visna virus and HIV-1 envelope products are also predicted (Fig. 6).

The overall size and structure of the two proteins are very similar. Visna virus and HIV-1 *env* exons code for 983 and 856 amino acids, respectively. There is a hydrophobic region (signal peptide) preceding the start of the glycosylated OMPs, gp135 and gp120, of visna virus (Fig. 6a) and HIV-1 (Fig. 6b), respectively. Upstream of the signal peptide in visna virus is a region that has been putatively identified as the first coding exon of *rev*. In visna virus, the hydrophobic signal peptide occupies the space between *rev* and gp135 and is the most amino-terminal sequence of HIV-1. More than likely, the putative *rev* region (Fig. 6a), although it is in the same reading frame as gp135, is probably not represented in the envelope protein of visna virus, and thus these amino acids should be subtracted from the overall calculated length of the mature *env* protein backbone. The first coding exon for *rev* in HIV-1 overlaps *env* but is in a different reading frame; thus it is not present in the HIV-1 profile.

There are 28 and 27 potential glycosylation sites overall for the visna virus and HIV-1 *env* gene products, respectively. The majority of these occur in the gp135 and gp120 domains and contribute to about half of the molecular weight of the sugar-modified backbone of the OMP. There is a hydrophobic stretch just following the cleavage site in the *env* precursor of visna virus and HIV-1 and another prominent one in the transmembrane domain of the TMPs (see also Fig. 8). There are five and six potential glycosylation sites in the TMPs (gp41) of visna virus and HIV-1, respectively. Figure 6 also indicates the probable sites of a number of antigenic determinants for visna virus and HIV-1. The hypervariable site in the carboxy terminus of the visna virus gp135 (Fig. 5) maps to a segment predicted to be antigenic, just prior to the gp135–gp41 cleavage sites (Fig. 6a). Interestingly, an immunoreactive site has been discovered in a similar location in HIV-1 gp120 (Palker *et al.*, 1987).

D. Homologies among Lentivirus env Proteins

Although the most conserved genes in the lentiviruses are *pol* and *gag*, sequence homology also extends to the *env* gene. Using the OMP–TMP processing site as the benchmark to align the *env* genes of visna virus, HIV-1, and EIAV, distant but significant amino acid and structural homology is revealed (Figs. 7 and 8, respectively). The homology extends for about 60 amino acids into the OMP over the carboxy-terminal region predicted to be an antigenic determinant (Fig. 6), which includes the hypervariable and immunoreactive sites in visna virus (Braun *et al.*, 1987) and HIV-1 (Palker *et al.*, 1987), respectively, before falling off. Retroviral TMPs generally contain three functional domains: amino-terminal extracellular, hydrophobic transmembrane, and cytoplasmic domains (Fig. 8). The homology between the *env* genes of the three lentiviruses is more extensive in the TMP than in the OMP and is about 20% in the extracellular domain and 43% in the transmembrane domain of the visna virus–HIV-1 TMP comparison (Figs. 7 and 8). Many nonidentical residues are chemically similar, which also contributes to the alignment score. This alignment also preserves important structural features in the TMPs, including potential N-linked glycosylation sites, two cysteine residues, and hydrophobicity profiles (Figs. 6–8).

E. Possible Role of RNA Secondary Structure in Genome Mutability

In search of the physicochemical mechanisms that would provide visna virus (and possibly other lentiviruses) with a hypermutable site in the *env* gene, Braun *et al.* (1987) proposed that there might be peculiar constraints in the secondary structure of the RNA genome that lead to a high error rate during reverse transcription. Figure 9 shows computer predictions, using the FOLD algorithms of Zuker and Stiegler (1981), of the most stable local secondary structure for 500 base pairs surrounding the hypervariable sites in the RNA genomes of various HIV-1 isolates and visna virus. The segment was centered on the putative processing site of the OMP–TMP. Interestingly, the hypervariable sites of these segments either fell on or overlapped a major (in one case minor) looped-out region of the predicted secondary structure.

Several explanations can be advanced for this finding. The apparent lack of base pairing in these regions may indicate that there are no secondary constraints that limit the rate of sequence divergence. Alternatively, this area may be more exposed to chemical mutagens, and single-stranded looped-out regions appear more mutable because mutations that occur there have little or no deleterious effect on the replicative ability of the virus. Finally, it may be more difficult for RT (or other polymerases) to replicate these regions accurately, although the lack of secondary structure might lead one to expect the opposite. In this regard, it is possible that the difficulty in replicating the unstructured areas may be that they form pseudoknots with other regions of the genome that were not included in the analysis. The correlation between hypervariable sites and regions predicted to have little intramolecular base pairing in visna virus and other lentiviruses may provide a clue to mechanisms promoting mutability. It is notable that higher rates of substitution in loops also have been observed in polio virus RNA (Currey *et al.*, 1986).

F. Rate of Sequence Divergence of Visna Virus Clones

The noted differences (Table 2) in the two visna virus sequences have permitted a rough estimate of the average rate of sequence divergence for visna virus variants (Braun *et al.*, 1987). Taking into consideration that 1.75 years of evolutionary time had elapsed between the time a sheep was infected with visna virus strain 1514 and the isolation of LV1-1 (Narayan *et al.*, 1978) and that 27 real substitutions had occurred, these data imply that 1.75×10^{-3} substitutions per nucleotide per year had occurred for the entire visna virus genome. Selection by the host immune system has undoubtedly favored some mutations, particularly in the *env* gene region (Fig. 5).

NH2
putative *rev* coding exon 1
signal peptide
100
gp135
150
200
250
300
350
400
450
hypervariable site
hydrophobic
700
gp41 gp135
cleavage site
750
800
hydrophobic transmembrane domain
850
950
HOOC

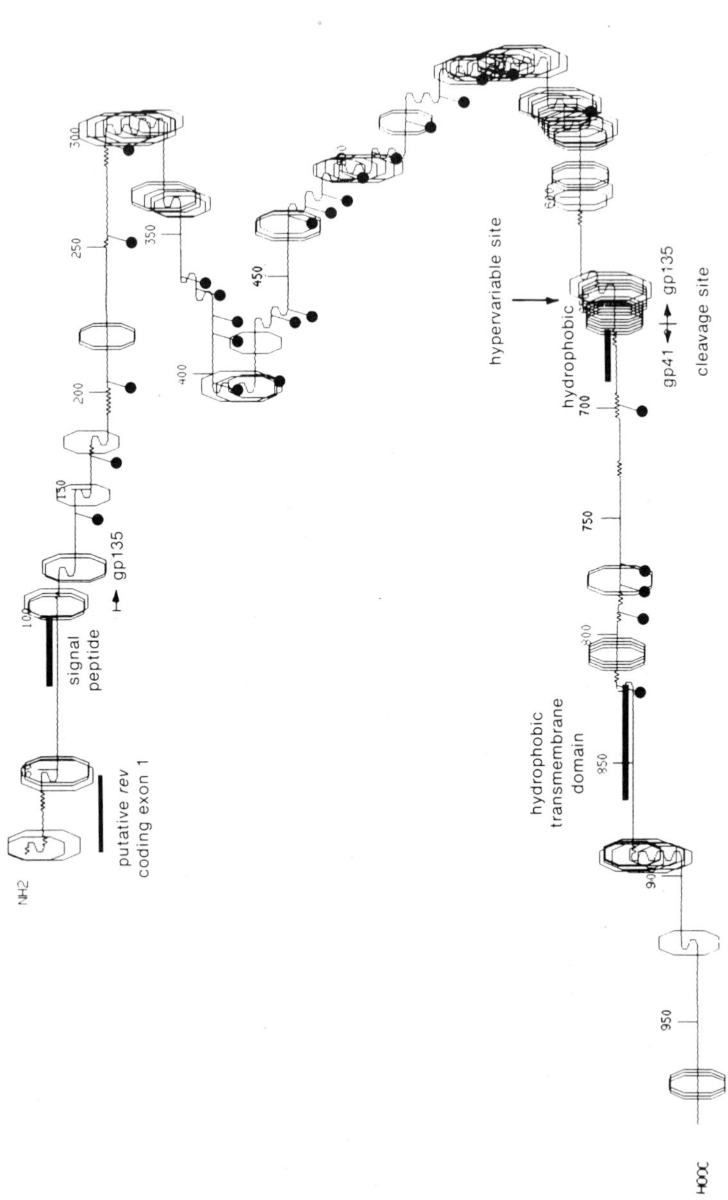

a

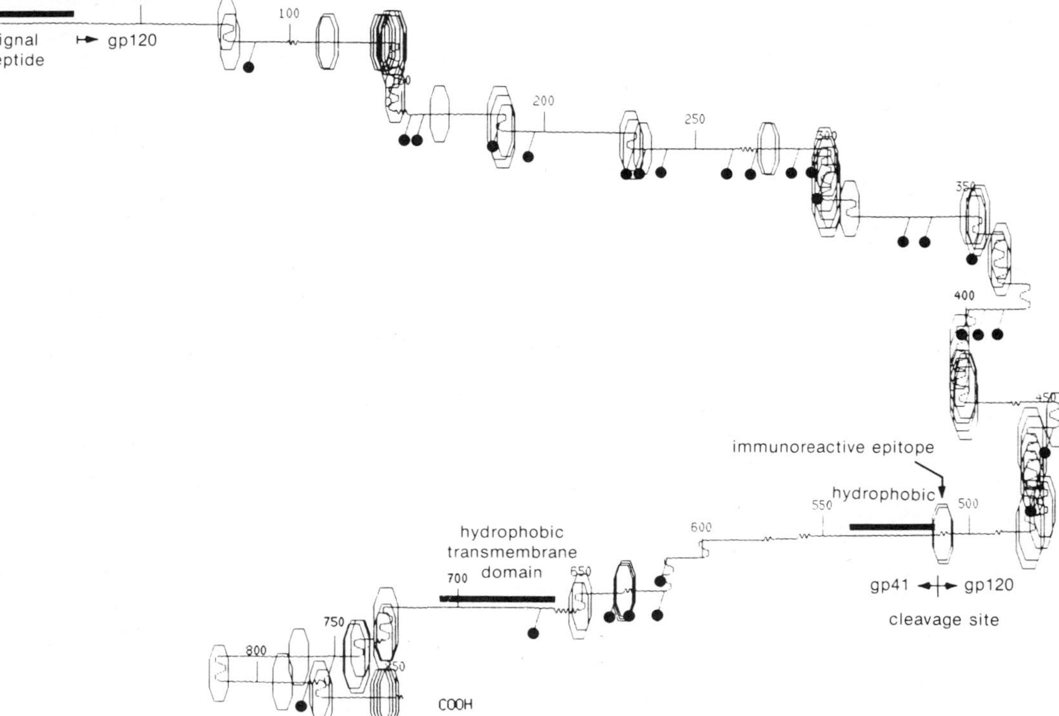

Figure 6. Computer-assisted analysis of visna virus and HIV-1 envelope structure. Chou—Fausman predictions of the secondary structure of the envelope proteins of (a) visna virus (Braun *et al.*, 1987) and (b) HIV-1 (Ratner *et al.*, 1985) were made using peptide structure and plot structure computer algorithms contained in the UWGCG DNA sequence analysis package (Devereaux *et al.*, 1984; Jameson and Wolf, 1988) under stringent conditions. The peptide sequences were derived by translating the DNA sequences starting at the first ATG in the *env* ORF; amino acids are numbered sequentially from the first *met* in *env*. Probable secondary structure, antigenic determinants, glycosylation sites, and select hydrophobic regions are shown. ⬡, potential antigenic sites; ●, sites of N-linked glycosylation; ▬, hydrophobic regions. OMPs (gp135, gp120 of visna virus and HIV-1, respectively) and TMPs (gp41) are indicated. The site of OMP and TMP cleavage is shown. The *rev* first coding exon which is in the *env* reading frame is designated in the visna virus profile (a). Hydrophobic signal peptide regions of the OMPs are labeled. The hydrophobic transmembrane and post OMP–TMP domains of the TMPs are shown for comparison (see also Fig. 8). The hypervariable site of visna virus and immunoreactive site of HIV-1 in the carboxy-terminus of the OMPs near the OMP–TMP cleavage site are indicated. Antigen index > = 1.2.

OMP / TMP cleavage

```
VISNA  RKKRGIGLVIVLAIMAIIAA----AGAGLGVANAVQQSYTRTAVQSLA[NAT]AAQQE   52
HIV    REKRAVGIGALF--LGFLGA----AGSTMGAASMTLTVQARQLLSGIVQQQNNLLR   50
EIAV   RHKRDFGISAIV--AAIVAATAIAASATMSYVALTEVNKIMEVQ[NHT]FEVE[NST]LN   54
              NH2-terminal hydrophobic domain
VISNA  VLEASYAMVQHIAKGIRILEARVARVEALVDRMMVYQELDCWHY--QHYCVTSTRS  106
HIV    AIEAQQHLLQLTVWGIKQLQARILAVERYLKDQQLLGIWGCSGK--LICTTAVPW[N  104
EIAV   GMDLIERQIKILYAMILQTHADVQLLKERQQVEETFNLIGCIERTHVFCHTGHPW[N  110

VISNA  EVANYV[NWT]RFKD[NCT]WQQWEEEIEQHEG[NLS]LLLREAALQVHIAQRDARRIPDAW  162
HIV    A S]W S[NKS]LEQIWN[NMT]WMEWDREIN[NYT]SLIHSLIEESQNQQEKNEQELLEL-DKW  159
EIAV   M S]WGHL--------[NES]TQWDDWVSKMEDLNQEILTTLHGARNNLAQSMITF-NTP  157

VISNA  KAIQEAF-----[NWS]SWFSWL--------KYIPWIIMGIVGLMCFRILMCVISMCL  205
HIV    ASL---W-----NWF[NIT]NWL--------WYIKLFIMIVGGLVGLRIVFAVLSVVN  199
EIAV   DSI---AQFGKDLWSHIGNWIPGLGASIIKYIVMFLLIYLLLTSSP----------  200
                          transmembrane domain
```

Figure 7. Alignment of the extracellular and transmembrane domains of the transmembrane *env* proteins of visna virus strain 1514 (Braun *et al.*, 1987), HIV (Ratner *et al.*, 1985), and EIAV (Rushlow *et al.*, 1986). The alignment was performed using ALIGN program (Dayhoff *et al.*, 1983) with a gap penalty of 25, matrix bias of 0, and 300 random comparisons. Identical residues are marked by dots. The alignments shown are those optimal with HIV; the visna virus–EIAV alignment can be improved slightly by moving gaps. Potential N-linked glycosylation sites are boxed. Arrows indicate conserved cysteine residues. (From Braun *et al.*, 1987, with permission of ASM.)

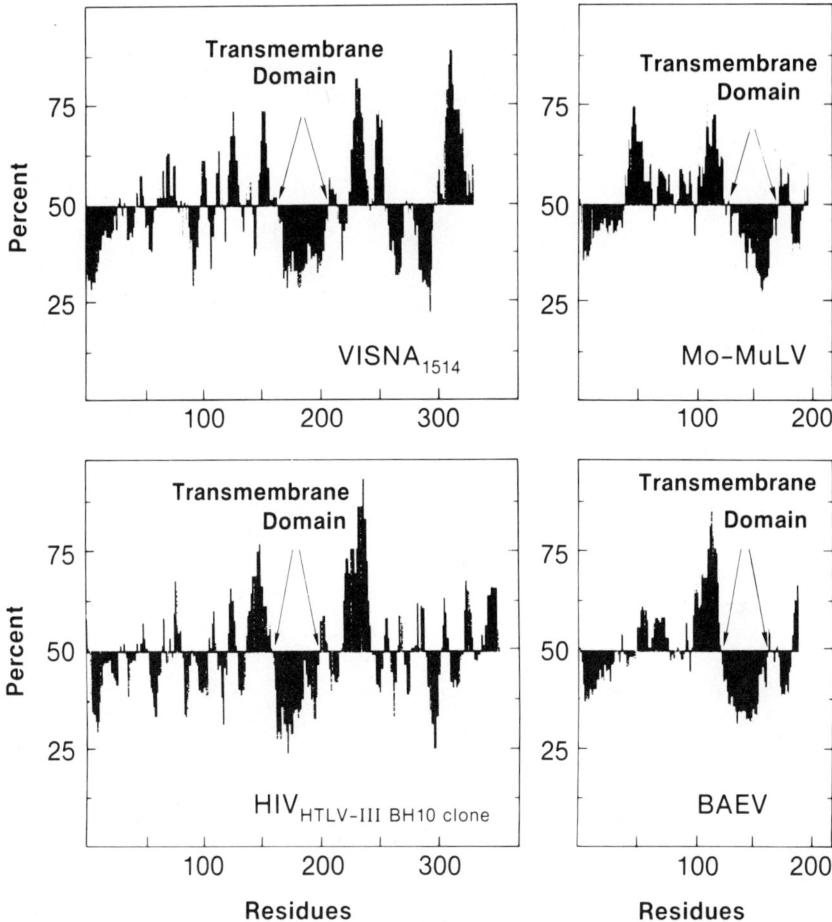

Figure 8. Hydrophobicity profiles of two lentivirus transmembrane *env* proteins (TMP) (visna virus and HIV), a murine type C TMP and a type D TMP (baboon endogenous virus; BAEV). Sequences used in the analysis were obtained from GenBank with the exception of BAEV. BAEV is a type C virus particle based on morphology, but its *env* gene shows strong sequence homology (80% identity in TMP) with type D viruses (Cohen and Braun, unpublished), and therefore it has been used here to represent type D TMPs. Hydrophobic regions below the mean, hydrophilic regions above. The analysis was performed by the method of Hopp and Woods (1981) with a step size of one and an averaging window of seven for the lentivirus TMPs and 10 for the type C and D TMPs. The transmembrane domains, predicted by the method of Klein *et al.* (1985), are indicated. (From Braun *et al.*, 1987, with permission of ASM.)

G. Reverse Transcriptase Fidelity and Genomic Heterogeneity

Studies on sequential HIV-1 isolates from persistently infected individuals suggest that HIV-1 evolves at a rate approximately a million times greater than DNA genomes; in fact, the substitution rates calculated for HIV were 10^{-4} and 10^{-3} for the *gag* and *env* genes, respectively (Hahn *et al.*, 1986). These values are in agreement with the estimate for visna virus (Braun *et al.*, 1987). The fidelity of the RT of HIV-1 has been assessed *in vitro* (Preston *et al.*, 1988; Roberts *et al.*, 1988). The misincorporation of nucleotides was estimated to be 1:2,000 to 1:4,000, suggesting that HIV-1

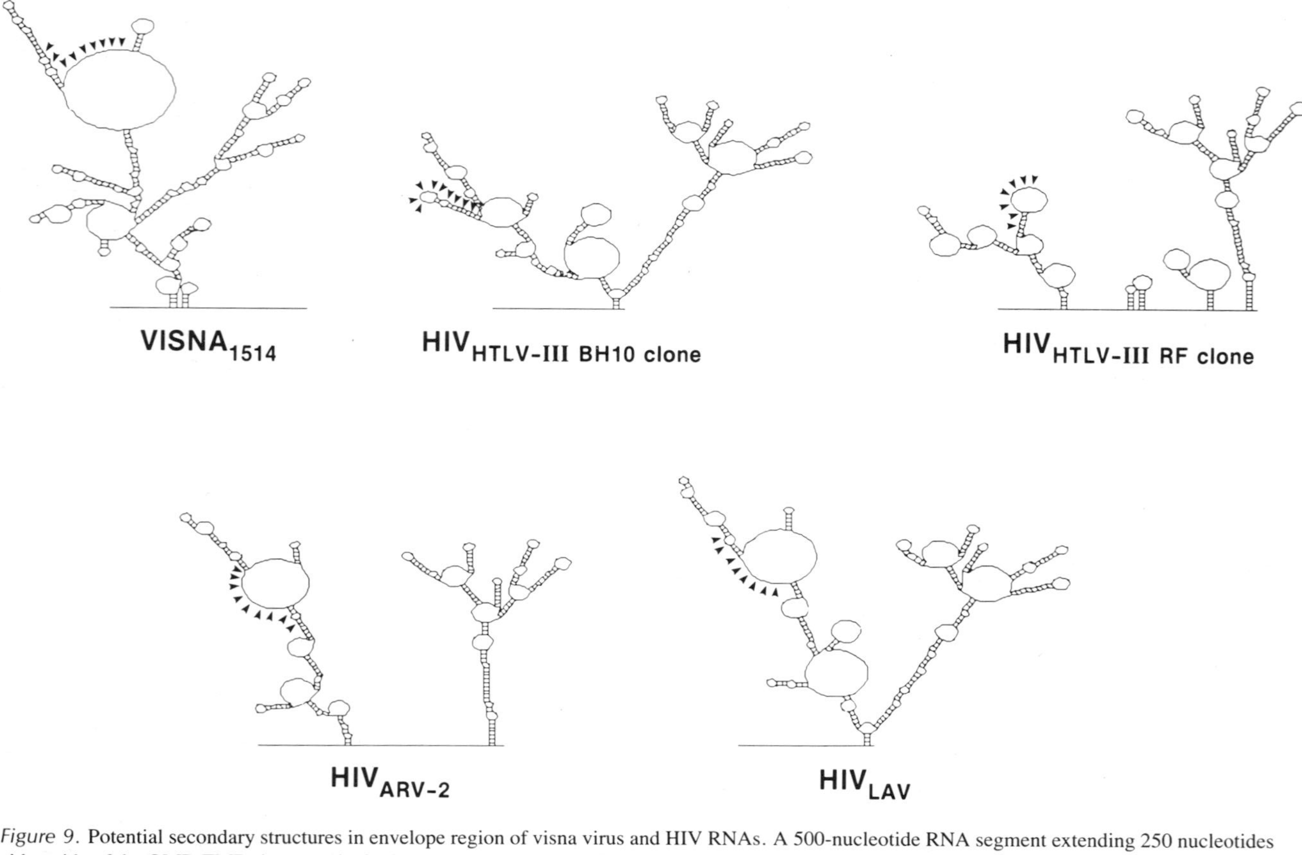

Figure 9. Potential secondary structures in envelope region of visna virus and HIV RNAs. A 500-nucleotide RNA segment extending 250 nucleotides either side of the OMP/TMP cleavage site in the *env* genes of visna virus strain 1514 (Braun *et al.*, 1987) and four strains of HIV (Ratner *et al.*, 1985; Sanchez-Pescador *et al.*, 1985; Wain-Hobson *et al.*, 1985; Starcich *et al.*, 1986) were used in the analysis. Secondary structure predictions were made using the program FOLD of Zuker and Stiegler (1981). Arrowheads mark the position of a hypervariable site in each segment. The hypervariable site for HIV strains lies at nucleotides 7170–7196 of the BH10 isolate (Ratner *et al.*, 1985) and was identified by Hahn *et al.*, (1986); that for visna virus lies at nucleotides 7864–7890 (Braun *et al.*, 1987). (From Braun *et al.*, 1987, with permission from ASM.)

has a ten-fold higher error rate than other retroviruses (1:17,000 and 1:30,000 for avian myeloblastosis virus and Mo-MuLV, respectively). In addition to substitutions, addition and deletion errors were also detected, and there were notable hotspots for these errors (Roberts *et al.*, 1988). More than likely, similar error rates will be found for the RT of visna virus.

V. CHALLENGES FOR DEVELOPING A VACCINE FOR LENTIVIRUSES

The possible consequences of retrovirus-host interactions can be quite different (Fig. 10). Nevertheless, retroviruses have developed several strategies, some of which are common to all retroviruses, to perpetuate their continued existence. Probably the greatest factor in viral persistence is their ability to integrate into the host genome and remain latent for long periods of time. Oncogenic viruses, with the exception of HTLV-1 and -2, have taken latency one step further in that some of these viruses have integrated into the germline of the host so that they are passed on as endogenous elements from one generation to the next. Thus, they have ensured their existence without the need for an infectious cycle.

Lentiviruses are exogenous viruses that initiate a slow, progressive infection of the immune and central nervous systems (Haase, 1986; Narayan *et al.*, 1987; Haseltine, 1988). Exogenous retroviruses must rely on an infectious cycle and therefore are subject to scrutiny and elimination by the host's immune system. With lentiviruses, there is a pattern of abundant early replication followed by periods of restricted virus expression (Narayan *et al.*, 1987; Haseltine, 1988). The major target cells appear to be monocytes and macrophages (Haase, 1986; Narayan *et al.*, 1987; Gartner *et al.*, 1986;

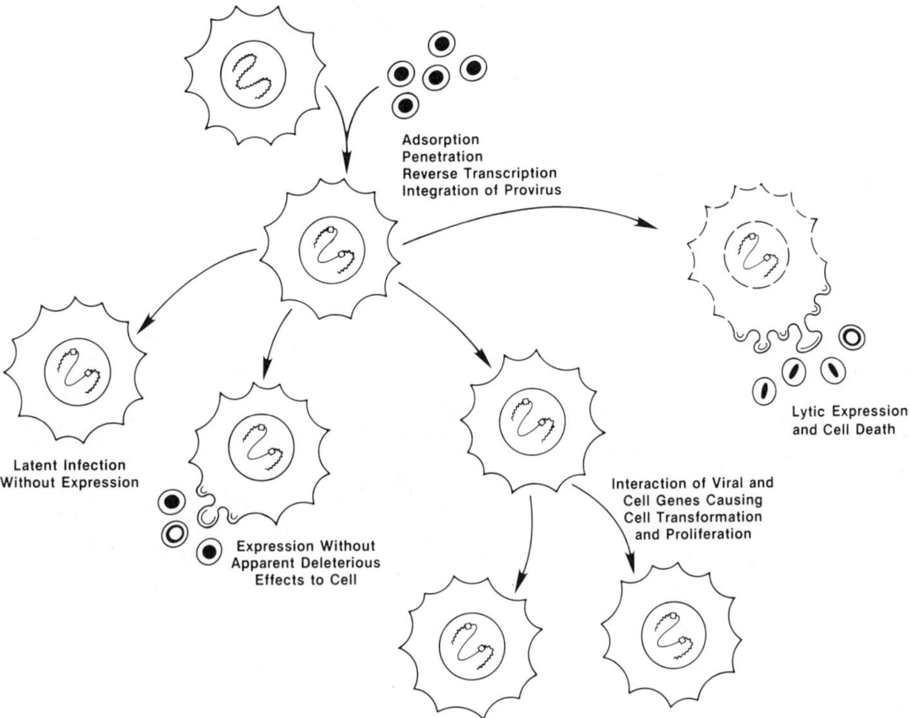

Figure 10. Possible consequences of retrovirus infection. (From Gonda *et al.*, 1989, with permission.)

Ho *et al.*, 1986; Asjo *et al.*, 1987) and, in the case of HIV, cells that carry T4 protein (Gallo *et al.*, 1984; Popovic *et al.*, 1984; Klatzman *et al.*, 1984). Orenstein *et al.* (1988) demonstrated that the viruses bud to the interior vesicles of macrophages and are not usually present on the surface of the cell. Thus, replication of lentiviruses in macrophages, at least in the case of HIV-1, is restricted at the level of cellular release. This is further complicated by the fact that monocytes and macrophages are antigen-presenting cells and may add to the infection process by cell-to-cell contact, without virus release. *In vitro,* lentiviruses are also cytopathic on cells that display an abundant supply of receptors; they may exhibit some of this property *in vivo,* adding to the pathogenesis of the virus.

Probably the greatest contributor to viral persistence within an individual is a consequence of the lentivirus life cycle whereby viral RNA is converted into double-stranded proviral DNA that is inserted into the host's genome and becomes part of the cell that it infects. Once a cell becomes infected, the viral genetic information becomes part of the host DNA and remains in the cell as long as the cell lives. Controlled replication and the latent state is due at least in part to the recently discovered elaborate regulatory gene network contained in the central region of the lentivirus genome (Haseltine, 1988).

Even with latency, the target cells have only a finite lifespan; thus, integration would have only a temporary effect on the survival of the viral genome. Virus production and dissemination to other target cells is necessary to enhance and expand the initial infection. This would put the virus in direct contact with the immune system where neutralizing antibodies could bind (Fig. 11). Lentiviruses have developed two additional mechanisms to avoid elimination during their exposed extracellular phase: masking of neutralizing epitopes probably by the acquisition of a large amount of sugar residues (Huso *et al.*, 1988) and, as shown here for visna virus, rapid and frequent production of point mutations in the *env* gene, which creates antigenic variation in potential neutralizing epitopes (Stanley *et al.*, 1987). In the first case, elimination is prevented because there is a lack of neutralizing antibodies or they are of such weak affinity that much of the virus escapes neutralization; epitope accessibility may also be a factor. Lack of neutralizing antibodies is best typified in the CAEV infection (Narayan *et al.*, 1984). In the second scenario, mutations arise in the neutralizing epitopes that offer a selective advantage in the face of preformed antibodies of the parental stock. Lentiviruses do not have to use either of the scenarios exclusively; there may be a blend of the two motifs.

Other studies have suggested that a humoral response to infection may not be compatible with protection. In fact, the presence of preformed antibodies to the virus may enhance the infection

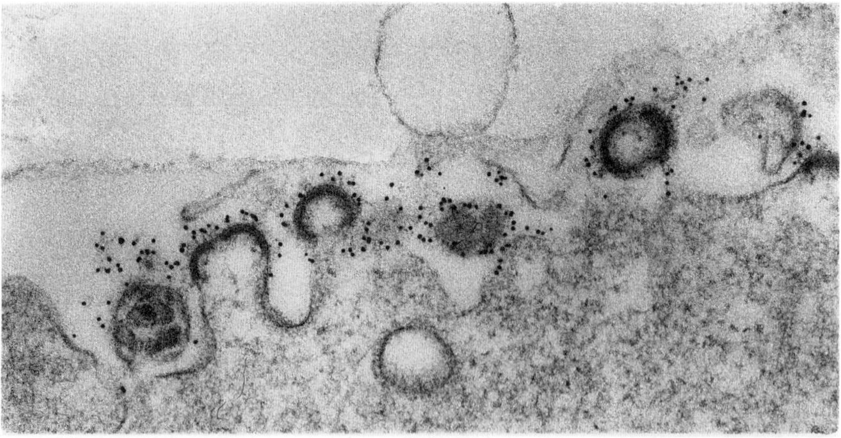

Figure 11. Immunoelectron microscopy of neutralizing sera to visna virus strain 1514. Virus particles are covered with immunogold marker indicating the strong binding of antiviral antibodies. 50,000×.

process in monocytes and macrophages. This is thought to occur through Fc-mediated mechanisms (Takeda *et al.*, 1988). Preformed antibodies may also potentiate the severity of the disease on virus challenge in some cases (Nathanson *et al.*, 1981; McGuire *et al.*, 1986). These features of lentivirus biology help explain why a vaccine for HIV and other lentiviruses has not been obtained. All of the above factors must be taken into consideration in developing novel approaches for preparing vaccines for lentivirus infections.

VI. CONCLUDING REMARKS

At the start of the AIDS epidemic, it would have been fair to say that much of what we came to know of HIV paralleled lentrivirus research over the last several decades. However, the pace of HIV research has far surpassed this early knowledge and it now provides a framework for comparative lentivirus research, particularly in studying the molecular genetics of this group of viruses. Although we have come far in our understanding of the HIV infection process and the host's response to HIV infection, we are still in search of an effective vaccine. *In vivo* models for AIDS-related research are sorely needed in this effort (Koff and Hoth, 1988). The close evolutionary relationship of visna virus and other lentiviruses to HIV makes them excellent candidates as models of HIV infection. A more thorough understanding of the molecular mechanisms of persistence and pathogenesis of the lentiviruses will undoubtedly yield new insights into curative therapies and eventually a vaccine.

ACKNOWLEDGMENTS

The author acknowledges Dr. M. Braun, a former colleague of my laboratory who worked extensively on the sequence of visna virus, and Drs. J. Clements and O. Narayan for their contributions and gifts of special reagents for some of the research presented in this article. I also thank K. Nagashima, D. Krell, and D. Chisholm for their technical assistance, K. McNitt, J. Smythers, M. S. Oberste, and N. Gunnell for their help with computer analysis, and J. Hopkins for preparation of the manuscript. This project has been funded at least in part with federal funds from the Department of Health and Human Services under contract number NO1-CO-74102. The content of this publication does not necessarily reflect the views or policies of the Department of Health and Human Services, nor does mention of trade names, commercial products, or organizations imply endorsements by the U.S. government.

REFERENCES

Alizon, M., Wain-Hobson, S., Montagnier, L., and Sonigo, P. (1986) *Cell* **46:** 63–74.

Asjo, B., Ivhed, I., Gidlund, M., Fuerstenberg, S., Fenyö, E. M., Nilsson, K., and Wigzell, H. (1987) *Virology* **157:** 359–365.

Barré-Sinoussi, F., Chermann, J. C., Rey, F., Negeyre, M. T., Chamaret, S., Gruest, J., Dauguet, C., Montagnier, L., Axler-Blin, C., Vézinet-Brun, F., Rouzioux, C., and Rozenbaum, W. (1983) *Science* **220:** 868–871.

Braun, M. J., Clements, J. E., and Gonda, M. A. (1987) *J. Virol.* **61:** 4046–4054.

Braun, M. J., Lahn, S., Boyd, A. L., Kost, T. A., Nagashima, K., and Gonda, M. A. (1988) *Virology* **167:** 515–523.

Charman, H. P., Bladen, S., Gilden, R. V., and Coggins, L. (1976) *J. Virol.* **19:** 1073–1079.

Chiu, I-M., Callahan, R., Tronick, S. R., Schlom, Jr., and Aaronson, S. A. (1984) *Science* **223:** 364–370.

Chiu, I-M., Yaniv, A., Dahlberg, J. E., Gazit, A., Skuntz, S. F., Tronick, S. R., and Aaronson, S. A. (1985) *Nature* **317:** 366–368.

Clavel, F., Guetard, D., Brun-Vézinet, F., Chamaret, S., Rey, M-A., Santos-Ferreira, M. O., Laurent, A. G., Dauguet, C., Katlama, C., Rouzioux, C., Klatzmann, C., Champalimaud, J. L., and Montagnier, L. (1986) *Science* **233:** 343–346.

Coffin, J. M. (1986) *Cell* **46:** 1–4.

Cohen, E. A., Terwilliger, E. F., Sodroski, J. G., and Haseltine, W. A. (1988) *Nature* **334:** 532–534.

Crawford, T. B., Adams, D. S., Cheevers, W. P., and Cork, L. C. (1980) *Science* **207:** 997–999.

Currey, K. M., Peterlin, B. M., and Maizel, J. V., Jr. (1986) *Virology* **148:** 33–46.

Cutlip, R. C., and Laird, G. A. (1976) *Am. J. Vet. Res.* **37:** 1377–1382.

Daniel, M. D., Letvin, N. L., King, N. W., Kannagi, M., Sehgal, P. K., Hunt, R. D., Kanki, P. J., and Desrosiers, R. C. (1985) *Science* **228:** 1201–1204.

Dayhoff, M. O., Barker, W. C., and Hunt, L. T. (1983) *Meth. Enzymol.* **91:** 524–545.

Devereaux, J., Haebesti, P., and Smithies, O. (1984) *Nucl. Acids Res.* **12:** 387–395.

Fukasawa, M. Miura, T., Hasegawa, A., Morikawa, S., Tsujimoto, H., Miki, K., Kitamura, T., and Hayami, M. (1988) *Nature* **333:** 457–461.

Gallo, R. C., Salahuddin, S. Z., Popovic, M., Shearer, G. M., Kaplan, M., Haynes, B. F., Palker, T. J., Markham, P. D., Redfield, T. J., Oleske, J., Safai, B., White, G., and Foster, P. (1984) *Science* **224:** 500–503.

Gallo, R., Wong-Staal, F., Montagnier, L., Haseltine, W. A., and Yoshida, M. (1988) *Nature* **333:** 504.

Gartner, S., Markovits. P., Markovitz, D. M., Kaplan, M. H., Gallo, R. C., and Popovic, M. (1986) *Science* **233:** 215–219.

Gonda, M. A. (1988) *J. Elect. Microsc. Tech.* **8:** 17–40.

Gonda, M. A., Charman, H. P., Walker, J. L., and Coggins, L. (1978) *Am. J. Vet. Res.* **39:** 731–740.

Gonda, M. A., Wong-Staal, F., Gallo, R. C., Clements, J. E., Narayan, O., and Gilden, R. V. (1985) *Science* **227:** 173–177.

Gonda, M. A., Braun, M. J., Clements, J. E., Pyper, J. M., Gallo, R. C., Wong-Staal, F., and Gilden, R. V. (1986) *Proc. Natl. Acad. Sci. USA* **3:** 4007–4011.

Gonda, M. A., Braun, M. J., Carter, S. G., Kost, T. A., Bess, J. W., Jr., Arthur, L. O., and Van Der Maaten, M. J. (1987) *Nature* **330:** 388–391.

Gonda, M. A., Boyd, A. L., Nagashima, K., and Gilden, R. V. (1989) *Arch. AIDS Res.* **3:** 1–42.

Haase, A. T. (1975) *Curr. Top. Microbiol. Immunol.* **72:** 101–156.

Haase, A. T. (1986) *Nature* **322:** 130–136.

Hahn, B. H., Shaw, G. M., Taylor, M. E., Redfield, R. R., Markham, P. D., Salahuddin, S. Z., Wong-Staal, F., and Gallo, R. C. (1986) *Science* **232:** 1548–1553.

Harris, J. D., Blum, H., Scott, J., Traynor, B., Ventura, P., and Haase, A. (1984) *Proc. Natl. Acad. Sci. USA* **81:** 7212–7215.

Haseltine, W. A. (1988) *J. AIDS* **1:** 217–240.

Henderson, L. E., Sowder, R. C., Copeland, T. D., Benveniste, R. E., and Oroszlan, S. (1988) *Science* **241:** 199–201.

Hess, J. L., Clements, J. E., and Narayan, O. (1985) *Science* **229:** 482–485.

Ho, D. D., Rota, T. R., and Hirsch, M. S. (1986) *J. Clin. Invest.* **77:** 1712–1715.

Hopp, T. P., and Woods, K. R. (1981) *Proc. Natl. Acad. Sci. USA* **8:** 3824–3828.

Huso, D. L., Narayan, O., and Hart, G. W. (1988) *J. Virol.* **62:** 1974–1980.

Jameson, B. A., and Wolf, H. (1988) *CABIOS* **4:** 181–186.

Kennedy, R. C., Eklund, C. M., Lopez, C., and Hadlow, W. J. (1968) *Virology* **35:** 483–484.

Klatzmann, D., Champagne, E., Chamaret, S., Gruest, J., Guetard, D., Hercend, T., Gluckman, J-C., and Montagnier, L. (1984) *Nature* **312:** 767–768.

Klein, P., Kanehisa, M., and DeLisi, C. (1985) *Biochim. Biophys. Acta* **815:** 468–476.

Koff, W. C., and Hoth, D. F. (1988) *Science* **241:** 426–432.

Laurence, J. (1988) *AIDS Res. Hum. Retroviruses* **4:** vii–viii.

Luciw, P. A., Cheng-Mayer, C., and Levy, J. A. (1987) *Proc. Natl. Acad. Sci. USA* **84:** 1434–1438.

Matsuda, Z., Chou, M. J., Matsuda, M., Huang, J-H., Chen, Y-M., Redfield, R., Mayer, K., Essex, M., and Lee, T-H. (1988) *Proc. Natl. Acad. Sci. USA* **85:** 6968–6972.

Maurer, B., and Flügel, R. M. (1988) *AIDS Res. Hum. Retroviruses* **4:** 467–473.

Mazarin, V., Gourdou, I., Querat, G., Sauze, N., and Vigne, R. (1988) *J. Virol.* **62:** 4813–4818.

McClure, M. A., Johnson, M. S., Feng, D-F., and Doolittle, R. F. (1988) *Proc. Natl. Acad. Sci. USA* **85:** 2469–2473.
McGuire, T. C., Adams, S., Johnson, G. C., Klevjer-Anderson, P., Barbee, D. D., and Gorham, J. R. (1986) *Am. J. Vet. Res.* **47:** 537–540.
McGuire, T. C., O'Rourke, K., and Cheevers, W. P. (1987) In *Contr. Microbiol. Immunol.* Vol. 8 (Lewis, R. E., Jr., and Cruse, J. M., eds.), pp. 77–89, Karger, Basel.
Modrow, S., Hahn, B. H., Shaw, G. M., Gallo, R. C., Wong-Staal, F., and Wolf, H. (1987) *J. Virol.* **61:** 570–578.
Molineaux, S., and Clements, J. E. (1983) *Gene* **23:** 137–148.
Narayan, O., Griffin, D. E., and Chase, J. (1977) *Science* **197:** 376–378.
Narayan, O., Griffin, D. E., and Clements, J. E. (1978) *J. Gen. Virol.* **41:** 343–352.
Narayan, O., Sheffer, D., Griffin, D. E., Clements, J. E., and Hess, J. (1984) *J. Virol.* **49:** 349–355.
Narayan, O., Clements, J. Kennedy-Stoskopf, S., Sheffer, D., and Royal, W. (1987) *Contr. Microbiol. Immunol.* **8:** 60–76.
Nathanson, N., Martin, J. R., Georgsson, G., Palsson, P. A., Lutley, R. E., and Petursson, G. (1981) *J. Comp. Pathol.* **81:** 185–191.
Orenstein, J. M., Meltzer, M. S., Phipps, T., and Gendelman, H. E. (1988) *J. Virol.* **62:** 2578–2586.
Palker, T. J., Matthews, T. J., Clark, M. E., Cianciolo, G. J., and Randall, R. R., Langlois, A. J., White, G. C., Safai, B., Snyderman, R., Bolognesi, D. P., and Haynes, B. F. (1987) *Proc. Natl. Acad. Sci. USA* **84:** 2479–2483.
Payne, S., Parekh, B., Montelaro, R. C., and Issela, C. J. (1984) *J. Gen. Virol.* **65:** 1395–1399.
Pedersen, N. C., Ho, E. W., Brown, M. L., and Yamamoto, J. K. (1987) *Science* **235:** 790–793.
Popovic, M., Sarngadharan, M. G., Read, E., and Gallo, R. C. (1984) *Science* **224:** 497–500.
Preston, B. D., Poiesz, B. J., and Loeb, L. A. (1988) *Science* **242:** 1168–1171.
Pyper, J. M., Clements, J. E., Molineaux, S. M., and Narayan, O. (1984) *J. Virol.* **51:** 713–721.
Pyper, J. M., Clements, J. E., Gonda, M., and Narayan, O. (1986) *J. Virol.* **58:** 665–670.
Ratner, L., Haseltine, W., Patarca, R., Livak, K. J., Starcich, B., Josephs, S. J., Doran, E. R., Antoni Rafalski, J., Whitehorn, E. A., Baumeister, K., Ivanoff, L., Petteway, S. R., Jr., Pearson, M. L., Lautenberger, J. A., Papas, T. S., Ghrayeb, J., Chang, N. T., Gallo, R. C., and Wong-Staal, F. (1985) *Nature (Lond.)* **313:** 277–284.
Roberts, J. D., Bebenek, K., and Kunkel, T. A. (1988) *Science* **242:** 1171–1173.
Rushlow, K., Olsen, K., Stiegler, G., Payne, S. L., and Montelaro, R. C. (1986) *Virology* **155:** 309–321.
Salinovich, O., Payne, S. L., Montelaro, R. C., Hussain, K. A., Issel, C. J., and Schnorr, K. L. (1986) *J. Virol.* **57:** 71–80.
Sanchez-Pescador, R., Power, M. D., Barr, P. J., Steimer, K. S., Stempien, M. M., Brown-Shimer, S. L., Gee, W. W., Renard, A., Randolph, A., Levy, J. A., Dina, D., and Luciw, P. A. (1985) *Science* **227:** 484–492.
Sigurdsson, B. (1954) *Br. Vet. J.* **110:** 255–270.
Sigurdsson, B. (1957) *J. Neuropathol. Exp. Neurol.* **16:** 389–403.
Sigurdsson, B., and Palsson, P. A. (1958) *Br. J. Exp. Pathol.* **39:** 519–528.
Sonigo, P., Alizon, M., Staskus, K., Klatzmann, D., Cole, S., Danos, O., Retzel, E., and Wain-Hobson, S. (1985) *Cell* **42:** 369–382.
Sonigo, P., Barker, C., Hunter, E., and Wain-Hobson, S. 91986) *Cell* **45:** 375–385.
Stanley, J., Bhaduri, L. M., Narayan, O., and Clements, J. E. (1987) *J. Virol.* **61:** 1019–1028.
Starcich, B. R., Hahn, B. H., Shaw, G. M., McNeely, P. D., Modrow, S., Wolf, H., Parks, E. S., and Parks, W. P. (1986) *Cell* **45:** 637–648.
Stephens, R. M., Casey, J. W., and Rice, N. R. (1986) *Science* **231:** 589–594.
Stone, L. B., Scolnick, E., Takemoto, K. K., and Aaronson, S. A. (1971) *Nature (Lond.)* **229:** 257–258.
Strebel, K., Klimkait, T., and Martin, M. A. (1988) *Science* **241:** 1221–1223.
Takeda, A., Tuazon, C. U., and Ennis, F. A. (1988) *Science* **242:** 580–583.
Thormar, H., and Palsson, P. A. (1967) In *Perspect. Virol.*, Vol. V (M. Pollard, ed.), pp. 291–308, Academic Press, New York.
Vallée, H., and Carré, H. (1904) *C. R. Acad. Sci.* **139:** 331–333.
Van Der Maaten, M. J., Boothe, A. D., and Seger, C. L. (1972) *J. Natl. Cancer Inst.* **49:** 1649–1657.
Varmus, H. (1988) *Science* **240:** 1427–1435.
Wain-Hobson, S., Sonigo, P., Danos, O., Cole, S., and Alizon, M. (1985) *Cell* **40:** 9–17.
Weiss, R. A. (1988) *Nature* **333:** 497–499.

Wong-Staal, F., Chanda, P. K., and Ghrayeb, J. (1987) *AIDS Res. Hum. Retroviruses* **3:** 33–39.
Yaniv, A., Dahlberg, J. E., Tronick, S. R., Chiu, I-M., and Aaronson, S. A. (1985) *Virology* **145:** 340–345.
Yokoyama, S., Chung, L., and Gojobori, T. (1988) *Mol. Biol. Evol.* **5:** 237–251.
Yu, X-F., Ito, S., Essex, M., and Lee, T-H. (1988) *Nature* **335:** 262–265.
Zuker, M., and Stiegler, P. (1981) *Nucl. Acids Res.* **9:** 133–148.

6

In Vivo and *In Vitro* Selection of Equine Infectious Anemia Virus Variants

Susan Carpenter, Leonard H. Evans, Martin Sevoian, and Bruce Chesebro

I. INTRODUCTION

Lentiviruses are a subfamily of retroviruses associated with slow, chronic infections in animals and humans. Included in the lentivirus family are visna virus, caprine arthritis-encephalitis virus (CAEV), equine infectious anemia virus (EIAV), ovine progressive pneumonia virus (OPPV), and immunodeficiency viruses of human (HIV), simian (SIV), feline (FIV), and bovine (BIV) species. These viruses share several structural and biologic features, including a complex genome structure characterized by several small open reading frames in both the *pol*/*env* intragenic region and in the 3′ end of the viral RNA genome (Chiu *et al.*, 1985; Gonda *et al.*, 1985; Rushlow *et al.*, 1986). In addition, lentiviruses are tropic for cells of the macrophage/monocyte lineage, and viral replication in these cell types is thought to play an important role in some aspects of lentiviral pathogenesis (Gendelman *et al.*, 1985, 1986; Gartner *et al.*, 1986; Narayan *et al.*, 1982, 1983).

Clinical manifestations of lentiviral infections vary widely, both among the different members of the lentivirus family as well as among individuals inoculated with the same virus. Both host and viral factors are thought to contribute to this variability. For example, the age of the host at the time of infection has been shown to affect the length of viral latency in HIV-infected individuals (Curran *et al.*, 1988), and there is evidence to suggest that certain host genotypes may influence the clinicopathologic outcome of lentiviral infections (Mann *et al.*, 1987).

Viral determinants of virulence and pathogenicity are less well defined. Lentiviruses are noted for a high degree of genetic heterogeneity, especially in the envelope region of the viral genome (Hahn *et al.*, 1986; Payne *et al.*, 1987; Starcich *et al.*, 1986; Wong-Staal *et al.*, 1985), and the biologic flexibility afforded by such heterogeneity may be an important factor in viral pathogenesis. For example, genetic mutations in the viral *env* gene could alter viral antigens and enable some lentiviruses to persist in the face of a virus-specific immune response. This may be the case with EIAV, where episodes of clinical disease are associated with replication of antigenically variant viruses (Montelaro *et al.*, 1984; Kono *et al.*, 1973; Thormar *et al.*, 1983). In addition, the target cell

---

Susan Carpenter, Leonard H. Evans, and Bruce Chesebro • Laboratory of Persistent Viral Diseases, Rocky Mountain Laboratories, Hamilton, Montana, 59840. *Martin Sevoian* • Department of Veterinary and Animal Sciences, University of Massachusetts, Amherst, Massachusetts 01002. *Present address of S.C.:* Department of Veterinary Microbiology and Preventive Medicine, Iowa State University, Ames, Iowa 50011.

specificity of lentiviruses has recently been associated with certain clinical and pathologic outcomes of infection (Levy, 1988; Gartner *et al.*, 1986). Recent studies indicate that HIV isolates associated with neuropathologic lesions in AIDS patients are "macrophage-tropic" *in vitro* and biologically distinct from "T-tropic" viruses isolated from PHA stimulated lymphocytes (Gartner *et al.*, 1986). These findings suggest that differences in the clinical outcome of lentiviral infections may be attributable to the presence of biologically distinct variant viruses. Moreover, multiple biologic and genetic variants are present in a single virus isolate from HIV-infected individuals (Fisher *et al.*, 1988; Sakai *et al.*, 1988), and studies with animal lentiviruses suggest coexistence of antigenically distinct viral isolates (Lutley *et al.*, 1983; Thormar *et al.*, 1983). Thus, understanding the mechanisms of variant generation and variant selection should aid our efforts to understand viral pathogenesis.

It is likely that lentivirus variants are generated as a result of errors in reverse transcription that occur during the spread of infection *in vivo* (Coffin, 1986). In addition, heterogeneous populations of variants may be transmitted during natural infections. In defining the role variation may play in viral pathogenesis, therefore, it is important to consider what factors contribute to the selection of a given variant from such heterogeneous populations. In certain animal lentivirus infections, such as EIAV and visna virus, there is evidence to suggest that the host immune response plays a critical role in selection of viral variants (Clements *et al.*, 1988).

The mechanism of immune selection of lentiviral variants is not fully understood, although several reports suggest that neutralizing antibody selects for antigenically novel variants (Clements *et al.*, 1988; Narayan *et al.*, 1981; Salinovich *et al.*, 1986). Thus, sera collected during chronic episodes of EIA contain antibody that will neutralize previously isolated virus variants, but will not neutralize virus variants isolated later during infection (Montelaro *et al.*, 1984; Kono *et al.*, 1973). In visna virus infections, however, there was no correlation between the neutralizing antibody titer to the inoculum virus and the occurrence of antigenic variants (Lutley *et al.*, 1983). Moreover, antigenically variant viruses were found to coexist with the inoculum virus (Lutley *et al.*, 1983; Thormar *et al.*, 1983), indicating that immunological elimination of the inoculum virus was not required for the emergence of antigenic variants. Finally, HIV variants selected by *in vitro* cultivation in the presence of neutralizing antibody genetically varied from the inoculum virus in regions of the viral genome completely unrelated to hypervariable regions previously identified in isolates recovered from HIV-infected individuals (Reitz *et al.*, 1988; Starcich *et al.*, 1986). Thus, neutralizing antibody is not likely to be the only selective pressure important in variant selection during natural lentiviral infections. To better define mechanisms important in the selection of lentivirus variants, we have studied *in vivo* and *in vitro* variation of EIAV.

II. VARIATION OF EQUINE INFECTIOUS ANEMIA VIRUS

In horses, EIAV causes a persistent disease characterized by a variable course. In acute cases, rapid viral replication leads to early death, generally within 1–4 weeks after infection. In the more classic, chronic form of EIAV, cycles of fever, weight loss, and anemia are interspersed with periods of clinical quiescence. Both the frequency and severity of the clinical episodes decline with time after infection (McGuire *et al.*, 1987), and clinical recovery eventually occurs. Finally, cases of EIA may also be classified as aclinical, where horses are seropositive but no clinical signs of disease are observed.

Antigenic variation of EIAV was first described by Kono *et al.* (1973), who found that virus isolated from successive febrile cycles could be antigenically distinguished by neutralizing antibody present after, but not before, variant isolation. Biochemical and molecular analyses indicated that antigenic variants of EIAV differ markedly in viral envelope proteins, but less so in viral core proteins (Montelaro *et al.*, 1984; Payne *et al.*, 1987; Salinovich *et al.*, 1986). These studies clearly demonstrated that recrudescence of clinical EIA was associated with the emergence of antigenic and genetic variants of EIAV. However, it was less clear whether there was, in fact, a causal relationship between the development of variant-specific neutralizing antibody and clinical quiescence.

Previous studies relied on endpoint dilution as a quantitative estimate of EIAV (Montelaro *et al.*, 1984; Kono *et al.*, 1973). However, this method does not have the accuracy of a focal infectivity, or plaque, assay and, moreover, required the use of a large amount of virus and *in vitro* passage in order to score infectious virus. Thus, nearly complete neutralization (>99.0%) was necessary to detect the presence of neutralizing antibody. We previously developed a focal immunoassay (FIA) for *in vitro* quantitation of murine retroviruses (Sitbon *et al.*, 1985a,b) and, more recently, for detection of HIV infectivity (Chesebro and Wehrly, 1988). Importantly, the FIA allowed us to quantitate neutralizing antibody using a small, but constant, amount of infectious virus and serial dilutions of antisera (Wiley *et al.*, 1986). The advantage of developing a similar assay for *in vitro* quantitation of EIAV is that it would allow us to quantitate the development of variant-specific antibody in relation to the clinical disease course.

A. *In Vivo* Selection of EIAV Variants

1. *In Vitro* Quantitation of EIAV

The FIA developed for *in vitro* quantitation of cell-free EIAV relies on the immunologic detection of EIA viral antigens using broadly reactive polyclonal horse sera as the first antibody and fluorescein isothiocyanate (FITC)-conjugated antihorse Ig (Cooper Biomedical, West Chester, PA) as the second antibody. When the FIA was used to assay equine dermal (ED) cells inoculated with near-limiting dilutions of cell-free EIAV, foci of fluorescent cells were observed (Fig. 1A). The number of foci was plotted against the amount of virus inoculum and a "one-hit" linear relationship was obtained (Fig. 1B), indicating that each focus resulted from a single infectious virion. Thus, the FIA could be used as a quantitative assay of infectious, focus-forming units (FFUs) of EIAV.

The FIA was initially used to titrate the presence of neutralizing antibody in serum samples collected from field-infected horses. Earlier studies have indicated little or no cross-neutralization among a variety of field isolates of EIAV (Kono *et al.*, 1971); however, low levels of cross-neutralizing antibody are often detected in the sera of AIDS patients (Prince *et al.*, 1987; Robert-Guroff *et al.*, 1985). Twofold serial dilutions of horse sera were mixed with 500–1000 FFUs of the prototype Wyoming isolate of EIAV (Malmquist *et al.*, 1973), and sera that gave a two-thirds or greater reduction in foci, as compared to normal horse serum or to diluent controls, were considered to contain neutralizing antibody. Interestingly, almost all of the sera tested were found to contain neutralizing antibody to the Wyoming isolate of EIAV (Table 1). At present, it is not known whether the neutralization is due to recognition of an epitope common to all isolates of EIAV or, rather, indicates that the field-infected horses were exposed to virus antigenically similar to the Wyoming isolate.

2. Immune Selection of EIAV Variants

The availability of a sensitive, quantitative assay allowed us to examine the development of neutralizing antibody in experimentally infected horses. For these studies, a horse was inoculated with whole blood from a field-infected, EIA-seropositive horse. Virus isolates from the first (MA-1) and fourth (MA-4) febrile cycles were antigenically characterized using serum samples collected throughout the course of disease. Interestingly, the sequential appearance of MA-1 and MA-4 occurred independent of the development of variant-specific neutralizing antibody (Carpenter *et al.*, 1987). MA-1 virus was isolated during the first febrile cycle at 13 days postinoculation (DPI), and MA-4 was isolated from blood collected during the fourth febrile cycle at 46 DPI (Fig. 2,top).

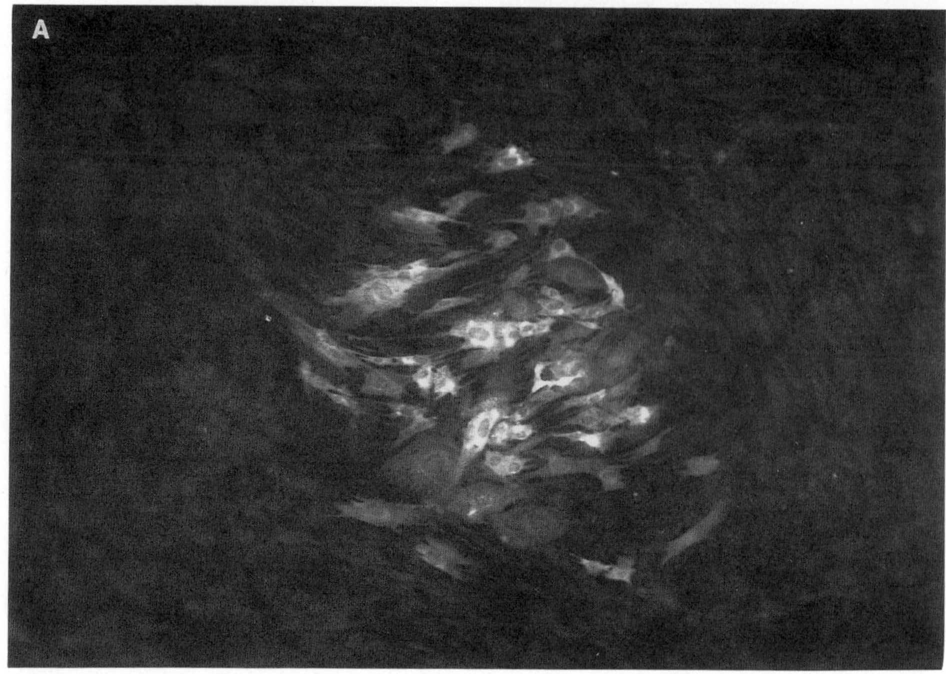

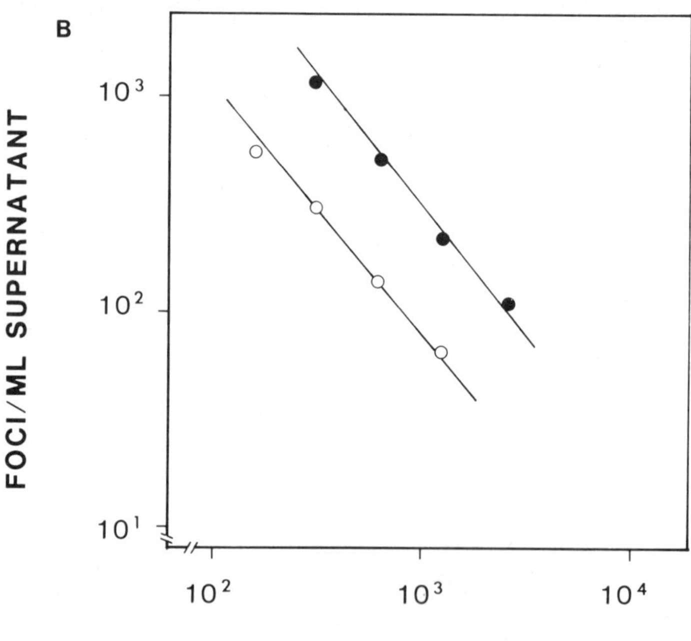

Figure 1. In vitro quantitation of EIAV infectivity using a focal immunoassay of methanol fixed cells. (A) Focus of EIAV-infected ED cells 5 days postinoculation with cell-free EIAV. (B) Linear one-hit kinetic relationship between the amount of virus inoculum and the number of foci detected. EIAV-Wyoming, open circles; MA-1, closed circles.

Table 1. Quantitation of Neutralizing Antibody to EIAV-Wyoming in Sera from Horses Infected with Field Isolates of EIAV

Serum sample	Neutralizing antibody titer to EIAV-Wyoming[a]
TO83	>256
NJ-M	>256
NH-O	>256
MA-G	16
VA-C	128
CT-B	32
CT-D	4
J-165	>128
CT-W	<4

[a]Neutralizing antibody titer represents the inverse of the highest serum dilution that gave 70% or greater reduction in foci as compared to normal horse serum or to diluent controls.

However, neutralizing antibody to MA-4 was detectable prior to neutralizing antibody to MA-1 (Fig. 2, bottom). In addition, multiple febrile cycles occurred in the interval between the isolation of MA-1 and the detection of MA-1 neutralizing antibody. Thus, the short period of clinical quiescence following the first febrile cycle was not due to the development of MA-1-specific neutralizing antibody. In fact, the delay in the appearance of MA-1 neutralizing antibody may indicate the continued presence of MA-1 during periods characterized by multiple febrile cycles and the emergence of antigenically variant viruses.

MA-1 and MA-4 were antigenically distinguished by membrane immunofluorscence. Serum collected in the interval between the first and the fourth febrile cycles had a higher antibody titer to MA-1-infected cells than to MA-4-infected cells (Fig. 3). Serum collected after the isolation of MA-4, had an increase in reactivity to MA-4-infected cells, but not to MA-1-infected cells. It thus appeared that antigenic variation of EIAV was expressed on the surface of virus-infected cells and, furthermore, that a variant-specific immune response occurred prior to the development of neutralizing antibody.

The presence of variant-specific antigens on the surface of EIAV-infected cells was confirmed by absorption studies. The reactivity of early, MA-1-specific, immune serum was significantly ($P < 0.05$) reduced by absorption with MA-1-infected cells, but not with MA-4-infected cells or with uninfected cells (Fig. 4). The demonstration of variation of virus-specific cell surface antigens suggested that immune selection of EIAV variants could occur through destruction of infected cells, either through antibody-mediated cytotoxicity (ADCC) or, perhaps, through cell-mediated cytotoxicity. Interestingly, it was recently shown that HIV epitopes susceptible to neutralizing antibody differ from those epitopes involved in ADCC (Lyerly *et al.*, 1988). A similar finding in EIA might explain the lack of neutralizing antibody in early serum shown to contain antibody reactive with MA-1-specific antigens on the surface of live cells.

We were not able to demonstrate a role for neutralizing antibody in immune selection of EIAV variants during early stages of chronic EIA. Others reports have indicated that febrile cycles and viral variation can occur prior to the development of neutralizing antibody (Kono, 1969; Montelaro *et al.*, 1984). Thus, our results suggest that during periods of rapid cycling early in the course of EIA, antigenic variation is not due to selective pressure of neutralizing antibody. However, immune pressure could select the EIAV variants through elimination of certain virus-infected cells or,

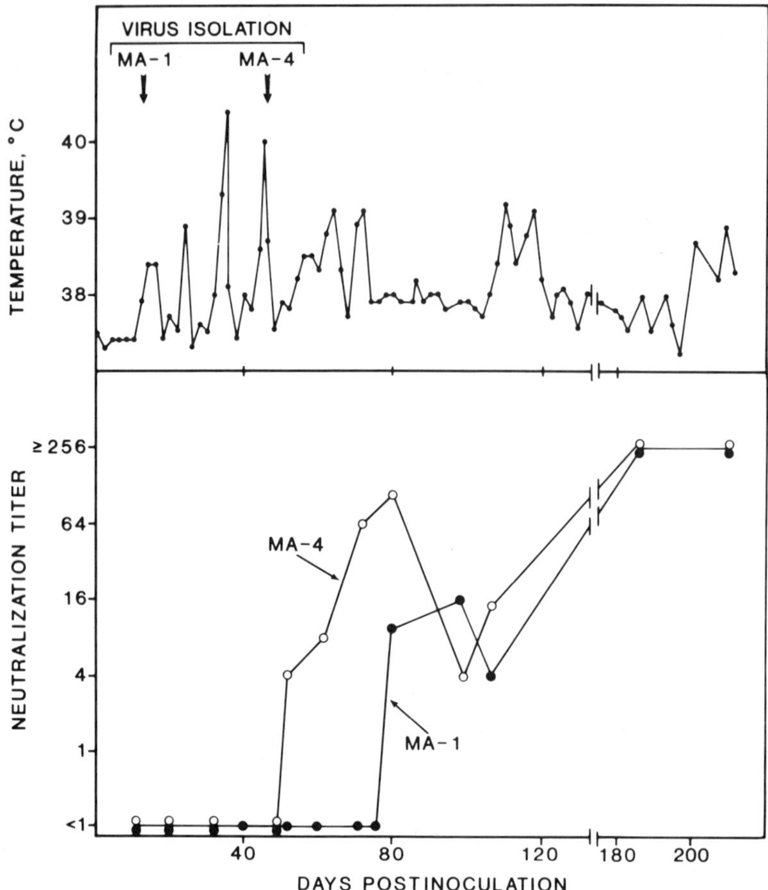

Figure 2. (Top) Clinical disease course in a horse experimentally inoculated with whole blood from a subclnical carrier of EIAV. (Bottom) Development of neutralizing antibody to MA-1 and MA-4 during the course of clinical disease in a horse experimentally inoculated with EIAV. Virus isolates were recovered from peripheral blood leukocytes recovered at day 13 (MA-1) and day 46 (MA-4) postinoculation and were biologically cloned before their use in neutralization assays. Neutralization titers are expressed as the reciprocal of the serum dilution that neutralized at least 70% of viral infectivity in a focus reduction assay (Carpenter *et al.*, 1987).

conversely, the variant-specific immune response might be a reactive phenomenon secondary to the sequential appearance of viral variants selected by other biological mechanisms.

3. Structural Analyses of MA-1 and MA-4

There is ample evidence to suggest that antigenic variation of EIAV is associated with genetic heterogeneity in the viral *env* gene (Payne *et al.*, 1984, 1987; Salinovich *et al.*, 1986; Montelaro *et al.*, 1984). However, both *env* and *gag* gene products are expressed on the surface of cells infected with some murine retroviruses (Tung *et al.*, 1976; Buetti and Diggelman, 1980), and it was possible that the antigenic differences between MA-1 and MA-4 were due to differences in the *gag* gene region. Radioimmunoprecipitation (RIP) assays were used to determine which viral protein(s) expressed variant specific epitopes. After metabolic labeling with [^{35}S]methionine, lysates of

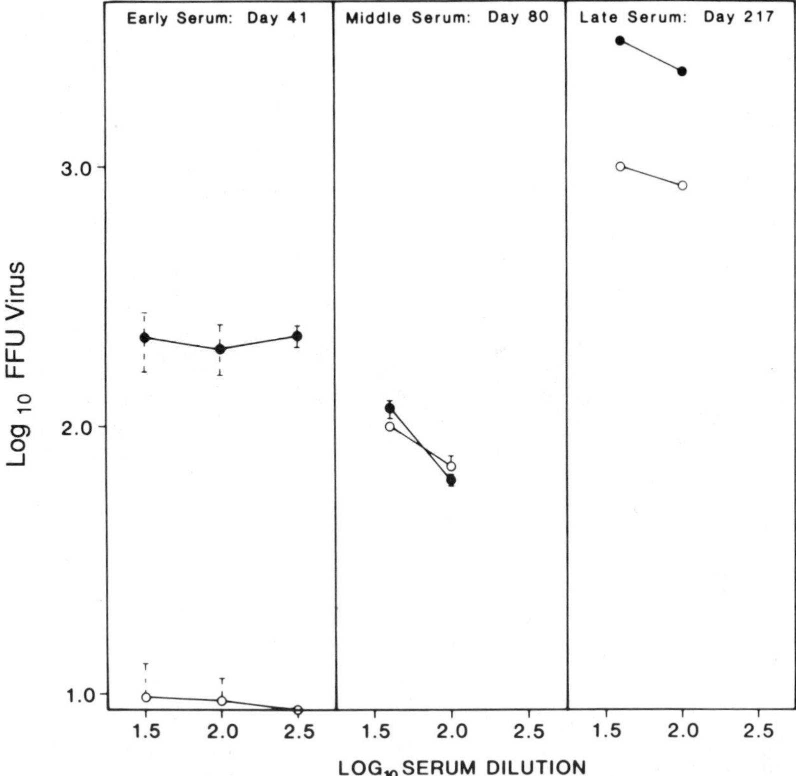

Figure 3. Detection of antigenic variants of EIAV by membrane immunofluorescence on live cell monolayers. ED cells were infected with serial dilution of biologically cloned stocks of MA-1 (closed circles) or MA-4 (open circles). Immune horse sera collected on days 41, 80, and 217 postinoculation were serially diluted and used as the first antibody in a focal immunofluorescence assay as previously described (Carpenter *et al.*, 1987).

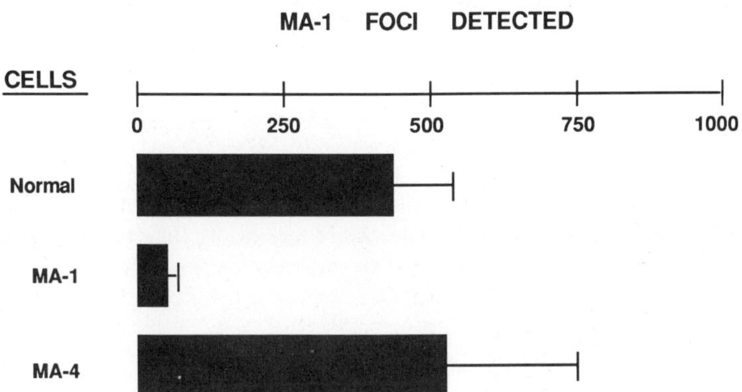

Figure 4. Early immune serum (day 41) was absorbed with normal ED cells or with ED cells chronically infected with either MA-1 or MA-4 virus and used as the first antibody in the FIA on ED cells inoculated with serial dilutions of MA-1 virus. Results are expressed as the number of focus-forming units of MA-1 virus per 0.5 ml of undiluted inoculum ± the standard error of the mean from replicate dishes. Serum absorbed with MA-1-infected cells differed significantly ($p < 0.05$) from serum absorbed with normal cells and from serum absorbed with MA-4-infected cells (Carpenter *et al.*, 1987).

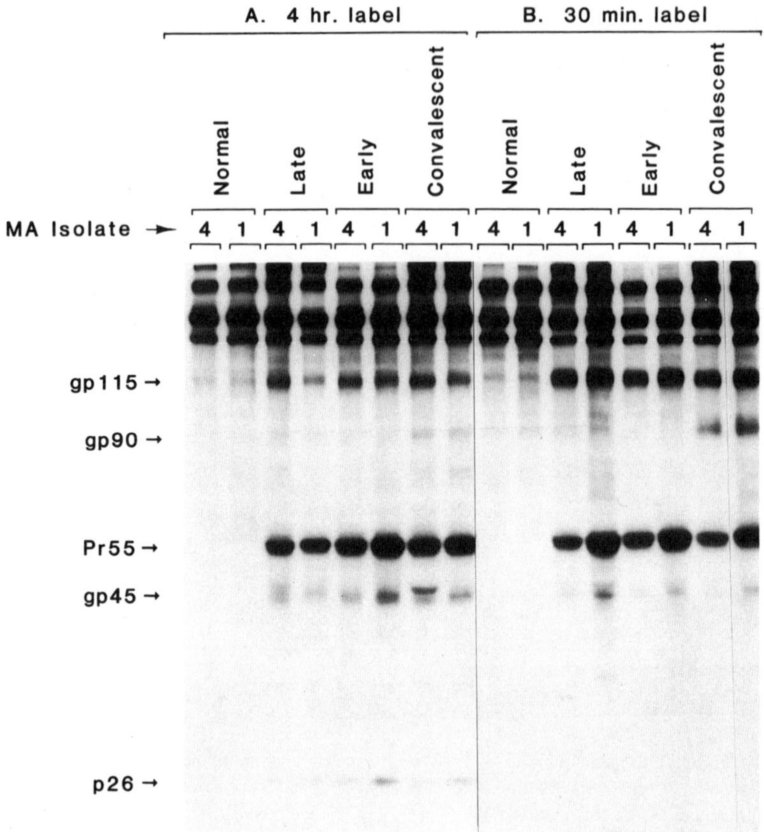

Figure 5. ED cells chronically infected with MA-1 or with MA-4 were labeled with [^{35}S]methionine for 4 hr (A) or 30 min (B), immunoprecipitated, and analyzed as previously described (Carpenter *et al.*, 1987). EIAV-specific proteins include the envelope glycoproteins gp115, gp90, and gp45 and the *gag*-specific proteins Pr55 and p26.

MA-1- and MA-4-infected cells were precipitated with early (MA-1-specific) or late (MA-1- and MA-4-specific) sera and the precipitates were analyzed by SDS–polyacrylamide gel electrophoresis (Fig. 5). Early immune serum was not MA-1-specific by RIP. Rather, cross-reactive epitopes were present on all viral *gag* and *env* proteins detected and, furthermore, antibody to broadly reactive determinants prevented detection of antibody to variant-specific determinants. Similar findings were recently reported for HIV-infected cells where variant-specific epitopes were detectable by membrane immunofluorescence on live cells, but were not distinguishable by RIP analyses (Cloyd and Holt, 1987). Together, these findings further emphasize that the exclusive use of assays that rely on detergent treatment of viral antigens may not allow detection of subtle antigenic changes with potential biologic relevance.

Further structural comparisons between MA-1 and MA-4 were done using RNase T_1-resistant oligonucleotide fingerprint analysis. The two isolates were found to be remarkably similar: of the almost 200 oligonucleotides resolved, five oligonucleotide differences were detected (Fig. 6). To approximate the location of the genomic differences between MA-1 and MA-4, the oligonucleotides were mapped relative to the 3′ end of the viral genome. Interestingly, the structural differences appeared randomly distributed (Fig. 7), suggesting that the genetic variants may have been gener-

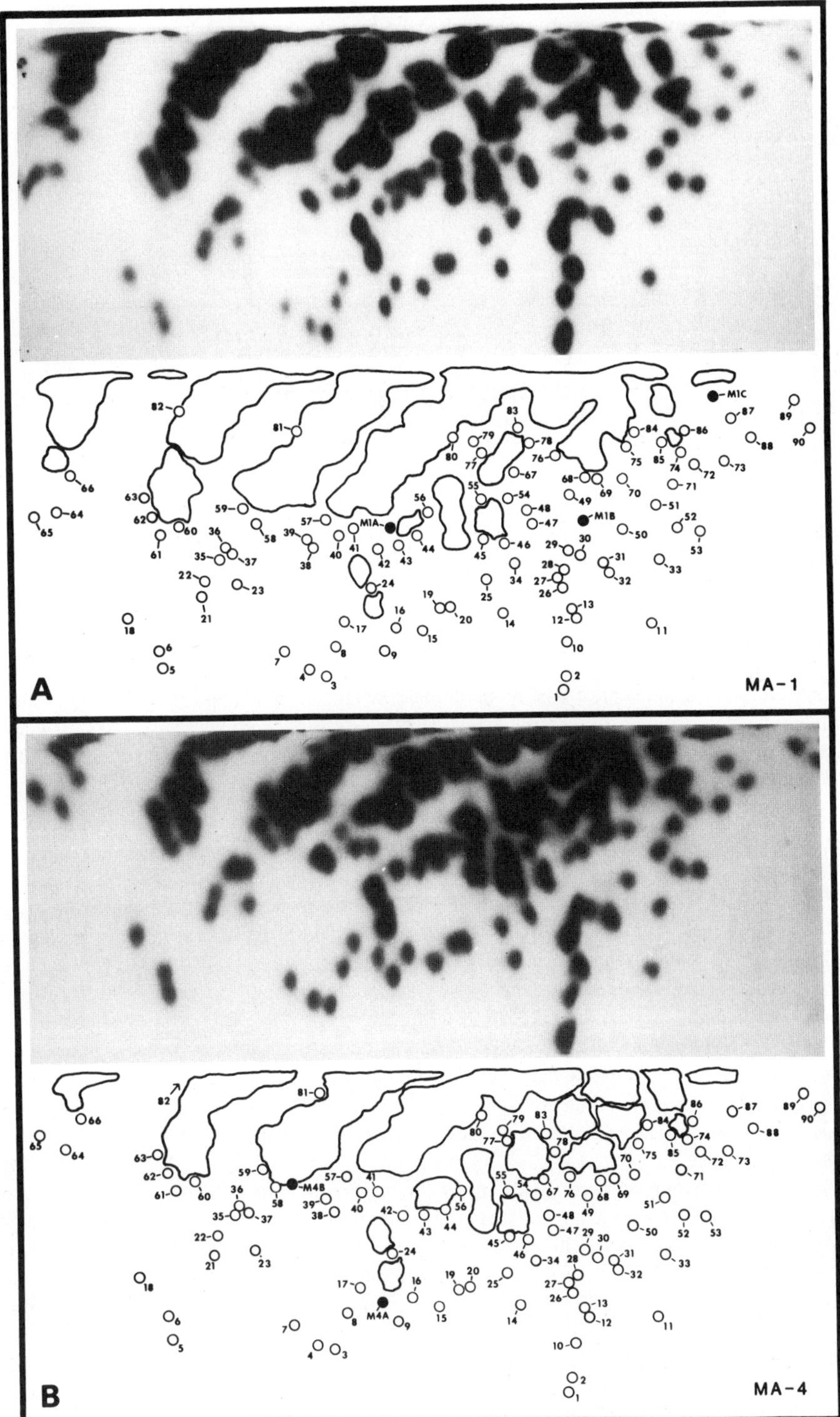

Figure 6. RNase T_1-resistant oligonucleotides of MA-1 (A) and MA-4 (B). Open circles are used to designate oligonucleotides common to both isolates, while closed circles indicate oligonucleotides unique to that particular isolate. The 70S [^{32}P]RNAs of each isolate were digested with RNase T_1 and fingerprinted as previously reported (Evans and Cloyd, 1984; Evans *et al.*, 1979) and oligonucleotides were assigned numbers. Electrophoresis was from left to right, and homochromatography was from top to bottom (Carpenter *et al.*, 1987).

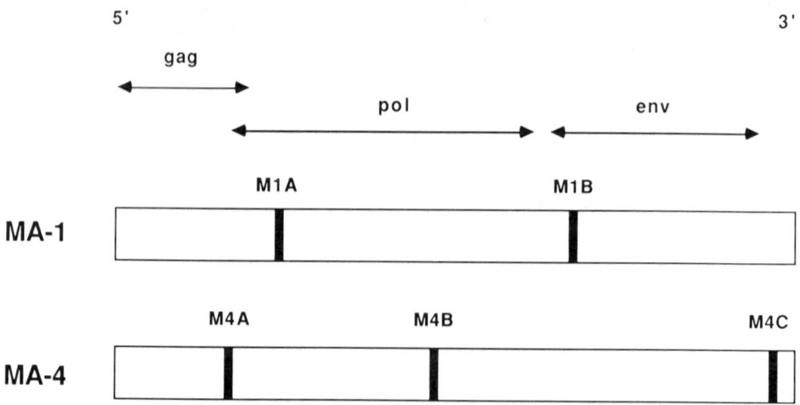

Figure 7. Approximate location of the observed oligonucleotide differences between MA-1 and MA-4. Ordering of the oligonucleotide differences was determined relative to the 3′ polyadenylate terminus of each isolate as previously described (Evans and Cloyd, 1984; Evans *et al.*, 1979). The positions indicated for the oligonucleotides are approximate and may differ (±10%) from their actual locations, particularly in the 5′ one-half of the genomes (Carpenter *et al.*, 1987).

ated and selected independent of host immune pressure. If so, variant-specific antibody might be a secondary event resulting from immune recognition of the predominant virus type selected by nonimmunologic factors. Thus, mechanisms other than immune pressure may potentially contribute to the selection of lentiviral variants.

B. In Vitro Variation of Equine Infectious Anemia Virus

It has recently been demonstrated that heterogeneous populations of HIV can be biologically distinguished by *in vitro* host range (Cheng-Mayer *et al.*, 1988; Folks *et al.*, 1986; Fisher *et al.*, 1988; Sakai *et al.*, 1988). Whether such variations are important *in vivo* is unclear. However, it can be imagined that genetic alterations that confer a broader *in vivo* host range may contribute to changes in viral pathogenesis. In this regard, it is interesting to note that progression of disease in acquired immunodeficiency disease (AIDS) patients could be correlated with recovery of HIV isolates that were more cytopathic and exhibited a broader *in vitro* host range (Cheng-Mayer *et al.*, 1988). To determine whether target cell specificity contributes to variant selection and viral pathogenesis *in vivo,* we initiated studies on the host range of EIAV isolates recovered from experimentally infected horses.

Similar to other lentiviruses, EIAV replicates *in vitro* in cells of the monocyte/macrophage lineage. In fact, horse macrophage cultures (HMC) are the only known *in vitro* culture system that readily supports replication of field strains of EIAV. However, HMC cultures are technically difficult. Consequently, studies in the last 15 years have almost exclusively utilized ED or equine kidney cell cultures together with the Wyoming isolate of EIAV which was adapted to replicate in these cultures (Malmquist *et al.*, 1973). Long-term replication of EIAV-Wyoming in ED cultures was accompanied by an eventual decrease in *in vivo* viral virulence (Orrego *et al.*, 1982; Gutekunst and Becvar, 1979), and studies of EIAV pathogenesis have been limited. Moreover, the almost exclusive use of ED-adapted virus is of concern owing to the increasing evidence that host cell tropism may play a role in lentivirus pathogenesis. HIV isolates associated with neuropathologic lesions in AIDS patients will replicate *in vitro* in human macrophage cultures, but are at least ten-fold less infectious for CD4+ lymphocytes *in vitro* (Cheng-Mayer *et al.*, 1988). To better define the effect of host cell tropism on the virulence and pathogenesis of EIAV, a field isolate of EIAV was studied during *in vitro* adaptation to growth in an ED cell line.

1. *In Vitro* Replication of Field Isolates of EIAV

Field strains of EIAV were isolated from peripheral blood monocytes collected from experimentally inoculated horses during febrile cycles of EIA. HMC were established *in vitro* as previously described (Evans *et al.*, 1984; Kobayashi and Kono, 1967) and were tested for the presence of EIAV by reverse transcriptase (Rushlow *et al.*, 1986; Willey *et al.*, 1988), immunofluorscence (Evans *et al.*, 1984; Crawford *et al.*, 1971), and immunodiffusion (Coggins and Norcross, 1970). EIAV replication was detectable in 5/7 HMC, but in none of ED cell cultures (Table 2). These results agreed with the previous reports of EIAV tropism for macrophages (Kobayashi and Kono, 1967) and further demonstrated the *in vitro* host cell restriction of field isolates.

We selected for an ED-adapted isolate of EIAV by continuous *in vitro* passage of ED cells inoculated with Th-1, one of the field isolates shown to replicate on HMC. ED cells were inoculated with Th-1, the culture was split 1:5 at 3- to 4-day intervals and assayed repeatedly for the presence of EIAV by indirect immunofluorescence on methanol-fixed cells. A single focus of infected cells was detectable following the third *in vitro* passage (Fig. 8). By the sixth *in vitro* passage, the infection had spread throughout the culture and bright cytoplasmic and membrane fluorescence was observed in 50–60% of the cells.

A virus stock, designated MA-1/11, was made from supernatant collected following the fourth *in vitro* passage. This stock was used to biologically clone an ED-adapted isolate using methods similar to those previously described for murine leukemia viruses (Sitbon *et al.*, 1985a). Briefly, the FIA was done on live monolayers of ED cells inoculated with limiting dilution of MA-1/11. Individual foci of infected cells were identified, picked from the monolayer, and expanded *in vitro*. The biologically cloned isolate of EIAV selected for *in vitro* replication in ED cells is referred to as MA-1. MA-1 replicates to a titer of 5–10 $\times$ 10^5 focus-forming units per ml of ED supernatant with no apparent cytopathic effects.

Table 2. In Vitro Replication of Field Isolates of EIAV

Horse	Number of virus isolations[a]	*In vitro* replication	
		HMC[b]	ED[c]
111	2	1	0
112	4	3	0
11	1	1	0

[a]Number of HMC established from whole blood collected during febrile cycles in horses experimentally with field isolates of EIAV.

[b]The presence of EIAV in HMC was determined by reverse transcriptase (Goff *et al.*, 1981; Willey *et al.*, 1988) or EIAV precipitating antigen (Coggins and Norcross, 1970).

[c]ED cultures were assayed for EIAV by membrane immunofluorescence on methanol-fixed cells.

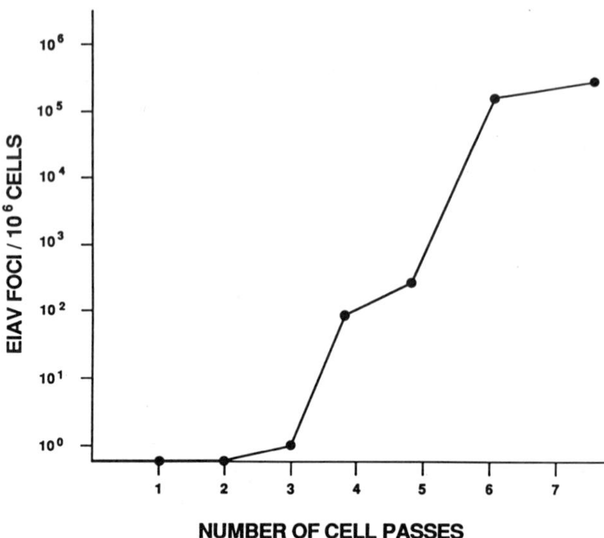

Figure 8. Replication of wild-type EIAV in ED cells. ED cells were inoculated with EIAV-positive HMC supernatant, passaged in duplicate, and at each pass confluent monolayers of ED cells were fixed with methanol and assayed for the presence of EIAV (Carpenter and Chesebro, 1989).

2. *In Vitro Host Range of EIAV Isolates*

The *in vitro* host range of Th-1, MA-1/11, and MA-1 was tested by titration of viral infectivity in both ED and HMC. The presence of Fc receptors on equine macrophages made the use of the FIA difficult; therefore, a reverse transcriptase (RT) assay (Rushlow *et al.*, 1986; Willey *et al.*, 1988) was used to titrate viral infectivity in both ED and HMC (Fig. 9). As expected, Th-1 replicated to relatively high titers in HMC, but was restricted in ED. Viruses present in MA-1/11, the uncloned virus stock, replicated in both ED and HMC, supporting previous studies which demonstrated that EIAV replication in HMC resulted in an expanded host range (Kono and Kobayashi, 1970). Surprisingly, however, MA-1, the biologically cloned, ED-adapted isolate, did not replicate to detectable levels in HMC, despite the high titer of MA-1 for ED. It was possible that the failure to detect MA-1 replication in HMC was due to the sensitivity of the RT assay. Therefore, supernatant from ED and HMC inoculated with serial dilutions of MA-1 were also tested for EIAV by the FIA on ED cells (Fig. 10). MA-1 was infectious for ED cells at dilutions of 10^4 or greater, while no virus was detectable in HMC cultures inoculated with a dilution greater than 10^1. In fact, kinetic studies indicated that the low levels of MA-1 detectable in HMC supernatant at day 5 were due to residual virus not removed by the washing procedures (data not shown). Thus, MA-1 was found to be free of HMC-tropic virus.

Although the MA-1 stock was found to contain a homogeneous virus population, both Th-1 and MA-1/11 appeared to contain mixtures of virus strains. The Th-1 stock was found to be predominately HMC-tropic; however, it also contained at least a small subpopulation of ED-tropic virus which was able to expand during *in vitro* growth on ED cells. The Wyoming isolate of EIAV has been shown to replicate in both ED and HMC (Malmquist *et al.*, 1969), and the dual tropism of the MA-1/11 stock may be due to the predominance of a virus population similar to the Wyoming isolate in that it is capable of *in vitro* replication in both HMC and ED. A mutation in viral gene(s) controlling host cell tropism could potentially lead to a loss of tropism for HMC, resulting in an ED-tropic virus such as MA-1. Alternatively, the MA-1/11 stock could contain mixtures of viruses with

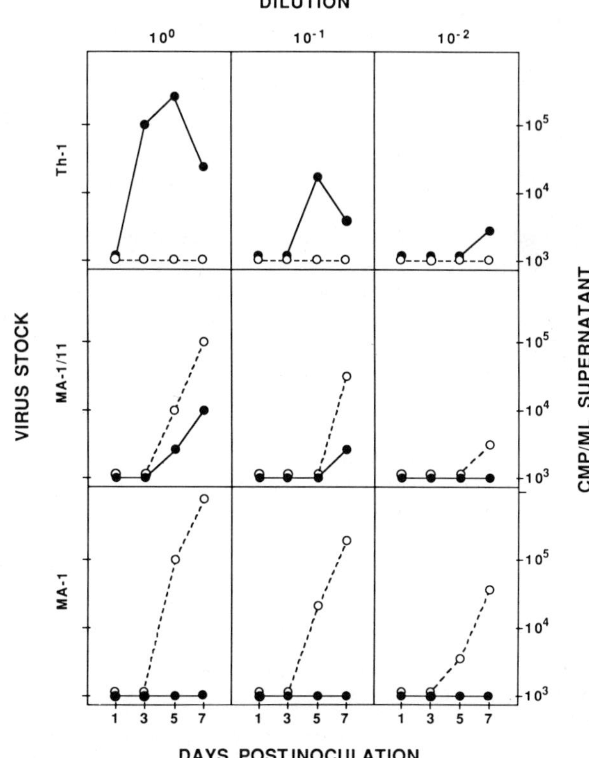

Figure 9. Adaptation of EIAV to *in vitro* growth in ED cells. Tenfold serial dilutions of Th-1, MA-1/11, and MA-1 were inoculated in duplicate onto HMC (closed circles) or ED (open circles). Culture supernatant fluid was collected every 2 days and assayed for the presence of RT. Autoradiographic results were converted to counts per minute (CPM)/ml supernatant by extrapolation from a standard curve established by exposing known CPMs to film under similar experimental conditions (Carpenter and Chesebro, 1989).

varied host cell tropism, including phenotypic mixtures of viruses in which genomes of one virus are encapsulated by the envelope of another virus.

3. *In Vivo* Replication of MA-1

The macrophage is believed to be the *in vivo* target cell for EIAV replication, and it was possible that the loss of tropism for the macrophage *in vitro* might affect viral replication and clinical

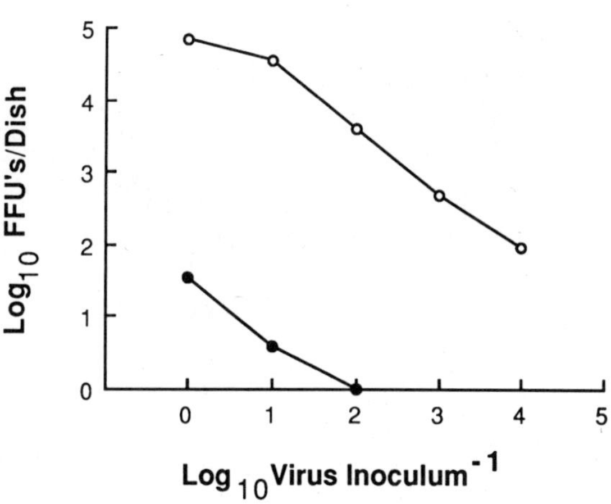

Figure 10. Detection of EIAV in HMC (closed circles) and ED (open circles) inoculated with serial dilutions of MA-1. HMC and ED cells inoculated with serial, tenfold dilutions of MA-1 culture supernatant were collected 5 days postinoculation, clarified, and inoculated onto ED cells. The ED cells were methanol-fixed and assayed for EIAV by FIA as described in the text.

disease *in vivo*. In fact, MA-1 was found to be avirulent *in vivo*. Horses inoculated intravenously either with 10^5 focus-forming units of MA-1 cell-free virus or with 10^7 MA-1-infected ED cells showed no clinical signs of EIA (Fig. 11). However, both horses developed neutralizing antibody to MA-1, indicating that MA-1 did replicate *in vivo*. Repeated attempts to reisolate virus from peripheral blood mononuclear cells were negative, however, indicating that MA-1 was not able to replicate to high titers *in vivo*. Thus, the restriction in *in vitro* replication in HMC may have contributed to the absence of clinical disease *in vivo*.

MA-1 was biologically cloned by picking infected cell foci at near-limiting dilution and should represent the most abundant virus type present after *in vitro* growth on HMC and ED. Our results suggest that even after a very few passages *in vitro*, an avirulent virus population becomes dominant. Studies of lentivirus pathogenesis may require particular attention to the choice of *in vitro* target cells and to the potential for biologic variation following virus replication *in vitro*.

III. CONCLUSIONS

It has long been thought that neutralizing antibody was the primary selective pressure for antigenic variation of EIAV (Kono *et al.*, 1973; Montelaro *et al.*, 1984; Clements *et al.*, 1988; McGuire *et al.*, 1987). Our studies clearly demonstrated, however, that neutralizing antibody does not contribute to variant selection during periods of rapid cycling early after infection. These findings extend previous studies which suggested that recrudescence of clinical EIA following administration of immunosuppressive drugs was due to ablation of cell-mediated, rather than

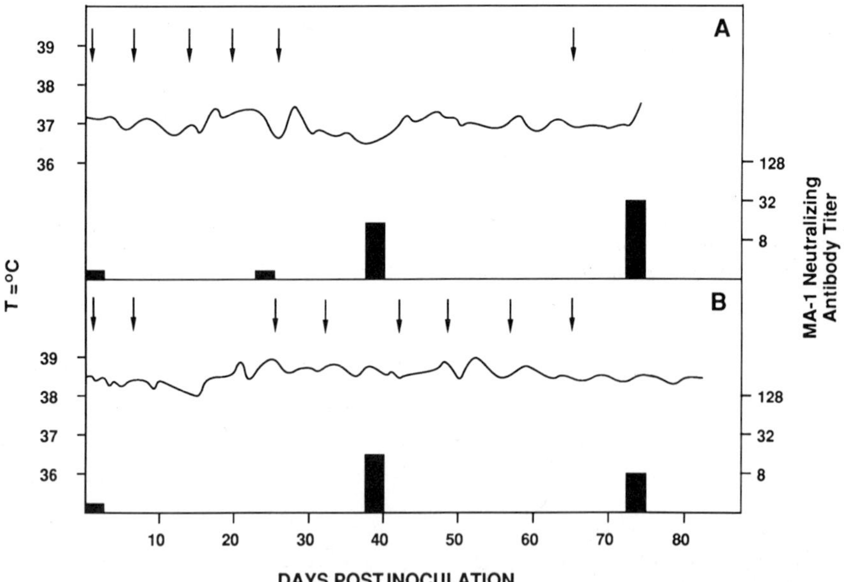

Figure 11. Clinical disease course in two horses inoculated with (A) 10^5 FFUs MA-1 or (B) 10^7 MA-1-infected cells. Arrows indicate days on which whole blood was collected for isolation of virus. A focus reduction assay was used to detect the presence of MA-1 neutralizing antibody as previously described (Carpenter *et al.*, 1987). Neutralization titers are expressed as the reciprocal of the highest serum dilution that neutralized at least 70% of viral infectivity as compared to immune serum and diluent controls (Carpenter and Chesebro, 1989).

antibody-mediated, immunity (Kono, 1972). Moreover, in spite of the genetic and biologic diversity found among isolates of HIV recovered sequentially from individuals (Cheng-Mayer *et al.*, 1988; Hahn *et al.*, 1986), there is no indication, to date, that HIV variants are selected by autologous neutralizing antibody. Although variants of some animal lentiviruses can be antigenically distinguished by monoclonal antibodies (Chesebro and Wehrly, 1988; Hussain *et al.*, 1987) and, in some cases, immune sera (Kono *et al.*, 1973; Montelaro *et al.*, 1984; Narayan *et al.*, 1977), there is no direct evidence that neutralizing antibody selects viral variants *in vivo*.

Immune selection of lentiviral variants could occur through recognition and destruction of variant-infected cells. For both EIAV and HIV, serum from infected individuals recognized variant-specific epitopes on virus-infected cells (Carpenter *et al.*, 1987; Cloyd and Holt, 1987). Moreover, in sequential serum samples from EIAV-infected horses, antibody to virus-infected cells appeared earlier than virus-neutralizing antibody. However, it is not clear whether there was a correlation between the development of an immune response to variant-specific cells and the clinical course of disease. Thus, it remains unclear whether the specificity of the immune response acts *in vivo* to select resistant variants and thereby contributes to viral pathogenesis.

The biologic diversity recently noted among HIV variants (Cheng-Mayer *et al.*, 1988; Fisher *et al.*, 1988; Sakai *et al.*, 1988) raises the possibility that mechanism of *in vivo* selection could also include host cell tropism. Our *in vitro* studies of EIAV isolated from an experimentally infected horse indicated that heterogeneous populations of virus are present during febrile cycles and, moreover, that these populations exhibit diverse *in vitro* host cell tropism. *In vitro* culture on ED cells quickly selected for a minor virus population that was found to be avirulent *in vivo*. These results further support the correlation between host cell tropism and lentiviral pathogenesis.

Only recently have we come to appreciate the diversity among lentivirus variants found to coexist *in vivo*. Such heterogeneous virus populations are likely to contribute to the myriad pathogeneses associated with lentiviruses and, perhaps, confound our efforts at preventive and therapeutic treatments. *In vivo* selection of lentivirus variants likely involves complex interactions among a variety of virus and host factors, including immune recognition, host cell tropism, and pathogenicity. Our understanding of lentiviral pathogenesis depends, in part, on our ability to define these interactions.

ACKNOWLEDGMENTS

The authors acknowledge the assistance of Gary Hettrick and Robert Evans in preparation of the figures and Cathy Steffensen and Karen Hethcote for preparation of the manuscript.

REFERENCES

Buetti, E., and Diggelman, H. (1980). *J. Virol.* **33,** 936–944.

Carpenter, S., Evans, L. H., Sevoian, M., and Chesebro, B. (1987). *J. Virol.* **61,** 3783–3789.

Carpenter, S., and Chesebro, B. (1989). *J. Virol.* **63,** 2492–2496.

Cheng-Mayer, C., Seto, D., Tateno, M., and Levy, J. A. (1988). *Science* **240,** 80–82.

Chesebro, B., and Wehrly, K. (1988). *J. Virol.* **62,** 3779–3788.

Chiu, I. M., Yaniv, A., Dahlberg, J. E., Gazit, A., Skuntz, S. F., Tronick, S. R., and Aaronson, S. A. (1985). *Nature* **317,** 366–368.

Clements, J. E., Gdovin, S. L., Montelaro, R. C., and Narayan, O. (1988). *Annu. Rev. Immunol.* **6,** 139–159.

Cloyd, M. W., and Holt, M. J. (1987). *Virology* **161,** 286–292.

Coffin, J. M. (1986). *Cell* **46,** 1–4.

Coggins, L., and Norcross, N. L. (1970). *Cornell Vet.* **60,** 331–334.

Crawford, T. B., McGuire, T. C., and Henson, J. B. (1971). *Arch. Gesamte Virusforsch.* **34,** 332–339.

Curran, J. W., Jaffe, H. W., Hardy, A. M., Morgan, W. M., Selik, R. M., and Dondero, T. J. (1988). *Science* **239,** 610–616.

Evans, K. S., Carpenter, S. L., and Sevoian, M. (1984). *Am. J. Vet. Res.* **45,** 20–25.

Evans, L. H., and Cloyd, M. W. (1984). *J. Virol.* **49,** 772–781.

Evans, L. H., Duesberg, P. H., Troxler, D. H., and Scolnick, E. M. (1979). *J. Virol.* **31,** 133–146.

Fisher, A. G., Ensoli, B., Looney, D., Rose, A., Gallo, R. C., Saag, M. S., Shaw, G. M., Hahn, B. H., and Wong-Staal, F. (1988). *Nature* **334,** 444–447.

Folks, T., Kelly, J., Benn, S., Dinter, A., Justement, J., Gold, J., Redfield, R., Sell, K. W., and Fauci, A. S. (1986). *J. Immunol.* **136,** 4049–4053.

Gartner, S., Markovits, P., Markovitz, D. M., Kaplan, M. H., Gallo, R. C., and Popovic, M. (1986). *Science* **233,** 215–219.

Gendelman, H. E., Narayan, O., Molineaux, S., Clements, J. E., and Ghotbi, Z. (1985). *Proc. Natl. Acad. Sci. USA* **82,** 7086–7090.

Gendelman, H. E., Narayan, O., Kennedy-Stoskopf, S., Kennedy, P.G.E., Ghotbi, Z., Clements, J. E., Stanley, J., and Pezeshkpour, G. (1986). *J. Virol.* **58,** 67–74.

Goff, S., Traktman, P., and Baltimore, D. (1981). *J. Virol.* **38,** 239–248.

Gonda, M. A., Wong-Staal, F., Gallo, R. C., Clements, J. E., Narayan, O., and Gilden, R. W. (1985). *Science* **227,** 173–177.

Gutekunst, D. E., and Becvar, C. S. (1979). *Am. J. Vet. Res.* **40,** 974–977.

Hahn, B. H., Sahw, G. M., Taylor, M. E., Redfield, R. R., Markham, P. D., Salahuddin, S. Z., Wong-Staal, F., Gallo, R. C., Parks, E. S., and Parks, W. P. (1986). *Science* **232,** 1548–1553.

Hussain, K. A., Issel, C. J., Schnorr, K. L., Rwambo, P. W., and Montelaro, R. C. (1987). *J. Virol.* **61,** 2956–2961.

Kobayashi, K., and Kono, Y. (1967). *Nat. Inst. Anim. Health Q.* **7,** 1–7.

Kono, Y. (1969). *Nat. Inst. Anim. Health Q.* **9,** 1–9.

Kono, Y. (1972). In *Proceedings of the 3rd International Conference on Equine Infectious Diseases* (J. T. Bryans, and H. Gerber, eds.), pp. 175–186, Karger, Basel.

Kono, Y., and Kobayashi, K. (1970). *Nat. Inst. Anim. Health Q.* **10,** 106–112.

Kono, Y., Kobayashi, K., and Fukunaga, Y. (1971). *Arch. Gesamte Virusforsch.* **34,** 202–208.

Kono, Y., Kobayashi, K., and Fukunaga, Y. (1973). *Arch. Gesamte Virusforsch.* **41,** 1–10.

Levy, J. A. (1988). *Nature* **333,** 519–522.

Lutley, R., Petursson, G., Palsson, P. A., Georgsson, G., Klein, J., and Nathanson, N. (1983). *J. Gen. Virol.* **64,** 1433–1440.

Lyerly, H. K., Pratz, J. E., Tyler, D. S., Matthews, T. J., Longlois, A. J., Bolognesi, D. P., and Weinhold, K. J. (1988). In *Vaccines 88 (R. Chanock et al.*, eds)., pp. 323–326, Cold Spring Harbor Laboratory, Cold Spring Harbor, NY.

Malmquist, W. A., Burnett, D., and Becvar, C. S. (1973). *Arch. Gesamte Virusforsch.* **42,** 361–370.

Mann, D. L., Murray, C., Goedert, J. J., Blattner, W. A. and Robert-Guroff, M. (1987). *III International Conference on AIDS, June 1–5, Abstracts Volume,* p. 206, Washington, D.C.

McGuire, T. C., O'Rourke, K., and Cheevers, W. P. (1987). *Contrib. Microbiol. Immunol.* **8,** 77–89.

Montelaro, R. C., Parekh, B., Preggo, A., and Issel, C. J. (1984). *J. Biol. Chem.* **259,** 10539–10544.

Narayan, O., Griffin, D. E., and Chase, J. (1977). *Science* **197,** 376–378.

Narayan, O., Clements, J. E., Griffin, D. E., and Wolinsky, J. S. (1981). *Infect. Immun.* **32,** 1045–1050.

Narayan, O., Wolinsky, J. S., Clements, J. E., Strandberg, J. D., Griffin, D. E., and Cork, L. C. (1982). *J. Gen. Virol.* **59,** 345–356.

Narayan, O., Kennedy-Stoskopf, S., Sheffer, D., Griffin, D. E., and Clements, J. E. (1983). *Infect. Immun.* **41,** 67–73.

Orrego, A., Issel, C. J., Montelaro, R. C., and Adams, N. W. (1982). *Am. J. Vet. Res.* **43,** 1556–1560.

Payne, S., Parekh, B., Montelaro, R. C., and Issel, C. J. (1984). *J. Gen. Virol.* **65,** 1395–1399.

Payne, S. L., Fang, F-D., Lui, C.-P., Dhruva, B. R., Rwanbo, P., Issel, C. J., and Montelaro, R. C. (1987). *Virology* **161,** 321–331.

Prince, A. M., Pascual, D., Kosolapov, L. B., Kurokawa, D., Baker, L., and Rubenstein, P. (1987). *J. Infect. Dis.* **156,** 268–272.

Reitz, M. S., Jr., Wilson, C., Naugle, C., Gallo, R. C., and Robert-Guroff, M. (1988). *Cell* **54,** 57–63.

Robert-Guroff, M., Brown, M., and Gallo, R. C. (1985). *Nature* **316,** 72–74.

Rushlow, K., Olsen, K., Stiegler, G., Payne, S. L., Montelaro, R. C., and Issel, C. J. (1986). *Virology* **155,** 309–321.

Sakai, K., Dewhurst, S., Ma, X., and Volsky, D. J. (1988). *J. Virol.* **62,** 4078–4085.

Salinovich, O., Payne, S. L., Montelaro, R. C., Hussain, K. A., Issel, C. J., and Schnorr, K. L. (1986). *J. Virol.* **57,** 71–80.

Sitbon, M., Nishio, J., Wehrly, K., Lodmell, D., and Chesebro, B. (1985a). *Virology* **141,** 110–118.

Sitbon, M., Nishio, J., Wehrly, K., and Chesebro, B. (1985b). *Virology* **140,** 144–151.

Starcich, B. R., Hahn, B. H., Shaw, G. M., McNeely, P. D., Modrow, S., Wolf, H., Parks, E. S., Parks, W. P., Josephs, S. F., Gallo, R. C., and Wong-Staal, F. (1986). *Cell* **45,** 637–648.

Thormar, H., Barshatzky, M. R., Arnesen, K., and Kozlowski, P. B. (1983). *J. Gen. Virol.* **64,** 1427–1432.

Tung, J-S., Yoshiki, T., and Fleissner, E. (1976). *Cell* **9,** 573–578.

Wiley, C. A., Schrier, R. D., Nelson, J. A., Lampert, P. W., and Oldstone, M. B. A. (1986). *Proc. Natl. Acad. Sci. USA* **83,** 7089–7093.

Willey, R. L., Smith, D. H., Lasky, L. A., Theodore, T. S., Earl, P. L., Moss, B., Capon, D. J., and Martin, M. A. (1988). *J. Virol* **62,** 139–147.

Wong-Staal, F., Shaw, G. M., Hahn, B. H., Salahuddin, S. Z., Popovic, M., Markham, P., Redfield, R., and Gallo, R. C. (1985). *Science* **229,** 759–762.

II

Genome and Antigenic Variability of Myxoviruses and Paramyxoviruses

Epidemiology and Control

7

Evolutionary Lineages and Molecular Epidemiology of Influenza A, B, and C Viruses

Peter Palese and Makoto Yamashita

I. INTRODUCTION

Influenza viruses are classified into three types—A, B, and C—on the basis of their type-specific nucleoprotein and matrix protein antigens. The proteins of viruses belonging to different types do not readily cross-react in serologic tests. However, many structural and biochemical features are shared by type A, B, and C viruses. All influenza viruses possess a segmented genome consisting of single-stranded RNAs of negative polarity. Influenza A and B viruses have eight RNA segments. In the case of the A viruses, these RNAs code for at least ten proteins: the three polymerase proteins (PB2, PB1, and PA), the hemagglutinin (HA), the nucleoprotein (NP), the neuraminidase (NA), two matrix proteins (M1 and M2), and two nonstructural proteins (NS1 and NS2). Ten proteins have also been described for influenza B viruses; although an M2 protein does not appear to be expressed in B-virus-infected cells, an additional protein (NB) coded for by the neuraminidase gene is found. In contrast to influenza A and B viruses, influenza C viruses contain only seven RNA segments encoding eight known proteins, and the C viruses lack the gene coding for the neuraminidase.

Only influenza A viruses are further grouped into antigenic subtypes according to their specific HA and NA surface proteins. To date, 13 distinct HA subtypes and nine NA subtypes have been described; these subtypes are recognized in the nomenclature system for influenza A viruses recommended by the World Health Organization. All three influenza viruses have hemagglutinin spikes on their virions, but no further subgroups or subtypes are identified for influenza B or influenza C viruses (for reviews, see Lamb, 1983; Krug, 1983).

Influenza A viruses have been isolated from humans and a wide variety of other mammals, including horses, pigs, and seals, as well as from avian species. For influenza B viruses the human appears to be the only known reservoir and, with a few exceptions, influenza C viruses are also only found in the human population in nature. For a summary of some of the differences among influenza A, B, and C viruses, see Table 1.

Peter Palese • Department of Microbiology, Mount Sinai School of Medicine, New York, New York 10029. *Makoto Yamashita* • Bioscience Research Laboratories, Sankyo Co. Ltd., Shinagawa-Ku, Tokyo, Japan.

Table 1. Differences among Influenza A, B and C Viruses

	A	B	C
Severity of disease	+++	++	+
Frequency of isolation	+++	++	+
Human reservoir	Yes	Yes	Yes
Animal reservoir	Yes	No	No
Appearance of subtypes	Yes	No	No
RNA segments	8	8	7
Variation	+++	++	+

II. EVOLUTIONARY RELATIONSHIP OF INFLUENZA VIRUS GENES

A. The P Genes

The two longest RNA segments of influenza A, B, and C viruses possess single long open reading frames which encode the PB2 and PB1 proteins of the polymerase complex. Amino acid sequence analysis suggests both a functional and an evolutionary relationship among the PB1 proteins and among the PB2 proteins of the A, B, and C viruses. As a matter of fact, the PB1 proteins share the greatest level of sequence identity found among all the proteins of these three virus types (Table 2). The PB1 proteins contain the asp-asp polymerase sequence motif, which is found in many RNA polymerases and even in DNA polymerases, and they also possess a core of conserved amino acids (40%). RNA segment 3 of influenza A, B, and C viruses encodes a protein that is also part of the viral polymerase complex. The polymerase proteins coded for by segment 3 of influenza A and B viruses are called PA (acid polymerase) due to the acid charge properties of the proteins at neutral pH. The deduced protein sequence encoded by RNA segment 3 of influenza C virus, however, has a positive charge at neutral pH. Therefore, we suggested that this protein of influenza C viruses should be called P3 protein rather than PA protein (Yamashita *et al.*, 1989). Despite differences in the intrinsic features of these proteins, the sequence homology found among the PA/P3 proteins of the three virus types again suggests a common function. It should be noted, however, that the precise role of this protein in the polymerase complex is still unknown.

B. Genes Coding for Surface Proteins

RNA segment 4 of all influenza virus types encodes the spike glycoprotein on virion surfaces which is required for the attachment of the virus to host cells. RNA segment 6 of influenza A and B viruses codes for another surface protein, the neuraminidase, which has a receptor-destroying activity. An important characteristic of influenza C viruses is that they do not contain the RNA segment which corresponds to RNA segment 6 of influenza A and B viruses, and therefore the C viruses lack an NA activity. However, the spike protein of influenza C viruses possesses a receptor-destroying/esterase activity (as well as a receptor-binding/hemagglutinin activity). Because of its combined hemagglutinin/esterase activity, we suggested that this protein should be called the HE protein (Vlasak *et al.*, 1987). The HAs of influenza A and B viruses recognize sialic acid on cell surfaces as a receptor and the NAs destroy the HA receptors by removing *O*-ketosidically bound sialic acid. In the case of influenza C viruses, the HE protein binds to 9-0-acetyl sialic acid on cell surfaces and the same protein has the ability, through its esterase activity, to remove the 9-acetyl group from sialic acids, thereby destroying the receptor (Rogers *et al.*, 1986).

Overall sequence identities between the HA proteins of A and B viruses are about 30% (Table

Table 2. Amino Acid Identities among the Proteins of Influenza A, B, and C Viruses

	A[a] vs. B[b] (%)	A vs. C[c] (%)	B vs. C (%)
PB2	39.5	27.7	27.5
PB1	60.6	39.3	42.2
PA/P3	38.0	28.5	30.7
HA1/HE1	25.2	16.2	18.3
HA2/HE2	41.0	23.8	25.5
NA	32.3	—	—
NP	36.3	22.1	21.8
M1	31.8	21.4	22.1
NS1	22.0	15.3	18.4
NS2	29.5	26.7	21.7

[a]A/PR8/34 sequences are used for the comparison (Fields and Winter, 1982; Winter and Fields, 1980, 1981, 1982; Winter *et al.*, 1981; Fields *et al.*, 1981; Baez *et al.*, 1980).

[b]HA, NA, NP, M1, and NS sequences of B/Lee/40 virus and PB2, PB1, and PA sequences of B/Ann Arbor/1/66 virus are used for the comparison (DeBorde *et al.*, 1988; Krystal *et al.*, 1982; Briedis and Tobin, 1984; Shaw *et al.*, 1982; Briedis *et al.*, 1982; Briedis and Lamb, 1982).

[c]PB2, PB1, P3, and M1 sequences of C/JJ/50 virus and HE, NP, and NS sequences of C/Cal/78 virus are used for the comparison (Yamashita *et al.*, 1988a, 1989; Nakada *et al.*, 1984a,b, 1985, 1986). The alignment was done using program THREE v 1.5 written for Macintosh computer (Johnson and Doolittle, 1986).

3). The HA protein of B viruses was examined in relation to the three-dimensional structure of the HA protein of A viruses and it was found that several of the important structural features were conserved: (1) Five of the six disulfide bonds in the A virus HAs are also conserved in the B virus HAs; (2) the sialic acid binding site involving at least 11 amino acid residues in the A virus HA is essentially conserved in the B virus HA protein; (3) the hydrophobic residues that form the stalk of the HA molecule of A viruses are also preserved in the B virus HA protein; and (4) specific amino acids defining loops and helices that can be recognized by antibodies in the A virus HA structure are also found in the HA sequences of the B virus (Krystal *et al.*, 1982).

On the other hand, only 16% and 18% sequence identities were detected between A and C virus HAs and B and C virus HAs, respectively (Table 2). Nevertheless, the following structural motifs are found among all of the A, B, and C viral HA/HE proteins: (1) a cleavage site between the HA1/HE1 and HA2/HE2 chains; (2) a conserved hydrophobic region (fusinogen) at the amino terminal of the small HA2/HE2 subunit; and (3) several conserved cysteine and proline residues distributed over the entirety of the HA/HE molecules (Nakada *et al.*, 1984a).

RNA segment 6 of influenza A and B viruses encodes neuraminidases, which are also virion surface glycoproteins. Sequence identities of A and B virus NAs are relatively high (Table 2). Both A and B type NAs have potential membrane-spanning domains near the amino terminal regions. Also, many of the cysteine residues—as well as the catalytic site—remain conserved in the B virus NAs, suggesting that the crucial structural and functional roles of the neuraminidases are the same in

Table 3. Influenza Viruses RNAs and Proteins

RNA segment	Nucleotide length (base)			Encoded protein			Amino acid length			M.W. (kDa)		
	A[a]	B[b]	C[c]	A	B	C	A	B	C	A	B	C
1	2341	2396	2365	PB2	PB2	PB2	759	770	774	86.0	88.0	87.8
2	2341	2369	2363	PB1	PB1	PB1	757	752	754	86.6	84.3	86.0
3	2233	2308	2183	PA	PA	P3	716	726	709	82.6	83.2	81.9
4	1778	1882	2071	HA1	HA1	HE1	327	346	432	36.7	37.6	48.3
				HA2	HA2	HE2	222	223	209	25.1	24.1	22.4
5	1565	1841	1809	NP	NP	NP	498	560	565	56.2	61.8	63.6
6	1413	1557	—	NA	NA	—	454	466	—	50.1	51.4	—
					NB							
7(6)[d]	1027	1191	1180	M1	M1	M1	252	248	242	27.9	27.6	27.0
				M2	M2 (?)	M2 (?)	97			11.0		
				M3 (?)								
8(7)	890	1096	934	NS1	NS1	MS1	230	281	286	25.9	32.1	32.5
				NS2	NS2	NS2	121	122	121	14.4	14.2	13.8
	13,588	14,640	12,905									

[a]Influenza A/PR/8/34 virus for all RNA segments (for references see Table 2).
[b]Influenza B/Lee/40 virus for RNA segments 4, 5, 6, 7, and 8, and influenza B/Ann Arbor/1/66 virus for RNA segments 1, 2, and 3 (for references see Table 2).
[c]Influenza C/JJ/50 virus for RNA segments 1, 2, 3, and 6, and C/Cal/78 virus for RNA segments 4, 5, and 7 (for references see Table 2).
[d]Parentheses show the number of the RNA segment for influenza C virus.

influenza A and B viruses. An important difference, however, is that the B virus NA-specific mRNA is bicistronic and that an additional protein, the NB protein, is translated from this RNA (Shaw *et al.*, 1983).

C. Genes Coding for the Type-Specific NP Proteins

The fifth largest RNA segments of influenza A, B, and C viruses encode the NP proteins (Table 3). These are major structural proteins that interact with the viral RNA segments to form ribonucleoprotein complexes. Overall, the sequence identity values between A and C and B and C virus NP proteins are low (Table 2). However, several stretches are found that show a great degree of sequence conservation. When all three NP proteins are aligned, conserved regions are clustered in the C terminal third of the molecule and the sequence identity between B and C virus NPs increases to 37% in positions 339–427. Specifically, two motifs are conserved among all NP proteins: One has been identified as the nuclear signal (positions 346–355) and it has been suggested that the other plays a role in RNA/protein or protein/protein interactions (positions 416–427) (Nakada *et al.*, 1984b; Davey *et al.*, 1985).

D. The M Genes

The M proteins of A and B viruses are coded for by RNA segment 7, whereas the M protein of C viruses is encoded by RNA segment 6 (Table 3). It has been shown that the M gene of A viruses codes for at least two matrix proteins, the M1 and the M2 (Table 2, Fig. 1). The latter protein is derived from a spliced mRNA, while the M1 protein is coded for by an mRNA that is colinear with

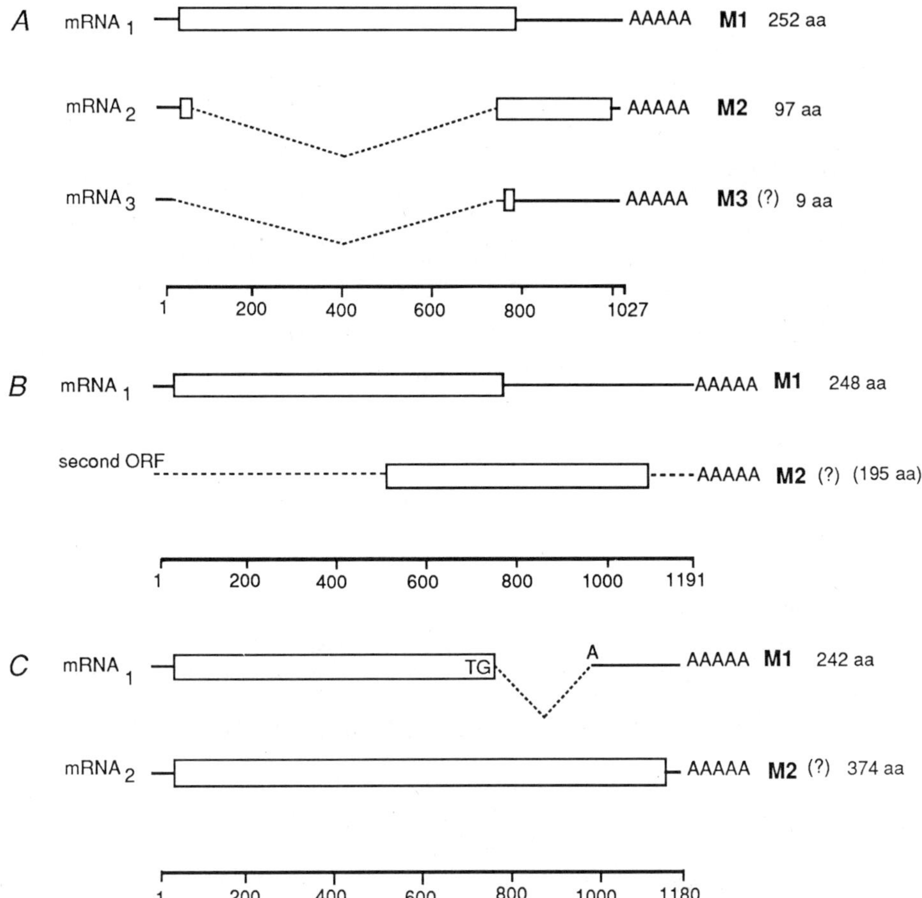

Figure 1. Coding strategies for the M proteins of influenza A, B, and C viruses. Solid lines, open boxes, and V-shaped broken lines indicate untranslated sequences, open reading frames, and intervening (spliced out) sequences, respectively. (Data from Lamb and Lai, 1981, Briedis *et al.*, 1982, and Yamashita *et al.*, 1988a.)

the M gene. In the case of the B and C viruses, only one protein, the M1 protein, has as yet been detected in virus-infected cells. Like that of the A viruses, the M1 protein of influenza B viruses is derived from an unspliced mRNA. In contrast, the coding strategy for the M1 protein of influenza C viruses is different. The latter protein is coded for by a spliced mRNA in which the splicing event results in the formation of a stop codon. The putative M2 protein would share the amino terminal sequence with the M1 protein and then extend to the entire length of the open reading frame in the M gene (Fig. 1; Yamashita *et al.*, 1988a).

Notwithstanding the differences in the coding strategies of their genes, it is interesting that the M1 proteins of all influenza virus types have a similar length of approximately 250 amino acids. However, the degree of sequence identities among the M1 proteins is low (Table 2). The greatest conservation is between A and B type M1 proteins and amounts to approximately 30%. Between A and C type and between B and C type M1 proteins, sequence identity is only about 20% (Table 2). The conserved amino acids are spaced throughout the molecules, but no stretch of more than three amino acids is conserved between A and C or B and C type M1 proteins. Nevertheless, there is a clear evolutionary relationship between the M genes of influenza C virus and those of A and B viruses. For example, eight different regions in the influenza C virus M1 proteins show identities with sequences either in the influenza A or in the influenza B virus M1 proteins. Hydrophobicity patterns of the M1 proteins of type A and B viruses superimpose but that of the C virus does not (Yamashita *et al.*, 1988a).

E. The NS Genes

The NS genes of A, B, and C viruses each code for two proteins—NS1 and NS2 (Table 3). The NS1 proteins are coded for by colinear mRNAs and the NS2 proteins by spliced mRNAs. In the case of the NS2 proteins of influenza A and B viruses, only ten amino acids at the N-terminus are shared with the NS1 proteins and the remainder of the NS2 proteins is derived from different coding frames; in contrast, the N-terminal 62 amino acids of the C virus NS1 and NS2 proteins are shared and only the remaining 59 amino acids of the C-terminus are unique to the NS2 protein of influenza C virus. The sequence identities among A, B, and C virus NS1 proteins are the lowest among all the influenza virus proteins (Table 2); in addition, none of the hydrophobicity patterns of these proteins may be superimposed (Nakada *et al.*, 1986). The NS2 proteins of these viruses show a slightly higher degree of relatedness (Table 2).

F. Summary

Another interesting feature common to all three types of influenza viruses is the presence of 12–15 conserved nucleotides at both the 3′ and the 5′ terminals of the RNA segments (Desselberger *et al.*, 1980). For influenza A viruses, it has also been demonstrated that these conserved ends form a stable panhandle in mature virus particles (Hsu *et al.*, 1987).

All data mentioned above—amino acid sequence homology, structural relationships among proteins, similar coding strategies of proteins, the presence of segmented RNA genomes, and nucleotide sequence similarities of both terminals of the RNA segments—show the close evolutionary relationship among influenza A, B, and C viruses and suggest that they arose from a common ancestor. However, in terms of evolutionary relationship, it appears that influenza C viruses have diverged from the more closely related influenza A and B viruses (Table 4).

III. EVOLUTION OF HUMAN INFLUENZA A VIRUSES

We examined variation in influenza A viruses by comparing the nucleotide sequences of the NS genes of 15 human viruses isolated over a 53-year period (1933–1985). The 15 sequences are easily aligned for analysis because of the size conservation (890 bases) of the NS segments. Substitutions in influenza A virus NS genes are generally retained in strains isolated at later times. This suggests that influenza A viruses emerge through a series of successive variants whereby, in general, only a single dominant one survives in a seasonal period. The sequences of the NS genes were analyzed by the maximum parsimony method to determine the phylogenetic tree of minimum length. The tree is shown in Fig. 2, suggesting a uniform and rapid rate of evolution.

In Fig. 3, the number of nucleotide substitutions between the origin of the tree and the tip of each branch is plotted against the date of isolation of the virus whose NS gene is represented by the tip. The major line, derived by linear regression analysis, shows that these sequences are evolving at the steady rate of $1.94 \pm 0.09 \times 10^{-3}$ substitutions per nucleotide site per year. Figure 3 also shows that the group of H1N1 subtype strains, which reappeared in the human population in 1977 after a 27-year absence, is evolving at the same rate. In summary, it appears that the evolution of influenza A viruses in the human population is unique in terms of its rapidity and uniformity, with the possible exception of the patterns shown by HIV (Starcich *et al,* 1986) and enterovirus 70 (Miyamura *et al.*, 1986).

Table 4. Overall Conservation (%) among the Proteins of the Three Influenza Virus Types

A vs. B	38.76[a]
A vs. C	26.48
B vs. C	27.68

[a]Analysis is based on sequence comparisons listed in Table 1. Between A and B, A and C, and B and C virus proteins, respectively, 1775 amino acids out of a total of 4249 amino acids, and 1192 amino acids out of a total of 4306 amino acids were conserved. For comparisons involving influenza C virus proteins the neuraminidases of either influenza A or influenza B viruses were omitted for analysis.

IV. EVOLUTION OF INFLUENZA B VIRUSES

We determined 12 NS and five HA genes of B viruses and compared them with two NS genes and five NA (HA1 domain) genes that had previously been sequenced. Pairwise comparisons of NS genes as well as of HA genes showed that viruses isolated at different times did not show differences strictly proportional to the time between isolations. A similar pattern was observed in the deduced amino acid sequences of these NS and HA1 genes (Yamashita *et al.*, 1988b). This feature is basically the same as that for influenza C viruses (see Section V). However, some replacements are

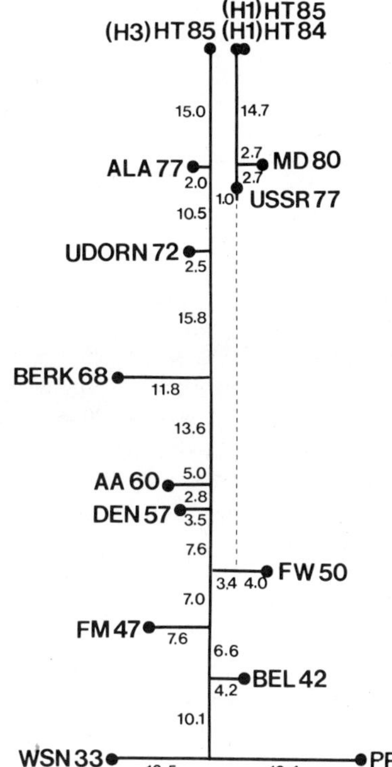

Figure 2. Most parsimonious evolutionary tree for 15 influenza A virus NS genes. The nucleotide sequences (Buonagurio *et al.*, 1986) were analyzed by the method of Fitch (1971). The lengths of the trunk and side branches of the evolutionary tree are proportional to the minimal number of substitutions required to account for the differences in sequence. The broken line represents the predicted number of additional substitutions between the NS genes of FW/50 and USSR/77 based on the calculated evolutionary rate. (From Buonagurio *et al.*, 1986.)

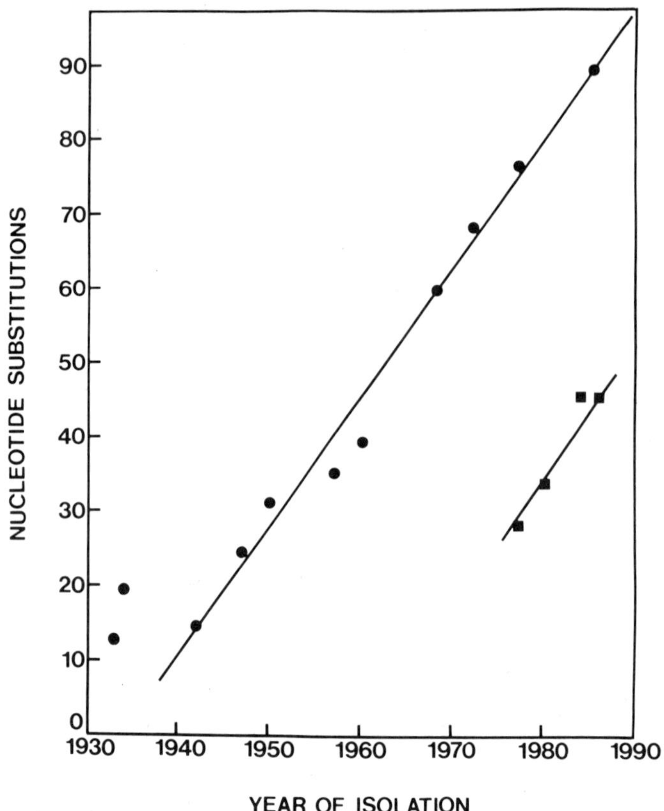

Figure 3. Linearity with time of number of substitutions in the NS genes of influenza A viruses. The abscissa represents the year of isolation of the influenza A virus used in th analysis. The ordinate indicates the number of substitutions observed in their NS genes between the first branching point formed by the WSN/33 and A/PR/34 sequences in Fig. 2 and the tips of all branches of the evolutionary tree. A line, generated by linear regression analysis, is drawn through the points. The slope of the line is 1.7 ± 0.1 substitutions per year. In addition to the sequences found on the trunk of the evolutionary tree (●), the NS genes of the four new H1N1 viruses are also represented in this graph (■). (From Buonagurio *et al.*, 1986.)

conserved in the genes of several strains, suggesting ancestor–descendant relationships. A maximum parsimony analysis of the data revealed a pattern of multiple B virus lineages circulating at any one time and a broad evolutionary tree (Yamashita *et al.*, 1988b).

V. COMPARISON OF EVOLUTIONARY PATTERNS OF INFLUENZA A, B, AND C VIRUSES

Survival of multiple lineages is also a hallmark of influenza C viruses in humans. Variation between isolates obtained in the same year may be as marked as between viruses obtained over a two-decade period. An additional characteristic of influenza C viruses is that viral genes (and most likely the virus as a whole) can survive unchanged for decades. For example, NS genes of C viruses isolated 19 years apart have shown no nucleotide changes (Buonagurio *et al.*, 1986). These findings suggest (1) that influenza C viruses have multiple genetic variants belonging to different evolutionary lineages which coexist in nature, and (2) that variation of influenza C viruses proceeds more slowly than that of influenza B or of influenza A viruses. With respect to the latter viruses, it should also be stressed that the evolutionary rate of approximately 2×10^{-3} changes/site/year (see Section III) is about 10^6-fold faster than that of eukaryotic genes (Li *et al.*, 1985) and that one can follow evolutionary processes in decades rather than in millions of years. A schematic diagram in Fig. 4 reflects these notions about the different epidemiologic patterns of influenza A, B, and C viruses.

VI. MEASUREMENT OF THE MUTATION RATE OF INFLUENZA A VIRUS AND COMPARISON WITH THAT OF OTHER RNA VIRUSES

As discussed in Section V, influenza A, B, and C viruses show both different degrees of variation and different epidemiologic patterns in humans. One of the factors that influence these characteristics is the ability of the virus to undergo mutational changes during a replication cycle.

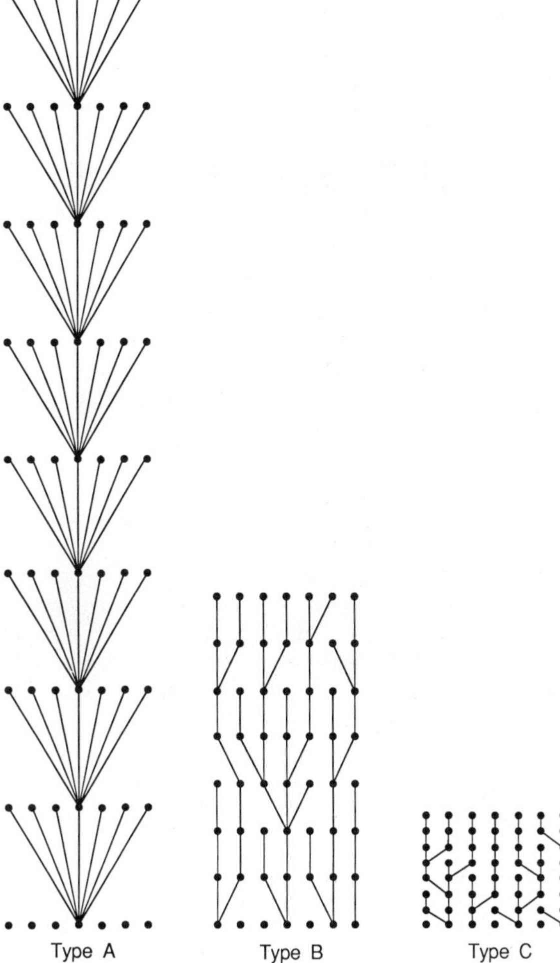

Figure 4. Evolutionary model for the propagation of influenza A, B, and C viruses in humans. Dots lying on a horizontal plane represent influenza virus variants arising in the same season (year). The left part of the diagram shows influenza A virus. The variants arise from only one lineage because of the dominating effects of favorable variants. The middle and right of the diagram show, respectively, the cocirculation of multiple influenza B and C virus lineages. Variation of influenza B viruses appeared to be slower than that of influenza A viruses and faster than that of C viruses. This is illustrated by the intermediate length of the evolutionary branches of the influenza B virus patterns. (The relative length of these branches is not to scale.) For all viruses, an arbitrary number of seven seasonal cycles is shown. (From Yamashita *et al.*, 1988b.)

Table 5. Comparison of Mutation Rates of RNA Viruses[a]

	Mutations/site per infectious cycle
Retrovirus (Rous sarcoma virus; *gag, pol, env, src* regions)	1.4×10^{-4}
Influenza virus (A/WSN/34 strain; NS gene)	1.5×10^{-5}
Poliovirus Type I (Mahoney; VPI gene)	$<2.1 \times 10^{-6}$

[a]Data from Parvin *et al.* (1986) and Leider *et al.* (1988).

Although we have not yet succeeded in measuring the mutation rates of all three types of influenza viruses, we have obtained results for an influenza A virus. One particle of the A/WSN/34 strain was used to produce a viral plaque; genes of progeny viruses in this plaque were sequenced in order to detect possible mutations. The NS gene was chosen for analysis because of its convenient size of about 900 nucleotides, and more than a hundred randomly selected viral clones were used for sequencing. Among the approximately 100,000 nucleotides sequenced, we identified seven point changes, which led us to a calculated mutation rate of 1.5×10^{-5} mutations/site per infectious cycle for influenza A virus (Parvin *et al.*, 1986). We are now attempting to measure the mutation rates for influenza B and C viruses. An approach identical to that used for influenza A/WSN/34 virus was taken to identify the mutation rate for the VPI gene of poliovirus type I (Parvin *et al.*, 1986). However, in this case, a much lower mutation rate was found: less then 2.1×10^{-6} mutations/site per infectious cycle. In order to determine the mutation rate of yet another RNA virus, Leider *et al.* (1988) employed a similar strategy to that used by Parvin *et al.* (1986) for the analysis of retroviruses. Progeny descended from a single particle were collected after a single replication cycle and the genomes were probed for point mutations in seven different regions using denaturing gel electrophoresis. Using this procedure, the mutation rate was determined to be 1.4×10^{-4} mutations/nucleotide per replication cycle.

In conclusion, the above results indicate that mutation rates for RNA viruses are high relative to those found for eukaryotic cells, but that marked differences exist between viruses. In fact, the mutation rates of retroviruses are faster by at least two orders of magnitude than are those of polioviruses, and influenza A viruses have a moderately rapid mutation rate (Table 5). Further experiments will show whether or not mutation rates are the primary factors in determining the ability of viruses to undergo rapid evolution in nature. Alternatively, other factors, such as the antigenicity of the virus, immune selection, and transmission and frequency of infection, may affect the evolutionary rates of viruses.

REFERENCES

Baez, M., Taussig, R., Zazra, J. J., Young, J. F., Palese, P., Reisfeld, A., and Skalka, A. M. (1980). *Nucl. Acids Res.* **8,** 5845–5858.

Briedis, D. J., and Lamb, R. A. (1982). *J. Virol.* **42,** 186–193.

Briedis, D. J., and Tobin, M. (1984). *Virology* **133,** 448–455.

Briedis, D. J., Lamb, R. A., and Choppin, P. W. (1982). *Virology* **116,** 581–588.
Buonagurio, D. A., Nakada, S., Fitch, W. M., and Palese, P. (1986). *Virology* **153,** 12–21.
Davey, J., Dimmock, N. J., and Colman, A. (1958). *Cell* **40,** 667–675.
Deborde, D. C., Donabedian, A. M., Herlocher, M. L., Naeve, C. W., and Maassab, H. F. (1988). *Virology* **163,** 429–443.
Desselberger, U., Racaniello, V. R., Zazra, J. J., and Palese, P. (1980). *Gene* **8,** 315–328.
Fields, S., and Winter, G. (1982). *Cell* **28,** 303–313.
Fields, S., Winter, G., and Brownlee, G. G. (1981). *Nature* **290,** 213–217.
Fitch, W. M. (1971). *Syst. Zool.* **20,** 406–416.
Hsu, M. T., Parvin, J. D., Gupta, S., Krystal, M., and Palese, P. (1987). *Proc. Natl. Acad. Sci. USA* **84,** 8140–8144.
Johnson, M. S. and Doolittle, R. F. (1986). *J. Mol. Evol.* **23,** 267–278.
Krug, R. M. (1983). In *Genetics of Influenza Viruses* (P. Palese and D. W. Kingsbury, eds.), pp. 70–98, Springer-Verlag, New York.
Krystal, M., Elliott, R. M., Benz, E. W., Young, J. F., and Palese, P. (1982). *Proc. Natl. Acad. Sci. USA* **79,** 4800–4802.
Lamb, R. A. (1983). In *Genetics of Influenza Viruses* (P. Palese and D. W. Kingsbury, eds.), pp. 21–69, Springer-Verlag, New York.
Lamb, R. A., and Lai, C. J. (1981). *Virology* **112,** 746–751.
Leider, J., Palese, P., and Smith, F. I. (1988). *J. Virol.* **62,** 3084–3091.
Li, W. H., Wu, C. I., and Luo, C. C. (1958). *Mol. Biol. Evol.* **2,** 150–174.
Miyamura, K., Tanimura, M., Takeda, N., Kono, R., and Yamazaki, S. (1986). *Arch. Virol.* **89,** 1–14.
Nakada, S., Creager, R. S., Krystal, M., and Palese, P. (1984a). *Virus Res.* **1,** 433–441.
Nakada, S., Creager, R. S., Krystal, M., Aaronson, R. B., and Palese, P. (1984b). *J. Virol.* **50,** 118–124.
Nakada, S., Graves, P. N., Desselberger, V., Creager, R. S., Krystal, M., and Palese, P. (1985). *J. Virol.* **56,** 221–226.
Nakada, S., Graves, P., and Palese, P. (1986). *Virus Res.* **4,** 263–273.
Parvin, J. D., Moscona, A., Pan, W. T., Leider, J. M., and Palese, P. (1986). *J. Virol.* **59,** 377–383.
Rogers, G. N., Herrler, G., Paulson, J. C., and Klenk, H. D. (1986). *J. Biol. Chem.* **261,** 5947–5951.
Shaw, M. W., Lamb, R. A., Erickson, B. N., Briedis, D. J., and Choppin, P. W. (1982). *Proc. Natl. Acad. Sci. USA* **79,** 6817–6826.
Shaw, M. W., Choppin, P. W., and Lamb, R. A. (1983). *Proc. Natl. Acad. Sci. USA* **80,** 4879–4883.
Starcich, B. R., Hahn, B. H., Shaw, G. M., McNeely, P. D., Modrow, S., Wolf, H., Parks, E. S., Parks, W. P., Josephs, S. F., Gallo, R. C., and Wong-Staal, F. (1986). *Cell* **45,** 637–648.
Vlasak, R., Krystal, M., Nacht, M., and Palese, P. (1987). *Virology* **160,** 419–425.
Winter, G., and Fields, S. (1980). *Nucl. Acids Res.* **8,** 1965–1974.
Winter, G., and Fields, S. (1981). *Virology* **114,** 423–428.
Winter, G., and Fields, S. (1982). *Nucl. Acids Res.* **10,** 2135–2143.
Winter, G., Fields, S., and Brownlee, G. G. (1981). *Nature* **292,** 72–75.
Yamashita, M., Krystal, M., and Palese, P. (1988a). *J. Virol.* **62,** 3348–3355.
Yamashita, M., Krystal, M., Fitch, W. M., and Palese, P. (1988b). *Virology* **163,** 112–123.
Yamashita, M., Krystal, M., and Palese, P. (1989). *Virology* **171,** 458–466.

8

Antigenic and Genetic Variation of Influenza A(H1N1) Viruses

Alan P. Kendal, Nancy J. Cox, and Maurice W. Harmon

I. INTRODUCTION

In 1977, type A(H1N1) influenza virus, related to strains that had circulated in 1950, reappeared and spread worldwide as a true pandemic virus. Summary reports to the World Health Organization showed that after detection in northern China in May 1977, the virus appeared in contiguous countries to the north and south in November of that year and spread across the temperate zones of the Northern Hemisphere during the following winter. Simultaneously, the virus appeared in tropical zones and spread early in the Southern Hemisphere winter season to temperate countries in the south Pacific and south Atlantic zones (Fig. 1).

In several countries, the first season of H1N1 virus coincided with ongoing activity of influenza type A(H3N2) virus (Fig. 2). This provided the opportunity for mixed infections to occur with the potential to yield reassortant viruses. Continued cocirculation of H3N2 and H1N1 viruses has occurred up to the present.

Several important antigenic variants have evolved since 1977, and regional epidemics have occurred as a result (Table 1). This includes the A/Chile/83-like variants and the A/Taiwan/86-like variants, derivatives of the latter being active as recently as the 1988–1989 winter in western Canada, Japan, and much of western Europe.

Because molecular biology and monoclonal antibodies were first applied to studying variation of influenza viruses at about the time the H1N1 strains reappeared, this period of history has afforded a unique opportunity to study influenza in a depth not previously possible. Here we review some of the noteworthy findings concerning genomic and antigenic variability of the H1N1 viruses in the past 12 years, expanding on findings from 1977–1980 that were summarized previously (Kendal *et al.*, 1981).

II. GENOME VARIATION

A. Reassortment

Although it has long been believed that new human influenza viruses could arise by genetic reassortment, the first definite proof of this came from analysis of H1N1 strains in 1978. Isolation of

Alan P. Kendal, Nancy J. Cox, and Maurice W. Harmon • Influenza Branch, Division of Viral Diseases, Center for Infectious Diseases, Centers for Disease Control, Atlanta, Georgia 30333.

Figure 1. Pandemic spread of H1N1 virus in 1977. 1, May–October 1977; 2, November, December 1977; 3, January 1978; 4, February 1978; 5, March 1978; 6, April 1978; 7, May 1978; 8, June 1978.

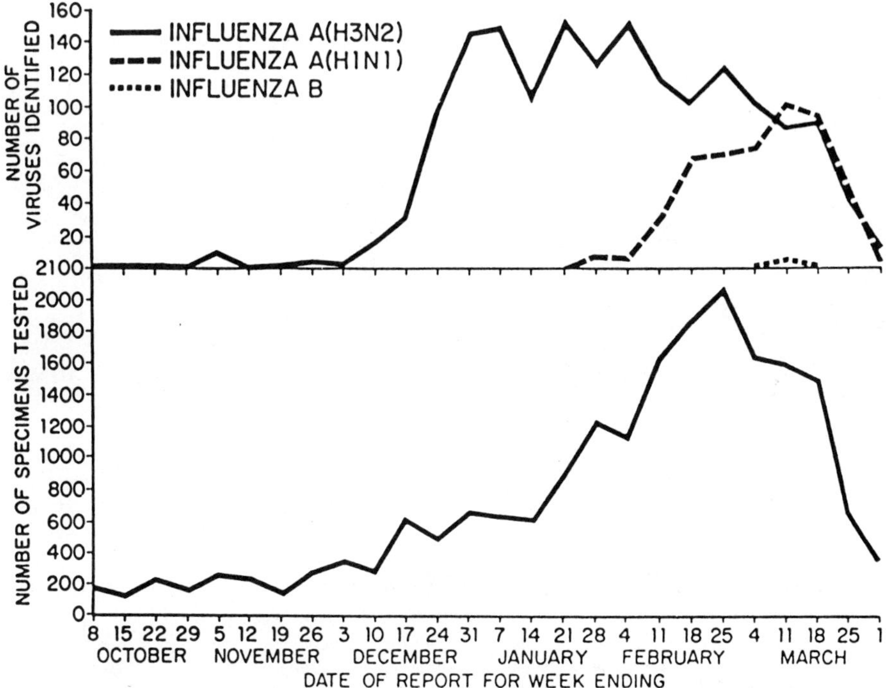

Figure 2. Appearance of influenza A(H1N1) viruses in midst of U.S. epidemic of H3N2 viruses in 1977–1978.

mixed populations of viruses from individual patients was reported from the United States, the United Kingdom, and Japan in early 1978, at a time when outbreaks of H3N2 and H1N1 viruses were occurring simultaneously. Genotyping of the isolates confirmed the presence of reassortants that were deriving different combinations of genes from H1N1 and H3N2 parents (Beare *et al.*, 1980, and Table 2).

The following winter (the second season of circulation of the H1N1 viruses after their reappearance), oligonucleotide mapping and nucleic acid hybridization techniques showed that the

Table 1. Periods of Major Epidemic Activity of Influenza A (H1N1) Viruses in the World since 1977

Reference strains[a]	Major epidemic year(s)	Major outbreak areas[b]
A/USSR/1/77	1977–1979	Global spread from N. China
A/Brazil/11/78[c]		
A/Chile/1/83	1983–1984	Many Pacific Basin countries, including United States, also parts of Europe
A/Victoria/7/83		
A/Taiwan/1/86	1986–1987	S.E. Asia, Japan, United States, Europe
A/Singapore/6/86	1988–1989	Australia, Fiji, Japan, W. Canada

[a]Pairs indicate reference strains with highly related hemagglutin genes.
[b]Partial listing.
[c]In some countries, isolates were predominantly reassortant viruses (see text).

Table 2. Derivation of Genes in Recombinant Viruses Recovered from Mixtures of Wild-Type Influenza A(H3N2) and A(H1N1) Strains

	Derivation in recombinant virus				
	A/Wyoming/1/78		A/Lakenheath/387/78[a]		
Gene	Clone 1 (H3N1)[b]	Clone 3 (H3N2)	Clone 4 (H3N1)	Clone 2 (H3N2)	A/Miyagi/7/78 clone 3 (H3N1)[c]
RNA 1	H1N1	H3N2	H1N1	H3N2	—
RNA 2	H3N2	H3N2	H3N2	H3N2	H1N1
RNA 3	H1N1	H1N1	H1N1	M	H1N1
HA	H3N2	H3N2	H3N2	H3N2	H3N2
Neuraminidase	H1N1	H3N2	H1N1	H3N2	H1N1
Nucleoprotein	H3N2	H3N2	H3N2	H3N2	H3N2
Matrix protein	H1N1	H1N1	H1N1	H1N1	H1N1
Nonstructural protein	H1N1	H1N1	H1N1	M	H1N1

[a]H1N1, Derived from influenza A (H1N1) virus; H3N2, derived from influenza A (H3N2) virus; M, possible mixture of genes from influenza A (H1N1) and A (H3N2) viruses; —, results could not be interpreted. Results were obtained initially by acrylamide gel electrophoresis of virion RNAs under several conditions, in which different recombinant clones from the same mixed infection were compared with each other and with wild-type H3N2 and H1N1 viruses. Confirmation was obtained by RNA–RNA hybridization. Functional assignments for RNAs 1, 2, and 3 have not been made for this, and the other, viruses.

[b]Determined by electrophoresis of virion RNAs through a urea-containing gel at 30°C.

[c]Determined by electrophoresis of virion RNAs through a urea-containing gel at 27°C.

majority of H1N1 viruses circulating in the United States were reassortant viruses. Most possessed four RNA segments derived from H3N2 parents (Young and Palese, 1979; Bean *et al.*, 1980; Nakajima *et al.*, 1981). An additional genotype that contained a further H3N2 gene was later identified. However, a prospective study of the genomes of future H1N1 isolates from countries around the world showed that the nonreassortant H1N1 viruses continued to circulate, and by about 1981 they became the only genotype recognized (Cox *et al.*, 1983).

B. "Silent" Evolutionary Pathways

Subsequent RNA oligonucleotide mapping of H1N1 viruses revealed a further large change in the pattern of virus isolates, coincident with the appearance of antigenic variants in 1983–1984. These viruses (reference strains A/Chile/1/83 and A/Victoria/7/83) caused waves of epidemic activity in the South Pacific and shortly thereafter in North America and Europe. Viruses from diverse areas of the world had similar oligonucleotide patterns, which diverged by a number of oligonucleotides considerably greater than would have been expected had the variants evolved directly from strains circulating the preceding season (Fig. 3). However, the mapping of individual gene segments and the partial sequencing of individual genes confirmed that all RNA segments of the A/Chile/1/83-like viruses were true H1N1 strains, with no genes derived from contemporary H3N2 strains (Cox *et al.*, 1989). The most probable explanation is that the A/Chile/1/83-like strains had been evolving over a 3- to 4-year period along a "silent" pathway. This pathway was not detected from the limited number of samples tested until the accumulated mutations resulted in a virus with antigenic properties and a genome constellation optimum for epidemic spread.

The possibility for such silent evolutionary paths was confirmed from studies of isolates from

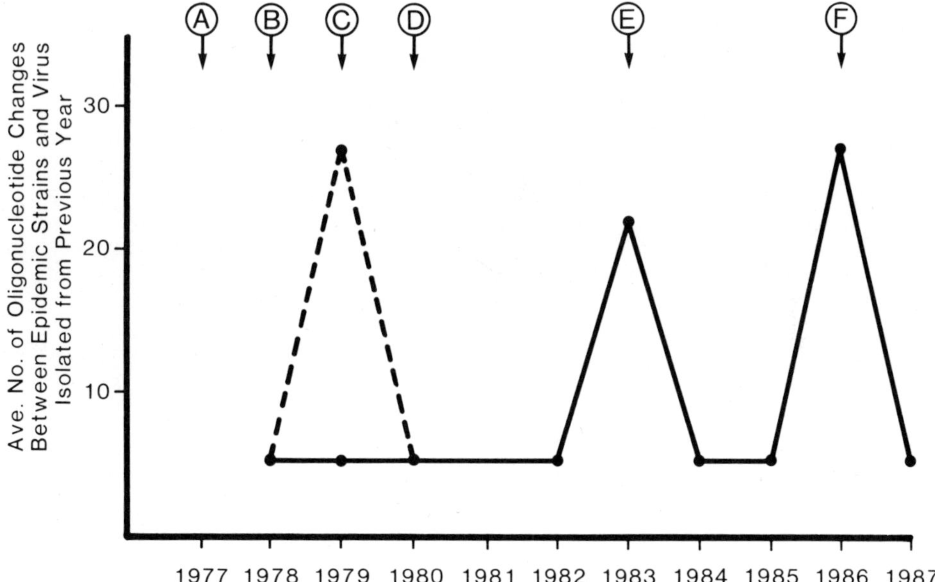

Figure 3. Average number of year-to-year changes in oligonucleotide patterns of influenza A(H1N1) epidemic viruses, 1978–1987. Reference strains for the years indicated are (A) A/USSR/90/77; (B) A/Brazil/11/78; (C) A/California/10/78; (D) A/England/333/80; (E) A/Chile/1/83; (F) A/Taiwan/1/86. Solid line indicates nonreassortant genome; dashed line indicates reassortant genome with four or five genes derived from H3N2 viruses.

1986–1987. At that time, major outbreaks occurred in Malaysia, Singapore, and Taiwan, and the reference strains A/Singapore/6/86 and A/Taiwan/1/86 exhibited marked antigenic variation from all previously identified epidemic strains (Table 3). Oligonucleotide maps of isolates from different outbreaks in different regions of the world showed a high degree of similarity, yet the viruses differed dramatically from the immediately preceding A/Chile/1/83-like strains (Fig. 3). Careful analysis of oligonucleotide patterns of viruses from the period 1980–1986, however, revealed that the strains from 1986 were quite closely related to strains from the pre-A/Chile/1/83 virus era (Table 4). Thus, epidemic strains of influenza A(H1N1) virus that appeared in 1986 were shown by analysis of their total genome composition to have been derived from strains prevalent 3 or more years earlier rather than from immediately preceding epidemic viruses.

C. Genome Diversity

The preceding analysis of the genome of H1N1 viruses in the period 1977–1987 in general showed a high degree of homogeneity in the isolates of a particular type (reassortant or nonreassortant) from different regions of the world when major outbreaks occurred. However, Brown (1988) suggested that in 1986–1987 a second type of genome pattern was present among A/Taiwan/1/86-like viruses isolated in western Canada. Ongoing analysis of isolates from the wave of epidemic activity in several countries in 1988–1989 supports the conclusion that isolates from contemporary epidemics of H1N1 virus may be similarly heterogeneous in their oligonucleotide patterns. As shown in Fig. 4, considerable genomic diversity exists between antigenically related virus isolates from the Americas, Asia, and Europe. The basis for this, and its relation to the observation by Brown, are presently being investigated.

Table 3. Hemagglutination–Inhibition Titers of Egg-Adapted Viruses Derived from an A(H1N1) Virus Isolated in MDCK Cells

	Ferret antisera							
Reference virus	A/USSR 90/77	A/Brazil 11/78	A/England 333/80	A/Chile 1/83	A/Victoria 7/83	A/Taiwan 1/86	Plaque 32a	Plaque 56c
A/USSR/90/77	*320*	80	160	20	<	10	40	40
A/Brazil/11/78	80	*160*	160	20	10	10	40	40
A/England/333/80	640	160	*320*	40	10	10	80	80
A/Chile/1/83	<	40	80	*80*	<	<	80	80
A/Victoria/7/83	160	80	160	80	*640*	40	160	160
A/Taiwan/1/86	<	20	20	10	<	*640*	<	<
MDCK parent virus[a]								
A/Chr/157/83	40	40	80	80	40	40	160	160
Egg-adapted MDCK plaques (group)[b]								
32a (1)	80	80	160	80	20	40	*160*	160
56c (2)	<	<	40	<	<	<	20	*80*
5a (3)	<	<	40	20	<	<	20	80

[a]Type A(H1N1) influenza virus from an outbreak at Christ's Hospital School was isolated directly from a clinical specimen in MDCK cells.

[b]Virus derived from individual plaques of parent virus on MDCK cells and adapted to growth in eggs. Strains 32a and 5a were passed once in the allantoic cavity; strain 56c required two passes in the amniotic cavity followed by one pass in the allantoic cavity. Group designation is based on monoclonal antibody reactivity according to Robertson *et al.* (1987).

Table 4. Oligonucleotide Differences between Pairs of H1N1 Influenza Viruses, 1978–1986

H1N1 virus genomes (epidemic period)	Interval of isolation (months)	Average no. of spot differences (range)	No. of pairs examined
H1N1 nonrecombinant	<12	5 (2–8)	17
vs. nonrecombinant (1978–1982)	12–23	8 (4–10)	17
	24–35	11 (10–13)	12
H1N1 recombinant	<12	4 (2–7)	9
vs. recombinant (1978–1981)	12–23	9 (7–10)	4
	24–35	11 (10–12)	4
H1N1 nonrecombinant	<12	27 (26–28)	3
vs. recombinant (1978–1980)	12–23	31 (26–37)	5
H1N1 nonrecombinant	<12	22 (20–23)	5
vs. 1983 variants (1980–1983)	12–23	22 (18–28)	6
H1N1 1983 variants	<12	6 (3–9)	6
vs. H1N1 1983 variants (1983–1984)			
H1N1 1986 variants	<12	27 (25–30)	3
vs. 1983-like variants (1983–1986)	12–23	27 (27)	2
	24–35	28 (27–30)	4
H1N1 1986 variants	36–47	14 (13–15)	2
vs. 1980–1982 nonrecombinants	48–59	15 (12–19)	5
	60–71	21 (17–25)	3
H1N1 1986 variants	<12	4 (3–6)	3
vs. H1N1 1986 variants			

III. THE HEMAGGLUTININ

A. Antigenic Heterogeneity

Although influenza epidemics are customarily ascribed to a so-called prototype strain as a point of antigenic reference, influenza isolates from outbreaks or epidemics usually exhibit minor heterogeneity. Heterogeneity can often be detected using animal antisera, and the sensitivity of monoclonal antibodies to a single amino acid change at one epitope increases the likelihood of detecting microvariant strains of virus. Different relationships may be observed between minor variants from the same epidemic when they are compared with polyclonal animal sera or monoclonal antibodies (Kendal *et al.*, 1981).

Attention has recently been directed at the likelihood that viruses isolated in different laboratory host systems may exhibit antigenic variation, which is a marker for amino acid changes selected under the influence of functional, rather than immunologic, pressure. Thus, it has been proposed that certain sites in the H1 hemagglutinin protein determine the host range of the virus, with different optimum sequences for the different species (Robertson *et al.*, 1987). This is analogous to the situation proposed for type B influenza or type A(H3N2) influenza (Robertson *et al.*, 1985; Katz and Webster, 1988), although different amino acid sites may be involved in different hemagglutinin types or subtypes.

These studies raise the question whether the antigenic heterogeneity among egg isolates is an artifact attributable to host selection, with antigenic variation occurring coincidentally. One way to examine this is by detailed comparisons of the antigenic relationship of "host-dependent variants" derived from single isolates, using the same polyclonal sera used to categorize the large numbers of field isolates sent in for reference analysis without regard to their passage histories.

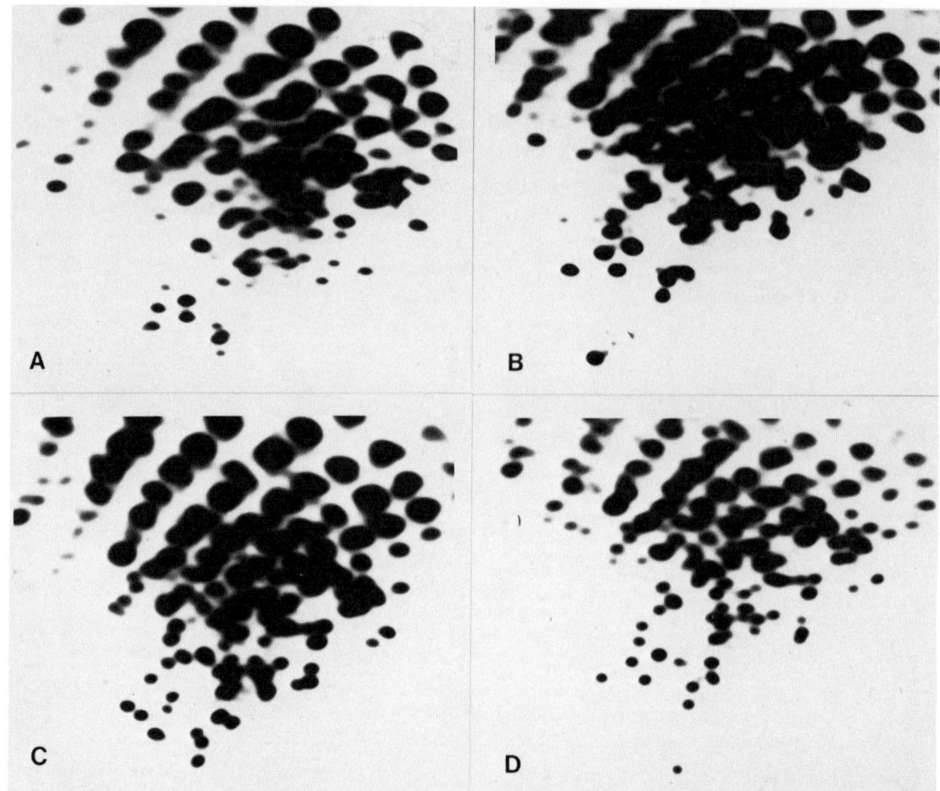

Figure 4. T_1 oligonucleotide maps of four 1988 influenza A (H1N1) viruses: (A) A/Sichuan/4/88; (B) A/Victoria/43/88; (C) A/South Carolina/6/88; (D) A/Czechoslovakia/1/88. Total viral RNA was extracted and digested with T_1 ribonuclease. The resulting oligonucleotides were labeled at the 5′ end with [$X^{32}P$]. Adenosine triphosphate using polynucleotide kinasese and separated by two-dimensional polyacrylamide gel electrophonesis.

Results shown in Table 4 were obtained with representative "clones" prepared in eggs by repassage of a tissue culture isolate from a single patient (Oxford *et al.*, 1987). The egg clones were shown by Oxford *et al.* to cluster in subgroups based on reactions with monoclonal antibodies. Our results, however, indicate that neither the egg clones nor the tissue culture isolate from the patient particularly resembles reference strains that are highly representative of the majority of field isolates in tests with polyclonal animal sera. Thus, whereas natural isolates, grown in eggs, tend to cluster in discrete patterns when analyzed with polyclonal sera, the laboratory-selected egg clones do not conform to these reaction patterns. We therefore believe that the use of egg clones from a single human isolate is not a reliable way to identify the type of variation seen among natural isolates. Although there is little doubt that host-related variation does occur in influenza isolates, alternative methods must be used to determine its significance relative to host-independent variation. As described in the following section, the sequencing of multiple isolates of field strains seems to be a successful way to approach this.

B. Sequence Analysis

The approach we have adopted for studying the evolution of the H1 hemagglutinin gene is identical to that we first developed for the H3 hemagglutinin, i.e., to sequence multiple isolates

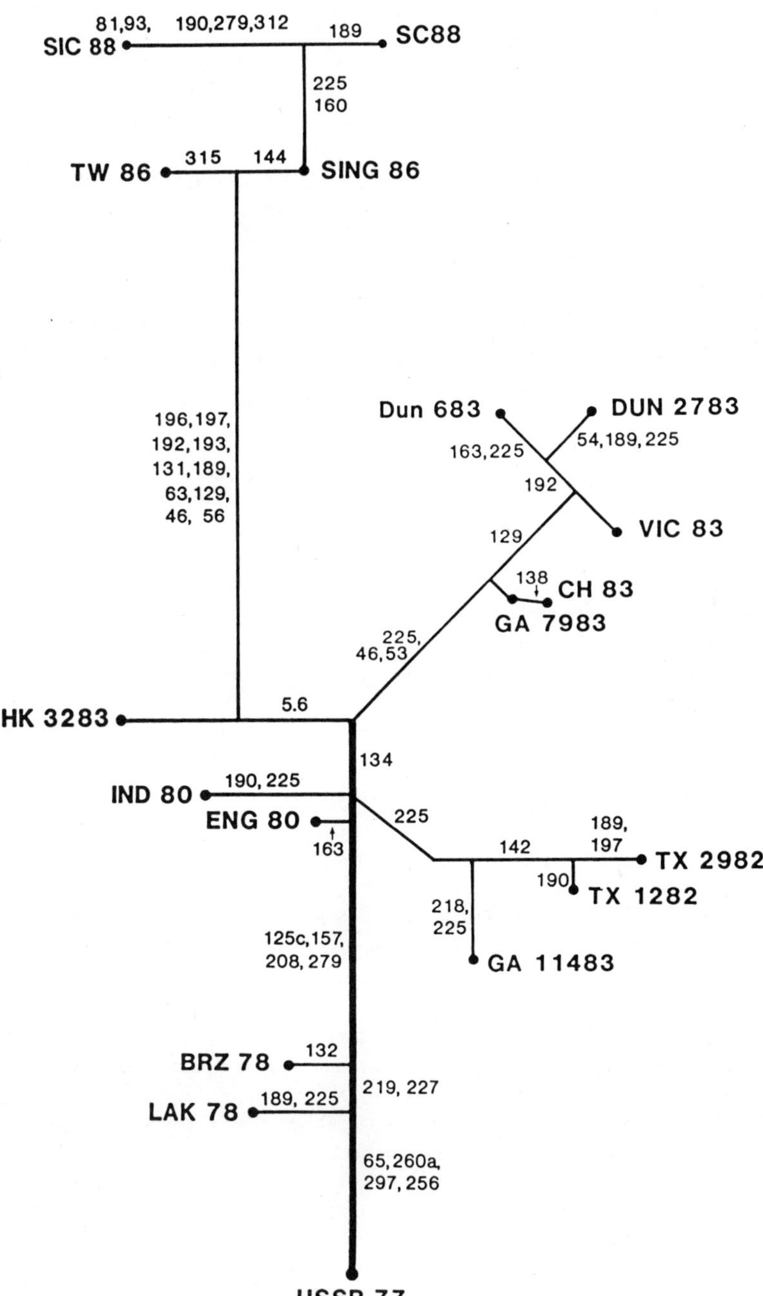

Figure 5. Evolutionary pathway of the influenza A H1 hemagglutinin (HA1 region) deduced for influenza isolates from 1977 to 1988. Distances between strains reflect the total number of oligonucleotide changes between virus sequences. Numbers refer to amino acid changes that are either conserved (shown on main branch) or strain-specific (shown on side branches). The letters are abbreviations for the place of virus isolation, and the last two digits indicate the year of isolation.

representing major and minor variants from each epidemic period. This permits us to identify mainstream mutations, which are conserved from year to year. Such mutations can be distinguished from isolate-specific sidestream mutations that are seen in individual isolates but not in other isolates from the same or later epidemic periods. Isolate-specific mutations may represent the effect of the host used for virus propagation, as well as the normal level of naturally occurring heterogeneity.

When this approach is applied to isolates from 1977 to the present, an evolutionary tree can be drawn. The epidemics in 1983–1984 (reference strain A/Chile/1/83) and 1986–1987 (reference strain A/Taiwan/1/86) were caused by viruses that seem to have evolved along separate branches (Cox *et al.*, 1989, and Fig. 5). This finding with the hemagglutinin gene is consistent with the results of total genome analysis by RNA fingerprinting described earlier. Despite the genomic heterogeneity described for viruses isolated during the H1N1 epidemics in 1988–1989, sequence analysis of their hemagglutinin genes shows that most of them are closely related and that they exhibit few mainstream mutations from the 1986 strains.

A further example of the intrinsic ability of the hemagglutinin to evolve along separate pathways comes from comparison with evolution in the earlier period of prevalence of H1N1 epidemic viruses. Thus, it is well documented by antigenic and genomic analysis that the H1N1 virus that reappeared in 1977 was virtually identical to reference strains available from about 1950 (Kendal *et al.*, 1978; Scholtissek *et al.*, 1978; Nakajima *et al.*, 1978). This was confirmed at a higher level of sensitivity by sequencing the hemagglutinin, which shows only a few differences between the 1950 and the 1977 reference strains (Raymond *et al.*, 1986).

Despite this remarkable similarity to an earlier virus when the H1N1 virus first reappeared in 1977, subsequent variants have followed a completely distinct pattern of evolution in terms of antigenic specificity and sequence changes from that which occurred from 1950 to 1957 (Raymond *et al.*, 1986).

IV. CONCLUSION

Analysis of the epidemic and laboratory properties of influenza A(H1N1) viruses has revealed the following points.

1. The genetic and antigenic similarity of type A(H1N1) virus isolates around the world after the reintroduction of this virus subtype into the human population in 1977 illustrates dramatically the ability of influenza virus to spread almost unchanged, from a single point throughout the entire human population.

2. Other instances of genetic uniformity of influenza A(H1N1) viruses isolated worldwide in subsequent epidemics support the contention that in many cases a point source exists for a variant that spreads worldwide.

3. Reassortment between influenza A viruses of different subtypes was conclusively demonstrated in 1978–1979, and for a short period such reassortant viruses were predominant in some regions. However, after 1981, reassortant viruses were not detected, indicating that genetic reassortment *per se* is not necessarily advantageous for long-term survival of the virus.

4. The influenza hemagglutinin can evolve along more than one pathway.

5. Silent pathways of evolution have occurred in the hemagglutinin and other genes, so that in some years epidemic viruses have been shown to be derived not only from the preceding epidemic strain, but from earlier viruses. Thus, at any given time, despite the great predominance of one virus genotype, low-level transmission of a different genotype may be occurring and may eventually become dominant.

6. As the epidemic era of H1N1 viruses continues, the complexity of evolutionary pathways may also be increasing. Nevertheless, it is possible that at intervals in the future all branches but one

will become extinct, so that a continuum of evolution will have occurred. Only because of the ongoing molecular monitoring of epidemic strains can we be aware of the existence of alternative evolutionary branches.

REFERENCES

Bean, W. B. Jr., Cox, N. J., and Kendal, A. P. (1980). *Nature* **284,** 638–640.

Beare, A. S., Kendal A. P., Cox, N. J., and Scholtissek, C. (1980). *Infect. Immuni.* **28,** 753–761.

Brown, E. G., (1988). *J. Clin. Microbiol.* **26** 313–318.

Cox, N. J., Bai, Z. S., and Kendal, A. P. (1983). *Bull WHO* **61,** 143–152.

Cox, N. J., Black, R. A., and Kendal, A. P. (1989). *J. Gen. Virol.* **70,** 299–313.

Katz, J. M., and Webster, R. G. (1988). *Virology* **165,** 446–456.

Kendal, A., Cox, N., Nakajima, S., Webster, R., Bean, W., and Beare, P. (1981). In *Genetic Variation among Influenza Viruses* (Nayak, D. P. and Fox, C. F., eds.), pp. 490–504, Academic Press, New York.

Kendal, A. P., Noble, G. R., Skehel, J. J., and Dowdle, W. R. (1978). *Virology* **89,** 632–636.

Nakajima, K., Desselberger, U., and Palese, P. (1978). *Nature* **274,** 334–339.

Nakajima, S., Cox, N. J., and Kendal, A. P. (1981) *Infect. Immuni.* **32,** 287–294.

Oxford, J. S., Corcoran, T., Knott, R., Bates, J., Bartolemei, O., Major, D., Newman, R. W., Yates, P., Robertson, J., Webster, R. G., and Schild, G. C. (1987). *Bull. WHO* **65,** 181–187.

Raymond, F. L., Caton, A. J., Cox, N. J., Kendal, A. P., and Brownlee, G. G. (1986). *Virology* **148,** 275–287.

Robertson, J. S., Bootman, J. S., Newman, R., Oxford, J. S., Daniels, R. S., Webster, R. G., and Schild, G. C. (1987). *Virology* **160,** 31–37.

Robertson, J. S., Naeve, C. W., Webster, R. G., Bootman, J. S., Newman, R., and Schild, G. C. (1985). *Virology* **143,** 166–174.

Scholtissek, C., Van Hoynigen, V., and Rott, R. (1978). *Virology* **86,** 613–617.

Young, J. F., and Palese, P. (1979). *Proc. Natl. Acad. Sci. USA* **76,** 6547–6551.

9

Antigenic Variation among Human Parainfluenza Type 3 Viruses

Comparative and Epidemiologic Aspects

Kathleen van Wyke Coelingh

I. INTRODUCTION

Parainfluenza type 3 (PIV3) was originally isolated from a child with lower respiratory disease (Chanock *et al.*, 1958) and has since been shown to be a significant cause of infantile pneumonia, bronchiolitis, and croup (Chanock *et al.*, 1961). The frequency of hospitalization of infants or children with PIV3 infections has been estimated at 1 in 400 (Murphy *et al.*, 1988). Infection with PIV3 frequently occurs during the first months of life, when infants still possess maternally derived neutralizing antibodies, and children typically experience repeated infections of decreasing severity during childhood. The factors contributing to the epidemiologic picture of partial immunity and reinfection are incompletely understood, but may reflect immunologic immaturity of the host, immunologic suppression by maternally derived antibodies, and production of nonfunctional antibodies (some of which may be obstructive).

PIV3 is a member of the family Paramyxoviridae. Other members of this family that infect humans include parainfluenza type 1 (PIV1), type 2 (PIV2), type 4a (PIV4a), type 4b (PIV4b), mumps, measles, and respiratory syncytial viruses. Paramyxoviruses that infect humans have antigenically related nucleoproteins (NP), but their matrix (M), hemagglutinin–neuraminidase (HN), and fusion (F) proteins have very little antigenic similarity (Ito *et al.*, 1987). Some members of the *Paramyxovirus* genus that infect animals [Sendai virus, bovine type 3 parainfluenza virus, and simian virus 5 [(SV5)] are antigenically related to human viruses: SV5 and human type 2 parainfluenza virus HN, F, NP, and phosphoproteins (P) are antigenically related. Likewise, the Sendai virus and human types 1 and 3 parainfluenza virus HN proteins are related, and the NP of Sendai virus shares antigenic features with all four serotypes of human parainfluenza viruses. Finally, the bovine and human type 3 parainfluenza viruses have antigenically related HN, F, NP, and M proteins (Rydbeck *et al.*, 1987). The HN of Newcastle disease virus (NDV), of avian origin, appears to be antigenically related to mumps virus HN (Kilham *et al.*, 1949) and the NDV NP may share antigenic features with the PIV3, mumps, and PIV2 NP (Ito *et al.*, 1987).

PIV3 viruses are pleomorphic enveloped viruses that contain a single-stranded RNA genome of negative polarity. Similar to other paramyxoviruses, the PIV3 genome is transcribed into six unique

Kathleen van Wyke Coelingh • Laboratory of Infectious Diseases, NIAID, National Institutes of Health, Bethesda, Maryland, 20892. *Present address:* Protein Design Labs, Inc., Palo Alto, California 94304.

mRNAs which encode seven proteins: the large nucleocapsid (L) polymerase protein, the abundant NP protein, the nucleocapsid P protein, the M protein, the F glycoprotein, and the HN glycoprotein (Spriggs and Collins, 1986). Infection of the host cell is initiated by attachment of the HN to sialic acid–containing cell receptors. After attachment, a second virion surface glycoprotein, F, mediates fusion of the virus envelope with the host cell membrane, delivering viral nucleocapsids into the cytoplasm. Incoming nucleocapsid RNA is transcribed into discrete mRNAs by the viral RNA-dependent RNA polymerase complex. The same complex is responsible for the synthesis of complementary, genome-length, positive-sense RNA strands, which in turn serve as templates for the synthesis of new negative-sense genomic RNA strands. Newly synthesized HN and F glycoproteins are transported to and inserted into the plasma membrane, where the F protein presumably mediates fusion of infected cells with uninfected cells. Newly assembled genomic RNA-containing nucleocapsid structures bud through the plasma membranes, which are relatively free of host cell proteins but contain the HN and F glycoproteins.

Due to the location of the HN and F on the virion surface and on the infected cell surface, antibodies induced by PIV3 infection are directed primarily to these proteins. Furthermore, since the HN and F initiate infection by attachment and penetration (Homma and Ohuchi, 1973; Scheid and Choppin, 1974) and mediate cell-to-cell spread of infection (Scheid and Choppin, 1974), antibodies directed toward these proteins neutralize virus infectivity (Mertz *et al.*, 1981) and play a key role in the development of immunity (Spriggs *et al.*, 1987a; Kasel *et al.*, 1984).

The eventual control of PIV3 illness requires an understanding of several aspects of PIV3 infection and immunity. These aspects include virus replication strategy, molecular epidemiology, antigenic, functional, and structural properties of the HN and F proteins, and the immune response of the host to infection. This chapter describes the structural properties of the HN and F glycoproteins and their antigenic and functional organization, the nature and extent of antigenic variation in the HN and F among clinical isolates of PIV3, the evolution of the F and HN genes in nature, and current approaches toward development of a PIV3 vaccine.

II. PIV3 HEMAGGLUTININ-NEURAMINIDASE PROTEIN

A. Functional and Structural Properties of PIV3 HN Protein

The HN is a multifunctional protein, performing receptor attachment (hemagglutination) and receptor release (neuraminidase). The size of the mature, glycosylated PIV3 HN is estimated to be 72 kDa (Guskey and Bergtrom, 1981). The HN is anchored in the virus envelope by a stretch of 20 hydrophobic amino acids located near the amino terminus of the polypeptide (Elango *et al.*, 1986) (see Fig. 1). This hydrophobic region also presumably contains the signal peptide, which is not cleaved from the mature protein. These features identify th PIV3 HN, the HN of other paramyxoviruses (Schuy *et al.*, 1984; Blumberg *et al.*, 1985b; Hiebert *et al.*, 1985), and the neuraminidase protein of influenza virus (Fields *et al.*, 1981) as type II glycoproteins.

The PIV3 HN protein is synthesized as a single polypeptide chain of 572 amino acids, containing four potential *N*-linked carbohydrate acceptor sites and 14 cysteines (Elango *et al.*, 1986) (see Fig. 1). The number and location of potential glycosylation sites are not conserved among the paramyxovirus HNs sequenced to date, but ten of the cysteins are strictly conserved among all the HN proteins, suggesting similar structural properties. The PIV3 HN is most homologous to the Sendai virus HN and is more distantly related to the HNs of NDV, SV5, and mumps viruses. Regions of the HNs that are conserved among all paramyxoviruses are likely to be functionally important. The amino acid sequence NRKSCS is a strictly conserved motif among all paramyxoviruses. For the PIV3 HN, this motif is located within a 90-residue region (amino acids 234–323) which has been postulated to be the HN sialic acid binding site (Jorgensen *et al.*, 1987). The amino acid sequence and the predicted secondary structure of this region are similar to a region of the

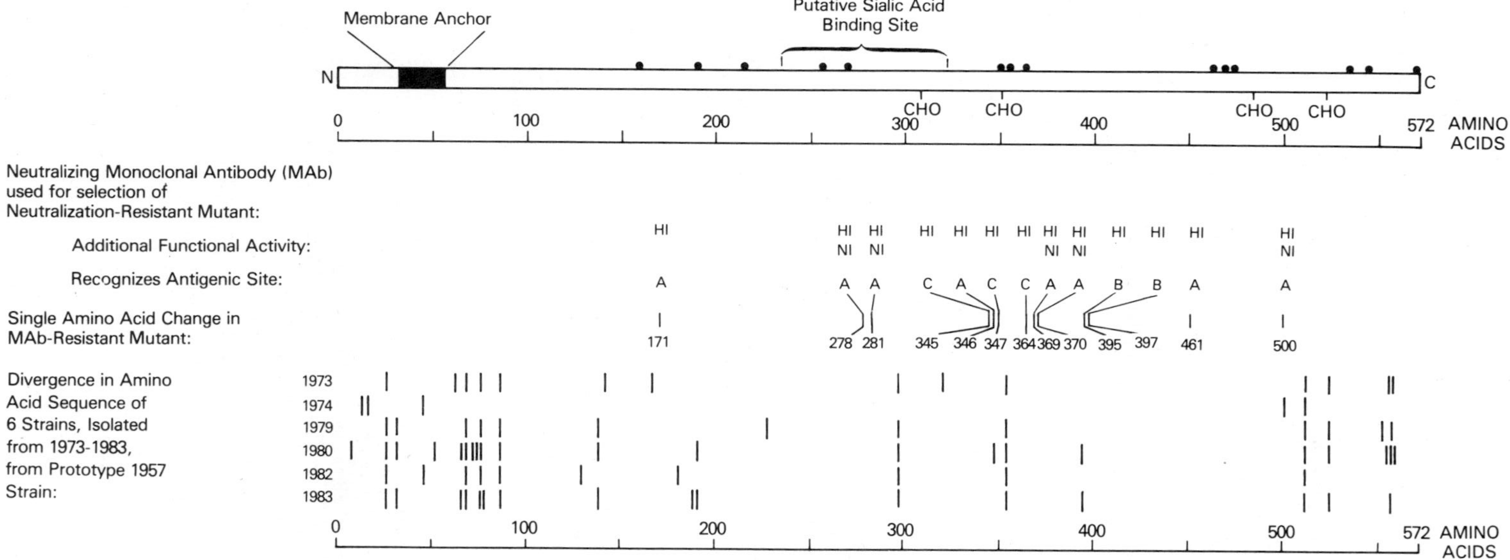

Figure 1. Schematic diagram of structural and antigenic features of PIV3 HN protein. A hydrophobic sequence near the N terminus anchors the protein in the virion envelope and also serves as the uncleaved signal peptide. The putative sialic acid binding site, cysteine residues (•), and potential *N*-linked carbohydrate acceptor sites (CHO) are indicated. Functional activities of neutralizing monoclonal antibodies (MAbs) [hemagglutination inhibition (HI) and neuraminidase inhibition (NI)] and their binding to antigenic sites A, B, or C are indicated. Position of the amino acid substitution in each MAb-resistant mutant selected with individual MAbs is shown. Position of amino acid divergence from the prototype 1957 strain is shown for six clinical strains.

influenza A virus neuraminidase (NA) protein (Blumberg *et al.*, 1985b; Jorgensen *et al.*, 1987), strengthening the argument that this region of the HN may be the sialic acid binding site. The remainder of the PIV3 HN molecule is predicted to be composed almost entirely of alternating β sheet structures and loops. Little experimentally derived information is available concerning the quaternary structure of the PIV3 HN; however, the HN undoubtedly forms oligomers, since Sendai, mumps, and NDV HN proteins exist as homodimers and homotetramers on mature virions (Nagai *et al.*, 1978; Markwell and Fox, 1980; Herrler and Compans, 1983).

B. Antigenic and Functional Organization of PIV3 HN Protein

To understand the antigenic and functional organization of the PIV3 HN, monoclonal antibodies (MAbs) specific for the HN and possessing certain functional properties were produced by immunizing mice with the prototype strain PIV3/Washington/47885/1957 (Coelingh *et al.*, 1985). The panel of 19 MAbs used for the studies described here possess one or more of the following functional activities: reactivity in ELISA, hemagglutination-inhibition (HI), neuraminidase-inhibition (NI), and virus neutralization tests. Viruses used included a variety of PIV3 strains isolated from clinically ill individuals and a panel of antigenic variants selected *in vitro* with neutralizing MAbs from a plaque-purified suspension of the prototype PIV3 strain. These reagents facilitated construction of a topologic map of the HN antigenic sites (regions of the HN containing one or clusters of proximal epitopes) and an operational map of individual epitopes (antibody-binding sites on the HN) within the antigenic sites. This information allowed a detailed examination of the functional organization of the PIV3 HN and of the nature and extent of variation in the individual HN epitopes of this monotypic virus.

Solid-phase competitive-binding radioimmunoassays of the MAbs in pairwise combinations were used to map the HN antigenic sites. The ability of two MAbs to compete for antigen binding indicates that the MAbs bind to epitopes that are in close proximity on the HN molecule. These assays indicated that the panel of MAbs define six antigenic sites (A–F) (Coelingh *et al.*, 1985, 1986) (see Table 1). Five of the antigenic sites are nonoverlapping, while one (site C) overlaps sites A and B. Three of the sites (A, B, and C) are recognized by neutralizing MAbs, whereas sites D, E, and F bind nonneutralizing MAbs.

To distinguish unique epitopes within the antigenic sites, all of the MAbs were tested in ELISA or HI assays against (1) a panel of laboratory-selected neutralization-resistant antigenic variants, and (2) a panel of PIV3 clinical strains isolated over a period of 26 years in Washington, D. C., Texas, and Australia. MAbs exhibiting unique patterns of reactivity with these viruses by definition recognize different HN epitopes. Epitopes distinguished by unique reactivities of antigenic variants were assigned numbers (I–X), and epitopes that could be further distinguished by their fine specificities with PIV3 clinical strains were also assigned letters (e.g., IIA, IIB, and IIC) (Table 1). These experiments indicated that there are seven unique neutralization epitopes in antigenic site A, two neutralization epitopes in site B, and two neutralization epitopes in site C. Sites D, E, and F each contain one nonneutralization epitope (Coelingh *et al.*, 1986).

The biologic activities that the MAbs inhibit are also shown in Table 1. The inhibitory properties of these MAbs do not completely dissociate the two biologic activities of the HN protein: Some of the MAbs (mapping in sites B and C) inhibit only hemagglutination, but all of the MAbs that inhibit neuraminidase activity also inhibit hemagglutination. Whether this means that one active site performs both functions is not known; however, MAbs that inhibit exclusively either hemagglutination or neuraminidase activity in Sendai virus HN and NDV HN have been produced (Portner, 1981; Iorio and Bratt, 1983), supporting a two-site model. The data in Table 1 suggest that antigenic sites A, B, and C also correspond to functional domains of the HN protein: all three sites are important for virus neutralization and binding to cell receptors (i.e., hemagglutination). Furthermore, it appears that site A may be located closer to the neuraminidase active site than are sites B and C. Antigenic sites D, E, and F do not appear important for neutralization, hemagglutination, or

Table 1. Antigenic and Functional Organization of the HN: Characterization Using MAbs and Antigenic Variants

MAb characterization			Variant characterization	
MAb	Antigenic site	Epitope	Biologic activity	Amino acid substitution in variant HN
170/7	A	I	N[a],HI[b],NI[c]	370 Pro → Thr
				370 Pro → His
271/7		I	N,HI,NI	281 Ala → Val
423/4		I	N,HI,NI	370 Pro → Thr
451/4		VI	N,HI,NI	278 Ser → Lys
166/11		VII	N,HI,NI	[d]
149/3		IIIB	N,HI,NI	[d]
128/9		IIIA	N,HI,NI	500 Lys → Glu
454/11		IIIA	N,HI,NI	369 Ser → Asn
101/1		IIIA	N,HI	461 Asn → Asp
429/5		IV	N,HI	171 Lys → Asn
447/12		IIC	N,HI	346 Glu → Gly
66/4	B	VA	N,HI	395 Lys → Asn
68/2		VB	N,HI	397 Trp → Leu
403/7	C	IIA	N,HI	364 Asn → Asp
61/5		IIA	N,HI	345 Asn → Asp
77/5		IIB	N,HI	347 Asn → Ser
155/2	D	VIII		[d]
44/1	E	IX		[d]
457/6	F	X		[d]

[a]N, Neutralization of virus infectivity.
[b]HI, Hemagglutination inhibition.
[c]NI, Neuraminidase inhibition.
[d]Mutants for sequence analysis could not be selected.

neuraminidase activity; however, in the presence of guinea pig complement, the MAb mapping to site D is capable of neutralizing PIV3 by facilitating virus lysis (Vasantha *et al.*, 1988).

To locate the HN epitopes within sites A, B, and C, the RNA sequences coding for the HN proteins of antigenic variants selected *in vitro* in the presence of neutralizing MAbs were analyzed. The HN gene of each variant contains a point mutation that results in an amino acid substitution (Table 1) (Coelingh *et al.*, 1986, 1987b). Each of these substitutions presumably accounts for the ability of the variants to escape neutralization by the selecting MAb, and the amino acid involved corresponds to the actual antibody binding site. Although strong evidence for this interpretation has been obtained for the influenza A virus hemagglutinin (Knowssow *et al.*, 1984), the possibility that distant amino acid substitutions may exert conformational alterations in the actual epitope must be kept in mind. The locations of all of the HN epitopes on the linear molecule are shown in Fig. 1. Most of the epitopes are located in the C-terminal half of the molecule, as expected for a protein anchored at its N-terminus. It is reasonable to expect that epitopes recognized by MAbs belonging to the same epitope group would cluster in the linear sequence of the HN molecule, and sometimes this is the case. For example, epitopes IIA, IIB, and IIC involve amino acids 345, 346, and 347, and epitopes VA and VB involve residues 395 and 397, which confirms the original site and epitope assignments for these antibodies. On the other hand, epitope IIIA (represented by MAbs 128/9, 454/11, and 101/1) involves widely scattered regions of the HN protein, suggesting that this epitope is conformational in nature, residues 369, 461, and 500 being juxtaposed on the folded molecule. A similar situation exists for epitope I, which involves residues 281 and 370. The conformational nature of all of the neutralizing HN epitopes is also supported by the fact that the MAbs do not react in Western blots or in ELISA with overlapping synthetic peptides covering the entire HN molecule (K. Hendrickson, personal communication).

Two of the HN neutralization epitopes (involving amino acids 278 and 281) are located in or near the above-described putative HN sialic acid binding domain (Fig. 1). The MAbs that recognize these epitopes inhibit neuraminidase activity. Similarly, Sendai virus HN epitopes recognized by MAbs that have NI activity are also located within this domain (amino acids 277 and 279). These results support the view that this domain of the HN protein is structurally conserved and functionally important for neuraminidase activity. That this region is important for enzymatic activity is reinforced by the finding that bovine PIV3 mutants with low neuraminidase activity have changes at amino acid position 193, which is 40 residues away from the putative HN sialic acid binding site.

The secondary structure of the PIV3 HN is predicted by the Garnier algorithm to resemble that of the influenza A virus neuraminidase (NA). The NA monomer has been shown by X-ray crystallography to be composed of alternating beta sheet structures and loops, which three-dimensionally resemble a six-bladed propeller (Varghese *et al.*, 1983) (Fig. 2). The rim of a large pocket, which binds sialic acid, contains 15 charged residues which are strictly conserved among all known NAs (Fig. 2A). Amino acids that vary in influenza antigenic drift strains and change in neutralization-resistant variants selected with MAbs are distributed among surface loops distal to the catalytic site (Fig. 2B).

The PIV3 HN sequences predicted to be beta sheets or loops were superimposed on the structural diagram of the influenza NA (Fig. 2, C and D). Four of the twelve charged residues that are conserved among paramyxovirus HN proteins were located within the putative HN sialic acid binding site, which Jorgensen *et al.* (1987) have suggested may correspond to the darkened NA loops proximal to the sialic acid binding site in Fig. 2C. Amino acids that change in neutralization-resistant PIV3 variants are located in surface loops distal to the putative sialic acid binding site by this comparison of the HN and NA (Fig. 2D). This comparison of the HN and the NA predicts that most amino acids that undergo substitutions in neutralization-resistant PIV3 variants are located in surface loops distal to the putative sialic acid binding site. Further, this comparison predicts that amino acids substitutions in several PIV3 variants, which were selected with MAbs having NI activity, would be within or adjacent to the putative sialic acid binding site. Although the nature of this comparison is speculative, it raises the possibility that the NA and HN may be structurally related.

III. PIV3 FUSION PROTEIN

A. Functional and Structural Properties of PIV3 F Protein

The PIV3 F protein functions in virus penetration and cell fusion, resulting in the formation of syncytia in infected cell monolayers. The PIV3 F protein is synthesized as a 539-amino-acid polypeptide. The mature, glycosylated F protein is estimated to be 60–63 kDa (Storey *et al.*, 1984) and contains four potential *N*-linked carbohydrate acceptor sites and 11 cysteines (see Fig. 3). Although the glycosylation sites are not conserved among F proteins of paramyxoviruses, the locations of the cysteins are highly conserved, suggesting overall structural similarities.

The PIV3 F protein is anchored in the virus membrane by a C-terminal sequence of 23 hydrophobic amino acids (Spriggs *et al.*, 1987b; Côté *et al.*, 1987) (see Fig. 3). At its N terminus is another stretch of 18 hydrophobic amino acids, which presumably functions as the signal sequence and which is proteolytically removed from the mature protein. These characteristics identify the PIV3 F protein and the F proteins of other paramyxoviruses (Paterson *et al.*, 1984; Collins *et al.*,

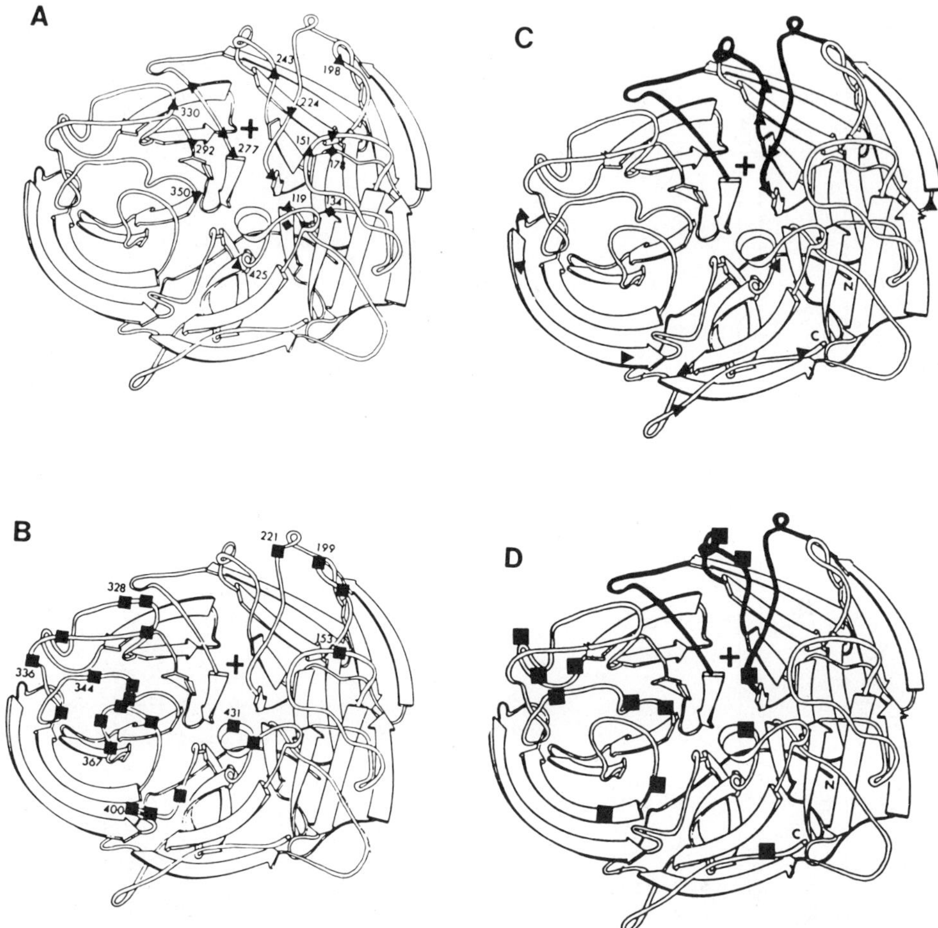

Figure 2. Drawing of the influenza A virus neuraminidase (NA) monomer, as determined by X-ray crystallography (from Coleman *et al.*, 1983, with permission). (A) Locations of charged amino acids (▲) which are strictly conserved among NA proteins and which rim the sialic acid binding site (+). (B) NA residues that change in influenza A antigenic drift strains and in neutralization-resistant influenza variants selected with monoclonal antibodies (MAbs) (▪). Similarities between the experimentally determined secondary structure of the influenza NA and predicted secondary structure of the PIV3 HN were used to superimpose the HN sequence on the structural diagram of the NA (C and D). HN sequences corresponding to the putative sialic acid binding site (Jorgensen *et al.*, 1987) are darkened. (C) Charged residues (▲) which are strictly conserved among paramyxovirus HN proteins. (D) Locations of amino acid substitutions in neutralization-resistant PIV3 variants selected with MAbs (▪).

1984; Blumberg *et al.*, 1985a; McGinnes and Morrison, 1986; Richardson *et al.*, 1986; Waxham *et al.*, 1987) as type I glycoproteins.

The F protein becomes functionally active after undergoing proteolytic cleavage by a trypsinlike host cell enzyme, yielding two disulfide-linked subunits (F1 and F2). This cleavage causes a conformational rearrangement in the F molecule which exposes a hydrophobic domain at the newly-generated N terminus of the F1 subunit (Hsu *et al.*, 1981). This hydrophobic domain of 26 amino acids is highly conserved among all paramyxovirus F proteins and shares sequence similarities with the fusogenic region of the transmembrane envelope proteins of several lentiviruses and retroviruses

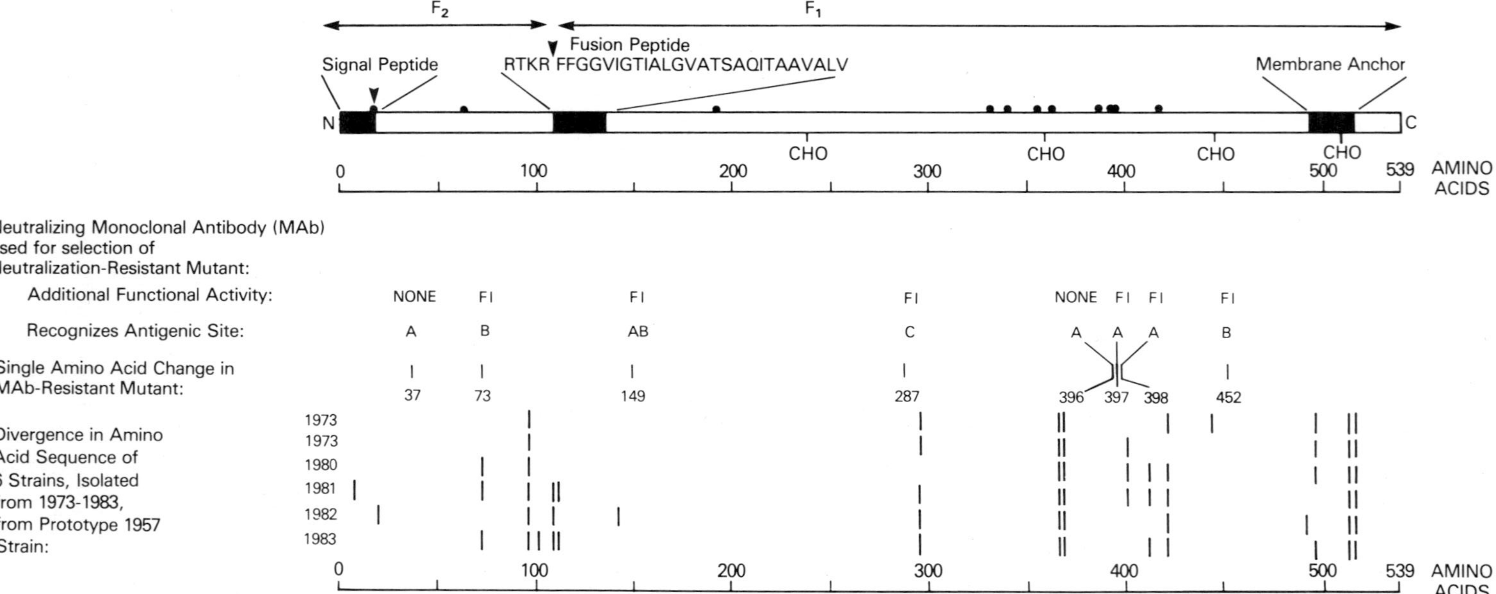

Figure 3. Schematic diagram of structural and antigenic features of PIV3 F protein. The F is synthesized as a single polypeptide which undergoes proteolytic cleavage at two sites (▼). Following its synthesis the N-terminal signal peptide is removed, and the F is cleaved at residue 109 into two subunits (F_2 and F_1) which remain disulfide-linked. Amino acid sequences shown flanking the cleavage site include the sequence RTKR, which is recognized by a trypsinlike enzyme. Also shown is the 26-residue hydrophobic fusion peptide at the newly-generated N terminus of the F_1 subunit. A hydrophobic sequence near the C terminus of the F_1 subunit anchors the F in the virion envelope. Cysteine residues (•) and potential *N*-linked carbohydrate acceptor sites (CHO) are indicated. Activity of neutralizing monoclonal antibodies (MAbs) in fusion-inhibition (FI) assays and their finding to antigenic sites A, B, AB, or C are indicated. Position of the amino acid substitutions in each MAb-resistant mutant selected with individual MAbs is shown. Position of amino acid divergence from the prototype 1957 strain is F is shown for six clinical strains.

(for review, see Gonzalez-Scarano *et al.*, 1987). On the basis of these similarities, it has been suggested that this domain is the fusogenic region (or fusion peptide) of the F molecule. This suggestion is reinforced by the observations that synthetic peptides mimicking the F1 fusion peptide can inhibit cell fusion by measles virus (Richardson *et al.*, 1980) and that the F1 N-terminal sequence of the SV5 paramyxovirus is capable of direct interaction with cell membranes (Paterson and Lamb, 1987).

Few experimental data are available concerning higher-order structure of the PIV3 F protein; however, it has been suggested that the F protein of mumps virus has a high level of ordered structure, with an estimated 82% of the protein predicted to participate in secondary structure interactions (Waxham *et al.*, 1987).

B. Antigenic and Functional Organization of PIV3 F Protein

Approaches similar to those used to characterize the HN protein were used to construct a topographic map of the F protein antigenic sites and an operational map of epitopes involved in MAb neutralization and/or binding and fusion inhibition. In the case of the PIV3 F, the MAb panel consisted of 14 neutralizing and 12 nonneutralizing MAbs.

Solid-phase competitive-binding assays indicated that there are seven nonoverlapping antigenic sites (A–F) and one bridge site (AB) (Coelingh and Tierney, 1989a) (see Table 2). Several of the site

Table 2. Antigenic and Functional Organization of the F: Characterization Using MAbs and Antigenic Variants

MAb characterization			Variant characterization	
MAb	Antigenic site	Epitope	Biologic activity	Amino acid substitution in variant F (subunit)
197	A	N1	N[a],FI[b]	398 Gly → Asp (F1)
267		N2	N,FI	397 Ile → Thr (F1)
214		N3	N	396 Gly → Asp (F1)
279		N4	N	396 Gly → Asp (F1)
284		N5	N	[d]
199		N6	N	[d]
191		N7	N,FI	397 Ile → Thr (F1)
102		N8	N	37 Gly → Glu (F2)
48		1		
145	AB	N9	N,FI,C′N[c]	[d]
108		N10	N,FI,C′N	149 Glu → Lys (F1)
128	B	N11	N,FI,C′N	73 Arg → Lys (F2) 452 Asp → Gly (F1)
110		N12	N,FI,C′N	73 Arg → Lys (F2)
215		N13	N,FI,C′N	73 Arg → Ser (F2)
640	C	N14	N,FI,C′N	287 Arg → Gly (F1)
591	D	2		[d]
4	E	3		[d]
14		4		
69	F	5		[d]
216	G	6		[d]

[a]N, Neutralization of virus infectivity.
[b]FI, Fusion inhibition.
[c]C′N, Complement-enhanced neutralization.
[d]Mutants for sequence analysis could not be selected.

A and site B MAbs participate in nonreciprocal competitive-binding reactions, suggesting that upon binding they induce conformational changes in other F epitopes. To distinguish epitopes within the antigenic sites, the MAbs were tested in ELISA and neutralization assays with a panel of PIV3 clinical strains and with a panel of laboratory-selected, neutralization-resistant antigenic variants. These analyses indicated that there are 14 neutralization epitopes and one nonneutralization epitope within sites A, B, C, and AB and five nonneutralization epitopes within sites D, E, F, and G. These results indicate that antigenic sites A, B, C, and AB are functional neutralization domains of the F protein.

To determine whether these neutralization domains are involved in fusion activity, the MAbs were tested in standard fusion-inhibition assays. Fusion-inhibiting activity is primarily associated with MAbs mapping to sites AB, B, and C. In addition, only site AB, B, and C MAbs exhibit enhanced neutralization following addition of guinea pig complement, providing additional evidence that the different antigenic regions of the PIV3 F protein correspond to distinct functional domains.

To more precisely define the locations of the neutralization epitopes on the F molecule, the RNA sequences of the F genes of antigenic variants selected *in vitro* were analyzed (Coelingh and Tierney, 1989b) (Table 2). The neutralization epitopes within sites A and B are scattered throughout hydrophilic regions of the F1 and F2 subunits (Fig. 3), indicating that distant regions of the linear polypeptide are folded into close proximity on the mature F molecule. For example, residues 37, 396, 397, and 398 are present in antigenic site A and residues 73 and 452 in site B.

None of the MAbs that inhibit fusion activity select variants with substitutions within the fusion peptide (residues 110–135), but site B and AB epitopes involve residues 73 and 149, respectively, which flank the fusion peptide. It is not currently understood why MAbs binding residues 397 and 398 inhibit fusion, while MAbs binding residue 396 are not inhibitory; however, the importance of this region for fusion function is supported by the fact that MAb-selected Sendai virus antigenic variants have a substitution at residue 400 and form very small plaques (Portner *et al.*, 1987).

IV. ANTIGENIC VARIATION IN THE HN AND F GLYCOPROTEINS AND MOLECULAR EPIDEMIOLOGY OF PIV3

To assess the nature and extent of antigenic variation in the surface glycoproteins among PIV3 strains, a group of 37 clinical isolates was tested with the HN MAbs in hemagglutination-inhibition tests or with the F MAbs in neutralization and ELISA tests. In addition, the HN and F genes of six clinical isolates were sequenced in order to examine the molecular epidemiology of PIV3.

A. Antigenic Variation in the HN Glycoprotein

Antigenic variation occurs in roughly half of the HN epitopes of PIV3 strains (Table 3). Of the 11 HN neutralization epitopes, three are variable and two are hypervariable (Coelingh *et al.*, 1985). Variation in the nonneutralization epitopes is less frequent (Coelingh *et al.*, 1986). The nature of the antigenic variation among PIV3 strains does not involve progressive accumulation of antigenic changes with time; antigenic changes detected in earlier isolates are not necessarily conserved in later strains. There is no strong correlation of the antigenic phenotype of the isolates with their geographic origin. Antigenic variation in PIV3 appears to represent genetic heterogeneity among strains and in this respect is similar to the epidemiology of influenza C viruses and strands in contrast to the progressive antigenic change in influenza A viruses (Buonagurio *et al.*, 1985).

The HN gene is 1882 nucleotides with one open reading frame of 1716 nucleotides. Nucleotide variability in the HN genes of clinical strains is most frequent in the 5′ and 3′ untranslated regions (where it ranges from 1 to 10%) and is less frequent in the coding region (about 1 to 5%) (Coelingh

Table 3. Antigenic Variability of the HN and F Epitopes of Clinical PIV3 Isolates

	HN protein variability			F protein variability			
MAb	Antigenic site	Epitope	Epitope variability[a]	MAb	Antigenic site	Epitope	Epitope variability[a]
170/7	A	I	C	48	A	1[b]	C
271/7		I	C	102		N8	V
423/4		I	C	197		N1	V
451/4		VI	C	214		N3	V
128/9		IIIA	C	279		N4	V
101/1		IIIA	C	284		N5	V
454/11		IIIA	C	199		N6	V
429/5		IV	C	191		N7	V
166/11		VII	C	267		N2	HV
447/12		IIC	HV	145	AB	N9	C
149/3		IIIB	HV	108		N10	C
66/4	B	VA	V	128	B	N11	V
68/2		VA	V	110		N12	V
77/5	C	IIB	C	215		N13	V
61/5		IIA	V	640	C	N14	C
403/7		IIA	V	591	D	2[b]	C
155/2	D	VIII[b]	C	14	E	4[b]	C
44/1	E	IX[b]	V	4		3[b]	V
457/6	F	X[b]	C	69	F	5[b]	C
				216	G	6[b]	V

[a]C, constant among all strains; HV, hypervariable among >70% of strains; V, variable among <50% of strains.
[b]Nonneutralization epitopes.

et al., 1988a). Changes in the coding region of the gene represent an overall 0.1–2.8% amino acid variability.

The clinical PIV3 isolates chosen for sequence analysis did not react with several HN MAbs (Coelingh *et al.*, 1985). In many instances, the amino acid changes identified in the clinical strains clearly account for their lack of reactivity with specific MAbs. For example, five of the six isolates that do not react with MAb 447/12 (which selects antigenic variants with mutations in residue 346) have amino acid changes at nearby residue 348. The 1974 isolate, which does react with MAB 447/12, has no alteration near position 346. A similar situation is observed for clinical strains that do not react with MAb 61/5; however, the reactivities of clinical strains with other MAbs cannot be unambiguously assigned to specific changes in the HN.

The variability in the HN of clinical strains is concentrated primarily at the N and C termini of the molecule in domains 1 and 3 (Table 4 and Fig. 1) and may reflect the functional features of the domains. For example, variability in the tail and transmembrane anchor regions of domain 1 would not be limited in degree, but would be restricted to changes maintaining the charged and hydrophobic characters of these regions. Similarly, changes in the proposed stalk region of domain 1 or in domain 3 would presumably be limited to substitutions maintaining helical structure. In contrast, variability in domain 2 would be restricted to changes that simultaneously maintain overall tertiary conformation, antigenicity, and biologic activity. In light of these considerations, the greater variability within domains 1 and 3 is not surprising.

What is surprising about the antigenic and genetic evolution of PIV3 strains is the high degree of conservation within domain 2. This is particularly striking in comparison to the high degree of

Table 4. Distribution and Frequency of Amino Acid Differences in HN Protein Domains of PIV3 Clinical Isolates

Domain characteristics	Domain 1 (residues 1–90)	Domain 2 (residues 91–509)	Domain 3 (residues 510–572)
Proposed structures and functions	N-terminus cytoplasmic tail	Hemagglutination	C-terminus
	Membrane anchor Stalk	Neuraminidase Neutralization epitopes	
Mean %[a] variation (range)[a]	6.5 (3.3–10.0)	1.0 (0.2–1.4)	5.8 (1.6–6.3)

[a]Mean percent variation and the range of this variation represent the amino acid differences in six clinical strains calculated with respect to the prototype Washington/47885/57 strain.

divergence observed, for example, in the hemagglutinin proteins of H1N1 influenza A drift strains isolated only 7 or 8 years apart (4.5–11.0%) (Raymond *et al.*, 1986).

There are several possible explanations for the slower evolution of PIV3 strains as compared to influenza A viruses. For example, the error frequency of the PIV3 polymerase could be less than that of influenza A. This seems unlikely, since the frequency of isolation of single-step mutants *in vitro* is similar for both viruses (Portner *et al.*, 1980; Coelingh *et al.*, 1985). Alternatively, there may be more stringent structure–function relationships for the PIV3 HN than for the influenza hemagglutinin. This explanation is also unlikely, since human and bovine PIV3 HN proteins maintain the same functions while conserving only 75% homology between them (Coelingh *et al.*, 1986). The most likely explanation for the different evolutionary rates between PIV3 and influenza A attachment proteins involves the nature of the immune response to infection with these viruses. In the case of influenza A, immunity is of long duration (Kendal *et al.*, 1979; Layde *et al.*, 1980), making it necessary for influenza viruses to evolve a structure able to accommodate rapid antigenic changes while maintaining essential attachment functions. On the other hand, the partial immunity induced by PIV3 infection (Chanock *et al.*, 1963) may allow the virus to survive without evolving an attachment protein capable of tolerating antigenic changes in regions also devoted to function.

B. Antigenic Variation in the F Glycoprotein

Antigenic variation occurs more frequently in the F glycoprotein than in the HN protein (see Table 3). Of the 14 F neutralization epitopes, ten are variable and one is hypervariable. In comparison, variation in the nonneutralization epitopes is less frequent (Coelingh and Tierney, 1989a). The nature of the antigenic variation observed for the F is similar to that of the HN: It does not involve cumulative change with time, nor is there a correlation with geographic origin of the isolate. An interesting feature of the epidemiology of PIV3 is the frequency with which viruses develop mutations that render them completely resistant to neutralization by MAbs, even though these MAbs bind to the F molecule very efficiently. This phenomenon also occurs with laboratory-selected variants.

The F gene is 1845 nucleotides with an open reading frame of 1617 nucleotides. The distribution of nucleotide changes in the F genes of clinical strains is very similar to that in the HN genes: Most of the variability (2.9–15%) occurs in the 5′ and 3′ untranslated regions (Coelingh and Winter, 1990). Nucleotide change in the coding region is lower (about 3.6–4.3%), resulting in an overall amino acid variability of 1.7–2.4%.

Specific amino acid changes in the F proteins of PIV3 clinical isolates correlate well with their resistance to neutralization by specific MAbs. For example, three strains that are not neutralized by site B MAbs (which select variants with mutations at amino acid 73) exhibit sequence divergence at position 73 (Fig. 3). Furthermore, it is clear that the nature of the amino acid at position 73 influences MAb binding and neutralization. For example, arginine 73 allows MAb 215 to bind and neutralize, but lysine 73 results in failure of MAb 215 to neutralize although it does allow MAb binding. A less conservative change to serine 73 abolishes MAb binding and consequently also renders the virus resistant to neutralization by this MAb. Some PIV3 clinical strains that are resistant to neutralization by site A MAbs have changes at residue 398 (which was also identified in antigenic variants selected with site A MAbs). In other strains resistant to neutralization by site A MAbs, changes in residues other than 398 are presumably responsible for the observed "binding without neutralization" phenomenon.

The distribution of the amino acid variability in the F proteins of clinical strains resembles that in the HN proteins. The highest concentration of change in F amino acid sequence occurs in the C-terminal transmembrane region (domain 3) (see Table 5). In comparison, the remainder of the molecule is strikingly conserved, particularly within a region from residue 149 to 366 in domain 2 (Coelingh and Winter, 1990). As in the case of the HNs of PIV3 clinical strains, the amino acid conservation in the F probably reflects the relative importance of functional constraints compared to the minor selective immunologic pressures imposed by an easily reinfected population.

V. PIV3 VACCINE DEVELOPMENT

Although PIV3 strains undergo genetic and antigenic variation in nature, the limited degree of this variation clearly indicates that these strains represent a monotypic group of agents. It is reasonable to conclude, therefore, that strain variability will not be a serious challenge to vaccine development.

Several vaccines against human PIV3 are currently being developed (for review, see Murphy *et al.*, 1988). Live virus vaccine candidates include cold-adapted PIV3 (Belshe and Hissom, 1982), bovine PIV3 (Coelingh *et al.*, 1988b), and vaccinia virus recombinants (Spriggs *et al.*, 1987a). In addition, various approaches are being taken to develop a subunit glycoprotein vaccine (Ray *et al.*, 1985; Ray and Compans, 1987; Coelingh *et al.*, 1987a).

Table 5. Distribution and Frequency of Amino Acid Differences in F Protein Domains of PIV3 Clinical Isolates

Domain characteristics	Domain 1 (residues 1–109)	Domain 2 (residues 110–493)	Domain 3 (residues 494–539)
Subunit	F_2	F_1	F_1
Proposed structures and functions	N-terminal signal peptide Cleavage activation site Neutralization epitopes	N-terminal fusion peptide Neutralization epitopes	C-terminal cytoplasmic tail Membrane anchor
Mean %[a] variation (range)[a]	2.1 (0.9–3.6)	1.5 (1.3–1.8)	5.8 (4.3–6.5)

[a]Mean percent variation and the range of this variation represent the amino acid differences in six clinical strains calculated with respect to the prototype Washington/47885/57 strain.

The major goal of immunization against human PIV3 is to prevent serious lower respiratory tract disease that requires hospitalization. Because natural infection with PIV3 induces only partial immunity in humans, it is unlikely that any vaccine candidate will completely prevent infection of humans with PIV3. It is possible, however, that a schedule of immunization with a vaccine could be developed that would induce sufficient immunity to restrict replication of human PIV3 and greatly reduce the occurrence of serious disease.

REFERENCES

Belshe, R. B., and Hissom, F. K. (1982). *J. Med. Virol.* **10,** 235–242.

Blumberg, B. M., Giorgi, C., Rose, K., and Kolakofsky, D. (1985a). *J. Gen. Virol.* **66,** 317–331.

Blumberg, B. M., Giorgi, C., Roux, L., Raju, R., Dowling, P., Chollet, A., and Kolakofsky, D. (1985b). *Cell* **41,** 269–278.

Bounagurio, D. A., Nakada, S., Fitch, W. M., and Palese, P. (1986). *Virology* **153,** 12–21.

Chanock, R. M., Parrott, R. H., Cook, K., Andrews, B. E., Bell, J. A., Reichelderfer, T., Kapikian, A. Z., Mastrota, F. M., and Heubner, R. J. (1958). *N. Engl. J. Med.* **258,** 207–213.

Chanock, R. M., Bell, J. A., and Parrott, R. H. (1961). In *Perspectives in Virology,* Vol. 2 (M. Pollard, ed.), pp. 126–139, Burgess, Minneapolis.

Chanock, R. M., Parrott, R. H., Johnson, K. M., Kapikian, A. Z., and Bell, J. A. (1963). *Amer. Rev. Respir. Dis.* **88,** 152–166.

Coelingh, K. V. W., and Tierney, E. L. (1989a). *J. Virol.* **63,** 375–382.

Coelingh, K. V. W., and Tierney, E. L. (1989b). *J. Virol.* **63,** 3775–3760.

Coelingh, K. V. W., and Winter, C. C. (1990). *J. Virol.* (in press).

Coelingh, K. V. W., Winter, C., and Murphy, B. R. (1985). *Virology* **143,** 569–582.

Coelingh, K. V. W., Winter, C. C., Murphy, B. R., Rice, J. M., Kimball, P. C., Olmsted, R. A., and Collins, P. L. (1986). *J. Virol.* **60,** 90–96.

Coelingh, K. V. W., Murphy, B. R., Collins, P. L., Lebacq-Verheyden, A-M., and Battey, J. M. (1987a). *Virology* **160,** 465–472.

Coelingh, K. V. W., Winter, C. C., Jorgensen, E. D., and Murphy, B. R. (1987b). *J. Virol.* **61,** 1473–1477.

Coelingh, K. V. W., Winter, C. C., and Murphy, B. R. (1988a). *Virology* **162,** 137–143.

Coelingh, K. V. W., Winter, C. C., Tierney, E. L., London, W. T., and Murphy, B. R. (1988b). *J. Infect. Dis.* **157,** 655–662.

Collins, P. L., Huang, Y., and Wertz, G. W. (1984). *Proc. Natl. Acad. Sci. USA* **81,** 7683–7687.

Colman, P. M., Varghese, J. N., and Laver, W. G. (1983). *Nature* **303,** 41–44.

Côté, M-J., Sotrey, D. G., Kang, C. Y., and Dimock, K. (1987). *J. Gen. Virol.* **68,** 1003–1010.

Elango, N., Coligan, J. E., Jambou, R. C., and Venkatesan, S. (1986). *J. Virol.* **57,** 481–489.

Fields, S., Winter, G., and Brownlee, G. G. (1981). *Nature* **290,** 213–217.

Gonzalez-Scarano, F., Waxham, M. N., Ross, A. M., and Hoxie, J. A. (1987). *Aids Res. Hum. Retroviruses* **3,** 245–252.

Guskey, L. E., and Bergtrom, G. (1981). *J. Gen. Virol.* **54,** 115–123.

Hiebert, S. W., Paterson, R. G., and Lamb, R. A. (1985). *J. Virol.* **54,** 1–6.

Herrler, G., and Compans, R. W. (1983). *J. Virol.* **47,** 354–362.

Homma, M., and Ohuchi, M. (1973). *J. Virol.* **12,** 1457–1463.

Hsu, M-C., Scheid, A., and Choppin, P. W. (1981). *Proc. Natl. Acad. Sci. USA* **79,** 5862–5866.

Iorio, R., and Bratt, M. A. (1983). *J. Immunol.* **133,** 2215–2219.

Ito, Y., Tsurudome, M., Hishiyama, M., and Yamada, A. (1987). *J. Gen. Virol.* **68,** 1289–1297.

Jorgensen, E. D., Collins, P. L., and Lomedico, P. T. (1987). *Virology* **156,** 12–24.

Kasel, J. A., Frank, A. L., Keitel, W. A., Taber, L. H., and Glezen, W. P. (1984). *J. Virol.* **52,** 828–832.

Kendal, A. P., Joseph, J. M., Kobayashi, G., Nelson, D., Reyes, C. R., Ross, M. R., Sarandria, J. L., White, R., Woodall, D. F., Noble, G. R., and Dowdle, W. R. (1979). *Am. J. Epidemiol.* **110,** 449–461.

Kilham, L., Jungherr, E., and Luginbuhl, R. E. (1949). *J. Immunol.* **63:**37–49.

Knowssow, M., Daniels, R. S., Douglas, A. R., Skehel, J. J., and Wiley, D. C. (1984). *Nature* **311,** 678–680.

Layde, P. M., Engelberg, A. L., Dobbs, H. I., Curtis, A. C., Craven, R. B., Graitcer, P. L., Sedmak, G. V., Erickson, J. D., and Noble, G. R. (1980). *J. Infect. Dis.* **142,** 347–352.

Markwell, M. A. K., and Fox, C. F. (1980). *J. Virol.* **33,** 152–166.
McGinnes, L. W., and Morrison, T. G. (1986). *Virus Res.* **5,** 543–556.
Mertz, D. C., Scheid, A., and Choppin, P. W. (1981). *Virology* **109,** 94–105.
Murphy, B. R., Prince, G. A., Collins, P. L., Coelingh, K. V. W., Olmsted, R. A., Spriggs, M. K., Parrott, R. H., Kim, H-W., Brandt, C. D., and Chanock, R. M. (1988). *Virus Res.* **11,** 1–15.
Nagai, Y., Yoshida, T., Hamaguchi, M., Iinuma, M., Maeno, K., and Matsumoto, T. (1978). *Arch. Virol.* **58,** 15–28.
Paterson, R. G., and Lamb, R. A. (1987). *Cell* **48,** 441–452.
Paterson, R., Harris, T., and Lamb, R. (1984). *Proc. Natl. Acad. Sci. USA* **81,** 6706–6710.
Portner, A. (1981). *Virology* **115,** 375–384.
Portner, A., Webster, R. G., and Bean, W. J. (1980). *Virology* **104,** 235–238.
Portner, A., Scroggs, R. A., and Naeve, C. W. (1987). *Virology* **157,** 556–559.
Ray, R., and Compans, R. W. (1987). *J. Gen. Virol.* **68,** 409–418.
Ray, R., Brown, V. E., and Compans, R. W. (1985). *J. Infect. Dis.* **152,** 1219–1230.
Raymond, F. L., Caton, A. J., Cox, N. J., Kendal, A. P., and Brownlee, G. G. (1986). *Virology* **148,** 275–287.
Richardson, C. D., Scheid, A., and Choppin, P. W. (1980). *Virology* **105,** 205–222.
Richardson, C., Hull, D., Greer, P., Hasel, K., Berkovich, A., Englund, G., Bellini, W., Rima, A., and Lazzarini, R. (1986). *Virology* **155,** 508–523.
Rydbeck, R., Lìve, A., Örvell, C., and Nórrby, E. (1987). *J. Gen. Virol.* **68,** 2153–2160.
Scheid, A., and Choppin, P. W. (1974). *Virology* **57,** 475–490.
Schuy, W. W., Garten, W., Linder, D., and Klenk, H-D. (1984). *Virus Res.* **1,** 415–426.
Spriggs, M. K., and Collins, P. L. (1986). *J. Virol.* **59,** 646–654.
Spriggs, M. K., Murphy, B. R., Prince, G. A., Olmsted, R. A., and Collins, P. L. (1987a). *J. Virol.* **61,** 3416–3423.
Spriggs, M. K., Olmsted, R. A., Venkatesan, S., Coligan, J. E., and Collins, P. L. (1987b). *Virology* **152,** 241–251.
Storey, D. G., Dimock, K., and Kang, C. Y. (1984). *J. Virol.* **52,** 761–766.
Vaghese, J. N., Laver, W. G., and Colman, P. M. (1983). *Nature* **303,** 35–40.
Vasantha, S., Coelingh, K., Murphy, B. R., Dourmashkin, R. R., Hammer, C. H., Frank, M. M., and Fries, L. F. (1988). *Virology* **167,** 433–441.
Waxham, M. N., Server, A. C., Goodman, H. M., and Wolinsky, J. S. (1987). *Virology* **159,** 381–388.

10

Genes Involved in the Restriction of Replication of Avian Influenza A Viruses in Primates

John Treanor and Brian Murphy

I. INTRODUCTION

The influenza viruses are enveloped viruses with a segmented genome of eight negative-sense stranded RNA segments (Murphy and Webster, 1989). Influenza viruses are classified into three types, A, B, and C, based on differences in the antigenicity of their nucleoprotein and matrix proteins. Influenza A viruses infect a variety of animal species, including swine, horses, marine mammals, birds, and humans. In humans, influenza A viruses cause respiratory infections with significant morbidity and mortality especially in the elderly and in individuals with chronic pulmonary or cardiac disease. The epidemiology of these infections is complex but is characterized by periodic worldwide pandemics with high attack rates. Such pandemics are associated with changes in the major protective antigens of the virus, the hemagglutinin and neuraminidase. This type of genetic change in influenza viruses is referred to as antigenic shift. The most recent pandemics of influenza A in humans occurred in 1957, when viruses of the H1N1 subtype were replaced by the H2N2 subtype, and in 1968, when the H2N2 virus was replaced by the H3N2 virus (Fig. 1). Since 1977, both the H1N1 and H3N2 subtypes of influenza A virus have cocirculated in humans.

The mechanism by which pandemic viruses are generated is unknown, but nucleotide sequence information suggests that antigenic shift cannot occur by the accumulation of a series of point mutations (Air, 1981). Rather, it has been suggested that new pandemic viruses arise via genetic reassortment of human influenza A viruses with avian influenza viruses, which donate novel hemagglutinin and/or neuraminidase genes to the human virus (Webster *et al.*, 1971). Such genetic reassortment has been observed in nature (Webster *et al.*, 1971; Cox *et al.*, 1983; Hinshaw *et al.*, 1980). Sequence data suggest that the RNA segments of the A/Singapore/57 (H2N2) virus encoding the polymerase proteins PB2 and PA, the nucleoprotein, matrix proteins, and nonstructural proteins were highly related to the corresponding gene segments of the previously circulating human H1N1 viruses (Lamb, 1989). However, the RNA segments encoding the hemagglutinin, neuraminidase, and polymerase PB1 were sufficiently different to suggest that they originate from an avian influenza A virus (Air, 1981; Treanor, Murphy, and Kawaoka, unpublished observations) (Table 3). Likewise,

John Treanor • Department of Medicine, Infectious Disease Unit, University of Rochester, Rochester, New York 14642. *Brian Murphy* • Laboratory of Infectious Diseases, National Institutes of Health, Bethesda, Maryland 20892.

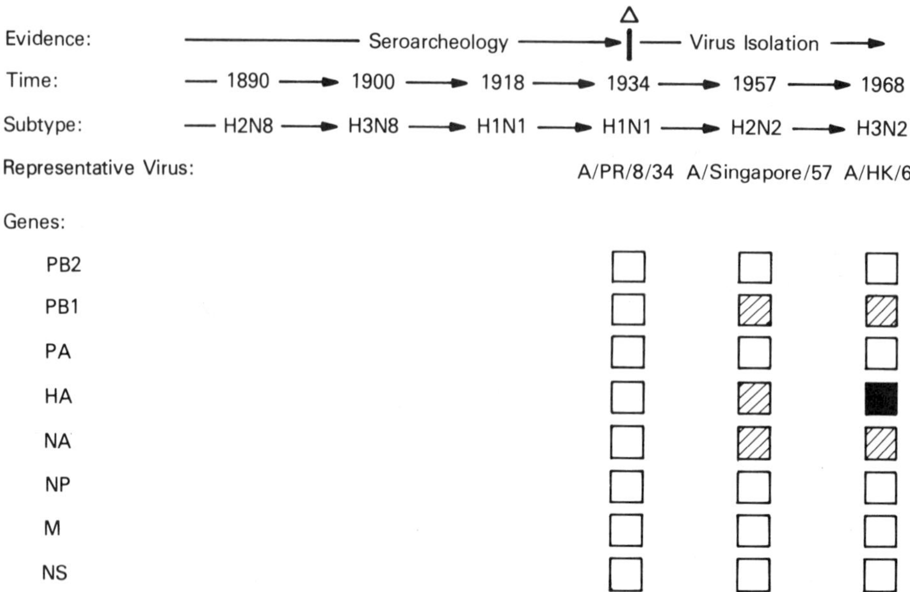

Figure 1. Epidemiology of influenza A in humans. The hemagglutinin and neuraminidase subtypes of pandemic influenza A viruses from 1890 to 1968 are shown, and the origin of the genes of the 1957 and 1968 pandemic viruses as deduced from nucleotide sequence analysis are indicated. □, Gene from the H1N1 virus; □, ■, genes from putative avian influenza A virus parent; △, influenza virus first isolated from humans in 1933.

seven gene segments of the A/Hong Kong/68 (H3N2) virus were homologous to those of preexisting human H2N2 viruses by hybridization, but the hemagglutinin gene segment was not (Scholtissek *et al.*, 1978). Instead, the H3 hemagglutinin of the human influenza virus is highly related to the H3 hemagglutinins of avian viruses (Kida *et al.*, 1987). The proposed origin of gene segments in the 1957 and 1968 human pandemic viruses is shown schematically in Fig. 1. Thus, the multiple influenza viruses that circulate in birds represent a large pool of influenza viruses which could infect humans or which could contribute novel genes to human influenza A viruses through genetic reassortment.

Since influenza in both the avian and human population occurs worldwide, the opportunities for such interactions are abundant. However, pandemics of influenza A occur rarely. There are several possible explanations for this paradox. Double infections may be rare because of host range restrictions of human and/or avian influenza viruses. For example, human influenza viruses are restricted in replication in the enteric tract of ducks, the major site of avian influenza virus replication (Hinshaw *et al.*, 1983). In addition, avian–human influenza reassortment viruses may not spread in the human population because they contain particular avian influenza virus genes or avian–human gene constellations that might limit their replication in humans. An understanding of the factors that determine the ability of avian influenza A viruses to replicate in primates would be of considerable interest in addressing these possibilities.

Nonhuman primates are a convenient model to evaluate these questions because various species including the squirrel monkey (*Saimiri sciureus*) develop febrile respiratory illnesses and shed virus in large quantities following infection with some human influenza A viruses (Murphy *et al.*, 1980). In contrast, avian influenza viruses manifest a spectrum of ability to replicate in both the upper and lower respiratory tract of squirrel monkeys, ranging from very low levels of replication to levels comparable with those of human viruses (Table 1) (Murphy *et al.*, 1982a).

Avian influenza viruses that manifest attenuated replication in the squirrel monkey are cur-

Table 1. Evaluation of Avian Influenza Viruses in Squirrel Monkeys[a,b]

Influenza A virus administered	Subtype	No. tested	Viral replication in: Nasopharynx [mean peak titer ($\log_{10}$ $TCID_{50}$/ml ± SE)[c]]	Trachea [mean peak titer ($\log_{10}$ $TCID_{50}$/ml ± SE)[c]]	Clinical response[d] (avg. severity of upper respiratory tract illness ± SE)
A/Pintail/Alb/268/78	H4N8	4	3.0 ± 0.2	2.4 ± 0.6	4.0 ± 0.9
A/Mallard/NY/6750/78	H2N2	4	2.5 ± 0.4	2.4 ± 0.4	1.0 ± 0.7
A/Pintail/Alb/358/79	H3N6	4	4.4 ± 0.1	2.8 ± 0.4	5.0 ± 0.7
A/Mallard/Alb/573/78	H1N1	4	2.5 ± 0.1	1.5 ± 0.0	4.3 ± 0.8
A/Mallard/Alb/827/78	H8N4	4	2.5 ± 0.5	2.9 ± 0.3	2.8 ± 1.6
A/Pintail/Alb/119/79	H4N6	4	2.8 ± 0.4	3.0 ± 0.2	4.0 ± 1.6
A/Turkey/MN/5/79	H10N7	4	3.0 ± 0.2	3.1 ± 0.4	9.8 ± 1.5
A/Mallard/Alb/88/76	H3N8	4	3.6 ± 0.5	4.3 ± 0.5	6.3 ± 2.7
A/Mallard/NY/6874/78	H3N2	4	4.4 ± 0.2	4.4 ± 0.5	14.3 ± 1.4
A/Pintail/Alb/121/79	H7N8	4	4.3 ± 0.3	5.0 ± 0.5	11.3 ± 1.0
A/Udorn/307/72	H3N2	11	5.0 ± 0.4	6.5 ± 0.2	12.5 ± 2.2

[a]Each monkey received $10^{7.0}$ $TCID_{50}$ of virus transtracheally in a 0.5-ml inoculum and was infected as determined by recovery of virus.

[b]Used by permission of Murphy *et al.* (1982a).

[c]The amount of virus in the nasopharyngeal swab or tracheal lavage from each monkey was determined and the maximum amount shed by each monkey was averaged. SE, standard error of the mean.

[d]Signs of upper respiratory illesss were scored daily on a scale of 1–3 and the total score achieved over 10 days of observation added for each animal. For eight uninfected control animals the average severity was 1.5 ± 0.8.

rently being evaluated as donors of genes to construct live avian–human influenza virus reassortant vaccines for humans (Murphy *et al.*, 1982b). Thus, we have been especially interested in determining the genetic basis of the attenuation phenotype of these viruses. The ability of influenza A viruses to exchange gene segments freely during mixed infection in cell culture provides an ideal system for investigating these questions. A schematic diagram of this approach is illustrated in Fig. 2. We have used this approach to generate reassortants that derive a specific gene or combination of genes from an avian influenza virus and all other genes from a human influenza virus. These reassortants can then be systematically evaluated in primates for the attenuation phenotype.

In the following review, the structure and function of the ten known gene products of the eight gene segments of avian and human influenza A viruses are briefly reviewed. For reference, the genes and gene products of the eight RNA segments of influenza A virus are presented in Table 2 (Lamb, 1989; Murphy and Webster, 1989). Where possible, differences in the nucleotide and amino acid sequences of those genes which may play a role in adaptation to a particular host are highlighted. The overall nucleotide sequence homology of the internal genes of the most extensively characterized avian influenza virus in this system, A/Mallard/NY/6750/78, and representative avian and human influenza A viruses is presented in Table 3. Data implicating the role of each gene segment alone or in combination in influencing the ability of avian influenza A viruses to replicate in primates are then presented.

II. POLYMERASE PROTEINS

The three largest RNA segments of influenza A virus encode the proteins of the influenza polymerase complex, denoted PB1, PB2, and PA on the basis of their mobility in two-dimensional

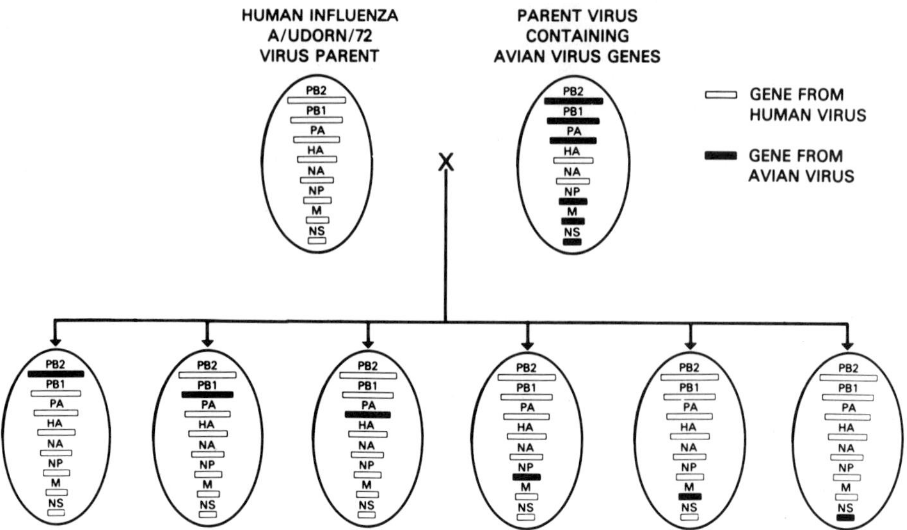

Figure 2. Generation of avian–human influenza A reassortant viruses to examine the genetic basis of attenuation specified by the avian parental virus.

electrophoresis (Kingsbury, 1989). Briefly, the three polymerase proteins function as a complex in the nucleus of the infected cell, where transcription and replication of influenza virus RNA occurs. It is thought that the PB2 protein recognizes and binds to methylated cap structures on host cell mRNA that are used to prime transcription following endonucleolytic cleavage, while PB1 is responsible for subsequent initiation and elongation of the message. The role of the PA protein, and the roles of the polymerase proteins in the synthesis of virion RNA, are less clear.

Table 2. Genes and Gene Products of Influenza A Virus

Gene segment	Protein product	Approximate no. molecules per virion	Proposed functions
1	PB1		
2	PB2	30–60	Internal proteins associated with RNA transcriptase activity
3	PA		
4	HA	500	Virus hemagglutinin responsible for attachment of virus to cells and fusion of virus and cell membranes; cleavage into HA1 and HA2 is required for infectivity
5	NA	100	Neuraminidase, responsible for release of virus from infected cells
6	NP	1000	Internal protein associated with RNA and polymerase proteins
7	M1	3000	Structural protein involved in virus assembly
	M2	20–60	Present on cell membranes and, perhaps virions, function unknown
8	NS1	Not applicable	Nonstructural proteins of unknown function
	NS2		

Table 3. Percent Nucleotide Homology of the Internal Genes of Avian Influenza A/Mallard/NY/6750/78 Virus with Representative Avian and Human Influenza A Viruses

	Percentage homology with:			
	Avian influenza virus:	Human influenza virus		
Gene	A/FPV/34 (H7N7)	A/PR/8/34 (H1N1)	A/AA/6/60 (H2N2)	A/NT/60/68 (H3N2)
PB1	N.A.[a]	84.1	93.8	93.4
PB2	85.5	86.9	85.3	84.7
PA	90.8	84.6	83.7	83.9
NP	90.1	85.6	83.6	84.3
M	91.6	92.1	90.9	89.9
NS	94.2	93.0	91.4	89.1

[a]N.A., Not available for comparison.

The polymerase proteins of a small number of avian and human influenza A viruses have been sequenced (Lamb, 1989). A comparison of the homology of the avian virus A/Mallard/NY/6750/78 with other avian and human viruses is given in Table 3. The few data available support the avian origin of the PB1 gene of H2N2 viruses, since the PB1 gene of the human A/Ann Arbor/6/60 (H2N2) virus shows greater sequence homology to PB1 genes of the avian viruses than it does to the PB1 gene of the human A/PR/8/34 (H1N1) virus.

The potential role of the polymerase genes in specifying restriction of replication of avian influenza viruses in primates has been studied by the use of avian–human influenza A reassortant viruses (Tian *et al.*, 1985; Snyder *et al.*, 1987). The results of these studies are summarized in Table 4. Reassortants were constructed using as parents the A/Mallard/NY/6750/78 virus, an avian virus that replicated poorly in primates, and the human A/Udorn/72 virus, which was known to replicate well in primates. Single gene reassortants that derived only the PB1 or PA gene segment from the avian virus parent and all other genes from the human virus parent were isolated and found to replicate as well in the lower respiratory tract of primates as the A/Udorn/72 virus. However, a reassortant that derived both PB1 and NS genes from the avian virus and all other genes from the human virus was restricted in replication in the upper and lower respiratory tract. The single gene reassortant with the avian PA gene was also slightly restricted in replication in the upper respiratory tract.

Additional studies were then performed using a different avian influenza virus parent, A/Pintail/119/79, and a different virulent human virus, A/Washington/80. In this case, a single gene avian–human influenza reassortant virus deriving the PB2 gene from the avian virus was isolated and was found not to be restricted in primates, while the single gene reassortant deriving the PA gene from the avian parent was slightly restricted in the lower respiratory tract. Reassortants deriving only the PB1 and NS gene from the avian parent were not isolated, so the influence of this gene combination on replication could not be evaluated.

A surprising finding of this study was that reassortants containing a specific constellation of avian and human polymerase gene segments were attenuated in primates and, in addition, manifested a difference in host range in cell culture. Four independently derived reassortants containing the PB2 and PA genes of A/Washington/80 and the PB1 gene of A/Pintail/79 [denoted the "HAH" genotype based on the parental origin of gene segments 1 (PB2), 2(PB1), and 3(PA)] had an infectivity titer in MDCK canine kidney cell monolayers that was significantly lower than that in primary chick kidney cultures. This was unexpected because neither of the parental viruses man-

Table 4. A Specific Constellation of Avian and Human Influenza A Polymerase Genes Attenuates Human Influenza A Virus for Squirrel Monkeys

Avian influenza virus parent	Human influenza virus parent	Avian–human influenza reassortant	No. of genes from avian parent	Parental origin of genes in reassortant						Reduction in virus shedding[a]				Restricted growth in MDCK
										Trachea		Nasopharynx		
				PB2	PB1	PA	NP	M	NS	Mean peak titer	Duration of virus shedding	Mean peak titer	Duration of virus shedding	
A/Mallard/78	A/Udorn/72	19[b]	1	H[c]	A	H	H	H	H	–	–	–	–	–
		10	1	H	H	A	H	H	H	–	–	+	+	–
		12	2	H	A	H	H	H	A	+	+	–	+	–
		3	2	H	H	A	H	H	A	–	–	–	–	–
A/Pintail/79	A/Wash/80	278	1	A	H	H	H	H	H	–	–	–	–	–
		15	1	H	H	A	H	H	H	++	–	–	–	–
		24	2	H	H	A	H	H	A	–	–	–	–	–
		228	2	A	H	H	H	H	A	–	–	–	–	–
		49	2	H	A	H	H	A	H	++	++	++	++	++
		22	3	H	A	H	A	A	H	++	++	++	++	++
		39–20	3	A	A	H	H	A	H	+	–	–	–	+

[a] –, No reduction compared to human influenza A parent; +, significant reduction compared to human influenza A parent; ++, reduction greater or equal to that of avian influenza A parent.
[b] Clone designation.
[c] H, gene derived from human parent; A, gene derived from avian parent. The HA and NA gene segments were derived from the human influenza A virus parent.

ifested this phenotype. Two of these reassortants, including one with no other attenuating avian genes, were administered to squirrel monkeys and were found to be markedly restricted in replication in both the upper and lower respiratory tract. Reassortants that derived both the PB2 and PB1 genes from the avian parent and the PA gene from the human parent (the "AAH" genotype) were also restricted in growth in MDCK cells, but only moderately so, and were only slightly restricted in the respiratory tract of primates. This indicated that the degree of host range restriction in cell culture and restriction of replication in primates were correlated. Further study of the biochemical mechanism of the host range phenotype manifested by some reassortants may be useful in understanding the interaction of the proteins of the polymerase complex with each other and with host factors.

It is interesting to note that the new pandemic influenza virus in 1957 appeared to be an avian–human reassortant influenza A virus that derived its PB1 gene from an avian parent and its PB2 and PA genes from a human parent; i.e. it was of the "HAH" genotype (Fig. 1). The 1957 virus was highly virulent in humans while the A/Pintail/79 x A/Washington/80 "HAH" reassortant was highly attenuated in primates. This illustrates that the specific parents that contribute genes to reassortants have an important effect on the properties exhibited by the resulting reassortant viruses.

Importantly, these data suggest that some, but not all, combinations of avian and human influenza virus genes can cooperate effectively in the respiratory tract of primates. Avian–human influenza A reassortant viruses with the phenotype of restricted replication in primates would not be able to spread efficiently from human to human, and therefore viruses with these gene constellations would not be expected to give rise to pandemic human influenza viruses. This represents one possible obstacle to the emergence of new pandemic influenza A viruses in humans, namely, the presence of avian–human influenza gene constellations that restrict viral replication in primates.

III. HEMAGGLUTININ

RNA segment four encodes the hemagglutinin (HA) of influenza A virus. The hemagglutinin monomer is cleaved into two polypeptide chains termed HA-1 and HA-2 with the removal of between one and six amino acids from the carboxy terminus of HA-1. HA-1 and HA-2 remain linked by a single disulfide bond and are anchored in the lipid membrane of the virus by an anchor sequence near the carboxy terminus of HA-2. Each HA spike consists of a trimer of three HA-1 and HA-2 monomers (Wiley and Skehel, 1987).

A major function of the influenza HA in virus replication is attachment to cell surface sialyloligosaccharides which serve as virus receptors. Depending on the sequence of amino acids in the receptor binding pocket of the HA molecule, the HA exhibits specificity for either neuraminic acid 2,3-galactose or 2,6-galactose linkages on cell surface glycoproteins. The receptor specificities of human and avian HAs differ, suggesting that the distribution of receptors in mammals and birds may be a determinant of host range (Rogers and Paulson, 1983).

Human influenza A viruses fail to infect the intestinal tract of birds after oral inoculation, and studies with avian–human influenza A reassortant viruses suggest that both the HA and NA of avian viruses are critical in allowing transmit through the digestive tract and replication in intestinal epithelial cells of ducks (Hinshaw *et al.*, 1982, 1983). For example, a reassortant that contained an HA gene segment derived from an H3 human parent and all other genes from an avian parent was unable to replicate in the intestinal tract of ducts. Mutants of this virus that had gained the ability to replicate in the intestinal tract of ducks were then isolated, and backcross mating showed that the genetic determinant of this phenotype was the HA gene. Sequence analysis of the HA genes of the parent and a number of such mutant viruses demonstrated that a single amino acid change at aa 226 from leucine to glutamine changed receptor specificity of the human HA to that of 2,6-galactose linkages but did not fully restore the ability of viruses bearing this HA to replicate in ducks. An additional mutation at aa 228 from serine to glycine, however, allowed full infectivity for avian viruses bearing the mutant human HA in the intestinal tract of ducts after oral inoculation (Naeve *et al.*, 1984). These studies demonstrate the importance of the HA in the tissue tropism of influenza A viruses in birds, and that a relatively small number of mutations can alter the tissue tropism specified by the human H3.

A second important function of the HA is fusion of the viral membrane with the cell membrane and subsequent entry of the virus into the cell (Wiley and Skehel, 1987). Cleavage of the HA by cellular trypsinlike proteases is necessary for viral infectivity and liberates a hydrophobic, highly conserved amino terminus of HA-2 which is homologous to the amino terminus of the F_1 subunit of the fusion glycoprotein of paramyxoviruses. Under conditions of low pH such as exist in endosomal vesicles, a conformational change occurs in the HA which results in the amino terminus of HA-2 being exposed on the outside of the molecule.

The ability of the HA to be cleaved by cellular proteases has been correlated with virulence of avian influenza viruses in birds (Webster and Rott, 1987). Avian influenza virus infection in fowls may cause two types of illness, either a primarily respiratory illness with low mortality or a highly lethal disseminated infection with encephalitis known as fowl plague. Of the 13 known HA subtypes, each of which has been identified on influenza A viruses isolated from birds, only viruses with the H5 and H7 HAs have been associated with disseminated disease. Both virulent and avirulent viruses with the H5 or H7 HA have been isolated, and the ability to cause disseminated disease has been associated with the ability of the HA to be cleaved by cellular proteases without added trypsin. Presumably, this allows virulent viruses to undergo cleavage activation throughout the tissues of the host, while avirulent viruses can only be activated by proteases within the cells of the respiratory and gastrointestinal tracts. Other genes also appear to influence the full expression of virulence of these viruses in birds.

Comparison of the HAs of virulent, cleaved and avirulent, uncleaved H5 and H7 viruses has revealed that both the presence of a series of basic amino acids and the overall length of the connecting peptide at the carboxy terminus of HA-1 are important determinants of cleavability (Kawaoka and Webster, 1988). Studies utilizing the technique of *in vitro* mutagenesis have confirmed the importance of the length and sequence of the connecting peptide and also the importance of additional structural features that may influence recognition of the cleavage site by cellular proteases. Recently, a spontaneous, highly virulent mutant H5 avian influenza virus which differed from its avirulent parent only by the loss of a glycosylation site near the cleavage activation site was isolated from chickens dying of influenza (Deshpande *et al.*, 1987). These findings indicate that mutations that alter the accessibility of the cleavage site to cellular proteases may also influence the virulence of avian influenza viruses in birds.

Relatively less is known about the potential role of avian HAs in the virulence of influenza A viruses in primates. As mentioned earlier, the HA serves as a neutralization antigen of the influenza virus, and a major component of protective immunity to influenza is antibody to the HA. Pandemics then result from the emergence of influenza viruses with novel HAs to which antibody has not developed in the human population. The avian HA gene therefore represents a gene that can increase the virulence of influenza viruses in humans by allowing viruses to escape from preexisting immunity. Avian HAs may also have properties apart from immunogenicity which effect their ability to replicate in primates, but this has not been examined in detail. Reassortants with avian influenza HA genes on a background of human influenza genes have not been studied because such reassortants would be unsafe to work with under standard laboratory conditions since they could escape from the laboratory and might initiate new pandemics of influenza. Therefore, only direct evidence is available concerning the ability of avian HA genes to influence replication in primates.

Avian influenza viruses representing 7 of the 13 HA subtypes were evaluated for their ability to replicate in the respiratory tract of primates, and a spectrum of level of replication was observed (Murphy *et al.*, 1982a). The level of replication of these viruses in the upper and lower respiratory tract of squirrel monkeys is shown in Table 1. Clearly, viruses with the avian H3 and H7 HA gene are able to replicate to high levels in these animals. The contribution that the HA gene makes to the restricted replication of the other avian influenza viruses in primates is more difficult to evaluate owing to the presence of these viruses of other potentially attenuating genes. The HA is not the sole determinant of restricted replication of these viruses, however, because reassortants bearing human HA and NA genes and all other genes from an avian virus are as restricted in replication in primates as their avian parent (Murphy *et al.*, 1982b). Avian viruses with the highly cleavable HAs associated with virulence in birds have not been evaluated in primates for the reasons discussed above.

Recent outbreaks of lethal influenza caused by viruses antigenically and genetically related to avian influenza A viruses in seals (A/Seal/Mass/1/80, H7N7) and mink (A/Mink/Sweden/84, H10N4) indicate that some avian influenza viruses have definite pathogenic potential for mammals (Webster *et al.*, 1981a; Klingborn *et al.*, 1985). The A/Seal/80 virus also caused conjunctivitis in human laboratory workers and was evaluated further for its ability to replicate in squirrel monkeys (Murphy *et al.*, 1983; Webster *et al.*, 1981b). The A/Seal/80 virus achieved the same level and duration of replication in both the upper and lower respiratory tract as the human A/Udorn/72 virus. In addition, one monkey infected with a human eye isolate of the A/Seal/80 virus died of pneumonia, and influenza virus was subsequently isolated from the spleen and liver but not the brain. It is unclear whether the systemic spread of this virus is an intrinsic property of the virus or an isolated consequence of overwhelming pulmonary infection. The genetic basis of the ability of a virus of apparently avian origin to replicate to high titer in primates also has not been determined. Of note, the H7 of A/Seal/80 had a connecting peptide sequence with some of the features of the virulent H7, but did not have the highly cleavable phenotype (Naeve and Webster, 1983). Thus, the potential for a human influenza virus with a highly cleavable HA derived from an avian influenza virus to emerge and cause disseminated disease in humans is not known.

On the other hand, a consistent finding of influenza epidemiology is the recycling of hemagglutinin subtypes in the human population. Three hemagglutinins, H1, H2, and H3, caused pandemics in humans in 1917, 1957, and 1968. In addition, serologic evidence indicates that H2 viruses were present in humans between about 1890 and 1900, and that H3 viruses were present between 1900 and 1917 (Masurel and Marine, 1973) (Fig. 1). The evidence that only three HA subtypes have been present on human influenza A viruses causing pandemics and epidemics over the past century while viruses with all 13 HA subtypes are widely distributed in birds indirectly suggests that the other HA subtypes restrict replication of influenza A viruses in primates. If such restrictions do exist, they are unlikely to be solely based on the receptor specificity of the HA given the ease with which this phenotype can change. However, more complex properties of the HA that are presently unidentified may limit the spread of avian influenza viruses in humans.

IV. NEURAMINIDASE

RNA segment 5 (or 6, depending on the gel system used) encodes the viral neuraminidase (NA) (Colman and Ward, 1985). Three potential functions of the NA in viral replication have been recognized. These include penetration of the sialic acid–rich mucus barrier to infection in the respiratory tract, trimming of terminal sialic acid from viral glycoproteins to prevent viral aggregation, and release of virus from infected cells.

Relatively little is known about the potential role of the NA in specifying the ability of avian influenza viruses to grow in primates. The NA of the human influenza A virus WSN has been implicated in the neurovirulence of this virus for mice (Sugiura and Ueda, 1980). In birds, the NA has been found to play a secondary role in virulence (Ogawa and Ueda, 1981). No information is available in primates regarding a direct role of the NA of avian viruses in specifying virulence. The NA is an important component of immunity to influenza, and antibody to the NA ameliorates influenzal disease in humans. Comments made previously regarding antigenic shift of the HA are therefore also applicable to the NA. In addition, a restricted number of NAs (N1, N2, and N8) have been identified on influenza A viruses that infect humans despite the presence of nine NA subtypes in birds (Fig. 1). Avian influenza viruses with the N2 or N8 NA were also able to replicate to high levels in the respiratory tract of squirrel monkeys (Table 1). As discussed earlier, the contribution of the NA to the restricted replication of the other avian viruses is difficult to evaluate because of the presence of other attenuating genes in the avian viruses. However, viruses with the other NA subtypes may not be able to gain access to the human population because of a restricting effect of this viral surface glycoprotein.

V. NUCLEOPROTEIN

RNA segment 6 (or 5 depending on the gel system used) encodes the nucleoprotein (NP). The NP interacts with RNA as a structural component of the virion and also associates with the three polymerase proteins to form the transcriptase complex (Lamb, 1989). The NP has also been implicated in the switch from mRNA to vRNA synthesis in the infected cell. Finally, the NP associates with M1 or M2 at the surface of the cell during virus assembly.

The NP genes of influenza A viruses from avian and human sources were first shown to be distinctly different by RNA–RNA hybridization. Five classes of NP genes were defined by this technique, including one human, two equine, and two avian classes (Bean, 1983). The genetic divergence of avian and human influenza virus NP genes was also indicated by the observation that human viruses were unable to complement a temperature-sensitive avian virus with a defect in the NP gene, while avian viruses functioned efficiently in complementation (Scholtissek *et al.*, 1985). Finally, sequence analysis has defined the presence of distinct avian and human NP genes (Buckler-White and Murphy, 1986).

A number of avian and human influenza virus NP gene sequences have been determined (Table 2). Analysis of these sequences confirmed the division of avian and human influenza NP genes into two classes (Buckler-White and Murphy, 1986). This analysis showed an approximately 91–95% nucleotide homology within each class but only about 85% homology between classes. In addition, there was evidence of a linear evolution of the NP genes of human viruses, while the avian viruses existed in multible sublines evolving in a nonlinear manner. The degree of nucleotide divergence within the avian group was greater than that within the human group, but the degree of amino acid divergence was less. This and the fact that the ratio of coding to noncoding changes in the NP gene is low suggested that there are functional constraints on the evolution of the NP gene in birds. Analysis of the deduced amino acid sequences also allowed determination of residues at which a specific amino acid was present in all avian viruses analyzed, while a different amino acid was present in all the human viruses. Fifteen such avian–human-specific amino acids were seen in the NP gene, again suggesting divergent evolution of the NP gene in humans and birds.

The existence of separate classes of the NP gene that correlate with host species suggested that the NP gene may play a role in the host range of influenza A viruses. This hypothesis has been tested by the construction of avian–human influenza A reassortant viruses that derive their NP gene from an avian virus and other genes from a human virus that replicates well in primates (Tian *et al.*, 1985; Snyder *et al.*, 1987). These studies are summarized in Table 5. The NP gene of the avian influenza A/Mallard/NY/6750/78 or A/Pintail/Alb/119/79 virus was shown to independently restrict the growth of virulent human influenza A/Udorn/72 or A/Washington/80 virus. One-hundred-fold or greater reductions in the level of virus replication of either of these single avian NP gene substitution reassortants compared to their respective human parents was seen in both the upper and lower respiratory tract of squirrel monkeys. Of note, the attenuating effect of the NP genes of the avian strains was seen not only when the NP gene was the only avian influenza viral gene present in the reassortant, but also in reassortants that derived the NP and other genes from the avian parent,

Table 5. The NP Genes of both the Influenza A/Mallard/78 and A/Pintail/79 Viruses Attenuate Human Influenza A Viruses for Squirrel Monkeys

Avian influenza virus parent	Human influenza virus parent	Avian–human influenza reassortant	No. of genes from avian parent	Parental origin of genes in reassortant						Reduction in virus shedding[a]			
										Trachea		Nasopharynx	
				PB2	PB1	PA	NP	M	NS	Mean peak titer	Duration of virus shedding	Mean peak titer	Duration of virus shedding
A/Mallard/78	A/Udorn/72	17[b]	1	H[c]	H	H	A	H	H	++	++	++	++
		15	2	H	A	H	A	H	H	+	+	–	+
		18	2	H	H	H	A	H	A	++	++	++	++
		19–1	3	H	H	A	A	H	A	++	++	++	++
A/Pintail/79	A/Wash/80	264	1	H	H	H	A	H	H	++	++	++	++
		251	2	H	H	H	A	A	H	++	++	++	++
		314	3	A	H	H	A	H	A	++	–	++	++
		226	4	A	H	H	A	A	A	++	++	++	++

[a] –, No reduction compared to human influenza A parent; +, significant reduction compared to human influenza A parent; ++, reduction greater than or equal to that of avian influenza A parent.

[b] Clone designation.

[c] H, Gene derived from human parent; A, gene derived from avian parent. The HA and NA gene segments were derived from the human influenza A virus parent.

including the polymerase and matrix proteins genes. Thus, the effects of the NP gene does not appear to act through a gene incompatibility mechanism as seen with the polymerase genes, but rather to be an independent effect in primate respiratory epithelium. In contrast to the polymerase gene reassortants, reassortants containing the avian NP gene manifested restricted replication in primates but did not have a host range phenotype in cell culture. Thus, this represents another level of obstruction to the emergence of new influenza A pandemic viruses, namely, the presence of avian influenza genes that independently restrict viral replication in primates.

The nucleotide sequences of the NP genes of A/Mallard/78 and A/Pintail/79 were subsequently found to be highly (98.8%) related and may represent one of the two classes of NP genes present in avian viruses (unpublished observations). Further studies to determine the ability of the NP genes of other avian viruses, including viruses that are not restricted in replication in primates, to attenuate human influenza viruses are underway. It is possible that the attenuation phenotype specified by the NP gene is allelic, or that variation exists within a single allele. Such studies and studies of the avian NP gene reassortants after prolonged replication in primates may help pinpoint the specific structural determinants of the attenuation phenotype.

VI. MATRIX PROTEINS M1 AND M2

The seventh RNA segment of influenza A virus encodes the two matrix proteins, M1 and M2 (Lamb, 1989). The M1 and mRNA is a linear transcript of the RNA while a second protein, M2, is encoded by segment 7 via a spliced mRNA. Although M1 is the most abundant protein in virions, little direct evidence is available regarding its function. The most likely function of M1 is structural, providing a support for the viral lipid envelope. The M2 protein is abundantly expressed on the cell surface in infected cells and small amounts of M2 also appear to be present in the virion.

The nucleotide sequences of RNA 7 of avian and human influenza A viruses are highly conserved, as are the predicted amino acid sequences of the avian and human M1 protein (Buckler-White *et al.*, 1986). There is an approximately 4% difference in the sequence of M1 between the most divergent avian and human viruses. A relatively higher degree of divergence, about 15%, is seen between the M2 proteins. Despite these similarities, some consistent differences have been noted. When the predicted amino acid sequences are aligned, three avian-human specific amino acids (defined in section V) are found in the M1 protein, and seven in the smaller M2 protein. The relatively greater divergence of M2 than M1 may indicate different pressures affecting the evolution of the two genes in birds.

Studies with avian–human single gene reassortant viruses have shown that RNA segment 7 of the avian influenza virus A/Mallard/NY/6750/78 can also independently restrict the replication of the virulent human A/Udorn/72 virus in the respiratory tract of squirrel monkeys (Table 6) (Tian *et al.*, 1985). However, RNA segment 7 of a second avian virus, A/Pintail/Alb/119/79, did not attenuate the human virus A/Washington/80 in the same system (Snyder *et al.*, 1987). In order to rule out the possibility that a suppressor gene was present in the A/Washington/80 virus, a third reassortant was made which derived segment 7 from the A/Pintail/79 virus and all other genes from the A/Udorn/72 virus. This reassortant, however, was also able to replicate like the human parent in both the upper and lower respiratory tract. Thus, the effect on virus replication of the avian M gene segment is clearly virus dependent.

The nucleotide sequences of the seventh RNA segment of the A/Pintail/79 and the A/Mallard/78 viruses were then compared to further investigate the structural basis of the attenuation phenotype. The two M genes were found to be highly related, with only three amino acid differences detected (unpublished observations). Two of these changes, proline to serine at aa90 and threonine to isoleucine at aal67 of M1, were unique to the A/Mallard/78 sequence. The third change, from serine to glycine at aa89 of M2, was present both in the A/Mallard/78 virus and in the sequence of the human influenza A/PR/8/34 virus. Since the M gene segment of A/PR/8/34 does not appear to

Table 6. The M Gene of A/Mallard/78 Virus but Not That of A/Pintail/79 Virus Attenuates Human Influenza A Virus for Squirrel Monkeys

Avian influenza virus parent	Human influenza virus parent	Avian–human influenza reassortant	No. of genes from avian parent	Parental origin of genes in reassortant						Reduction in virus shedding[a]			
										Trachea		Nasopharynx	
				PB2	PB1	PA	NP	M	NS	Mean peak titer	Duration of virus shedding	Mean peak titer	Duration of virus shedding
A/Mallard/78	A/Udorn/72	6[b]	1	H[c]	H	H	H	A	H	+	+	++	++
		13	1	H	H	H	H	A	H	++	++	−	+
		11	2	H	H	H	H	A	A	++	++	+	+
A/Pintail/79	A/Wash/80	204	1	H	H	H	H	A	H	−	−	−	−
		218	2	A	H	H	H	A	H	−	−	−	−
		208G	4	A	H	A	H	A	A	+	−	−	+
A/Pintail/79	A/Udorn/72	227	1	H	H	H	H	A	H	−	−	−	−

[a]−, No reduction compared to human influenza A parent; +, significant reduction compared to influenza A parent; ++, reduction greater than or equal to that of avian influenza A parent.
[b]Clone designation.
[c]H, Gene derived from human parent; A, gene derived from avian parent. The HA and NA gene segments were derived from the human influenza A virus parent.

be an attenuating gene in primates, this change is unlikely to be a basis for the attenuation phenotype of A/Mallard/78 virus. Thus, despite the apparently more divergent evolution of M2 in birds, the attenuation phenotype mediated by the A/Mallard/78 gene segment 7 for primates appears to be specified by the M1 protein.

VII. NONSTRUCTURAL PROTEINS NS1 AND NS2

RNA segment 8 encodes the two nonstructural proteins NS1 and NS2 (Lamb, 1989). The NSI mRNA is a colinear transcript of virion RNA, while the NS 2 mRNA is a spliced transcript such that the amino terminal 10 amino acids of NS2 are shared with NS1 and the remainder are translated in a +1 reading frame. NS1 accumulates in the nucleus early in infection and occasionally in the cytoplasm in paracrystalline arrays. The functions of NS1 and NS2 are not known, although studies with temperature-sensitive mutants of RNA segment 8 have implicated the NS proteins in the shutoff of host cell protein synthesis and in the switch from mRNA to vRNA synthesis.

The sequences of the NS genes of a large number of avian and human influenza A viruses are known (Buonagurio *et al.*, 1986; Nakajima *et al.*, 1987; Baez *et al.*, 1981). Two distinct alleles of the NS genes of avian influenza A viruses have been defined in studies utilizing RNA–RNA hybridization (Scholtissek and Von Hoyningen-Huene, 1980) or sequence analysis (Treanor *et al.*, 1989). One group of avian viruses had NS gene sequences with 90% or greater homology to the NS gene of A/Fowl Plague/Rostock/34 (allele A) while another group was homologous to the NS gene of A/Duck/Alb/76 (allele B). The homology between the two alleles was about 70%. In contrast, the NS genes of influenza A strains isolated from humans since 1933 belong to a single group in which new mutations have accumulated linearly over time. The sequences of the NS genes of early human strains are about 93% homologous to the NS genes of allele A, but only about 73% homologous to allele B and thus belong to allele A.

The ability of the eighth gene segment of avian influenza viruses to attenuate human influenza viruses for primates appears to correlate with the NS allele (Treanor *et al.*, 1989). In studies of avian–human single gene substitution reassortant influenza viruses, the presence of an NS gene of the avian influenza virus A/Mallard/NY/6750/78 did not significantly restrict the replication of the A/Udorn/72 virus in the respiratory tract of squirrel monkeys (Tian *et al.*, 1985). Similarly, the presence of an NS gene of the A/Pintail/119/79 virus did not restrict the replication of the A/Washington/80 virus (Snyder *et al.*, 1987) (Table 7). Both of these NS genes were subsequently found to belong to allele A by nucleotide sequence analysis. In contrast, the allele B NS genes of A/Mallard/88/76 and A/Pintail/121/79 viruses were both able to independently exert a modest, but definite attenuating effect on the replication of the A/Udorn/72 virus in the upper and lower respiratory tract of squirrel monkeys (Treanor *et al.*, 1989). Evidence of host range restriction in cell culture was not seen. Thus, the more divergent evolution of the B NS allele in avian hosts may have resulted in gene products that cannot function as efficiently in primates or in the genetic background of other influenza genes adapted to growth in primates. The specific amino acids responsible for this phenotype remain to be identified.

VIII. DISCUSSION

RNA hybridization and nucleotide sequence data suggest that influenza A viruses that cause pandemics in humans may contain novel hemagglutinin and neuraminidase genes derived from avian influenza A viruses by genetic reassortment. Gene segments of avian and human influenza A viruses reassort freely *in vitro,* and there is evidence to suggest that such reassortment may take place *in vivo* as well. Ample opportunities for such reassortment to occur exist, since both avian and

Table 7. The Allele B NS Genes of Two Avian Influenza A Viruses Attenuate a Human Influenza A Virus for Squirrel Monkeys

											Reduction in virus shedding[a]			
											Trachea		Nasopharynx	
Avian influenza virus parent	Human influenza virus parent	Avian–human influenza reassortant	No. of genes from avian parent	NS allele	Parental origin of genes in reassortant						Mean peak titer	Duration of virus shedding	Mean peak titer	Duration of virus shedding
					PB2	PB1	PA	NP	M	NS				
A/Mallard/78	A/Udorn/72	2[b]	1	A	H[c]	H	H	H	H	A	–	–	–	–
		3	2	A	H	H	A	H	H	A	–	–	–	–
A/Pintail/79	A/Wash/80	82–7	1	A	H	H	H	H	H	A	–	+	+	–
		24	2	A	H	H	A	H	H	A	–	–	–	–
		228	2	A	A	H	H	H	H	A	–	–	–	–
A/Mallard/76	A/Udorn/72	5	1	B	H	H	H	H	H	A	–[d]	+	++	++
A/Pintail/79	A/Udorn/72	25	1	B	H	H	H	H	H	A	–[d]	+	++	++

[a]–, No reduction compared to human influenza A parent; +, significant reduction compared to human influenza A parent; ++, reduction greater than or equal to that of avian influenza A parent.

[b]Clone designation.

[c]H, Gene derived from human parent; A, gene derived from avian parent. The HA and NA genes were derived from the human influenza A virus parent.

[d]Statistically significant reduction from human parent when results of these two groups were pooled.

human influenza A infections are widespread. Yet, these events must take place rarely, since only three pandemic human influenza A viruses have been generated in this century. A consideration of the factors responsible for this observation is of some interest.

The generation of avian–human influenza A reassortant viruses capable of transmission and spread within the human population would require several components. First, a double infection of a cell with a human and an avian influenza A virus would have to occur in some host. This requires that a human virus be able to replicate within the gut or respiratory tract of a bird, or that an avian virus be able to replicate within the respiratory tract of a human, or that both viruses be able to replicate in some third host. Second, the reassortant virus generated would have to be able to replicate to a reasonably high level within the respiratory tract of humans, in order to cause disease and to be transmitted. Third, the surface glycoproteins of the reassortant would need to be sufficiently distinct from those of previous human viruses to allow spread in the face of preexisting immunity. Finally, the transmissibility and antigenic novelty of the reassortant influenza A virus would have to be great enough to allow the new virus to compete with the currently circulating influenza A viruses in humans. Otherwise the new reassortant influenza A virus might only be capable of causing isolated human infections or localized outbreaks of influenza, as seen previously with swine influenza A viruses in humans (Hinshaw *et al.*, 1978; Patriarca *et al.*, 1984). In this chapter, we reviewed the extent and genetic mechanisms of the host range restriction of avian and human influenza A viruses in an attempt to understand some of the factors that might affect the potential of avian–human influenza A reassortant viruses to fulfill these requirements.

Several obstacles to the formation of such reassortant viruses were discussed. First, the degree of host range restriction of many avian and human influenza viruses is such that successful double infections may not be as common as predicted by the relative frequency of individual infections. Successful infection of humans with avian viruses manifesting restricted replication in primates would be unlikely. Similarly, after oral inoculation human influenza viruses are unable to replicate in the intestinal tract of ducks, the major site of influenza virus replication in these birds. These findings indicate that host range differences that restrict replication of the majority of avian and human influenza viruses in the heterologous host likely contribute to a decreased frequency with which double infections can occur in the cells of these two species. This, in turn, could be partly responsible for a lower-than-expected emergence of pandemic human influenza A viruses.

Second, specific combinations of avian and human influenza gene segments are incompatible with efficient virus replication in the respiratory tract of humans. Studies of avian–human single and multiple gene substitution reassortant influenza A viruses have demonstrated three different ways in which avian influenza gene segments may contribute to the restricted growth or attenuation of these reassortants in primates. First, the avian nucleoprotein, matrix proteins, or allele B of the nonstructural protein gene is able to confer the attenuation phenotype on a human influenza virus independently. Second, combinations of avian influenza virus genes, such as the PB1 and NS gene of A/Mallard/78, may specify this phenotype. Finally, specific constellations of avian and human genes, such as the A/Pintail/76 x A/Washington/80 polymerase complex, may confer attenuation. Thus, not all combinations of avian and human influenza virus genes allow viruses to replicate in primates well enough to spread throughout the human population. This indicates that a second factor contributing to the infrequent emergence of pandemic viruses is the restriction of replication in primates of many of the avian–human influenza A reassortant viruses generated during double infection. These observations are also compatible with the particular constellation of avian genes seen in the human 1957 pandemic virus, in which the potentially attenuating NP, M, and NS genes were derived from the preexisting human viral parent. On the other hand, two of ten avian influenza viruses tested and the majority of avian–human influenza A reassortant viruses evaluated in these studies appear to function efficiently in primates. Thus, the paradox of infrequent emergence of pandemic influenza viruses may not be totally explained by restriction of growth in primates imposed by certain avian influenza virus internal genes.

A third possibility is that the hemagglutinin could also potentially contribute to restricted growth of avian viruses in humans. The human population can be considered to be immunologically virgin with reference to the majority of HA subtypes on influenza A viruses circulating in birds. However, only three hemagglutinin subtypes have been observed in human influenza A viruses during the past century, and current evidence suggests that these HA subtypes recirculate when the level of immunity to the particular HA subtype in the population declines. Only indirect evidence is available regarding the potential restriction of viruses bearing the other HA subtypes in primates. However, the HA did not appear to be solely responsible for the restriction of replication in primates of the avian viruses that manifested this phenotype. Further studies of reassortants bearing avian HA genes under carefully controlled conditions may be useful in investigating this possibility further. Ultimately, a full understanding of the determinants of the host range of avian influenza viruses will have important implications for influenza biology, epidemiology, and vaccine development.

REFERENCES

Air, G. M. (1981). *Proc. Natl. Acad. Sci. USA* **78,** 7639–7643.

Baez, M., Zazra, J. J., Elliott, R. M., Young, J. F., and Palese, P. (1981). *Virology* **113,** 397–402.

Bean, W. J. (1983). *Virology* **133,** 438–442.

Buckler-White, A. J., and Murphy, B. R. (1986). *Virology* **155,** 345–355.

Buckler-White, A. J., Naeve, C. W., and Murphy, B. R. (1986). *J. Virol.* **57,** 697–700.

Buonagurio, D. A., Nakada, S., Parvin, J. D., Krystal, M., Palese, P., and Fitch, W. M. (1986). *Science* **232,** 980–982.

Colman, P. M., and Ward, C. W. (1985). *Curr. Topics Microbiol. Immunol.* **114,** 177–255.

Cox, N. J., Bai, Z. S., and Kendal, A. P. (1983). *Bull WHO* **61,** 143–152.

Deshpande, K. L., Fried, V. A., Ando, M., and Webster, R. G. (1987). *Proc. Natl. Acad. Sci. USA* **84,** 36–40.

Hinshaw, V. S., and Webster, R. G. (1982). In *Basic and Applied Influenza Research* (A.S. Beare, ed.), pp. 79–104, CRC Press, Boca Raton, FL.

Hinshaw, V. S., Bean, W. J., Webster, R. G., and Easterday, B. C. (1978). *Virology* **84,** 51–62.

Hinshaw, V. S., Bean, W. J., Webster, R. G., and Sriram, G. (1980). *Virology* **102,** 412–419.

Hinshaw, V. S., Webster, R. G., Naeve, C. W., and Murphy, B. R. (1983). *Virology* **128,** 260–263.

Kawaoka, Y., and Webster, R. G. (1988). *Proc. Natl. Acad. Sci. USA* **85,** 324–328.

Kida, H., Kawaoka, Y., Naeve, C. W., and Webster, R. G. (1987). *Virology* **159,** 109–119.

Kingsbury, D. W., (1989). In *Virology* (B. N. Fields *et al.*, eds.), Raven Press, New York (in press).

Klingborn, B., Englund, L., Rott, R., Juntti, N., and Rockborn, G. (1985). *Arch. Virol.* **86** 347–351.

Lamb, R. A. (1989). In *The Influenza Viruses* (R. M. Krug, ed.), pp. 1–67, Plenum Press, New York.

Masurel, N., and Marine, W. M. (1973). *Am. J. Epidemiol.* **97,** 44–49.

Murphy, B. R. and Webster, R. G. (1989). In *Virology* (B. N. Fields *et al.*, eds.), Raven Press, New York (in press).

Murphy, B. R., Sly, D. L., Hosier, N. T., London, W. T., and Chanock, R. M. (1980). *Infect. Immun.* **28,** 688–691.

Murphy, B. R., Hinshaw, V. S., Sly, D. L., London, W. T., Hosier, N. T., Wood, F. T., Webster, R. G., and Chanock, R. M. (1982a). *Infect. Immun.* **37,** 1119–1126.

Murphy, B. R., Sly, D. L., Tierney, E. L., Hosier, N. T., Massicot, J. G., London, W. T., Chanock, R. M., Webster, R. G., and Hinshaw, V. S. (1982b). *Science* **218,** 1330–1332.

Murphy, B. R., Harper, J., Sly, D. L., London, W. T., Miller, N. T., and Webster, R. G. (1983). *Infect. Immun.* **42,** 424–426.

Naeve, C., and Webster, R. G. (1983). *Virology* **129,** 298–301.

Naeve, C., Hinshaw, V. S., and Webster, R. G. (1984). *J. Virol.* **51,** 567–569.

Nakajima, K., Nobusawa, E., Ogawa, T., and Nakajima, S. (1987). *Virology* **158,** 465–468.

Ogawa, T., and Ueda, M. (1981). *Virology* **113,** 304–313.

Patriarca, P. A., Kendal, A. P., Zakowski, P. C., Cox, N. J., Trautman, M. S., Cherry, J. D., Auerbach, D. M., McCusker, J., Belliveau, R. R., and Kappau, K. D. (1984). *Am. J. Epidemiol.* **119,** 152–158.

Rogers, G. N., and Paulson, J. C. (1983). *Virology* **127,** 361–373.

Scholtissek, C., and Von Hoyningen-Huene, V. (1980). *Virology* **102,** 13–20.

Scholtissek, C., Rohde, W., Von Hoyningen, V., and Rott, R. (1978). *Virology* **87,** 13–20.

Scholtissek, C., Burger, H., Kistner, O., and Shortridge, K. F. (1985). *Virology* **147,** 287–294.

Snyder, M. H., Buckler-White, A. J., London, W. T., Tierney, E. L., and Murphy, B. R. (1987). *J. Virol.* **67,** 2857–2863.

Sugiura, A., and Ueda, M. (1980). *Virology* **101,** 440–449.

Tian, S-F., Buckler-White, A. J., London, W. T., Reck, L. J., Chanock, R. M., and Murphy, B. R. (1985). *J. Virol.* **53,** 771–775.

Treanor, J. J., Snyder, M. H., London, W. T., and Murphy, B. R. (1989). *Virology* **171,** 1–9.

Webster, R. G., and Rott, R. (1987). *Cell* **50,** 665–666.

Webster, R. G., Campbell, C. H., and Granoff, A. (1971). *Virology* **44,** 317–328.

Webster, R. G., Geraci, J. R., Petursson, G., and Skirnisson, K. (1981a). *N. Engl. J. Med.* **304,** 911.

Webster, R. G., Hinshaw, V. S., Bean, W. J., Van Wyke, K. L., Geraci, J. R., St. Aubin, D. J., and Petursson, G. (1981b). *Virology* **113,** 712–724.

Wiley, D. C., and Skehel, J. J. (1987). *Annu. Rev. Biochem.* **56,** 365–394.

11

Newcastle Disease Virus Variations

P. H. Russell, A. C. R. Samson, and D. J. Alexander

Variation among Newcastle disease variations (NDV) strains has traditionally been by functional tests, e.g., virulence, thermostability of the hemagglutinin, plaquing ability. Variation has recently been described at the genetic level by gene cloning and at the antigenic level by monoclonal antibodies (MAb); in this chapter we discuss these two factors separately.

I. GENETIC VARIATION

A. Introduction

Nucleic acid sequences from cloned NDV genes were first published in 1986 (Chambers *et al.*, 1986a,b; McGinnes and Morrison, 1986). Since then the fusion (F) and hemagglutinin–neuraminidase (HN) gene sequences from several different strains have been published together with one or two sequences from the remaining NDV genes; NP, NAP(P), M, and L. In addition to information gleaned from sequence studies using cloned NDV genes, further sequence information has been obtained using specific oligonucleotides to prime DNA sequencing reactions with the NDV RNA genome as template.

This section deals only with sequence information obtained from the genes that code for the two virion surface glycoproteins F and HN, for which there now exist sufficient data for a fruitful analysis of strain-to-strain variation. In addition, sequence changes within the HN genes of certain NDV strains that result from the acquisition of resistance to neutralizing monoclonal antibodies are discussed.

B. Fusion Protein F

For all NDV strains there is an absolute requirement for a specific cleavage event to biologically activate the fusion protein. This posttransitional cleavage event is mediated by a trypsinlike protease and is a prerequisite for virus–cell membrane fusion and hence normal infection. (For a review of this topic, see Choppin and Scheid, 1980; Samson, 1988).

P. H. Russell • Department of Veterinary Pathology, The Royal Veterinary College, London NW1 0TU, England. *A. C. R. Samson* • Department of Biochemistry and Genetics, University of Newcastle upon Tyne, Newcastle upon Tyne NE2 4HH, England. *D. J. Alexander* • Poultry Department, Central Veterinary Laboratory, Weybridge, New Haw, Surrey KT15 3NB, England.

F gene sequencing studies (e.g., Chambers *et al.*, 1986b; McGinnes and Morrison, 1986) established that the predicted amino acid sequences for the unprocessed F protein contain three runs of about 25 hydrophobic amino acids; one within the N-terminal signal sequence (1–32), one at the N-terminus of what after F_0 cleavage becomes the F_1 polypeptide (117–142) and is thought to promote virus–cell fusion (denoted "fusor" sequence in Fig. 1), and one close to the C-terminus of F_1, which anchors the protein in the virus membrane (500–525, see Fig. 1).

Close examination of the predicted amino acid sequences for F, representing the seven NDV strains for which there are published complete sequences, reveals the following common features in addition to the runs of hydrophobic amino acids noted above. All seven F proteins are the same length (553 residues). Six well-conserved potential glycosylation sites (NXT/S) are found at the following residue numbers (identifying the N residues in Fig. 1); 85, 191, 366, 447, 471, and 541. All but the first are located in the larger polypeptide F_1. The conserved potential glycosylation site at 541 lies beyond the C-terminal anchor sequence, and if indeed this site is glycosylated, it would suggest that this part of the C-terminal region would be on the outside of the virus membrane. The only variation with respect to the potential for glycosylation seen so far is an overlapping site at 192 and an additional site at 497 both in the Miyadera strain.

Also well conserved among the seven strains are the positions of cysteine residues, many of which are likely to play a vital role in secondary and tertiary protein structure. Twelve cysteine residues are conserved in all seven strains; 25, 76, 199, 338, 347, 370, 394, 399, 401 (but see Ulster sequence), 424, and 523. Cysteine residues occur in some strains at 27 and 514. The cysteine residue at position 76 is the only cysteine found in mature F_2 and therefore must be involved in the disulfide bridge linking it to F_1 (Scheid and Choppin, 1977).

It has been suggested by Chambers *et al.* (1986b) that cysteine 523, which is located within the hydrophobic anchor region, is a candidate for attachment of fatty acid to the F protein. The complete conservation of this residue lends further support to this suggestion.

The region of the F amino acid sequence that exhibits most strain-to-strain variation is within the signal sequence (1–32). This sequence is thought to be cleaved off the F protein soon after the association of the nascent peptide with membrane vesicles, it therefore does not form part of the functioning F protein and thus may be subject to fewer evolutionary constraints.

By far the most significant strain-to-strain differences within the F protein are concerned with the region bordering the F_2/F_1 cleavage site. Recent sequencing studies have shown unequivocally that there are significant amino acid sequence differences in this region and that these correlate with the virulence of the NDV strain (Toyoda *et al.*, 1987; Glickman *et al.*, 1988; Millar *et al.*, 1988).

In Table 1 relevant regions from cloned F gene sequences (from Fig. 1) are combined with sequences for this region obtained using specific oligonucleotides to prime reverse transcriptase–driven DNA sequencing reactions using NDV RNA genomes as templates (Toyoda *et al.*, 1987; Glickman *et al.*, 1988). As pointed out by these workers, all virulent NDV strains have the amino acid sequence RRQK(orR)RF [viz. ArgArgGlnLys(orArg)ArgPhe] between residues 112 and 117, whereas all avirulent NDV strains have the amino acid GR(orK)QGRL [viz. GlyArg(orLys)GlnGly-ArgLeu] at these positions. In both cases cleavage occurs just before residue 117.

The efficient proteolytic cleavage of F protein in virulent strains in a variety of cell lines (Nagai *et al.*, 1976) thus correlates with the presence of two pairs of basic amino acids K or R (lysine or arginine) separated by a Q (glutamine) residue. This motif is immediately followed by an F (phenylalanine) residue, which marks the N-terminus of the F_1 polypeptide. In contrast, none of the avirulent strains possesses pairs of basic amino acids, and in all cases residues 112 and 115 are both G (glycine), and the N-terminus of such F_1 polypeptides is L (leucine) (Toyoda *et al.*, 1987, as revised by Nagai, personal communication; Glickman *et al.*, 1988).

C. Hemagglutinin–Neuraminidase Protein HN

HN is the larger of the two NDV glycoproteins and is responsible for the attachment of virions to host cell receptors and for receptor destroying activity (neuraminidase) Scheid and Choppin,

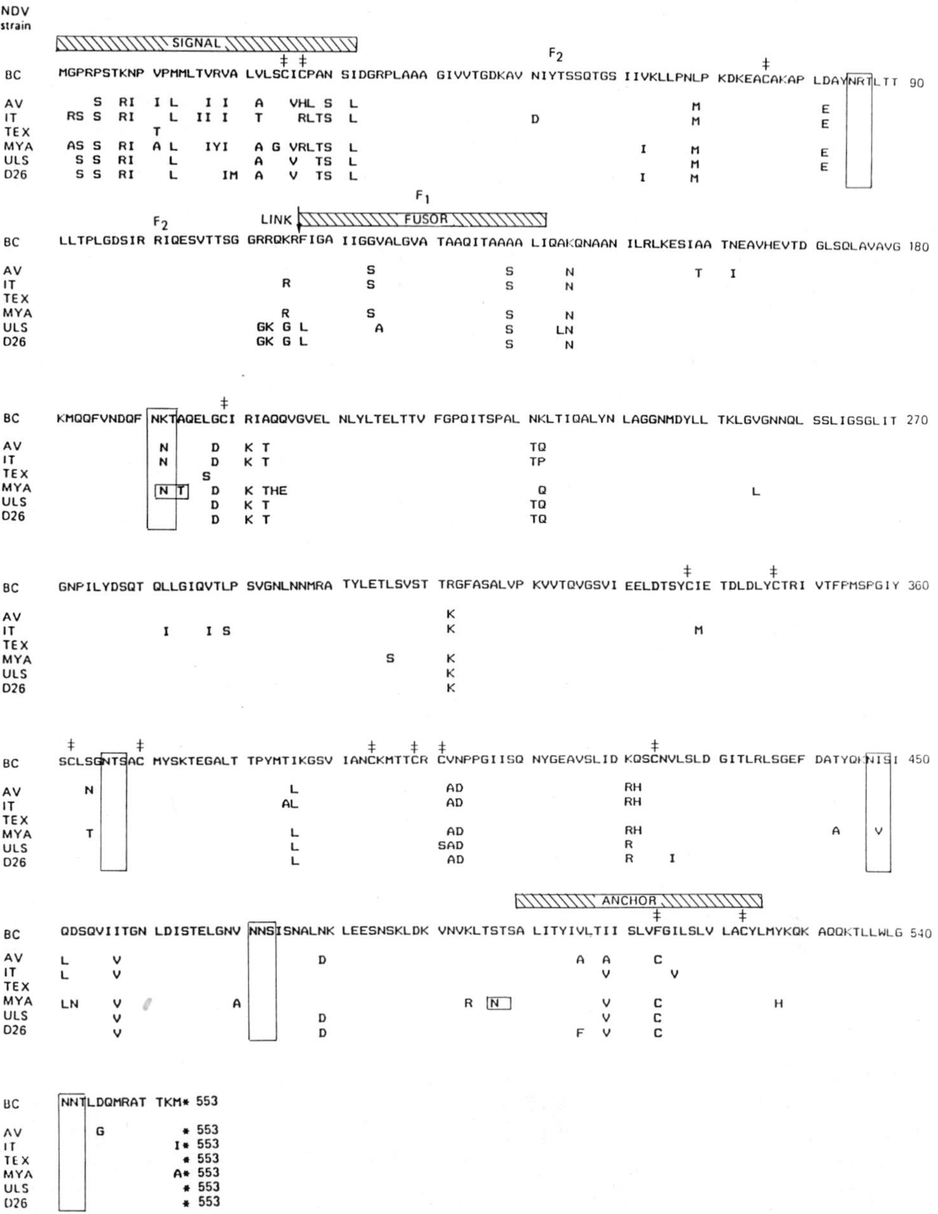

Figure 1. Comparison of the predicted amino acid sequences of the F proteins from seven NDV strains. The complete single letter code amino acid sequence for the Beaudette C (BC) strain is given and only amino acid differences are shown for the other strains. The three hydrophobic regions signal, fusor, and anchor are shown as hatched bars above the sequence. The F_2, F_1, and linker regions are also indicated above the sequence together with the site of cleavage between F_1 and F_2 (↓). Potential glycosylation sites (NXT/S) are shown boxed within the sequences. Positions of cysteine residues for any strain are indicated with a ‡ above the sequence. The position of the first stop codon is shown as an * in each sequence. Sources of sequence information are: BC (Beaudette C from Chambers *et al.*, 1986b), AV (Australia–Victoria from McGinnes and Morrison, 1986), IT (Italien from Espion *et al.*, 1987), TEX (Texas from Schaper *et al.*, 1988), MYA (Miyadera from Toyoda *et al.*, 1987), ULS (Ulster from Millar *et al.*, 1988), and D26 (Sato *et al.*, 1987a).

Table 1. Comparison of Predicted Amino Acid Sequences around the F_2/F_1 Boundary among Virulent and Avirulent NDV Strains[a]

NDV strain	Amino acid sequence differences F_1/F_2 boundary																			
Virulent	107				F_2					↓				F_1						126
Beaudette C[b]	T	T	S	G	G	R	R	Q	K	R	F	I	G	A	I	I	G	G	V	A
Australia–Victoria[b]																		S		
Italien[b]									R									S		
Texas[b]																				
Miyadera[b]									R									S		
California RO[c]									R						L					
Israel HP[c]									R									S		
Italy–Milano[c]									R									S		
Iowa–Salsbury[c]																		V		
NJ-Roakin[c]																				
L-Kansas[c]																				
Herts[d]									R											
Avirulent																				
Ulster[b]						G	K		G		L									
D26[b]						G	K		G		L									
NJ-La Sota[c]						G			G		L									
B1-Hitchner[c]						G			G		L									
Wisconsin-Appleton[c]						G			G		L									
England-F[c]						G			G		L									
Queensland[d]						G	K		G		L									

[a]Amino acid residues 107–126 inclusive are shown for the Beaudette C strain; only amino acid differences are shown for other strainns. The site of cleavage between F_2 and F_1 is shown by a vertical arrow.
[b]See Fig. 1 for reference details.
[c]Taken from Glickman *et al.* (1988).
[d]Taken from Toyoda *et al.* (1987) and Nagai (personal communication).

1973). In some avirulent NDV strains the HN protein is made as an inactive precursor polypeptide (HN_0) which is cleaved to the biologically active HN *in ovo* and in certain cell lines. In most NDV strains there is no such precursor (Nagai *et al.*, 1976; Nagai and Klenk, 1977). HN gene sequencing studies have shown that the lengths of the predicted open reading frames vary from strain to strain whereas the HN gene length appears similar if not the same (see Millar *et al.*, 1988). Moreover, the sequence of codons beyond the stop codon for those NDV strains which do not code for a precursor HN_0 protein bears a remarkably close resemblance to those found at the same positions for strains which do (see Millar *et al.*, 1988, for details). This strongly suggests that strains that produce HN and not HN_0 have evolved from HN_0-producing strains via the acquisition of upstream stop codons thereby shortening the open reading frame (Chambers *et al.*, 1986a; Millar *et al.*, 1988). Heterogeneity in HN polypeptide length is thus due to differences in the relative position of the first stop codon in the HN gene. The open reading frame of the Australia–Victoria strain, however, is shortened still further by the deletion of the codon for tyrosine 185 relative to the other sequences.

Close examination of the predicted amino acid sequences for HN and HN_0, taken from the eight NDV strains for which there are published sequences, reveals the following common features (Fig. 2). There is a run of 38 mainly hydrophobic amino acids between residues 8 and 45 which is the presumed anchor sequence for the HN glycoprotein (Millar *et al.*, 1986). An N-terminal anchor sequence had earlier been reported for the neuraminidase protein of influenza virus (Fields *et al.*, 1981) and for the HN proteins of other paramyxoviridae (Blumberg *et al.*, 1985; Hiebert *et al.*, 1985).

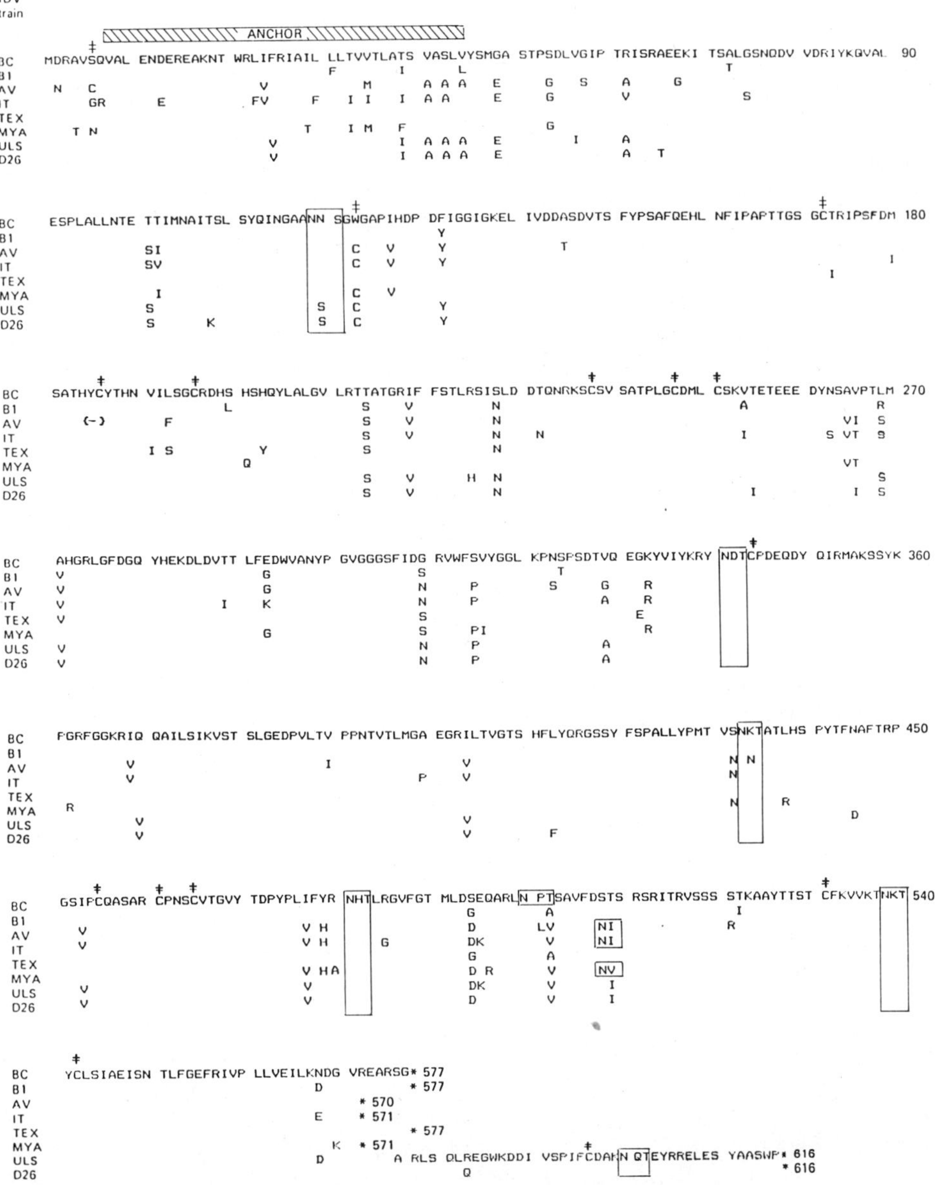

Figure 2. Comparison of the predicted amino acid sequences of the HN proteins from eight NDV strains. The complete single letter code amino acid sequence for the Beaudette C (BC) strain is given and only amino acid differences are shown for the other strains. The sequence of amino acids for the avirulent strain Ulster (ULS) is shown beyond residue 577 together with any changes in strain D26. The hydrophobic anchor region is shown as a hatched bar above the sequence. Potential glycosylation sites (NXT/S) are shown boxed within the sequences. Positions of cysteine residues for any strain are indicated with a ‡ above the sequence. The position of the first stop codon is shown as an * in each sequence. The deletion of codon 185 from the AV strain is shown as a (−). Sources of sequence information are: BC (Beaudette C from Millar *et al.*, 1986), B1 (B1-Hitchner from Jorgensen *et al.*, 1987), AV (Australia–Victoria from McGinnes *et al.*, 1987), IT (Italien from Wemers *et al.*, 1987), TEX (Texas from Schaper *et al.*, 1988), MYA (Miyadera from Toyoda *et al.*, 1987), ULS (Ulster from Millar *et al.*, 1988, cf Gotoh *et al.*, 1988) and D26 from Sato *et al.* (1987b).

Five well-conserved potential glycosylation sites are found at the following asparagine residue numbers: 119, 341, 433, 481, and 538 within HN and a sixth site at 599 within HN_0. A sixth potential glycosylation site within HN has been reported for the Beaudette C strain at residue 500 (Millar *et al.*, 1986) and for Australia–Victoria, Italien, and Miyadera strains at residue 508 (McGinnes *et al.*, 1987; Wemers *et al.*, 1987; Gotoh *et al.*, 1988).

Twelve cysteine residues are conserved in all eight NDV strains sequenced so far: 172, 186, 196, 238, 247, 251, 344, 455, 461, 465, 531, and 541 within HN and a further site at 596 within HN_0. Cysteine residues occur in some strains but not others at residues 6 (Australia–Victoria) and 123 (Australia–Victoria, Italien, Ulster, and D26). The Cys 123 residue has been shown to be responsible for the disulfide-linked dimer form of the HN protein in a number of NDV strains (Sheehan *et al.*, 1987). These workers used oligonucleotides to prime reverse transcriptase–driven DNA sequencing reactions using NDV genome RNAs as templates in order to examine the sequences of several NDV strains within the vicinity of Cys 123. They found a strict correlation between the presence of Cys 123, from the predicted amino acid sequence, and reducing agent–sensitive dimerization of the HN protein. However, they found no causal relationship between disulfide-linked dimeric HN and virulence.

The locations of the two active sites on the NDV HN glycoprotein are not yet defined. Jorgensen *et al.* (1987) compared their NDV B1-Hitchner sequence to that of other paramyxovirus HN proteins and to the region around the known sialic acid binding site of influenza neuraminidase and concluded that the region 212–303 in NDV HN most resembled this site. In addition Millar *et al.* (1986) identified a site at residues 399–404 which was conserved among paramyxoviruses and was similar to the influenza hemagglutinin sialic acid binding site.

Further work with NDV and other paramyxovirus HN proteins is needed before these active sites can be unequivocally defined.

D. Location of Antigenic Sites and Epitopes within the HN Protein

The ability to prepare monoclonal antibodies (MAbs) against a chosen antigen has been exploited by virologists ever since the classic studies of Köhler and Milstein in 1975. MAbs that neutralize viruses can be used to select escape mutants (MAb-R) which are now resistant to the neutralizing effect of these antibodies. The genomes from these MAb-R mutants can then be subjected to rapid DNA sequencing by using oligonucleotides (chosen and synthesized on the basis of the previously known wild-type gene sequence), to act as specific primers for reverse transcriptase–driven DNA sequencing reactions using mutant genomes as templates.

Individual MAbs from a panel of MAbs can be grouped on the basis of competition binding studies. MAbs are said to bind to the same antigenic site if they compete with one another in a binding assay. Antigenic sites may be subdivided into epitopes on the bases of the sensitivity or resistance of MAb-R mutants to other MAbs. MAbs reacting with different sites or epitopes can be used to probe the biologic activities of the wild-type virus protein to which they bind. Coupled with the MAb-R sequencing studies this approach allows a correlation between a specific change in a gene sequence (and hence its product) and the perturbation of a particular biologic activity by the MAb. In addition, valuable insight into the structure of antigenic sites can be gained.

Several groups have isolated MAbs raised against several different NDV strains (see Table 3). More recently the locations of antigenic sites and epitopes on the HN protein recognized by certain MAbs has been determined by partial digest analysis of HN protein (Samson *et al.*, 1988), by synthesis of cloned HN gene fragments in *Escherichia coli* (Chambers *et al.*, 1988), and by rapid gene sequencing of MAb-R escape mutants (Gotoh *et al.*, 1988; Yusoff *et al.*, 1988).

Sequencing studies by Gotoh *et al.* (1988) using escape mutants selected from the avirulent D26 strain of NDV identified the locations of three of the four antigenic sites described in earlier work from that group (Nishikawa *et al.*, 1986). Mutants at the immunodominant site I give rise to a change from glutamic acid (E) at amino acid position 347 to mainly lysine (K) or in one case glycine (G). Antibodies used to select these mutants inhibit neuraminidase activity and map closer to the sialic acid binding site proposed by Jorgensen *et al.* (1987) than mutants at other sites. A mutant at site II had a change from glutamic acid (E) at position 495 to valine (V). Two mutants at site IV suffered a change from an asparagine (N) at position 481 to aspartic acid (D), a change that eliminates a potential glycosylation site (see Table 2). Gotoh *et al.* (1988) found that the mobility of the HN_0 protein is increased in site IV mutants compared with that of the wild type. Nishikawa *et al.* (1986) had shown previously that site IV was sensitive to treatment with endoglycosidase F. Gotoh

Table 2. Location of Amino Acid Changes Predicted from the Nucleotide Sequences of HN Genes from Monoclonal Antibody-resistant Mutants Compared to Sequence Data for 13 Field Isolates of NDV

Amino acid residue number[a]	D26 strain[b]		Beaudette C strain[c]	
	Antigenic site	Amino acid change	Epitope	Amino acid change
284			B1	Lys → Glu
284			B1	Lys → Asn
287			B2	Asp → Tyr
290			B2	Thr → Lys
293			B2	Glu → Lys[d]
325			B3	Pro → Ser
325			B3	Pro → Leu
347	I	Glu → Lys[e]	A1	Glu → Lys[e]
347	I	Glu → GLy	A1	Glu → Gly
349			A1	Asp → Tyr
454			A2	Pro → Leu
456			A2	Gln → His
457			A2	Ala → Asp
460			A3	Arg → Gly
481	IV	Asn → Asp[f]	C1	Asn → Asp
495	II	Glu → Val[g]		

[a]See Fig. 2.
[b]Taken from Gotoh *et al.* (1988).
[c]Taken from Yusoff *et al.* (1988).
[d]Glu 293 is present on sequenced viruses of our MAb-binding groups D, G, and Q compared to Lys or Gly on those of groups B, E, and R (see Sakaguchi *et al.*, 1988 and Table 4). MAb 445 to B2 correspondingly binds D, G, and Q but not B, E, and R isolates (Table 4).
[e]Glu 347 is present on sequenced viruses of our MAb-binding groups B, D, E, G, and Q compared to Lys on Ibaragi/85 of group R. MAbs 14 to site 1 and A1 correspondingly bind all the above groups except R.
[f]No sequence differences among all 13 isolates at aa residues 284, 287, 349, 454, 456, 457, 460, 481 (Sakaguchi *et al.*, 1988).
[g]Glu 495 is present on D26 compared to Val on La Sota or Lys on Italien with loss of binding to La Sota and Italien by MAb 21 to site II (Nishikawa *et al.*, 1987).

et al. (1988) suggest that the hydrophilic character of the glycoslyated site IV is diminished by the loss of carbohydrate at this site in the escape mutants.

Samson *et al.* (1988) used certain MAbs directed against the Ulster (Russell *et al.*, 1983) and Italien (LeLong *et al.*, 1986) strains, which neutralize the virulent Beaudette C strain, to select MAb-R mutants and to obtain an approximate position for the immunodominant group of MAbs (group la). This was achieved by Western blot analysis of partial digest fragments of HN. This immunodominant epitope, which is resistant to boiling in SDS in the presence of 2-mercaptoethanol (and is therefore a nonconformational or linear epitope), was identified more precisely by Western blot analysis of cloned fragments of the HN gene and synthetic oligonucleotides expressed in *E. coli*. In this study the peptide DEQDYQIR (residues 346–353) was identified as the linear epitope. Furthermore, sequence analysis of a MAb-R mutant revealed a glutamic acid (E) to lysine (K) change at position 347 (Chambers *et al.*, 1988).

More recently an extended panel of neutralizing MAbs has been used to select MAb-R mutants from the Beaudette C strain. So far seven different epitopes have been found among three antigenic sites (sites A, B, and C) (Yusoff *et al.*, 1988). Epitope Al is the immunodominant linear epitope referred to earlier and previously named group la. It is the only linear epitope found so far in the Beaudette C strain. [All antigenic sites described by Gotoh *et al.* (1988) for D26 were conformational.]

Table 2 compares the amino acid replacement data for the D26 strain (Gotoh *et al.*, 1988) with that for the Beaudette C strain (Yusoff *et al.*, 1988), and it can be seen that the two immunodominant sites (site I in D26 and epitope Al in Beaudette C) map to the same amino acid residue Glu 347). However, the environment of these sites/epitopes must be different because site I in the avirulent D26 is sensitive to reducing agent (and SDS) but epitope Al in the virulent Beaudette C strain is not.

The Asn to Asp change for site IV of D26, which eliminates a potential glycosylation site, is also found for the Cl epitope in Beaudette C (residue 481). The Beaudette C mutant also appears to produce an HN protein with an increased electrophoretic mobility. We reported earlier that no change in electrophoretic mobility had been seen in this mutant (Yusoff *et al.*, 1988).

One Beaudette C mutant produced an HN protein that had a lower electrophoretic mobility than the wild type. Examination of the sequence of this mutant showed it had a change from proline (325) to serine, which creates an additional potential glycosylation site; moreover, the electrophoretic mobility of both mutant and wild-type nonglycosylated (tunicamycin blocked) HN was the same (Yusoff *et al.*, 1988).

The only amino acid residue change reported for the D26 strain which was not found in the Beaudette C was at 495 (Glu to Val) at site II. This mutant was unusual in that it retained binding capacity to the antibody used for selection (Nishikawa *et al.*, 1983).

None of the amino acid substitutions found by Gotoh *et al.* (1988) fell within the 212–303 region proposed to contain the sialic acid binding site for neuraminidase proposed by Jorgensen *et al.* (1987). It will be recalled that changes at site I in D26 (residue 347), were generated by antibodies that inhibit neuraminidase. The only MAbs that inhibited neuraminidase in the Beaudette C strain selected mutants located at epitopes A2 and A3 (residues 454–460) but not at epitope Al (347).

Further exploitation of MAbs that inhibit the functions of HN should lead to a better definition of the active sites of this multifunctional protein.

Recently two papers have been published reporting sequence changes in escape mutants selected with monoclonal antibodies to the NDV fusion protein; Toyoda *et al.* (1988) for the Sato strain (together with the F gene sequence for that strain), and Neyt *et al.* (1989) for the Italien strain. Work in our laboratory with the Beaudette C strain has identified similar sequence changes in F escape mutants (Yusoff *et al.*, 1989). In addition, a new review article "Molecular Biology of Newcastle Disease Virus" by Nagai *et al.* (1989) has been published.

II. MONOCLONAL ANTIBODIES AND ANTIGENIC VARIATION OF FIELD ISOLATES

A. Introduction

NDV isolates are monotypic by hemagglutination inhibition (HI) testing using avian polyclonal antisera but show a spectrum of virulence. The most avirulent field isolates can be used as vaccines without further attenuation, because virus with cleaved fusion protein, i.e., infective virus, is produced in only a narrow range of cells—allantoic endoderm after *in ovo* inoculation and the surface epithelium of the respiratory tract (Hitchner B1) or gut (Queensland V4) after contact infection. Virulent viruses replicate throughout the egg and kill birds within a few days after a more extensive cell damage centered on the gut and viscera (viscerotropic, e.g., Herts '33), respiratory tract (pneumotropic, e.g., Essex '70), or brain (neurotropic, e.g., GB Texas). Virulent viruses also form plaques in cell culture without the addition of trypsinlike proteases.

B. Conserved and Variable Epitopes

Following the use of MAb to analyze antigenic drift of influenza virus and neurovirulence of rabies, several laboratories produced MAbs to NDV to search for antigenic variation that may be relevant to virulence or epizootiology. MAb to the HN were produced readily and could differentiate strains by HI testing. MAb to F were less easy to produce or to detect by immunoprecipitation. MAb to most internal viral proteins, M, NP, and P, but not to the large polymerase protein (L), have been reported. The results are summarized in Table 3.

Regions, i.e., distinct MAb-combining sites, on these molecules were analyzed by competition blocking between labeled and unlabeled MAbs. Several laboratories have subdivided HN into four regions using their own panels of MAbs. The relationship of these regions defined by separate laboratories is usually unknown because of the limited exchange of MAb and NDV isolates (Nishikawa *et al.,* 1986).

A region can be subdivided into epitopes by several criteria: nonreciprocal blocking, i.e., one epitope may overlap another; nonreciprocal neutralization of MAb escape mutants; binding of MAb to different field isolates, which obviously depends on choice of isolates.

The definition of a conserved epitope also depends on choice of isolate. The HN-1 epitope of Ulster 2C was considered to be largely conserved because it bound 36/40 isolates and the four exceptions were lentogenic and from feral ducks (Russell and Alexander, 1983). However, the virus strain responsible for the panzootic in pigeons and which caused outbreaks of ND in British poultry flocks in 1984 also lacked HN-1 (Alexander *et al.*, 1985a,b). The causal virus was, however, shown to differ from the duck viruses using other MAb and the suggestion that the four feral duck viruses were irrelevant exceptions was therefore disproved.

Another HN epitope, defined by Meulemans' MAb 8C11, was more truly conserved and bound to all viruses tested including the above duck and pigeon isolates with HI titers of 40–10,240 (Meulemans *et al.,* 1987; Alexander, unpublished). This MAb is suitable for diagnosis of NDV isolates by HI testing.

MAb to F were most readily obtained by the inoculation of infectious purified whole virus. The epitopes these MAb bind appear to be more conserved than those on HN: 50% compared to 38%. Two of the MAb produced by Meulemans' laboratory—1C3 and 12C4—bind all NDV isolates so far examined (Meulemans *et al.,* 1987).

Variation among epitopes also exists on the internal proteins: 40% on matrix, 75% on P, and 25% on NP. MAb to these proteins do not neutralize and such variation has been ascribed to spontaneous variation (Nishikawa *et al.,* 1987). Spontaneous variation rather than immunoselection may also explain antigenic variation of the envelope proteins which each carry at least one immunodominant conserved epitope.

Table 3. Regions and Epitopes on NDV

Virus isolate	Protein	No. of MAbs made	Methods used[a]	No. of regions	No. of epitopes	No. of conserved epitopes	Key ref. (includes earlier refs.)
Australia–Victoria (A-V)	HN	17	a,b,c,e	4	9/10	6	Iorio *et al.* (1986)
JapDuck 26 (D26)	HN	11	b,c,d,e	4	4	1	Nishikawa *et al.* (1986)
Sato	HN	11	a,b,d,f	2	6	2	Abenes *et al.* (1986b)
La Sota	HN	18	c,d	?	4	1	Meulemans *et al.* (1987)
Miyadera	HN	3	c,e	?	3	1	Ishida *et al.* (1985)
Taka	HN	1		?	1	0	
Ulster	HN	4	a,b,c,d,e	2	2	0	Russell and Alexander (1983)
Ulster	F	1	c	1	1	0	Russell and Alexander (1983)
La Sota	F	1	a,c	?	5?	2	Meulemans *et al.* (1987)
Italien	F	4		?			
Sato	F	12	a,b,c,f	4	6	4	Abenes *et al.* (1986a)
JapD26	M	8	b,c	2	6	4	Nishikawa *et al.* (1987)
A-V	M	5	b,c	3	3	2	Faaberg and Peebles (1988)
Ulster	M	1	c	1	1	0	Russell and Alexander (1983)
Ulster	P	1	c	1	1	0	Russell and Alexander (1983)
JapD26	P	5	b,c	4	4	1	Nishikawa *et al.* (1987)
JapD26	NP	3	b,c	2	2	2	Nishikawa *et al.* (1987)
Ulster	NP	2	b,c	1	2	1	Russell and Alexander (1983)

[a]Methods used to enumerate regions and epitopes. a, Binding of MAbs to their mutants; b, cross-blocking; c, binding to different field isolates; d, stability and glycosylation of the antigen; e, f, functional properties, hemagglutination inhibition, hemolysis inhibition, VN, and neuraminidase inhibition with neuraminlactose and fetuin as substrates (e) or fetuin only (f).

Epitopes on NP and P are less conformation dependent than those on HN or F because 8/9 on NP and 3/3 on P resist boiling in SDS and reducing agent compared to 1/12 on HN and 0/5 on F (Iorio and Bratt, 1983; LeLong *et al.*, 1986; Nishikawa *et al.*, 1986; Russell *et al.*, 1983).

C. MAb to Shared PMV Antigens

NDV is classified as avian paramyxovirus type 1 (PMV-1) and 8 other serotypes are recognized (PMV-2 to PMV-9). MAb that cross-react between NDV and other serotypes of avian paramyxovirus are rare; two have been reported to date. One is to the F of NDV (Sota), which cross-reacts to a wide spectrum of NDV isolates and to a single isolate of PMV-2, -3, and -4 but not to -6 or -7 (Abenes *et al.*, 1986a). The second is to the MAb 161 to the HN of NDV/pigeon isolates (group P, see Section II.D.1) and cross-reacts to PMV-3 viruses from caged exotic birds but not to PMV-3 viruses of poultry or any other PMV-1 isolate (Collins *et al.*, 1989). This unique HN antigen could indicate either that these two virus groups have some common evolutionary origin and all other members of their serotypes have lost this epitope, or that it has evolved spontaneously in both viruses at different times.

D. Variation of NDV Field Isolates with Regard to Epizootiology

The majority of MAb to NDV bind to epitopes that are shared by some, but not all, isolates independent of their virulence characteristics. By the continued use of one panel of seven MAbs to

NDV (Ulster) we have been able to group existing NDV isolates by HI and/or indirect immunoperoxidase binding assays (Russell and Alexander, 1983). How these groups relate to epizootiology will be discussed later.

The first panel of MAb to Ulster recognized two epitopes on HN and NP and single epitopes on P, M, and F (Russell and Alexander, 1983). MAb to HN and F stain the plasma membrane of formalin-fixed MDBK cells which had been infected overnight. The MAbs to NP and P stain intracytoplasmic inclusion bodies. The MAb to M stains the cytoplasm diffusely and the nucleus if the formalin level is reduced (Russell *et al.*, 1983); and this intranuclear staining is enhanced by posttreatment with 0.1% triton as modified from Faaberg and Peebles (1988).

This classification method has now been extended using MAbs produced against the NDV pigeon isolate 617/83 by the Central Veterinary Laboratory and using MAb donated to us by several laboratories, notably those of Meulemans, and results are collated in Table 4, which lists viruses into Groups A–Q. The numbers of viruses tested so far for each group and their main properties are

Table 4. Summary of the Newcastle Disease Virus Groups Based on the Binding of Monoclonal Antibodies against NDV Ulster 2C and 617/83 to Infected MDBK Cells

Group	Brief description	Examples	MAb[a] bound	MAb[a] not bound
A	1970–1974 panzootic	Essex '70	14[b], 38, 479, 162, 165	445, 481, 688, 424, 161.38/617
B	Other VVND[c] probably the first isolates	Herts '33	14, 481, 38, 479, 165, 162, 38/617[d]	445, 688, 424, 161
C1	Velogenic	Kuwait 256	14, 481, 38, 479, 424, 165, 162, 38/617	445, 688, 161
C2	Lentogenic duck	1092/81	14, 481, 38, 479, 424, 165	445, 688, 161, 162, 38/617
D	"Pneumoencephalitis"	Texas GB	14, 445, 38, 688, 424, 479, 165	481, 161, 162, 38/617
E	Lentogenic vaccines	B1	14, 38, 688, 424, 479, 165	445, 481, 161, 162, 38/617
F	Lentogenic	F	14, 38, 424, 479	445, 481, 688, 161, 165, 162, 38/617
G	Lentogenic waterfowl poultry	Ulster 2C and some Queensland V4	14, 445, 38, 481, 688, 424, 479, 165	161, 162, 38/617
H	Lentogenic aquatic birds	MC110	481, 38, 424, 479	14, 445, 688, 161, 162, 165, 38/617
L	Lentogenic aquatic birds	Loon/83	14, 481, 38, 688, 424, 479, 165	445, 161, 38/617
P	Pigeon isolates	617/83	481, 38, 479, 161, 162, 165, 38/617	14, 445, 688, 424
Q	Lentogenic waterfowl poultry	Some Queensland V4	14, 445, 38, 481, 424, 479, 165	161, 162, 38/617, 688
R		IBA/85	481, 38, 479, 162, 165, 38/617	14, 445, 688, 424, 161

[a]MAbs bound to these proteins: 14, 32, 86, 445, 161, 162, 165 to HN; MAb 481 to F; MAb 688 to P; MAb 38, 479, 38/617 to NP; MAb 424 to M.
[b]MAbs 32 and 86 were bound identically to 14.
[c]Viscerotropic velogenic Newcastle disease.
[d]Some isolates do not bind 38/617.

summarized in Table 5. Viruses from every group, now A–R, have been recovered from field samples during 1985–1989 as if viruses from all groups continue to circulate.

The three large panzootics of NDV, however, involved different groups of viruses. Group B emerged from the Far East in 1926 (Java) then spread slowly over the next 30 years. Group A arose in the Middle East in the 1960s and then spread more rapidly as a result of the increased trade in poultry products and pet birds to reach most countries by 1973. Group P arose in the Middle East (Iraq, 1977), then was spread via pigeons to reach England in 1983; outbreaks among all poultry resulted from pigeon carcases being ground up into food.

1. Group A

Viruses in group A include the velogenic viruses isolated during the 1970–1974 panzootic. Of the viruses examined from that time both the UK and Californian isolates show the same binding pattern. The introduction of the disease to California was suspected to have been by importation of smuggled psittacines, which have been shown to have the potential to become persistent excretors of this virus (Erikson *et al.*, 1978), from Central and South America. Velogenic viruses isolated from exotic psittacines held in quarantine in the United States as late as 1985 and from poultry in Mexico and Paraguay in 1987–1988 (Alexander *et al.*, unpublished) have been shown to give group A binding patterns, which suggests that the panzootic virus may still be enzootic in these areas despite its apparent absence from other parts of the world.

2. Group B

Group B encompasses the earliest isolates of viscerotropic NDV, which appear to have arisen in Asia in the 1920s. The group includes Herts '33, its attenuated derivative, H, and the laboratory attenuated Indian strain, Mukteswar. With the exception of the latter strains, which may be regarded as mesogenic, all viruses in group B have been velogenic for chickens.

A 1968 Iraqi isolate from Abu-Ghraib, AG68, also falls into group B, which suggests it is not an early isolate of the 1970–1974 panzootic, as suggested by Allan *et al.* (1978). Two viruses in group B, Texas 19 and Northants '72, isolated in 1970 and 1972, respectively, are separated from

Table 5. Summary of Viruses Placed in MAb Binding Groups[a]

Group	Number tested	Pathogenicity	Main host	Comments
A	18	Velogenic 100%	Chickens 66%	Panzootic 1970–74
B	48	Velogenic 90%	Poultry 55%	Pre-1970 Post-1974
C1	46	Velogenic 100%	Poultry 62%	Exotic birds 33%
C2	6	Lentogenic 100%	Waterfowl 83%	
D	12	Various	Chickens 75%	USA virus
E	71	Vaccinal 100%	Poultry 75%	B1 and La Sota
F	6	Lentogenic 100%	Chickens 66%	F strain
G and Q	26	Lentogenic 100%	Aquatic birds 81% poultry 19%	Ulster 2C V4
H	18	Lentogenic 100%	Waterfowl 94%	
L	4	Lentogenic 100%	Aquatic birds 100%	
P	351	Various	Pigeons 91%	Poultry 7%
R	1	Velogenic	Poultry	IBA/85

[a]Figures represent percentage of the viruses tested in that category.

the panzootic viruses of group A, which were prevalent at that time. This coincides with clinical typing of Northants '72 as distinct from the panzootic virus (Alexander and Allan, 1974). Group B viruses and even different stocks of Herts '33 vary in their binding to MAb 38/617 to the NP of the pigeon virus.

3. Group C

Group C has now been divided into C1 and C2 on the basis of binding or not of MAb 38/617 and 162/617, which are not bound by C2 viruses.

Viruses placed in group C1 have been exclusively velogenic for chickens. The first of the isolates tested was obtained from chickens in Kuwait in 1968, but the majority of the subsequent isolates up to the 1980s were from exotic birds—usually psittacine species, obtained during routine sampling of birds in quarantine. However, in recent years viruses of this group have been associated with epizootics and outbreaks of virulent Newcastle disease in poultry in Mauritius, Saudi Arabia, Morocco, Italy, and Australia (Alexander *et al.*, 1987).

To date six viruses have been tested that form group C2; five are lentogenic viruses from waterfowl and the sixth was isolated from a pheasant in England.

4. Group D

Group D encompasses the viruses of varying virulence that are enzootic in poultry in the United States and were responsible for the form of the disease originally termed "pneumoencephalitis." Thus, included in this group are the velogenic neutrotropic strain Texas GB, the mesogenic vaccine strain Roakin, and the laboratory-derived mesogenic strain Beaudette C.

5. Group E

Group E includes the live vaccine strains Hitchner B1 and La Sota, and the majority of the isolates tested to date (Table 5) which fall into this group presumably represent reisolation of vaccine. These viruses represent viruses of low virulence isolated originally in the United States. Hitchner B1 was the first virus for which the vaccine potential was realized, by Hitchner although the virus had originally been supplied to him by Beaudette as an isolate of infectious bronchitis virus (Hitchner and Johnson, 1948). Hitchner B1 and La Sota are both relatively thermolabile strains and replicate readily in the respiratory tract. They are the most widely used live vaccines and may be administered by aerosol, coarse spray, intraocularly, or in the drinking water. Both viruses retain some inherent pathogenicity for day-old susceptible chickens, and in stringent tests La Sota vaccines may give mortality of 5–33%, compared to 0–13% for Hitchner B1 (Borland and Allan, 1980). Because of this high pathogenicity the use of La Sota is banned in some countries, e.g., Great Britain.

The two viruses can be separated by Group IV MAb 7D4 produced by Meulemans *et al.* (1987). This MAb gives an HI titer of about 1/2000 to La Sota and 1/20 to Hitchner B1. Erdei *et al.* (1987) produced a MAb that uniquely bound to La Sota out of 300 viruses tested by ELISA. La Sota has also been distinguished from B1 using MAb produced by Nishikawa *et al.* (1983, 1987) to the M-II and HN-II sites of NDV strain D26.

6. Group F

This group of six lentogenic viruses includes the F strain isolated from chickens by Asplin in 1952. The F strain has been used as a vaccine.

7. Groups G and Q

Russell and Alexander (1983), using their panel of MAbs to Ulster 2C, placed the homologous virus and Queensland V4 in Group G with other isolates, mainly from waterfowl, which also bound all MAb used. However, subsequent tests with other laboratory-maintained isolates, also designated Queensland V4 or V4, indicated that these viruses bound MAb 688 weakly if at all. Some isolates of V4 showed binding of MAb 688 at a dilution of 10^{-3} compared to $10^{-4.5}$ for Ulster 2C, while others, including the commercially available vaccine, showed only weak binding at 10^{-2}. Field isolates have been made, mainly from waterfowl, which show binding of all MAb including 688 at the same titer as Ulster 2C, while others have shown no binding at the lowest dilution tested. These findings suggest that two distinguishable groups of these closely related viruses exist, and it is proposed that group Q is established for viruses not binding MAb 688 for which V4 is the prototype virus.

Some isolates designated V4 and others from feral birds that would be placed in group C or Q have also shown some differences from Ulster 2C in that, although showing binding of MAb to the HN-1 epitope of Ulster 2C designated by Russell and Alexander (1983), unlike Ulster 2C they are not inhibited in HI tests with these MAb (Alexander *et al.*, unpublished).

Nishikawa *et al.* (1983) reported MAb raised against D26 that differentiate between V4 and Ulster 2C, and Collins *et al.* (unpublished) have raised a MAb to the HN of Ulster 2C, 70/U, which has bound to all NDV strains and isolates tested to date with the exception of V4-designated and V4-like (i.e., group Q) viruses.

Viruses of groups G and Q, typified by Ulster 2C and V4, all appear to be lentogenic, thermostable, enterotropic viruses. Ulster 2C may grow to extremely high titers in embryonated fowl's eggs and has been used as the virus antigen source for inactivated NDV vaccines. In recent years V4 has been proposed as a suitable live virus vaccine for use in tropical countries due to its thermostability (Sagild and Haresnape, 1986).

Australia has been considered free of ND since 1932 apart from infections with V4-like viruses. Antigenic assessment of viruses isolated from feral birds in Western Australia using MAb confirmed the presence of group Q viruses in wading and passerine birds but also demonstrated that a second antigenically distinct virus was present in waterfowl, as nine viruses were isolated in 1979 which showed group H binding patterns (Alexander *et al.*, 1986).

8. Group H

Viruses forming group H have been isolated from waterfowl and wading birds in the United States, Europe, and Australia; all have proved to be of low virulence for chickens.

An NDV isolate of low virulence was obtained from chickens in Northern Ireland (McNulty *et al.*, 1988) and subsequently shown to be of group H (Alexander and Manvell, unpublished), which may be interpreted as evidence of spread from feral waterfowl.

Viruses of group H appear to share very few of the HN epitopes with other NDV strains and this variant nature has been detectable in HI tests with polyclonal sera (Hannoun, 1977; McNulty *et al.*, 1988).

9. Group L

Three lentogenic viruses showing binding of all MAb except 445 have been isolated from feral aquatic birds in the United States (Alexander *et al.*, 1987; Alexander, unpublished), from a "duck" in 1972, a coot (*Fulica* sp.) in 1977, and a loon *Gavia* sp.) in 1983.

Isolates showing similar binding patterns were obtained from commercial ducks in Great Britain in 1988.

The loon isolate bound MAb to the HN-1 epitope but was not inhibited in HI tests.

Lentogenic viruses from aquatic birds have been placed into five different groups, C2, G, H, L, and Q, and there has been some evidence that viruses from these antigenic groups may have spread to poultry. Kawamura *et al.* (1987) separated 19 lentogenic Japanese duck isolates into four groups using their eight MAb in HI tests; these groups also excluded other viruses of chickens such as Hitchner B1 and Miyadera.

10. Group P

In 1981 a disease of racing pigeons was described in Italy which resembled the neurotropic form of Newcastle disease in chickens and from that time the disease appeared to spread westward across Europe and subsequently beyond to become a true panzootic. There is strong evidence to suggest that the virus was present as early as 1977 in Iraq but misdiagnosed as pigeon encephalitis herpesvirus (Kaleta *et al.*, 1985).

In binding tests with the panel of monoclonal antibodies used by Russell and Alexander (1983), the NDV (PMV-1) isolates from pigeons gave a unique pattern and were designated group P. Group P viruses have been confirmed as affecting pigeons in the following countries: Great Britain, Belgium, Federal Republic of Germany, Denmark, Sweden, Norway, Italy, Portugal, Spain, Switzerland, Austria, Hungary, France, Holland, Czechoslovakia, Israel, Egypt, Sudan, Uganda, South Africa, Hong Kong, Japan, Canada, and the United States (Alexander *et al.*, 1984, 1985b, 1987). This worldwide spread has been associated with the international nature of pigeon races and the trade in racing and show birds.

The disease arrived in Great Britain in 1983 and rapidly spread among pigeons during the racing season, including spread to feral birds (Lister *et al.*, 1986). Great Britain had had a nonvaccination policy for domestic poultry since 1981 for NDV, and although no ND outbreak had been recorded since 1978, in 1984 23 outbreaks occurred (Alexander *et al.*, 1985a). MAb typing of the 21 isolates that were made from these outbreaks showed 20 of the viruses to be of group P, while the other was of group E and probably represented vaccine virus (Alexander *et al.*, 1985a). This unequivocal confirmation of the similarity between poultry isolates and the virus causing disease in pigeons greatly aided the epizootiologic tracing of the source of infection of food stores infested with diseased feral pigeons from which ingredients of food had been taken and fed untreated to chickens, and enabled remedial measures to be taken.

A national poultry flock fully susceptible to NDV and the practice of using untreated food were probably unique to Great Britain. Nevertheless, the use of MAb typing has also confirmed the presence of group P viruses in domestic chickens in Austria, Federal Republic of Germany, Saudi Arabia, Uganda, and Kenya, turkeys in Israel, ducks in Switzerland, various exotic birds, such as important psittacines in Federal Republic of Germany, semidomesticated birds, i.e., kestrels in Great Britain, a buzzard in Federal Republic of Germany, and a feral sparrow in Switzerland (Alexander *et al.*, 1987 and unpublished).

Collins *et al.* (1989) raised MAbs to a group P isolate, PMV-1/pigeon/England/617/83. Seven of the nine MAb tested bound to all group P viruses, but two, to the nucleoprotein/polymerase complex, were specific for 617/83 and bound to no other NDV isolates. The other seven produced a similar classification scheme to the original MAb raised against Ulster 2C.

The Kenyan and Ugandan poultry viruses do not bind MAb 14 in HI tests as expected, although they do bind MAb 14 weakly in IIP tests. This had earlier been reported for a Sudanese food-pigeon isolate (Alexander *et al.*, 1985) and more recently for pigeon viruses from Egypt but not South Africa (which imports European pigeons), as if the pigeon virus group shows some heterogeneity in Africa.

Group R

Group R contains one isolate, IBA/85, from Japanese poultry. This virus may be "intermediate" between groups B and P. It differs from group B by failing to bind MAb 14 and from group P by failing to bind MAb 161.

IBA/85 is one of the viruses used by Sakaguchi *et al.* (1988) to study the evolution of the HN gene by sequencing. Our MAb groups did correspond to the evolutionary tree of Sakaguchi *et al.* (1988) as based on K_s and K_a values of HN nucleotides (Fig. 3).

Sakaguchi used viruses from our MAb groups B, D, E, G, L, Q, and R but omitted virus from the two more recent NDV panzootics, groups A and P, which reached Britain and the United States in the 1970s and mid-1980s, respectively. MAb grouping implies the pigeon virus is more related to group B and group R than to group A, and gene sequencing would establish this.

MAb and sequence data both show (1) the divergence of American groups D and E and their +ve or −ve binding with MAb 445 to HN of Ulster (see Fig. 3 and Table 4) and (2) that three of the four epitopes on HN which vary both on field strains and after MAb selection contain the two negatively charged amino acids glutamic and aspartic next to each other in most viruses. Change of the glu (293, 347, or 495) is associated with loss of binding (Table 2 and Fig. 2). On the whole HN molecule five such pairs of glu–asp occur, and it is remarkable that three are variable epitopes.

Sequencing analysis of the F and M genes gives a similar evolutionary tree to the HN gene (Toyoda *et al.*, 1989). This demonstrates how mutations occur at almost the same rate in all three genes and underlines the lack of gene recombination between virulent and avirulent strains (Toyoda *et al.*, 1989).

E. MAb and Virulence

The search for MAb to label epitopes associated with virulence has not given clear-cut results. Certain MAbs bind predominantly to virulent or avirulent strains, but this may be coincidence

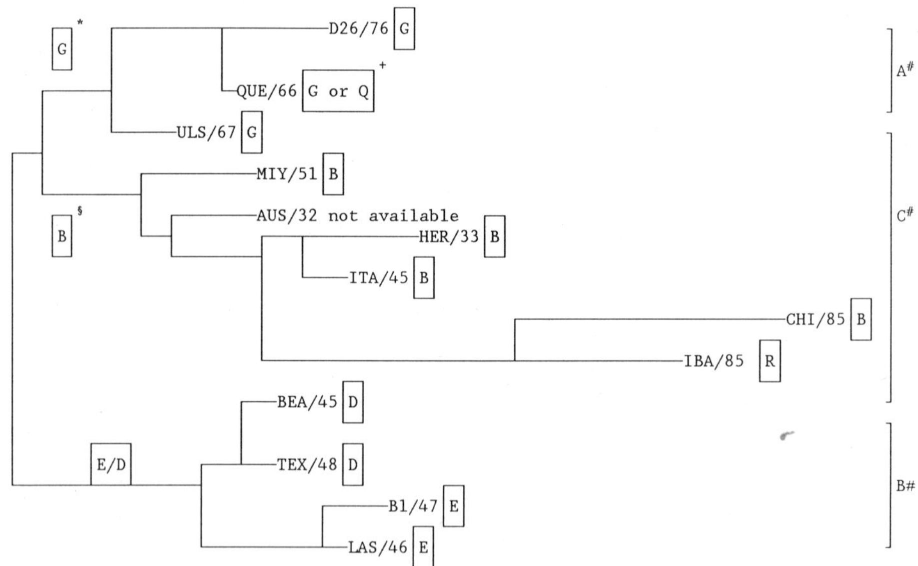

Figure 3. Evolutionary tree based on the K_s values in a coding region (nucleotides 92 to 1804) of HN. *,MAb binding group (Russell and Alexander). No group A, C, H, L, and P isolates sequenced. +,G or Q depending on isolate of Queensland V4 used. #,Lineage depending on presence of biologically active/inactive HN/HN_0 or F/F_0 in cell culture as related to gene expression and/or proteolytic cleavage respectively. A = F_0, $HN_{616}(HN_0)$; B = $F_0/_F$, HN_{577}; C = F/HN_{571}. $,Group B varies with respect to 38/617 binding. All the isolates shown were negative although Herts might be +ve depending on isolate used. From Sakaguchi *et al.* (1989).

because exceptions always occur. Meuleman's MAb 5A1 to HN group III bound to Italien, a virulent isolate, but did not cause HI of Italien or 13 other virulent viruses, whereas it was active by HI using 26 or 18 avirulent viruses with Ulster as one of the exceptions. Our MAb 424, now known to bind to matrix, did not bind velogenic viruses of Groups A, B, or P, but bound to those in Group C. Conversely, another MAb to matrix, the M-3 site, bound to 3/7 virulent strains but not to 5/5 avirulent strains (Faaberg and Peebles, 1987).

III. CONCLUDING REMARKS

This review has elucidated antigenic and structural variation of NDV using MAb and sequencing data. No new markers for virulence have been obtained and the cleavability of the fusion protein remains the single known restrictive element. Some lentogenic strains have an accompanying uncleaved HN_0 protein. Antigenic and virulence variation is widespread. Whether amino acid changes caused by immunoselection using MAb parallel those seen in the field, as reported for H3 molecules of influenza (Wiley *et al.*, 1981), awaits further comparisons although present evidence suggests that for NDV antigenic variation may have occurred spontaneously rather than as an escape from immunosurveillance. This would be linked to the continual supply of susceptible hosts because yolk sac IgG is not always able to prevent the replication of NDV, whether virulent or avirulent at the gut or respiratory tract surface (see Borland and Allan, 1980).

REFERENCES

Abenes, G. B., Kida, H., and Yanagawa, R. (1986a). *Arch. Virol.* **90,** 97–110.

Abenes, G. B., Kida, H., and Yanagawa, R. (1986b). *Jpn. J. Vet. Sci.* **48,** 353–362.

Alexander, D. J., and Allan, W. H. (1974). *Avian Pathol.* **3,** 269–278.

Alexander, D. J., Russell, P. H., and Collins, M. S. (1984). *Vet. Rec.* **114,** 444–446.

Alexander, D. J., Wilson, G. W. C., Russell, P. H., Lister, S. A., and Parsons, G. (1985a). *Vet. Rec.* **117,** 429–434.

Alexander, D. J., and 25 other co-authors. (1985b). *Avian Pathol.* **14,** 365–376.

Alexander, D. J., Mackenzie, J. S., and Russell, P. H. (1986). *Aust. Vet. J.* **63,** 365–367.

Alexander, D. J., Manvell, R. J., Kemp, P. A., Parsons, G., Collins, M. S., Brockman, S., Russell, P. H., and Lister, S. A. (1987). *Avian Pathol.* **16,** 553–565.

Allan, W. H., Lancaster, J. E., and Toth, B. (1978). *Newcastle Disease Vaccines,* FAO, Rome.

Blumberg, B., Giorgi, C., Roux, L., Rajin, R., Dowling, P., Chollet, A., and Kolakofsky, D. (1985). *Cell* **41,** 269–278.

Borland, L. J., and Allan, W. H. (1980). *Avian Pathol.* **9,** 45–59.

Chambers, P., Millar, N. S., Bingham, R. W., and Emmerson, P. T. (1986a). *J. Gen. Virol.* **67,** 475–486.

Chambers, P., Millar, N. S., and Emmerson, P. T. (1986b). *J. Gen. Virol.* **67,** 2685–2694.

Chambers, P., Nesbit, M., Yusoff, K., Millar, N. S., Samson, A. C. R., and Emmerson, P. T. (1988). *J. Gen. Virol.* **69,** 2115–2122.

Choppin, P. W., and Scheid, A. (1980). *Rev. Infect. Dis.* **2,** 40–61.

Collins, M. S., Alexander, D. J., Brockman, S., Kemp, P. A., and Manvell, R. J. (1989). *Arch. Virol.* **104,** 53–61.

Erdei, J., Erdei, J., Bachir, K., Kaleta, E. F., Shortridge, K. F., and Lomniczi, B. (1987). *Arch. Virol.* **96,** 265–269.

Erikson, G. A., Mare, C. J., Gustafson, G. A., Miller, L. D., Proctor, S. J., and Canbrey, E. A. (1978). *Avian Dis.* **21,** 642–654.

Espion, D., De Henau, S., Letellier, C., Wemers, C. D., Brasseur, R., Young, J. F., Gross, M., Rosenberg, M., Meulemans, G., and Burny, A. (1987). *Arch. Virol.* **95,** 79–95.

Faaberg, K. S., and Peebles, M. E. (1988). *J. Virol.* **62,** 586–593.

Fields, S., Winter, G., and Brownlee, G. G. (1981). *Nature (Lond.)* **290,** 213–217.

Glickman, R., Syddall, R. J., Iorio, R. M., Sheehan, J. P., and Bratt, M. A. (1988). *J. Virol.* **62,** 354–356.
Gotoh, B., Sakaguchi, T., Nishikawa, K., Inocencio, N. M., Hamaguchi, M., Toyoda, T., and Nagai, Y. (1988). *Virology* **163,** 174–182.
Hannoun, C. (1977). *Dev. Biol. Stand.* **39,** 469–472.
Hiebert, S. W., Paterson, R. G., and Lamb, R. A. (1985). *J. Virol.* **54,** 1–6.
Hitchner, S. B., and Johnson, E. P. (1948). *Vet. Med.* **43,** 525–530.
Iorio, R. M., and Bratt, M. A. (1983). *J. Virol.* **48,** 440–450.
Iorio, R. M., Borgman, J. B., Glickman, R. L., and Bratt, M. (1986). *J. Gen. Virol.* **67,** 1393–1403.
Ishida, M., Nerome, K., Matsumoto, M., Mikami, T., and Oya, A. (1985). *Arch. Virol.* **85,** 109–121.
Jorgensen, E. D., Collins, P. L., and Lomedico, P. T. (1987). *Virology* **156,** 12–24.
Kaleta, E. F., Alexander, D. J., and Russell, P. H. (1985). *Avian Pathol.* **14,** 553–557.
Kawamura, M., Nagata-Matsubara, K., Nerome, K., Yamane, N., Kida, H., Koderma, H., Izawa, H., and Mikami, T. (1987). *Microbiol. Immunol.* **31,** 831–835.
Köhler, G., and Milstein, C. (1975). *Nature (Lond.)* **256,** 495–497.
Lister, S. A., Alexander, D. J., and Hogg, R. A. (1986). *Vet. Rec.* **118,** 476–479.
LeLong, L., Portetelle, D., Ghysdael, J., Gonze, M., Burny, A., and Meulemans, G. (1986). *J. Virol.* **57,** 1198–1202.
McGinnes, L. W., and Morrison, T. G. (1986). *Virus Res.* **5,** 343–356.
McGinnes, L. W., Wilde, A., and Morrison, T. G. (1987). *Virus Res.* **7,** 187–202.
McNulty, M. S., Adair, B. M., O'Loary, C. J., and Allan G. M. (1988). *Avian Pathol.* **17,** 509–513.
Meulemans, G., Gonze, M., Carlier, M. C., Petit, P., Burney, A., and Long, L. (1987). *Arch. Virol.* **92,** 55–62.
Millar, N. S., Chambers, P., and Emmerson, P. T. (1986). *J. Gen. Virol.* **67,** 1917–1927.
Millar, N. S., Chambers, P., and Emmerson, P. T. (1988). *J. Gen. Virol.* **69,** 613–620.
Nagai, Y., and Klenk, H-D. (1977). *Virology* **77,** 125–134.
Nagai, Y., Klenk, H-D., and Rott, R. (1976). *Virology* **72,** 494–508.
Nagai, Y., Hamaguchi, M., and Toyoda, T. (1989). In *Prog. Vet. Microbiol. Immunol.,* Vol. 5, pp. 16–64, Karger, Basel.
Neyt, C., Geliebter, J., Slaoui, M., Morales, D., Meulemans, G., and Burny, A. (1989). *J. Virol.* **63,** 952–954.
Nishikawa, K., Isomura, S., Suzuki, S., Watnabe, E., Hamaguchi, M., Yoshida, T., and Nagai, Y. (1983). *Virology* **130,** 318–330.
Nishikawa, K., Morishima, T., Toyoda, T., Miyadai, T., Yokochi, T., Yoshida, T., and Nagai, Y. (1986). *J. Virol.* **60,** 987–993.
Nishikawa, K., Morishima, T., Yoshida, T., Hamaguchi, M., Toyoda, T., and Nagai, Y. (1987). *Virus Res.* **7,** 83–92.
Russell, P. H., and Alexander, D. J. (1983). *Arch. Virol.* **75,** 243–253.
Russell, P. H., Griffiths, P. C., Goswami, K. K. A., Alexander, D. J., Cannon, M. J., and Russell, W. C. (1983). *J. Gen. Virol.* **64,** 1069–2072.
Sagild, I. K., and Haresnape, J. M. (1986). *Avian Pathol.* **16,** 165–176.
Sakaguchi, T., Toyoda, T., Gotoh, B., Kuma, K., Miyata, T., and Nagai, Y. (1989). *Virology* **169,** 260–272.
Samson, A. C. R. (1988). In *Newcastle Disease* (D. J. Alexander, ed.), pp. 23–44, Kluwer, Boston.
Samson, A. C. R., Nesbit, M., Lyon, A. M., and Meulemans, G. (1988). *J. Gen. Virol.* **69,** 473–480.
Sato, H., Hattori, S., Ishida, N., Imamura, Y., and Kawakita, M. (1987a). *Virus Res.* **8,** 217–232.
Sato, H., Oh-hira, M., Ishida, N., Imamura, Y., Hattori, S., and Kawakita, M. (1987b). *Virus Res.* **7,** 241–255.
Schaper, U. M., Fuller, F. J., Ward, M. D. W., Mehrotra, Y., Stone, H. O., Stripp, B. R., and De Buysscher, E. V. (1988). *Virology* **165,** 291–295.
Scheid, A., and Choppin, P. W. (1973). *J. Virol.* **11,** 263–271.
Scheid, A., and Choppin, P. W. (1977). *Virology* **80,** 54–66.
Sheehan, J. P., Iorio, R. M., Syddall, R. J., Glickman, R. L., and Bratt, M. A. (1987). *Virology* **161,** 603–606.
Toyoda, T., Sakaguchi, T., Kunitoshi, I., Inocencio, N. M., Gotoh, B., Hamaguchi, M., and Nagai, Y. (1987). *Virology* **158,** 242–247.
Toyoda, T., Sakaguchi, T., Hirote, H., Gotoh, B., Kuma, K., Miyata, T., and Nagai, Y. (1988). *J. Virol.* **62,** 4427–4430.
Toyoda, T., Sakaguchi, T., Hirote, H., Gotoh, B., Kuma, K., Miyata, T., and Nagai, Y. (1988). *Virology* **169,** 273–282.

Wemers, C. D., De henau, S., Neyt, C., Espion, D., Letellier, C., Meulemans, G., and Burny, A. (1987). *Arch. Virol.* **97,** 101–113.

Wiley, D. C., Wilson, I. A., and Skehel, J. J. (1981). *Nature* **289,** 373–378.

Yusoff, K., Nesbit, M., McCartney, H., Emmerson, P. T., and Samson, A. C. R. (1988) *Virus Res.* **11,** 319–333.

Yusoff, K., Nesbit, M., McCartney, H., Meulemans, G., Alexander, D. J., Collins, M. S., Emmerson, P. T., and Samson, A. C. R. (1989). *J. Gen. Virol.* **70,** 3105–3109.

III

Variability of Picornaviruses and Rotaviruses

Molecular Epidemiology

12

Molecular Epidemiology of Wild Poliovirus Transmission

Olen M. Kew, Baldev K. Nottay, Rebeca Rico-Hesse, and Mark A. Pallansch

I. INTRODUCTION

Paralytic poliomyelitis is a serious and highly visible public health problem in most developing countries, where up to 400,000 children are stricken each year (Assaad and Ljungars-Esteves, 1984; Sabin, 1985). Large areas of Asia, Africa, North America, and South America are currently endemic for this disease. In many urban centers of the developing world, attack rates are particularly high. The epidemiologic situation in Bombay, where the annual incidence of poliomyelitis cases in that one city alone is over 100-fold higher than in the United States and Canada combined (Enterovirus Research Centre, Bombay, 1987; Pan American Health Organization, 1987; World Health Organization, 1988), underscores the magnitude of the problem that still awaits resolution.

Outbreaks of poliomyelitis are caused by wild polioviruses. In most developed countries, circulation of the indigenous wild polioviruses has been eliminated through immunization (Kim-Farley *et al.*, 1984). In these countries, poliomyelitis is a rare, virtually forgotten, disease. The few isolated cases that do occur are almost exclusively associated with use of the oral poliovaccines (Nkowane *et al.*, 1987). However, high immunization coverages must be maintained in developed countries to prevent recurrence of epidemics and the reestablishment of endemicity by imported wild polioviruses.

Clearly, the most desirable way to protect populations in developing and developed countries alike is through universal immunization. Since humans are the only natural hosts for polioviruses, an effective global immunization program should achieve eradication of wild strains. Indeed, a central component of the World Health Organization's Expanded Program on Immunization (Henderson, 1984) is the Poliomyelitis Eradication Program, which has set goals for eradication of indigenous wild polioviruses from the American Region by 1990 (Pan American Health Organization, 1985) and worldwide by the year 2000 (World Health Assembly, 1988). Major strides toward poliomyelitis control have been taken under this program in recent years, especially in the American Region, where several countries have attained the threshold of eradication (Pan American Health Organization, 1988).

Olen M. Kew, Baldev K. Nottay, Rebeca Rico-Hesse, and Mark A. Pallansch • Division of Viral Diseases, Center for Infectious Diseases, Centers for Disease Control, Atlanta, Georgia 30333. *Present Address of R. R-H*: Department of Epidemiology and Public Health, Yale University School of Medicine, New Haven, Connecticut 06510.

While the highly successful precedent of the Smallpox Eradication Program has shaped many of the current strategies for control of poliomyelitis, numerous fundamental clinical, epidemiologic, and virologic features distinguish the two diseases (Evans, 1984). These differences call for the Poliomyelitis Eradication Program to have a much greater reliance on laboratory support. Laboratory investigations yield crucial information regarding vaccine potencies, individual and community seroconversion rates, confirmation of suspected cases, and characterization of clinical isolates. As immunization programs proceed and the incidence of poliomyelitis declines, laboratory findings become even more important for guiding control activities toward those areas requiring reinforced efforts. Accordingly, the Poliomyelitis Eradication Program has sought close integration of laboratory and epidemiologic activities, in order to better utilize the limited health resources available in countries endemic for poliomyelitis and to accelerate progress toward control.

Molecular methods, based primarily on comparison of poliovirus genomic sequences, offer a powerful, independent approach for epidemiologic surveillance (Nottay *et al.*, 1981; Kew and Nottay, 1984a; Rico-Hesse *et al.*, 1987). Use of these methods has yielded new insights into the distribution and transmission of wild polioviruses in nature. The endemic origins of different wild polioviruses can be independently traced, aiding evaluation of the factors leading to epidemics, and contributing to the formulation of better methods for poliomyelitis control. Extension of these approaches has led to the development of new and highly sensitive techniques for the direct detection of specific wild viruses. Simplified, rapid procedures have been designed for use in the laboratories of poliomyelitis-endemic countries, so that detailed epidemiologic information may be obtained from routine diagnostic tests.

II. BASIC PRINCIPLES AND METHODS

A. Fundamental Epidemiologic Questions

A primary objective of epidemiologic surveillance is to determine the patterns of spread of wild viruses, so that improved strategies for interrupting transmission may be designed. In this context, the most critical questions center on (1) the relationships between cases and outbreaks occurring in different regions, (2) the local, regional, and global patterns of transmission of wild polioviruses, and (3) identification of the major reservoirs for maintenance of poliovirus circulation.

Prior to the application of molecular methods to epidemiologic surveillance, many of these questions could not be easily approached. Because a high proportion (>99%) of poliovirus infections are subclinical, the standard epidemiologic techniques of case-contact investigations were often unable to reveal the origins of wild viruses. Most difficult to trace were isolated, sporadic cases, potentially associated with either vaccine-related or wild polioviruses. However, recognition that genetic relatedness implies epidemiologic linkage opens an additional avenue for epidemiologic investigation, leading to the detection of links that could not be determined by any other means.

B. Rapid Evolution of Poliovirus Genomes Permits Detailed Molecular Epidemiologic Surveillance

The precision with which any molecular method can distinguish field isolates, and therefore monitor virus transmission, is proportional to the rate of viral genomic evolution during replication in their natural hosts. For DNA viruses and some RNA viruses (Smith and Inglis, 1987; Steinhauer and Holland, 1987), evolution rates are slow, and the epidemiologic resolution of molecular techniques is consequently low. Poliovirus genomes, in contrast, experience rapid variation upon replication in humans (Nottay *et al.*, 1981; Kew *et al.*, 1981; Minor *et al.*, 1982, 1986b; Kew and Nottay, 1984b). For example, oligonucleotide fingerprint (see next section) comparisons of serial isolates from an epidemic revealed an average rate of fixation of mutations over the entire genome of about one to two nucleotide substitutions per week (Nottay *et al.*, 1981). High evolution rates have also been reported for the genomes of other picornaviruses (Domingo *et al.*, 1980; Takeda *et al.*, 1984). Since most mutations are fixed cumulatively into polioviral genomes, chains of transmission may be inferred from the location and types of mutations found in the RNAs of clinical isolates. An important corollary to these considerations is that the extent of sequence divergence between related wild poliovirus isolates provides an approximate measure of the number of intervening infections separating the cases.

C. A Brief Description of Analytic Methods

We have used three different methods for characterizing poliovirus isolates. All sample viral genomic sequences, but in separate ways, and the choice of approach depends on the nature of the questions under study.

1. Genomic Sequencing

Genomic sequencing (Sanger *et al.*, 1977; Zimmern and Kaesberg, 1978; Biggin *et al.*, 1983; Rico-Hesse *et al.*, 1987) is the most definitive method for assessing the genetic relatedness among polioviruses. Comparisons of selected genomic regions, representing as little as 2% of the total RNA, provide detailed information on the genetic relationships among polioviruses. Sequence analyses detect distant evolutionary relationships (up to 15% base sequence divergence in the region examined), thus providing a broad overview of the distribution of wild poliovirus genotypes in nature.

2. Oligonucleotide Fingerprinting

Oligonucleotide fingerprinting (Kew *et al.*, 1984; Kew and Nottay, 1986) samples noncontiguous intervals of sequence information along the entire genome. Fingerprints are usually produced by digestion of the viral RNA with ribonuclease T_1 and electrophoretically separating the radiolabeled oligonucleotide fragments in two dimensions. Approximately 10–15% of the genome is represented by longer oligonucleotides (>12 nucleotides), derived from unique genomic sequences, that resolve into patterns of spots highly characteristic for each RNA sequence. Fingerprinting is extremely sensitive to mutations (Aaronson *et al.*, 1982); therefore, comparisons are most reliable when RNA molecules share >95% nucleotide sequence homology.

3. DNA Probe Hybridization

DNA probe hybridization (Meinkoth and Wahl, 1984; Studencki and Wallace, 1984) is a rapid method for determining sequence homologies within defined genomic intervals. We have prepared two classes of synthetic DNA hybridization probes: (1) those that base-pair with conserved genomic sequences, and thus hybridize to all poliovirus strains, and (2) those that bind to variable genomic regions characteristic of particular poliovirus strains. Probes specific for each of the Sabin strains have been made (Nottay *et al.*, in preparation), as have probes that recognize specific wild poliovirus genotypes (Rico-Hesse *et al.*, 1990; da Silva *et al.*, 1990).

The three methods are in many ways complementary. Isolates can be quickly screened for sequence homology to other polioviruses using DNA probes. Genomic sequencing reveals the most distant evolutionary (hence, epidemiologic) relationships. Fingerprinting surveys the entire genome and detects small differences not readily picked up by the other methods. Since fingerprinting is the most technically demanding of the molecular approaches, it is best used in specialized applications, such as comparing very closely related strains or screening for poliovirus isolates having recombinant genomes (Kew and Nottay, 1984b; Minor *et al.*, 1986b).

A fourth method, the polymerase chain reaction (PCR) (Saiki *et al.*, 1985, 1988), has recently been applied in our laboratory to achieve highly sensitive detection of specific viral genomic sequences (C-F. Yang, unpublished results). Application of this technique requires some knowledge of the nucleotide sequences of the target genome. It is likely that PCR, possibly in combination with probe hybridization, will find wide application in virus diagnostics. These techniques are especially suitable for use in laboratories of developing countries, as they require comparatively inexpensive equipment and materials.

4. Genomic and Antigenic Analyses Compared

An alternative, and more traditional, approach for characterizing poliovirus isolates has been to determine their reactivities with preparations of neutralizing antisera (Nakano *et al.*, 1978; van Wezel and Hazendonk, 1979). The precision and specificities of identifications based on antigenic properties have been enhanced by the development of panels of highly specific neutralizing monoclonal antibodies (Humphrey *et al.*, 1982; Minor *et al.*, 1982; Crainic *et al.*, 1983; Osterhaus *et al.*, 1983). Antibody reactivity patterns define the immune surface of poliovirions, which is encoded by approximately 0.5–1% of the genomic sequences (Minor *et al.*, 1986a; Page *et al.*, 1988), providing information not directly obtainable from the molecular methods described here.

Distant epidemiologic relationships among wild polioviruses have been accurately predicted from their antigenic properties (Osterhaus *et al.*, 1983). However, antigenic data provide at best qualitative evidence for evolutionary relationships among polioviruses, but yield no quantitative measures of the extent of genetic relatedness. A more critical limitation to identifications based exclusively on antigenic properties is the capacity of both vaccine-derived and wild polioviruses to undergo significant antigenic variation during replication in a single patient (Nakano *et al.*, 1963; Crainic *et al.*, 1983; Minor *et al.*, 1986b; Huovilainen *et al.*, 1988). Moreover, the antigenic variation of wild polioviruses observed to occur in individuals does not necessarily produce a progressive antigenic drift during transmission in human populations (Houvilainen *et al.*, 1988). Further complicating serologic identifications is the existence of unrelated polioviruses sharing similar antigenic properties (Nakano *et al.*, 1978).

III. GENOTYPIC RELATIONSHIPS AMONG POLIOVIRUS ISOLATES

The most direct method currently available for comparing poliovirus genomic sequences is by extension of synthetic DNA primers with reverse transcriptase in the presence of chain-terminating inhibitors (Sanger *et al.*, 1977; Zimmern and Kaesberg, 1978; Biggin *et al.*, 1983). The most useful primers for comparative studies are complementary to conserved intervals adjoining highly variable sequences. While many conserved intervals exist, most abundantly in the noncapsid and nontranslated regions (Toyoda *et al.*, 1984), an especially useful site for primer binding is located 100 nucleotides to the 3′-side of the VP1/2A (capsid/noncapsid) junction. Using a primer complementary to that site (Rico-Hesse *et al.*, 1987), we have compared the sequences of over 200 poliovirus isolates of all three serotypes.

The VP1/2A region was selected for several reasons. This interval spans two functional domains, those of the major capsid surface protein, VP1, and those of a viral protease, protein 2A, catalyzing an early morphogenetic cleavage of the polyprotein (Toyoda *et al.*, 1986). Capsid sequences are generally of greatest interest because polioviruses are primarily identified by serotype. The VP1 interval compared contains sequences characteristic of serotype. When additional sequence information is required to clearly distinguish isolates, determination of the 3′-

terminal sequences of VP1 opens the way for systematic sequencing of additional capsid sequences by recognition of other conserved primer binding sites. Sequences from protein 2A were also included in our comparisons to evaluate whether this noncapsid domain evolves differently from the VP1 domain.

A. Comparative Sequence Alignments

Partial genomic sequences (90 nucleotides from VP1, 60 from 2A) of independent type 1, type 2, and type 3 isolates are aligned in Figs. 1–3. All sequences within each serotype are colinear. However, because the surface loop near the carboxy terminus of VP1 (Hogle *et al.*, 1985) contains one more amino acid residue in type 1 strains than in types 2 and 3 (Toyoda *et al.*, 1984), an additional codon is included at the 5′-end of the alignments for types 2 and 3. In each alignment, the corresponding Sabin oral poliovaccine sequences are shown for reference. Alignments also include the sequences of a vaccine-derived isolate, which are typically very similar to the sequences of the reference vaccine strain.

Within each serotype, the sequences of the wild polioviruses differed substantially from those of the reference vaccine strains. Compared with the corresponding Sabin reference sequences, the wild isolates differed at up to 66% (type 1), 66% (type 2), and 74% (type 3) of codons.

All mutations observed within each serotype were nucleotide substitutions. For closely related pairs of viruses (<5% sequence divergence), over 80% of the substitutions were transitions. Most (>97%) of the observed nucleotide substitutions encoded changes to synonymous codons, which are thought to be phenotypically silent, not appearing through biological selection but arising via genetic drift. Thus, the resolving power of genomic comparisons is very high, because isolates having identical biological properties may easily be distinguished biochemically.

For many isolates, patterns of genomic similarities were evident from the aligned nucleotide sequences (Figs. 1–3). While close genetic similarities could easily be seen from the sequence alignments, more distant relationships were generally much more difficult to detect.

B. Dendrograms of Sequence Relatedness

Analyses of the relationships among a large set of sequences can be further refined and quantitated by performing all possible pairwise comparisons. The results of such comparisons may be summarized for easy visual comparison as dendrograms of sequence relatedness (Fitch and Margoliash, 1967; Rico-Hesse *et al.*, 1987). In the dendrograms showing intratypic sequence relatedness for type 1 (Fig. 4), type 2 (Fig. 5), and type 3 (Fig. 6) polioviruses, the percentage difference between any two strains is twice the distance along the abscissa to the connecting node. Because most of the amino acid residues within the VP1/2A region are conserved across serotypes, and few of the observed mutations encoded amino acid substitutions, the limit divergence obtained within each serotype (type 1, 23%; type 2, 23%; type 3, 26%) approached the 29% divergence across the same interval for the three Sabin poliovaccine strains (Toyoda *et al.*, 1984).

We have previously defined a genotype as a group of polioviruses having no more than 15% genomic divergence within the 150-nucleotide VP1/2A region (Rico-Hesse *et al.*, 1987). While this criterion for demarcation of genotypes is subjective, it is sufficiently restrictive to exclude nearly all spurious associations and yet include most linkages expected on epidemiologic grounds and all linkages confirmable by other means. As the divergence exceeds 15%, the reliabilities of the relationships established by the dendrogram diminish. At the extreme right sides of the dendrograms, distantly related genotypes appear to be nearly equally divergent, and no clear patterns for their evolutionary interrelationships can be discerned.

For the dendrograms in Figs. 4–6 to accurately reflect the evolutionary relationships among polioviruses, three conditions must be met: (1) sequence divergence must occur primarily by ac-

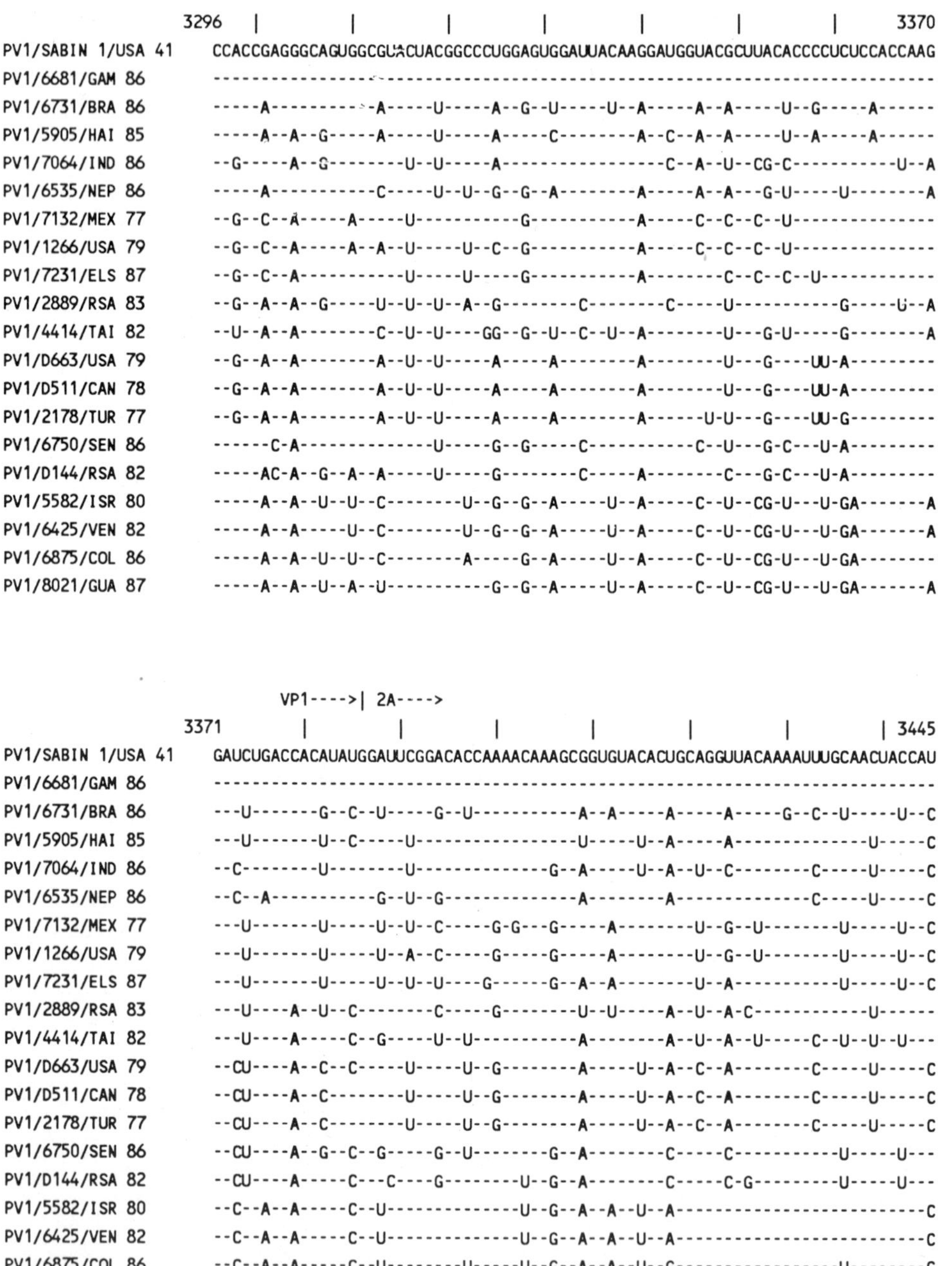

Figure 1. Comparison of nucleotide sequences (90 from VP1/60 from 2A) at the capsid/noncapsid junction region of 20 type 1 polioviruses. Sequence reactions were performed by extension of a synthetic DNA primer with reverse transcriptase in the presence of [α-^{35}S]dATP and dideoxynucleoside triphosphates (Rico-Hesse *et al.*, 1987). Nucleotide differences from the Sabin 1 reference sequence are shown; dashes indicate identities. Isolate PV1/6681/GAM 86 is vaccine-derived; the 18 other case isolates are wild. Nucleotide positions are numbered according to Nomoto *et al.* (1982).

```
                       3295 |         |         |         |         |         |         |    3369
PV2/SABIN 2/USA 54      CGACCUCCACGAGCAGUCCCAUACUUCGGACCAGGUGUUGAUUAUAAAGAUGGGCUCACCCCACUACCAGAAAAG
PV2/7837/PER 84         ---------------------------------------------------------------------------
PV2/II-996/SWE 77       --------C----UG--G-----U-AU-----G--G-----C--C--G-----AU-A--------------G--A
PV2/WHO6683/YUG 78      --------C----UG--G-----U-AU-----G--G--------C--G-----AU-G--------------G--A
PV2/LEDERLE 2           -----C--U----------------A---------G-----C--C--G-----A--A--------G--------A
PV2/W-2                 -----C--U----------------A---------G-----C--C--G-----A--A--------G-----G--A
PV2/MEF-1/EGY 42        -----C--U----------------A---------G-----C--C--G-----A--AG-------G-----G--A
PV2/II-317/EGY 55       --G--C--U----U---U--------AU--------G-----C-----G-----A--G-----G-----------A
PV2/II-209/VEN 61       --G--A--C-----------G--U-----G-----A--------C--------U--U--U-----------G---
PV2/II-580/COL 64       -----A--C-----------G--U--U--------A--------C-----C--U--U--U-----------G---
PV2/II-314/GUA 57       -----C-----------U-----U-----------------------G-----------U--------------A
PV2/6876/COL 86         -----C--U--------U--G--------------A-----C-----G--C-----U--U-----------G--A
PV2/7834/PER 83         -----------G-----------------------------C--------C--------------U---------
PV2/LANSING/USA 37      -----C-----------U-----U-A---G--U-----C--------G--C-----U-----U--U--G------
PV2/7079/IND 86         --U--C--U--U--G----------AU--------G-----C--C-----------U--U--CU----G-----A
PV2/II-637/IND 68       --------C--G--------G----AU--G-----G-----C--C--G--C--C--------C------------
PV2/II-710/KEN 71       --G--C--U--C--G--------------G--G--G--C-----C--------U--------CU-G--------A
PV2/15340/IND 82        --C--C--C--C--G----------A---G-----A--C--C--C--------A--A-----CU----------A
```

```
                                VP1---->| 2A---->
                       3370      |         |         |         |         |         |         | 3444
PV2/SABIN 2/USA 54      GGAUUAACGACUUAUGGAUUUGGACACCAAAACAAAGCUGUGUACACAGCUGGCUACAAAAUUUGCAAUUACCAC
PV2/7837/PER 84         ---------------------------------------------------------------------------
PV2/II-996/SWE 77       --GC-G--A--C--------C--U--------U--G-----------G--G-----U------------------
PV2/WHO6683/YUG 78      --GC-G--A--C--------C--U--U-----U--G-----------G--G--U--U------------------
PV2/LEDERLE 2           --C--G--A--C-----U-----C--------U--G--A--------G--A--U---------------------
PV2/W-2                 --C--G-UA--C-----U-----C--------U--G--A--------G--A--U---------------------
PV2/MEF-1/EGY 42        --C--G--A--C-----U-----C--------U--G--A--------G--A--U---------------------
PV2/II-317/EGY 55       --GC-------A--C--C--C--C--U--G-----G--A--------G--A--U---------------------
PV2/II-209/VEN 61       --U--------C-----------U--U--G--------A-----U-----A--U-----------------U--U
PV2/II-580/COL 64       --U--------C-----------U--U--G--------A--C--------A-----U-----C--U-----U--U
PV2/II-314/GUA 57       -----G-----C-----------G-----G-----G--C--------------U--U-----C------------
PV2/6876/COL 86         --U--G-----C-----------------------G--A-----U-----------U-----------------U
PV2/7834/PER 83         --------A--------C-----G-----------------------G--------U-----------C--U---
PV2/LANSING/USA 37      --G-----A--C--C-----------------U-----------------------U--------U--------U
PV2/7079/IND 86         --G--G--C--G--C--------G-----G--------A--U-----C--G--G--U-----A--U--C------
PV2/II-637/IND 68       --GC-G--C--G--C--------------------G--A--C--U--U--G--G--U-----A--U--------U
PV2/II-710/KEN 71       --------C--A------C----------G-----G--G--U-----U-----G--------A--U--C------
PV2/15340/IND 82        ---C----C-----C--------G--U--------G--A--C-----C--A--G--------A--U-----U--U
```

Figure 2. Comparison of nucleotide sequences at the VP1/2A junction region of 18 type 2 polioviruses. Nucleotides are numbered by alignment with the corresponding position in the reference Sabin 2 sequence (Toyoda *et al.*, 1984). In Figs. 2 and 3, an additional codon is included at the 5′-end of the string of nucleotides, in order to compensate for a codon deletion (relative to type 1 RNAs) occurring within the interval encoding the carboxy-terminal surface loop (Hogle *et al.*, 1985) of the VP1 proteins of poliovirus types 2 and 3 (Toyoda *et al.*, 1984). Isolate PV2/7837/PER 84 is vaccine-derived.

```
                      3287 |         |         |         |         |         |         |    3361
PV3/SABIN 3/USA 37    AGACCGCCGCGCGCGGUACCUUAUUAUGGACCAGGGGUGGACUAUAGGAACAACUUGGACCCCUUAUCUGAGAAA
PV3/6880/COL 86       ---------------------------------------------------------------------------
PV3/7105/IND 86       --U--U--A--G-UA--G--G-----C-----G-----------C-A-G-UGGG----CG-----G--G-----G
PV3/7095/IND 86       C----C-----G-UA--G--C--C--C--G--C--U--------C-A-G--GGG----CA---C----G--A--G
PV3/V-17/BAN 82       C----C-----G--A--G--A--C--C-----G-----A-----C-A-G-UGGG----CG--------A-----G
PV3/V-1 /EGY 82       --G--C-----G--U--G--G--C--C-----------------C-A-G--GGG--A-CA--UC----A-----G
PV3/V-2 /LIB 82       --G--C-----G--U--G--G--C-----G-----A-----U--C-A-G--GGG----CA--UC----A-----G
PV3/6184/FIN 84       -----C-----G-----A--G--C--C--G-----A--------C-AAG--GGG----CA--UC----A-----G
PV3/V-3 /SPA 83       -----C-----G-----G--G--C--C--G--G--A--------C-A-G--GGG---ACA--UC-G--A-----G
PV3/8178/VEN 88       -----C--U--G-----U--G--C--C--------A--------C-A-G-UGGG----CA--UC-G--A------
PV3/6879/COL 86       C----C-----G-----G--A--C--C--G--------U-------AAG-UGGGC---CA--G-----A--A--G
PV3/6699/ECU 86       C----C-----G--------A--C--C-----------U-----C-AAG--GGGC---CA--GC----A--A--G
PV3/7840/PER 86       C-G--C-----G--U--G--G-----C--G--------A--U--C-AAG-UGGG----CA--GC-G--A-----G
PV3/7377/BOL 86       C-G--C-----G--U--G--G-----C-----------A--U--C-AAG-UGGG----CA--GC-G--A-----G
PV3/USOL-D-bac        C----C-----G--U--G--G-----C--G-----------U--C-A-G--GGGC-A-CA-----G--A-U---G
PV3/4-22/POL 68       C----C-----G--U--G--G-----C--G-----------U--C-A-G--GGGC-A-CA-----G--A-----G
PV3/SAUKETT/USA 52    -----A--A-----------------------------------C-AAG--------A-------G----C----
PV3/V-7 /GUA 84       --G--A--A--U--A--G--A-----------G--A-----U--C-AAG-UGGUC--A----AC-G--C------
PV3/1565/USA 80       --G--A-----U--A--G--A-----------G--------U--C-AAG-U-GU--AA-U---C-G---------
PV3/7124/MEX 87       -----A-----U--A--G--A-----C--G--G-----A--U--C-AAG--------A-----C-G---------

                               VP1---->| 2A---->
                      3362     |         |         |         |         |         |         | 3436
PV3/SABIN 3/USA 37    GGUUUGACCACAUAUGGCUUUGGGCAUCAGAAUAAAGCUGUGUACACUGCUGGUUACAAGAUCUGCAACUACCAU
PV3/6880/COL 86       ---------------------------------------------------------------------------
PV3/7105/IND 86       --AC-C-----U-----G-----A--C--A--C--G--A-----U--A--C--G--------------------C
PV3/7095/IND 86       ---C-C-----U--C--G-----A--C--A-----G--A-----U--A--C--G--U-----U-----U-----C
PV3/V-17/BAN 82       ---C-U-----U--C--G-----A--C-----------A--------G-----G--U-----U-----U-----C
PV3/V-1 /EGY 82       --AC-C--A-----C--G-----U-----A-----G--A--C-----A--A-----------------------C
PV3/V-2 /LIB 82       --AC-C--G-----C--G-----A--C--A--C-----A--U--------A--A--------------U-----C
PV3/6184/FIN 84       --AC-C--A--G--C--G-----U--C--A-----------U-----A--G-----U-----------U-----C
PV3/V-3 /SPA 83       --AC-C--A--G--C--A--C--A--C--A--C-----A--------A--G--G-----A--A-----------C
PV3/8178/VEN 88       --AC-C--A-----C--G-----A-----A--------A-----U--G--A--C--U--A--U-----------C
PV3/6879/COL 86       --AC-C--A-----C--A-A---A--------C-----G--------G-----G--------U-----U-----C
PV3/6699/ECU 86       --CC-C--G--------G-A---A--C-----------------U--A--A--C-U---A--U-----U-----C
PV3/7840/PER 86       --GC-U--G--C-----A-------G------C--G--A--A-----A-----------A-----U--U------
PV3/7377/BOL 86       --GC-C--G--C-----A--------------C--G--A--A-----A--A--------A-----U--U------
PV3/USOL-D-bac        --GC-C--G--U-------A------C-----C--G--A--------A-----G--U--------U--U-----C
PV3/4-22/POL 68       --GC-C--G--U-------A------C-----C--G--A--------A-----G--U--------U--U-----C
PV3/SAUKETT/USA 52    -----------C-----------A-----------G-----U--U--C-----A--------U--U--------C
PV3/V-7 /GUA 84       --C-----A--U-----------------A--C-----G--------C--------------A--U-----U--C
PV3/1565/USA 80       --A-----A--U-----------A-----A--C--G-----------C--------U--A--U------------
PV3/7124/MEX 87       --------A--U--C--------------A--C--G--A--------C--A-----------U-----U--U--C
```

Figure 3. Comparison of nucleotide sequences at the VP1/2A junction region of 20 type 3 polioviruses. Nucleotides are numbered by alignment with the corresponding position in the reference Sabin 3 sequence (Stanway *et al.*, 1984; Toyoda *et al.*, 1984). Isolate PV3/6880/COL 86 is vaccine-derived.

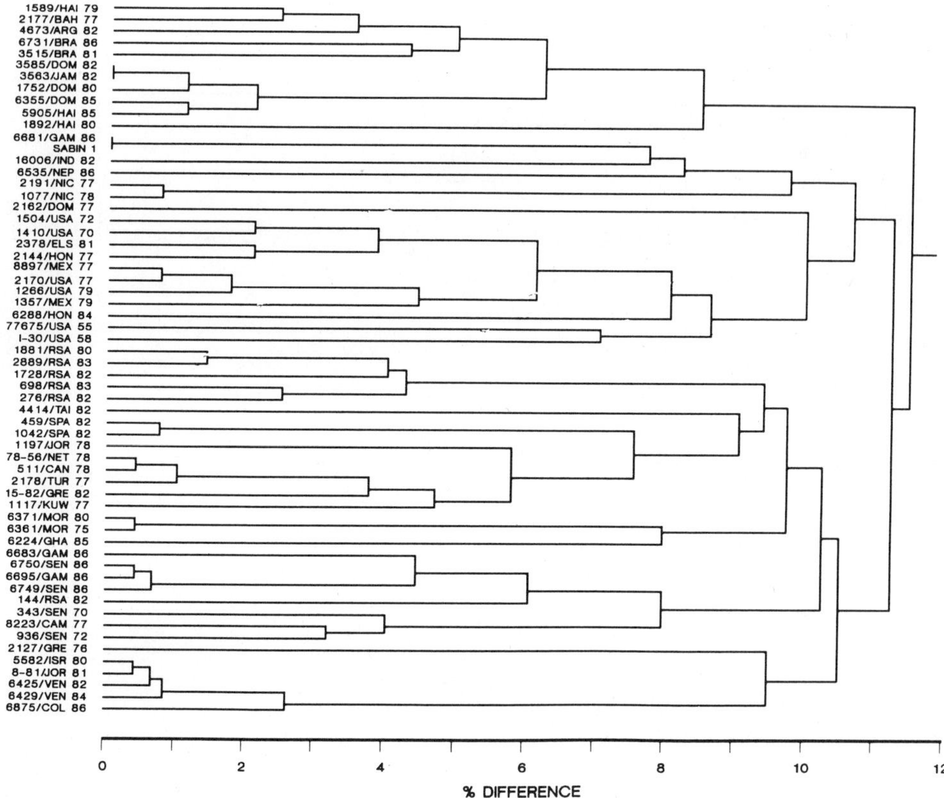

Figure 4. Dendrogram constructed according to sequence relatedness among 60 type 1 polioviruses across the interval of nucleotides 3296–3445 (VP1/2A region). All pairwise comparisons were performed using a computer program that assigned each nucleotide substitution an equivalent statistical weight. The percentage nucleotide sequence divergence between any two strains is twice the distance along the abscissa to the connecting node.

cumulation of substitution mutations, (2) the rates of divergence for homologous regions of different strains must not vary widely, and (3) the sequence interval selected for comparison must include only domains having a common lineage. In situations where the pathways of virus transmission were known, the experimental results were consistent with progressive fixation of mutations into poliovirus genomes (Nottay *et al.*, 1981; Rico-Hesse *et al.*, 1987, unpublished results). While it is clear that different regions of the poliovirus genome evolve at different rates (Toyoda *et al.*, 1984; Hughes *et al.*, 1986; La Monica *et al.*, 1986), the available evidence suggests that different poliovirus genomes evolve at similar rates (Nottay *et al.*, 1981; Minor *et al.*, 1986b; Rico-Hesse *et al.*, 1987; da Silva *et al.*, 1990). Thus, if all three conditions are satisfied, the branch structures of trees sampling different genomic intervals would be similar, even though the ranges of sequence variation indicated on the abscissa scales may differ.

If genetic exchange between distantly related polioviruses has occurred within the VP1/2A region, then the third condition described above is not satisfied. In fact, the genomes of a small number of isolates have sequence patterns expected for recombinants of poliovirus (and possibly other enterovirus) genotypes (Rico-Hesse *et al.*, 1987). If crossover sites map within the interval under comparison, recombinants may be separated in the dendrograms from both parental genotypes, thereby appearing as new genotypes. The best characterized of the recombinant genomes

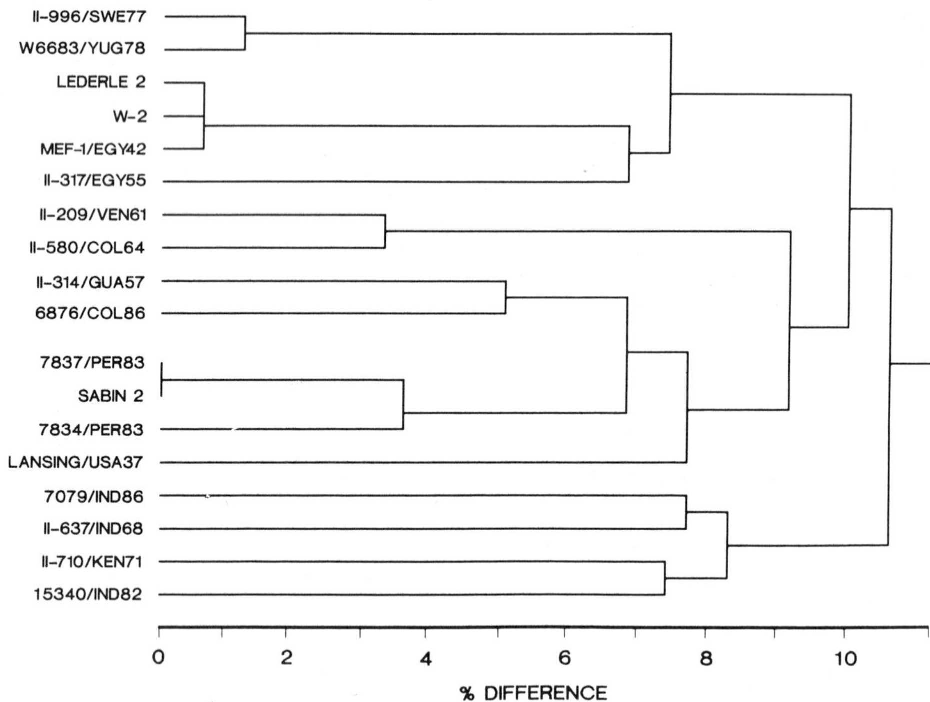

Figure 5. Dendrogram constructed according to sequence relatedness among 18 type 2 polioviruses across the interval of nucleotides 3295–3444 (VP1/2A region).

appear to have arisen by exchange of the entire capsid region sequences as a "cassette" (Rico-Hesse *et al.*, 1987, unpublished results). Recombinant genomes are recognized by having long intervals with high homology to a reference sequence sharply bounded by extended intervals with low homology.

When all three conditions are satisfied, variable region sequences representing as little as 2% of the total genome (such as the 150-nucleotide VP1/2A interval) provide a reasonable overview of the distribution and transmission of wild polioviruses in nature. Very detailed determinations of the patterns of spread of closely related strains (e.g., for the analyses of epidemics) may require larger sequence samples per isolate.

Although the dendrograms superficially resemble phylogenetic trees, they differ by lacking a coordinate for plotting divergence as a function of time. Construction of reliable phylogenetic trees for each serotype is impeded by the existence of many highly divergent genotypes, for which no single common ancestor can be identified. It is possible to build phylogenetic trees for viruses that have recently (<10 years) evolved from a recognized progenitor.

C. Relationships among Type 1 Polioviruses

The dendrogram for type 1 polioviruses (Fig. 7), the most common etiologic agents of poliomyelitis, illustrates the relationships among many cases occurring in many different regions of the world. Several distinct genotypes of type 1 poliovirus currently exist in nature. When the den-

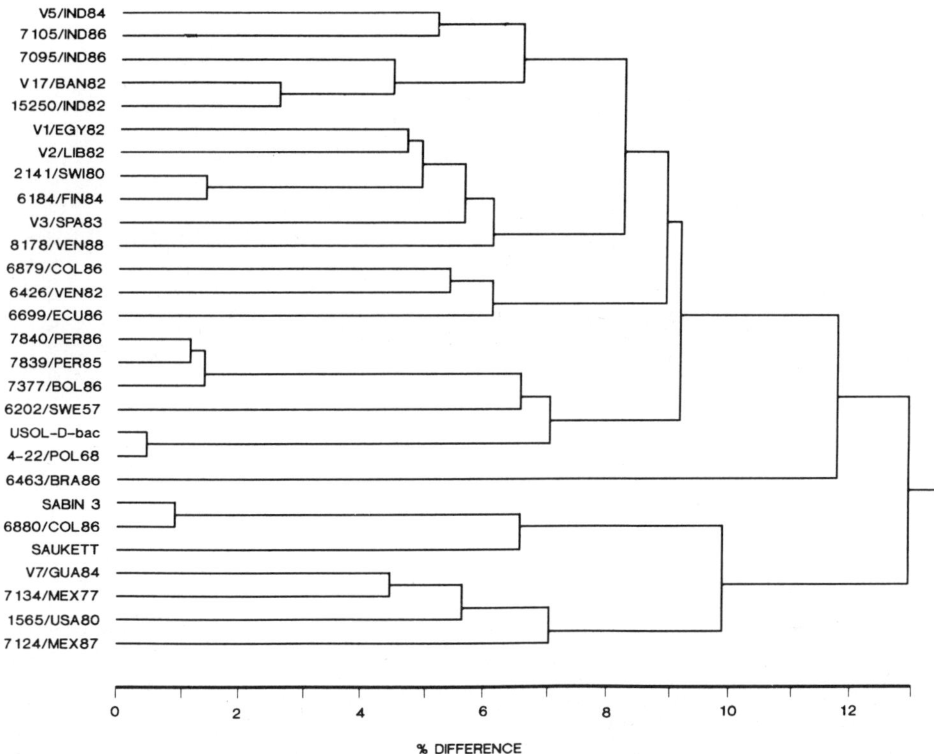

Figure 6. Dendrogram constructed according to sequence relatedness among 28 type 3 polioviruses across the interval of nucleotides 3287–3436 (VP1/2A region).

drogram groups poliovirus isolates by nucleotide sequence similarities, they are simultaneously grouped by geographic origins. Thus, each of the many genotypes has a regionally defined range of endemicity. The worldwide transmission patterns for wild polioviruses contrast sharply with those of influenza A viruses, which are spread in global pandemics, with the total number of distinct genotypes in concurrent circulation being small (Buonagurio *et al.*, 1986b). The global epidemiologic picture for polioviruses more closely resembles that of influenza C viruses, for which numerous gene lineages coexist (Buonagurio *et al.*, 1986a).

The relationships among wild type 1 polioviruses have already been described in detail (Rico-Hesse *et al.*, 1987), so only a few examples will be discussed here to illustrate how sequence data are used to detect links between cases and outbreaks.

The wild type 1 polioviruses indigenous to the United States are believed to have been eradicated sometime before 1970. All subsequent cases associated with wild polioviruses are thought to involve imported strains, a view fully supported by sequence data. The most frequent source of imported wild viruses has been the region of Mexico and Central America. Isolates from epidemics in Texas (1970) and Connecticut (1972) and from numerous sporadic cases were closely related to viruses endemic to that region (Fig. 4). The last case associated with a wild type 1 poliovirus imported into the United States from Mexico occurred in 1979. In that same year, an epidemic occurred within Amish communities in several states, preceded by cases among unimmunized members of related denominations in Canada (1978) and the Netherlands (1978). A direct

Figure 7. Geographic distribution of wild type 1 poliovirus genotypes. A genotype is defined here as a group of polioviruses sharing ≥85% nucleotide homologies in the VP1/2A junction compared in the alignments (Figs. 1–3) and dendrograms (Figs. 4–6). Each genotype is assigned a shade and number code according to the most populous country in which it is endemic. In all maps, only the genotypes of isolates from cases occurring after 1976 are shown. Countries believed not to be endemic areas are shaded and coded to identify the gentotypes of epidemic isolates: Argentina (1982), Kuwait (1978), the Netherlands (1978), and Taiwan (1982). The distribution maps for genotypes present in the American region are nearly complete and current (as of October 1988); the indigenous wild polioviruses appear to have been eliminated from Brazil, the Caribbean, and Central America (except Mexico). The genotype associated with cases in Turkey in 1977 and the Netherlands in 1978 was the agent of the 1978–1979 epidemics in Canada and the United States. Lightly shaded areas identify countries where poliomyelitis is considered to be endemic (World Health Organization, 1989), but where our current information on the distribution of wild genotypes is incomplete.

link between the Netherlands–Canada–US epidemics and cases occurring in Turkey in 1977 was strongly suggested by the close genetic relatedness among the isolates. The importation of a Middle Eastern genotype into North America underscores the potential of wild polioviruses to infect susceptible individuals and populations anywhere in the world.

Representatives of the same genotype as the 1977 Turkish virus were associated with outbreaks in other parts of the Middle East, including Kuwait (1977) and Jordan (1978). By 1980–1981, an entirely distinct genotype was the agent of cases occurring in Israel and Jordan. Sequence comparisons revealed the virtual identity of isolates from Israel (1980) and Venezuela (1981), an observation consistent with importation of an Andean virus to the Middle East around 1980. Viruses of this genotype are now endemic to both the Andean region and the Middle East and include the agents of the 1988 outbreak in Israel (unpublished results).

Another genotype having a wide distribution is endemic to West Africa and was responsible for the 1986 epidemic in Senegal and the Gambia. Viruses of the West African genotype have also been recently isolated in South Africa. However, the majority of isolates obtained throughout South Africa belong to a different genotype, suggesting that isolates of the West African genotype represent imported viruses.

In addition to revealing previously unsuspected links between cases and outbreaks, sequence comparisons can rule out potential epidemiologic links that might be considered on the basis of proximity of cases. For example, it is very unlikely that the type 1 poliovirus associated with an outbreak in Spain originated in Morocco, because the sequences of the viruses found in the two neighboring countries were very different (Fig. 4). Similarly, no polioviruses of Caribbean origin have been associated with cases in the United States for many years, despite frequent migration from endemic areas of the Caribbean to continental North America.

D. Relationships among Type 2 Polioviruses

Among wild polioviruses, type 2 strains are the least frequent cause of poliomyelitis, and few contemporary wild type 2 genotypes have been characterized biochemically. As a result, our selection of type 2 genotypes (Figs. 2 and 5) includes a smaller proportion of recent isolates.

India is probably the largest reservoir for wild type 2 polioviruses. The large sequence differences between recent isolates from the west (Bombay, represented by the 1986 isolate, 7079) and the south (Vellore, represented by the 1982 isolate, 15340) suggest that at least two independent foci for endemic transmission of wild type 2 polioviruses exist within the country.

The dendrogram (Fig. 5) clearly shows the close linkage between the epidemics in Sweden (1977) and Yugoslavia (1978) and suggests a distant link to cases in the Middle East. Interestingly, the experimental oral poliovaccine strain W-2, thought to be derived from a 1936 U.S. isolate (Carp *et al.*, 1963), is, like Lederle 2, an attenuated derivative of the 1942 Egyptian isolate, MEF-1. MEF-1, commonly used as a reference virus, is quite distinct from type 2 viruses from the United States, here represented by Lansing (Michigan, 1938) and Sabin 2, an attenuated variant of the clinical isolate, P712 (New Orleans, 1954; Sabin and Boulger, 1973).

E. Relationships among Type 3 Polioviruses

The dendrogram in Fig. 6 reveals the existence of numerous genotypes of wild type 3 polioviruses. Since the introduction of the poliovaccines, the relative proportion of poliomyelitis cases associated with wild type 3 strains appears to have increased, possibly because some poliovaccine formulations induce comparatively low immune titers and seroconversion rates to serotype 3 (Patriarca *et al.*, 1988).

As was the case for type 2 isolates, the wild type 3 strains from India were found to be genetically diverse, consistent with the existence of multiple pockets of endemicity within the large Indian population. The European isolates (Switzerland, 1980; Spain, 1983; Finland, 1984) were distantly related to viruses endemic to the Middle East (Egypt, 1982; Libya, 1982). The isolates from the northern Andean region (Venezuela, Colombia, and Ecuador), except for a 1988 isolate from Venezuela (8178) which appeared to be distinct, comprised a somewhat heterogeneous genotype. Viruses from the central Andean region, Peru and Bolivia, formed another genotype. Independent genotypes also exist in the region of Mexico and Central America and in Brazil.

Sabin 3, derived from a 1937 U.S. (Los Angeles) case isolate (Sabin and Boulger, 1973) was found to be only distantly related to Saukett (Pittsburgh, 1950), the type 3 component of the inactivated poliovaccine.

A large epidemic occurred in Poland following a 1967 field trial of an experimental type 3 oral poliovaccine, USOL-D-bac (Kostrzewski *et al.*, 1970). The close sequence correspondence between a clinical isolate from the epidemic, 4-22, and USOL-D-bac (Figs. 3 and 6) clearly implicates neurovirulent derivatives of the candidate vaccine as the agents of the epidemic. Thus, the sequence data are fully consistent with previous results obtained with oligonucleotide fingerprinting (Kew and Nottay, 1984a).

IV. GEOGRAPHIC DISTRIBUTION OF WILD POLIOVIRUS GENOTYPES

Maps showing the distributions of wild poliovirus genotypes (Figs. 7–9) reinforce the point that many distinct genotypes concurrently exist worldwide. The patterns of distribution, where each genotype has a geographically defined range of endemicity, suggest that local conditions, such as vaccine coverage rates and sanitation, are more important determinants of transmission than specific virologic properties, such as antigenic structure. This observation is highly encouraging to the prospects for poliomyelitis control, since it is possible to improve the critical conditions.

The maps are most complete for the American region, where we have received the largest number of isolates. We have identified three type 1, two type 2, and four type 3 genotypes that are currently endemic to the Americas. We expect to discover few, if any, additional indigenous American genotypes, because each of the many poliovirus specimens we have received from this region belongs to one of the genotypes already described. Our maps are still incomplete for much of Africa and Asia, because we lack recent specimens from most of the poliomyelitis-endemic countries of these continents.

Since entire countries are assigned shadings and codes according to their endemic genotypes, the maps are necessarily coarse and represent only a preliminary survey of the distributions of polioviruses in nature. The maps identify the most important reservoirs for wild polioviruses imported into nonendemic areas. For the United States, the principal reservoir is Mexico and Central America; for Europe, it is the eastern Mediterranean. Thus, the maps are consistent with common patterns of human migration within and from endemic areas and may assist in the identification of the major portals through which wild genotypes are imported.

Type 1 polioviruses are the most widely distributed in nature, followed by type 3, and then type 2. Some areas appear to be endemic only for type 1 polioviruses. Examples of such areas are the island of Hispaniola (Haiti and the Dominican Republic), Honduras, and possibly South Africa. No areas are known to be endemic solely for type 2 strains. Wild type 2 polioviruses have apparently been eradicated from North and Central America, as no wild isolates have been obtained for many years. The northern Andean region is the only part of the Americas known to still be endemic for type 2 strains. Since type 2 viruses are most effectively controlled by the poliovaccines, their presence may indicate serious deficiencies in vaccine coverages.

Figure 8. Geographic distribution of wild type 2 poliovirus genotypes. Sweden and Yugoslavia are shaded to show identities of 1977 and 1978 epidemic isolates. The sequence heterogeneity (Figs. 2 and 5) among recent isolates from India indicates that at least two genotypes of type 2 polioviruses are endemic to that country.

Figure 9. Geographic distribution of wild type 3 poliovirus genotypes. The type 3 poliovirus isolates obtained in Europe were associated with limited outbreaks (Spain, 1983; Finland, 1984) and an isolated case (Switzerland, 1980).

V. HIGH-RESOLUTION MOLECULAR EPIDEMIOLOGY

While the molecular epidemiologic results presented thus far offer a new, global perspective on patterns of wild poliovirus transmission, many of the more immediate and practical epidemiologic issues focus on modes of spread at the regional and community levels. To obtain a more detailed view of transmission within locales, both oligonucleotide fingerprinting and genomic sequencing are useful. Generally, the best approach is to compare more sequence information per isolate, thereby increasing chances for detecting small differences among genomes and, in effect, producing higher-resolution images of branch structures at the left sides of dendrograms. For example, complete VP1 region sequences were compared in order to monitor the spread of a wild type 1 poliovirus strain during the 1988 epidemic in Israel (unpublished results). An important question addressed by these studies was whether infections in one community were ancestral to subsequent infections and cases occurring elsewhere in the country.

Identification of the major regional reservoirs for poliovirus circulation should aid in the formulation of more effective control strategies. Using high-resolution molecular methods, it may be possible to determine the relationships among cases in neighboring urban and rural areas. If, for example, rural cases are directly linked to preceding urban infections, then intensified immunization programs in urban zones that are continuously endemic for poliomyelitis should yield immediate benefits to the countrysides. In addition, critical analyses of the evolutionary relationships among isolates within a specific region may provide more accurate estimates of the minimum population sizes required for unbroken poliovirus circulation in different environments.

VI. MOLECULAR EPIDEMIOLOGIC DATA FROM ROUTINE DIAGNOSTIC PROCEDURES

Epidemiologic data are most valuable when the findings are quickly made available to the health personnel in the affected areas, so that corrective measures can be taken. Building on the foundations of our genomic sequence database, we have developed diagnostic procedures for routine use in laboratories of developing countries. Local utilization of these methods should minimize problems of specimen transport and permit more rapid recognition of wild viruses. New biochemical reagents have been prepared that allow identification of viruses by genotype, so that the distribution of specific genotypes can be determined from the results of routine diagnostic tests.

The molecular methods most applicable for routine use are DNA probe hybridization and PCR technology. Synthetic DNA probes designed to recognize vaccine-related strains or specific wild genotypes (Fig. 10) are already in use in some countries in the American region (Pan American Health Organization, 1988). Since wild poliovirus genomes are constantly evolving, we expect the useful lives of our 25-nucleotide probes designed to complement the sequences of contemporary wild isolates to be on the order of 5–10 years (see Fig. 10; Edson da Silva, personal communication). Probes specific for the vaccine strains should be useful indefinitely, since these viruses are rarely transmitted beyond the immediate contacts of primary vaccines, so that opportunities for significant genomic evolution are reduced. Our original probes contained radioisotopes as reporters. However, the use of synthetic DNA probes with covalently attached alkaline phosphatase reporters (Jablonski *et al.*, 1986) permits sensitive detection of hybrids through standard histochemical reactions, potentially widening the clinical applications of probe technology by eliminating the need for radioisotopes.

The unparalleled sensitivities of PCR techniques, coupled with the simplicities of the procedures (Saiki *et al.*, 1985, 1988), have already revolutionized viral diagnostics. Genotype-specific primer pairs can be readily prepared to detect the equivalent of one infectious poliovirion (C-F. Yang, unpublished results). Viruses present in clinical samples may be identified within 8 hr, while bypassing the expense associated with virus isolation in cell culture for routine identifications.

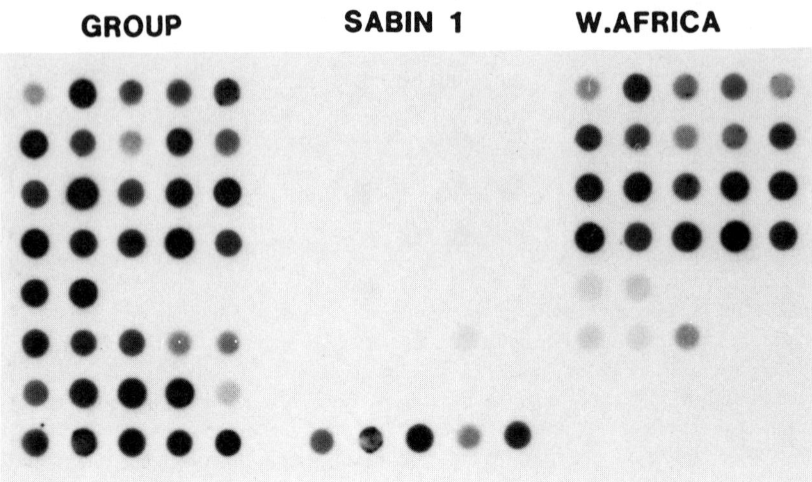

Figure 10. Autoradiogram demonstrating the use of specific synthetic DNA probes to identify vaccine-derived and wild poliovirus isolates by dot-blot hybridization. Clarified culture fluids from virus-infected cells were prepared as described by White and Bancroft (1982). Samples (100 μl) were filtered in three identical sets through nitrocellulose membranes. Each filter was subjected to hybridization with one of three ^{32}P-labeled probes (polio group-, Sabin 1-, or West Africa wild type 1-specific) and washed in parallel under standard conditions (Meinkoth and Wahl, 1984). Virus samples were blotted (left to right) as follows: Rows 1–4: 1986 epidemic isolates from Senegal and the Gambia. Row 5: Wild type 1 isolates from Senegal, 1970 and 1972. Row 6: Wild type 1 isolates from Africa: Cameroon 1977, Morocco 1980, RSA (Johannesburg) 1982, RSA (Letaba) 1982, Ghana 1985. Row 7: Other wild type 1 isolates: Spain 1982, Colombia 1986, Honduras 1985, Dominican Republic 1985, Brazil 1986. Row 8: Type 1 vaccine-related isolates: Sabin 1 (WHO reference), the Gambia 1986, Morocco 1985, Argentina 1985, Chile 1986. Genetic relationships detected by probe hybridization closely corresponded to those shown by the dendrogram (Fig. 4).

VII. SOME ADDITIONAL EPIDEMIOLOGIC LESSONS LEARNED

Molecular epidemiologic analyses have provided much more information on the dynamics of poliovirus transmission than was heretofore available. It is our hope that comprehensive understanding of the underlying mechanisms of these dynamic processes will point the way toward more effective control methods.

A. Genetic Parameters of Endemicity and Epidemicity

In regions endemic for wild polioviruses, cases that are temporally associated may not have direct epidemiologic links. For example, isolates obtained in the same years within South Africa (type 1, Fig. 4), India (type 3, Fig. 6), or the northern Andean region (type 3, Fig. 6) showed considerable sequence heterogeneity for each genotype. Such diversity among contemporary isolates indicates the existence of separate pockets for circulation and evolution of wild polioviruses. Even within the same country, separate endemic pockets may persist for many years. For example, the sequences of type 1 viruses isolated in 1986 in southern Mexico and El Salvador were closer to those of isolates obtained in the same region in 1978 than they were to sequences of other 1986 isolates from northern Mexico. Thus, separate southern and northern endemic pockets had been maintained for at least 8 years.

Epidemics generally involve the rapid, clonal expansion of a specific lineage. Very limited genetic heterogeneity (1–2% within the VP1/2A region) is usually observed among isolates from an epidemic. Prior to an epidemic, extensive genetic diversity, typical of multifocal endemic transmission, may exist. For example, at least two distinct variants of the West African type 1 genotype, differing by about 9% in the VP1/2A region, were present in the Gambia immediately before the 1986 epidemic (Fig. 4). All epidemic isolates shared high sequence homologies with only one of the preepidemic variants. The lineage of the other variant may have become extinct during the epidemic, since the pool of susceptibles was decreasing as the population of competing viruses was increasing. Thus, one possible outcome of the epidemic may be that the range of genetic diversity of the poliovirus population in the Gambia was reduced, only to broaden again with time as the progeny of the epidemic virus diverge.

B. Coexistence and Displacement of Genotypes

The structures of the dendrograms suggest that displacement and extinction of lineages is a frequent occurrence in the natural evolution of polioviruses. The observation that isolates obtained from distant locales show detectable sequence homologies is evidence for the past predominance of progenitor viruses over large geographic areas. If events bringing about the narrowing of genotypes were uncommon, then we would expect prolonged (>30 years) epidemiologic isolation to produce much greater genetic diversity among regional isolates than is observed.

We have already described the displacement of one genotype with an unrelated genotype in the Middle East. Similar selective processes appear to have recently taken place in Central America. For example, type 1 viruses isolated in Nicaragua in 1977–1978 were not closely related to any other poliovirus we have so far examined. The prevalence in Nicaragua of this seemingly unusual genotype may have declined by 1979, as isolates from that year were members of the genotype that is widely distributed in Central America and Mexico. Competition between genotypes may also be occurring in Guatemala. Type 1 isolates obtained in 1983 from Guatemala belonged to the major Central American genotype described above. More recent isolates, from 1986 and 1987, included members of both the Central American and northern Andean genotypes. This situation may represent the early stages of a displacement process which, if not terminated by eradication efforts, will lead to the predominance in Guatemala, as in Israel, of only the Andean genotype.

C. Eradication of Genotypes

While natural selection appears to frequently lead to the eradication of some poliovirus lineages upon the expansion of others, the goal of the Poliomyelitis Eradication Program is to render all wild poliovirus lineages extinct. In fact, many lineages and some entire genotypes have been eradicated through immunization. Extinct genotypes include the wild polioviruses of the United States and Canada and the wild type 2 poliovirus from Brazil.

The molecular approaches described here will aid in monitoring the disappearance of specific wild genotypes from nature. A good example of a specific application of these approaches is a study currently underway in Brazil. Using probes specific for the type 1 and 3 genotypes indigenous to that country, it has been possible to monitor the progressive and dramatic reduction in the geographic ranges of the wild polioviruses following the intensified national immunization program begun in 1980 (Pan American Health Organization, 1988; E. da Silva and H. Schatzmayr, personal communication).

VIII. ROLE OF MOLECULAR EPIDEMIOLOGIC SURVEILLANCE IN POLIOMYELITIS CONTROL

The most productive applications of molecular epidemiologic approaches vary with the stage of development of poliomyelitis control programs. In the early stages, when nearly all clinical isolates are wild, sequence characterization of a limited number of isolates, having diverse geographic origins and representing all indigenous serotypes, provides a valuable initial survey of the relationships among cases within a region. The sequence information also serves as the basis for design of hybridization probes and PCR primers for use in later studies.

As immunization programs proceed, and the incidence of poliomyelitis declines, an increasing fraction of clinical isolates may be vaccine-derived, so that the ability to distinguish vaccine-related and wild viruses becomes critical to accurate classification of cases. In addition, data on the relative importance of each wild poliovirus serotype to paralytic disease may justify modification of vaccine formulations and schedules to optimize protection against the most prevalent agents of poliomyelitis.

Upon full implementation of national immunization programs, intensification efforts can be directed toward the endemic reservoirs, identified through both case surveillance and sequence comparisons of wild isolates, in order to break the chains of transmission sustaining endemicity. A program stressing control at the endemic centers may achieve eradication most efficiently. Afterward, special immunization and surveillance activities should concentrate on communities most likely to be exposed to imported wild polioviruses.

At advanced stages of poliomyelitis control, when vaccine coverage levels are high, very few wild poliovirus infections may result in paralytic cases. Under these conditions, standard case surveillance and characterization of clinical isolates may not be sufficiently sensitive to detect continued circulation of wild polioviruses in populations that are largely immune to paralytic illness, though not to intestinal infection. It may be possible, through environmental samplings (Metcalf *et al.*, 1988; Pöyry *et al.*, 1988), to detect widespread subclinical wild virus infections, signaling the need to mobilize further control measures. At the final stage, a critical line of evidence for certification of eradication of indigenous wild polioviruses will be the consistent inability to detect their presence in either clinical or environmental specimens.

IX. SUMMARY AND CONCLUSIONS

Wild polioviruses, associated with up to 400,000 cases of poliomyelitis each year, are endemic to the developing countries of five continents. To determine the genetic relationships among wild polioviruses, case isolates were compared by partial nucleotide sequencing of their RNA genomes. Comparative sequence data were summarized as dendrograms of sequence relatedness, permitting panoramic visualization of the evolutionary and epidemiologic links among cases and outbreaks occurring worldwide. Viruses sharing $\geq 85\%$ nucleotide homologies in the VP1/2A region are considered to be members of a single genotype, derived from a common ancestral infection. Several independent genotypes, each having a geographically defined range of endemicity, exist within each serotype. Type 1 polioviruses are the most widely distributed, followed by types 3 and 2.

The rapid evolution of poliovirus genomes upon replication in humans allows us to estimate the proximity of epidemiologic links among cases by determining the nucleotide sequence relatedness among isolates. Consequently, previously unsuspected links among cases occurring in distant parts of the world could be unambiguously established. Epidemicity could be distinguished from endemicity by measuring the extent of sequence heterogeneity among contemporary isolates within a country or region. The details of epidemic spread can be resolved using sequence comparisons or oligonucleotide fingerprinting of isolates.

The genetic sequence information has been applied to the design of synthetic DNA probes and PCR primers complementary to the genomes of either vaccine-related polioviruses or specific wild genotypes. These new molecular reagents permit detailed epidemiologic information to be obtained from simple, routine diagnostic tests and open the way for improved laboratory surveillance for wild polioviruses, performed within the laboratories of those regions most affected.

We believe that the enhanced surveillance capabilities offered by molecular epidemiologic approaches will play an important role in monitoring the progress of poliomyelitis control programs. In addition, by tracking of the origins of case isolates, molecular studies may identify important endemic reservoirs requiring reinforced control efforts. Analysis of environmental samples may detect the presence of wild polioviruses within communities having no paralytic cases or, in the final stages of control, provide evidence for the eradication of indigenous polioviruses. It is hoped that the close integration of epidemiologic and laboratory activities implemented within the Poliomyelitis Eradication Program will have general applicability to the control of other vaccine-preventable viral diseases, and that these coordinated efforts will hasten the day when paralytic poliomyelitis becomes a relic of the past.

ACKNOWLEDGMENTS

We thank the many scientists throughout the world whose donation of poliovirus specimens made this study possible. We appreciate the support of many other colleagues: George Marchetti, Mary Flemister, and Beverly Hamby for preparation of virus isolates, Velma George and Francisco Candal for production of cell cultures, Brian Holloway for synthesis of oligonucleotide primers, Pete Seidel, Jim Dobbins, and Sandy Ford for preparation of graphics, and Larry Anderson and Fred Murphy for encouragement and constructive suggestions. R.R.H. was a National Research Council Postdoctoral Fellow.

REFERENCES

Aaronson, R. P., Young, J. F., and Palese, P. (1982). *Nucleic Acids Res.* **10,** 237–246.

Assaad, F., and Ljungars-Esteves, K. (1984). *Rev. Infect. Dis.* **6**(Suppl 2)**,** S302–S307.

Biggin, M. D., Gibson, T. J., and Hong, G. F. (1983). *Proc. Natl. Acad. Sci. USA* **80,** 3963–3965.

Buonagurio, D. A., Nakada, S., Fitch, W. A., and Palese, P. (1986a). *Virology* **153,** 12–21.

Buonagurio, D. A., Nakada, S., Parvin, J. D., Krystal, M., Palese, P., and Fitch, W. M. (1986b). *Science* **232,** 980–982.

Carp, R. I., Plotkin, S. A., Norton, T. W., and Koprowski, H. (1963). *Proc. Soc. Exp. Biol. Med.* **112,** 251–256.

Crainic, R., Couillin, P., Blondel, B., Cabau, N., Boue, A., and Horodniceanu, F. (1983). *Infect. Immun.* **41,** 1217–1225.

da Silva, E. E., Schatzmayr, H. G., and Kew, O. M. (1990). *Braz. J. Biol. Med. Sci.* (in press).

Domingo, E., Davila, M., and Ortin, J. (1980). *Gene* **11,** 333–346.

Enterovirus Research Centre, Bombay (1987). Annual Report of Poliomyelitis Surveillance. Indian Council of Medical Research, New Delhi.

Evans, A. S. (1984). *Prog. Med. Virol.* **29,** 141–165.

Fitch, W. M., and Margoliash, E. (1967). *Science* **155,** 279–284.

Henderson, R. H. (1984). *Rev. Infect. Dis.* **6**(Suppl 2)**,** S475–S479.

Hogle, J. M., Chow, M., and Filman, D. J. (1985). *Science* **229,** 1353–1365.

Hughes, P. J., Evans, D. M. A., Minor, P. D., Schild, G. C., Almond, J. W., and Stanway, G. (1986). *J. Gen. Virol.* **67,** 2093–2102.

Humphrey, D. D., Kew, O. M., and Feorino, P. M. (1982). *Infect. Immun.* **36,** 841–843.

Huovilainen, A., Kinnunen, L., Ferguson, M., and Hovi, T. (1988). *J. Gen. Virol.* **69,** 1941–1948.

Jablonski, E., Moomaw, E. W., Tullis, R. H., and Ruth, J. L. (1986). *Nucleic Acids Res.* **14,** 6115–6128.

Kew, O. M., and Nottay, B. K. (1984a). *Rev. Infect. Dis.* **6**(Suppl 2)**,** S499–S504.

Kew, O. M., and Nottay, B. K. (1984b). In *Modern Approaches to Vaccines* (R. Chanock and R. Lerner, eds.), pp. 357–362, Cold Spring Harbor Laboratory, Cold Spring Harbor, NY.

Kew, O. M., and Nottay, B. K. (1986). In *Concepts in Viral Pathogenesis,* Vol. 2 (A. L. Notkins and M. B. A. Oldstone, eds.), pp. 317–323. Springer-Verlag, New York.

Kew, O. M., Nottay, B. K., Hatch, M. H., Nakano, J. H., and Obijeski, J. F. (1981). *J. Gen Virol.* **56,** 337–347.

Kew, O. M., Nottay, B. K., and Obijeski, J. F. (1984). In *Methods in Virology,* Vol. 8 (K. Maramorosch and H. Koprowski, eds.), pp. 41–84, Academic Press, New York.

Kim-Farley, R. J., Bart, K. J., Schonberger, L. B., Orenstein, W. A., Nkowane, B. M., Hinman, A. R., Kew, O. M., Hatch, M. H., and Kaplan, J. E. (1984). *Lancet* **2,** 1315–1317.

Kostrzewski, J., Kulesza, A., and Abgarowicz, A. (1970). *Epidemiol. Rev.* **34,** 89–103.

La Monica, N., Meriam, C., and Racaniello, V. R. (1986). *J. Virol.* **57,** 515–525.

Meinkoth, J., and Wahl, G. (1984). *Anal. Biochem.* **138,** 267–284.

Metcalf, T. G., Jiang, X., Estes, M. K., and Melnick, J. L. (1988). *Prog. Med. Virol.* **35,** 186–214.

Minor, P. D., Schild, G. C., Ferguson, M., Mackay, A., Magrath, D. I., John, A., Yates, P. J., and Spitz, M. (1982). *J. Gen. Virol.* **61,** 167–176.

Minor, P. D., Ferguson, M., Evans, D. M. A., Almond, J. W., and Icenogle, J. P. (1986a). *J. Gen Virol.* **67,** 1283–1291.

Minor, P. D., John, A., Ferguson, M., and Icenogle, J. P. (1986b). *J. Gen. Virol.* **67,** 693–706.

Nakano, J. H., Gelfand, H. M., and Cole, J. T. (1963). *Am. J. Hyg.* **78,** 214–226.

Nakano, J. H., Hatch, M. H., Thieme, M. L., and Nottay, B. (1978). *Prog. Med. Virol.* **24,** 178–206.

Nkowane, B. M., Wassilak, S. G. F., Orenstein, W. A., Bart, K. J., Schonberger, L. B., Hinman, A. R., and Kew, O. M. (1987). *JAMA* **257,** 1335–1340.

Nomoto, A., Omata, T., Toyoda, H., Kuge, S., Horie, H., Kataoka, Y., Genba, Y., Nakano, Y., and Imura, N. (1982). *Proc. Natl. Acad. Sci. USA* **79,** 5793–5797.

Nottay, B. K., Kew, O. M., Hatch, M. H., Heyward, J. T., and Obijeski, J. F. (1981). *Virology* **108,** 405–423.

Osterhaus, A. D. M. E., van Wezel, A. L., Hazendonk, T. G., UytdeHaag, F. G. C. M., van Asten, J. A. A. M., and van Steenis, B. (1983). *Intervirology* **20,** 129–136.

Page, G. S., Mosser, A. G., Hogle, J. M., Filman, D. J., Rueckert, R. R., and Chow, M. (1988). *J. Virol.* **62,** 1781–1794.

Pan American Health Organization (1985). Eradication of Indigenous Transmission of the Wild Poliovirus from the Americas, Plan of Action. Washington, DC (July 1985).

Pan American Health Organization (1987). *EPI Newslett.* **9,** 1–5.

Pan American Health Organization (1988). Sixth Meeting of the EPI Technical Advisory Group on Polio Eradication in the Americas. Buenos Aires (November 1988).

Patriarca, P., Laender, F., Palmeira, G., Couto Oliveira, M. J., Lima Filho, I., de Souza Dantes, M. C., Tenorio Cordeiro, M., Risi, J. B., and Orenstein, W. A. (1988). *Lancet* **1,** 429–432.

Pöyry, T., Stenvik, M., and Hovi, T. (1988). *Appl. Environ. Microbiol.* **54,** 371–374.

Rico-Hesse, R., Pallansch, M. A., Nottay, B. K., and Kew, O. M. (1987). *Virology* **160,** 311–322.

Sabin, A. B. (1985). *J. Infect. Dis.* **151,** 420–436.

Sabin, A. B. and Boulger, L. R. (1973). *J. Biol. Standard.* **1,** 115–118.

Saiki, R. K., Scharf, S., Faloona, F., Mullis, K. B., Horn, G. T., Ehrlich, H. A., and Arnheim, N. (1985). *Science* **230,** 1350–1354.

Saiki, R. K., Gelfand, D. H., Stoffel, S., Scharf, S. J., Higuchi, R., Horn, G. T., Mullis, K. B., and Ehrlich, H. A. (1988). *Science* **239,** 487–491.

Sanger, F., Nicklen, S., and Coulson, A. R. (1977). *Proc. Natl. Acad. Sci. USA* **74,** 5463–5467.

Smith, D. B., and Inglis, S. C. (1987). *J. Gen. Virol.* **68,** 2729–2740.

Stanway, G., Hughes, P. J., Mountfort, R. C., Reeve, P., Minor, P. D., Schild, G. C., and Almond, J. W. (1984). *Proc. Natl. Acad. Sci. USA* **81,** 1539–1543.

Steinhauer, D. A., and Holland, J. J. (1987). *Annu. Rev. Microbiol.* **41,** 409–433.

Studencki, A. B., and Wallace, R. B. (1984). *DNA* **3,** 269–277.

Takeda, N., Miyamura, K., Ogino, T., Natori, K., Yamazaki, S., Sakurai, N., Nakazono, N., Ishii, K., and Kono, R. (1984). *Virology* **134,** 375–388.

Toyoda, H., Kohara, M., Kataoka, Y., Suganuma, T., Omata, T., Imura, N., and Nomoto, A. (1984). *J. Mol. Biol.* **174,** 561–585.

Toyoda, H., Nicklin, M. J. H., Murray, M. G., Anderson, C. W., Dunn, J. J., Studier, F. W., and Wimmer, E. (1986). *Cell* **45,** 761–770.

van Wezel, A. L., and Hazendonk, A. G. (1979). *Intervirology* **11,** 2–8.

White, C. M., and Bancroft, E. C. (1982). *J. Biol. Chem.* **257,** 8569–8572.

World Health Assembly (1988). Global Eradication of Poliomyelitis by the Year 2000. Resolution WHA41.28, Geneva (May 1988).

World Health Organization (1988). *Weekly Epidemiol. Rec.* **63,** 249–251.

World Health Organization (1989). *Expanded Programme on Immunization, Global Situation—Poliomyelitis, May 1989.*

Zimmern, D., and Kaesberg, P. (1978). *Proc. Natl. Acad. Sci. USA* **75,** 4257–4261.

13

Virus Variation and the Epidemiology and Control of Rhinoviruses

David Arthur John Tyrell

As a background to this chapter it seems important to give first a brief outline of current concepts of the basic structure, biology, and replication of rhinoviruses since these have been changing fast recently. How these affect the epidemiology and the possibility of control by drugs or vaccines will then be discussed. The aim will be to show how changes in the nucleotide sequences of the genetic material are reflected in the behavior of the virus at the biologic level, as it interacts with the host and environment, produces disease, and provides an immune response. We have not reached that point yet, but there is interesting and encouraging progress to report.

I. RHINOVIRUS STRUCTURE AND REPLICATION

Early work showed that rhinoviruses are members of the picornavirus group and so far 100 distinct serotypes that infect humans have been recognized (Collaborative report, 1987). The particles are about 30 nm in diameter, composed of protein and about 30% RNA, which is positive sense and single stranded. This RNA genome is organized very similarly to other picornaviruses, such as enteroviruses, like polioviruses, and foot-and-mouth disease virus. Analogies from these other viruses, which have in some cases been studied in more detail, have been helpful in clarifying some aspects of the strategy of the virus genome and the mechanism of replication. At the time of writing, full sequences were available for rhinoviruses type 14, 2, and 9, so we have full information on the genome of only a small proportion of the 100 serotypes of rhinovirus now recognized.

The genome arrangement is summarized in Fig. 1. There is first a 5′ noncoding region, which is long and myristilated yet seems to be performing some important function since it is unexpectedly well conserved—judging by the fact that cDNA for the RV14 5′ end hybridizes with RNA from most serotypes (Al-Nakib *et al.*, 1986). Indeed, there are two short sequences which seem to be completely conserved in the known sequences of rhinoviruses and enteroviruses. An oligonucleotide cDNA probe for these has been synthesized and has been found to hybridize well with all 57 rhinoviruses tested, which must therefore all contain very similar sequences (Bruce *et al.*, 1988). After the noncoding region come the sequences encoding the capsid proteins, which are translated as part of a single polyprotein and rapidly cleaved to VP0, VP1, VP3, and as the virus particle is assembled, VP0 is cleaved to give VP2 and VP4. The sequences here are very varied and there is no

David Arthur John Tyrrell • Medical Research Council Common Cold Unit, Harvard Hospital, Salisbury Wilts SP2 8BW, England

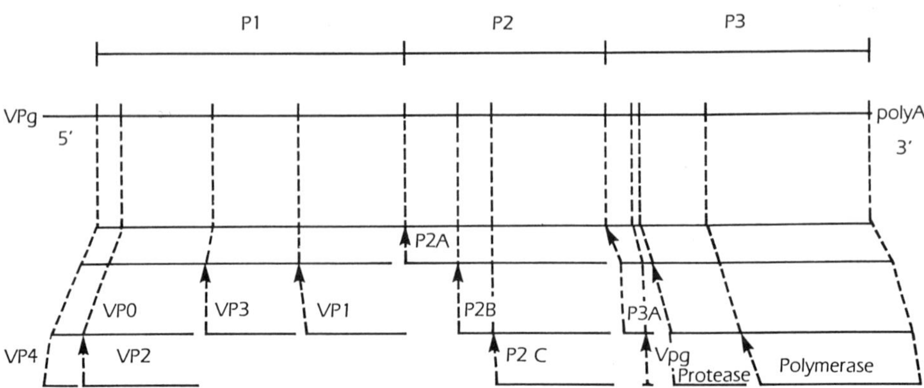

Figure 1. Arrangement of the rhinovirus genome. (By permission of W. S. Barclay.)

evidence for sequence homology; pulse chase experiments show that there are wide differences in the size of the cleavage products of different serotypes (Decock and Billiau, 1986).

The spatial arrangement of the capsid proteins in the assembled particle of RV14 has now been discovered by X-ray crystallography (Rossman *et al.*, 1985). It is similar to that of other picornaviruses, including polioviruses and even plant picornaviruses. In particular, the proteins form "β barrels," which provide the structural framework of the capsid of the virus particle, and the terminal sequences link one peptide with another and the interior of the particle, in which the RNA is contained (Fig. 2). There are also hydrophilic "loops" that protrude outward from the virus surface. Each triangular facet is formed by a uniform arrangement of the four peptides, so placed that the fivefold axis is composed of VP1 and surrounded by a cleft or "canyon"; it is suggested that this is the region with which the receptor site on the cell surface interacts. Further aspects of the loops and the "canyon" will be discussed later.

After the structural proteins the virus enzymes are encoded (Fig. 1), namely, the polymerase and protease. Their functions are on the one hand to transcribe the input RNA to form a negative strand and then further positive strand copies, and on the other to cleave the polyprotein as described above. Their functioning is a large subject and cannot be discussed further here.

It was anticipated that the enzyme proteins would be well conserved, and it has been shown that at the nucleotide and amino acid level there is homology between the putative polymerase-coding regions of poliovirus type 1 and rhinovirus type 2 (Skern *et al.*, 1984); hydrophobicity plots show that there are still considerable similarities between the polymerase of a typical poliovirus and a rhinovirus.

The genome ends with a short polyadenylated 3′ noncoding region, which by cross-hydridization shows no evidence of homology from serotype to serotype (W. Al-Nakib, personal communication).

Rhinoviruses bind to the cell surface by specific receptors which have now been identified (Tomassini and Colonno, 1986). In an initial absorption stage (Lonberg-Holm and Korant, 1972) VP1 is lost and there is a change in the structure of the particle to reveal C-type antigenic determinants (Neubauer *et al.*, 1987). The RNA then enters the cell and is translated and transcribed. Virus particles are formed in replicative complexes associated with internal membranes and are released when the cell disintegrates (see MacNaughton, 1982).

It has become clear in recent years that many viruses are genetically unstable, certainly at the nucleotide level (Reanney, 1984). There are several reasons for this; one is that they undergo numerous rapid cycles of replication, and another that they have less accurate methods for copying their genomes than cellular organisms. This is particularly true of RNA viruses, especially those that

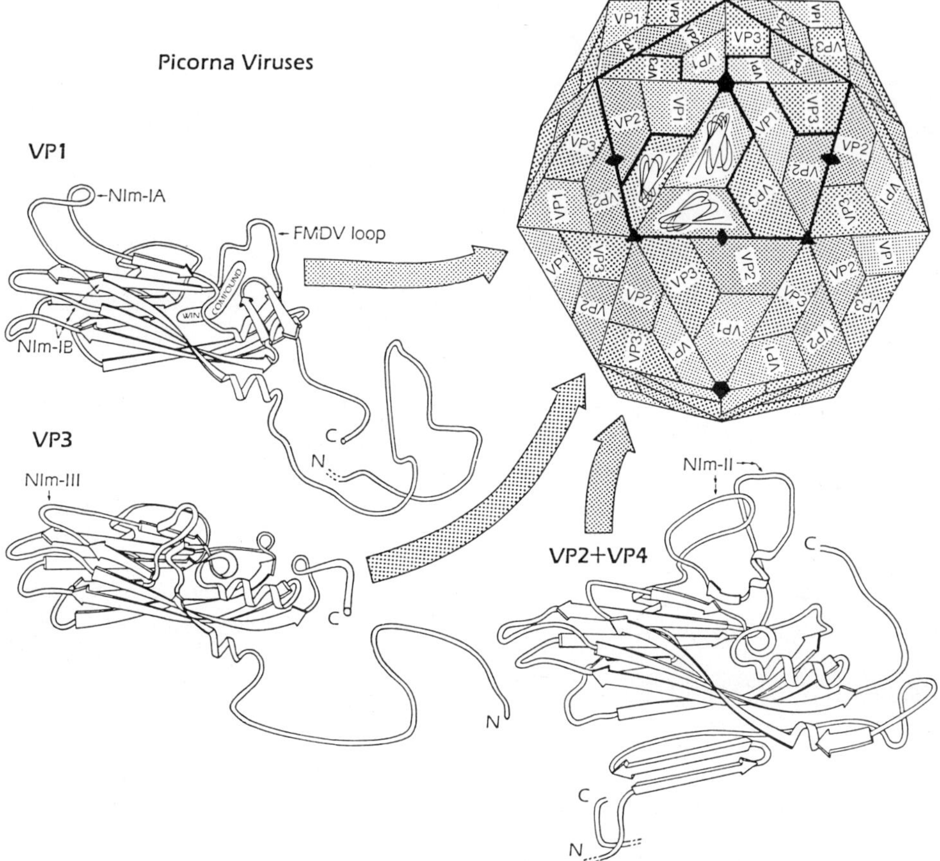

Figure 2. Schematic drawing of the way in which viral polypeptides are folded to form the icosahedral structure of a complete rhinovirus particle. (Reproduced by permission of M. Rossman.)

are single stranded. Indeed, if one applies a selective environment to any pool of rhinovirus (such as a monoclonal neutralizing antibody or a specific antiviral drug), one is likely to recover resistant viruses with a frequency of 10^{-4} or 10^{-5}, which indicates that there must be a high mutation rate.

II. RHINOVIRUSES AS PATHOGENS

These viruses were first recognized because they caused common colds, and early studies paid great attention to this as well as defining their basic biologic properties (see Tyrrell, 1968).

It was found that they were fastidious viruses to grow in the laboratory and very limited and specialized in their natural host range. A few viruses would grow in monkey kidney cells and are now called the minor receptor group (Mischak *et al.*, 1988) but most required carefully selected human cells (human embryo kidney or selected strains of human fibroblasts or HeLa cells). They were able to grow well only at reduced temperatures, typically 33°C, and they did not tolerate low pH, as enteroviruses do. Their cell specificity seems to be due to their use of special cell surface receptors (Tomassini and Colonno, 1986).

Experimental inoculations showed that they would cause colds in chimpanzees, but would scarcely replicate, let alone cause symptoms, in other apes or simians. Furthermore, even in humans they would not infect the pharynx or mouth, or the outside of the nose, and although infection could be produced by conjunctival inoculation, the virus replicated only in the nasal epithelium. Because they grew to quite high titer in the nose and stimulated sneezing and an excess of nasal secretion, they were able to spread by the respiratory route.

For practical reasons, not many viruses were studied in animals, but the laboratory characteristics were found to be quite consistent, and certainly all the isolates and serotypes studied have proved to cause common colds when given to nonimmune volunteers as nasal drops.

Thus, much of the early debate on taxonomy revolved around the paradox that these viruses seemed very much like enteroviruses on laboratory study, but that they had some small, consistent differences that seemed linked with differences in their natural history, in particular their fine adaptations to invade the upper respiratory tract and to spread by the respiratory route.

Nevertheless, there is no sharp distinction between enteroviruses and rhinoviruses. An early example was the discovery of strains of the Coe virus, which grew in tissue culture and had the basic properties of an enterovirus, but caused typical common colds (Parsons *et al.*, 1960). In due course, Coe viruses were found to be neutralized by antiserum against Coxsackie virus A21, though they grew to a very limited extent in suckling mouse muscle (the main biologic marker of type A Coxsackie viruses) and the 5′ noncoding sequence closely resembles that of poliovirus (Hughes *et al.*, 1987).

Thus, the rhinovirus serotypes detected by neutralization tests are very varied, and we want to know their basis, origin, and biologic significance—and at the same time the viruses are all of similar pathogenicity. We need to know how all this is related to the epidemiology of colds and the possibility of controlling them by vaccination.

The Basis and Origin of Serotypes

The basis of serotype specificity has now been well established for RV14. Sherry *et al.* (1986) prepared neutralizing monoclonal antibodies against this virus and then used these to select "escape" mutants by mixing a stock of virus with antibody and propagating the surviving particles. The RNA of these viruses was then prepared and sequenced. It was found that nucleotide substitutions were clustered, which indicated that the monoclonal antibody was reacting with a rather limited sequence of amino acids, though sometimes the regions involved were separated. These findings were "illuminated" by applying the sequenced data to the three-dimensional map of the particle mentioned above (Rossman *et al.*, 1985). This shows that the regions of amino acid substitution fell on the surface of the virus particle. In some cases the sequences formed loops emerging from the surface. In addition, it was seen that the sequences that seemed to be separated were actually adjacent when they were presented as part of the folded-up molecule. Rossman *et al.* also suggested that a "canyon" or cleft that surrounds the fivefold axis of symmetry is involved in virus entry and uncoating and went on to note that the immunogenic sites mapped onto the edge of the canyon and suggested that it was likely therefore that antibody found there would be well placed to interfere with the early stages of virus replication. It was thus possible to mark out on the surface of the virus the sites that acted as epitopes for neutralizing antibody (NIM).

It is likely that this broad picture is correct and that rhinoviruses and other picornaviruses each have several sites or epitopes on the surface of the particle to which antibody can attach and annul the virus infectivity. On the other hand, we are far from understanding the matter fully. For instance, antibody has been made against an amino acid sequence that is well conserved and located in the "canyon"—this antibody neutralizes virus of a number of different serotypes (McCray and Werner, 1987), which suggests that the simple idea based on the analysis of crystals of virus is incomplete: Perhaps the capsid is quite flexible at times so that the "canyon" opens up (perhaps when it attaches to a cell) and in this state antibody can gain access to the cleft. There is evidence from the study of the interaction between influenza virus neuraminidase and antibody that the model of a rigid "lock and key" interaction will not fit the facts and there must be some degree of flexibility (Colman *et al.*, 1987).

Complete models and coordinates are available only for RV14 at the moment but others will be available soon. It will then be possible to determine the apparent sequences of amino acids responsible for the characteristic serotypes of the virus. At the moment the information is fragmentary. We know from monoclonal antibody studies that the mouse immune system "sees" and responds to several different sites, but these are in fact quite varied. Competition experiments or escape mutant studies may show that there is an antigenic area which is spanned by a series of overlapping monoclonals (Fig. 3). On the other hand, a set of monoclonals may all map to a very limited antigenic site (G. Appleyard, personal communication).

It is also likely that studies with mouse monoclonals represent only part of the picture. Polyclonal sera apparently contain antibodies that are directed at additional sites. In some recent work we have found that human sera that neutralize RV2 to substantial titers do not apparently contain antibodies that block the mouse monoclonals directed at any of the three sites defined so far on that virus (W. Barclay, unpublished). Similar findings have emerged with poliovirus antibodies and human sera and also with foot-and-mouth disease virus and animal sera. This may, of course, mean that we should search further for mouse monoclonals with different specificities. Whether or not this is the right research strategy, the results suggest that the immune system of the inbred mouse that provides the monoclonals has a different antigenic repertoire from that of humans and the other laboratory animals whose sera have been examined.

It is important to remember that it is not enough to look at three-dimensional constructs of a virus particle and at which protruding pieces of peptide chain might prove immunogenic. We need to

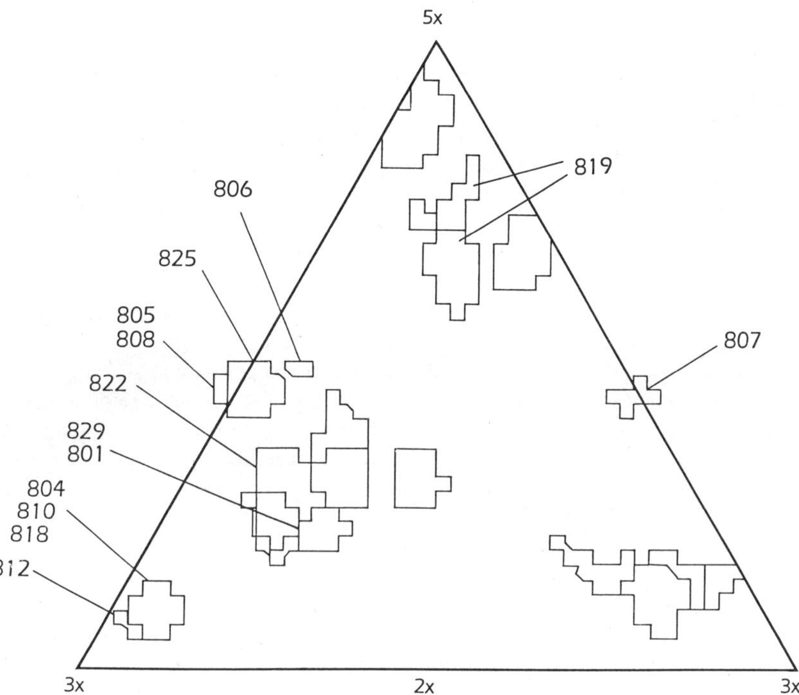

Figure 3. Tentative map of the sites of attachment of a set of monoclonals. The sites of amino acid substitutions are shown for monoclonal antibodies 801–829. (Reproduced by permission of G. Appleyard.)

understand the matter more deeply, for instance, the exact conformation of the "loops" and their relation with adjacent peptide chains, the presence of T-cell epitopes that can be "seen" by the T-helper cells which are needed to generate an immune response (Francis *et al.*, 1987b), and the specific ability of the host immune system to respond to certain sequences and not to others, as has been well recognized for years from classical work on the Ir genes. Thus, the immunologic variability of rhinoviruses can only be understood as being in some measure a reflection of the immunogenic variability of vertebrates, in particular of humans, and we already know that different individuals produce antibody directed against different epitopes of influenza (Wang *et al.*, 1986) and RV2 (Barclay, unpublished).

In my opinion, we must therefore admit that in spite of all the recent progress, we cannot yet understand fully at the molecular level the antigenicity of rhinoviruses. Nevertheless, a great deal of careful work was done previously using standard techniques, e.g., immunizing animals and testing their sera in virus neutralization tests. We showed in the very early days that human sera reacted with certain rhinoviruses and not with others, and it was quickly found that sera from immunized animals could be very specific. As a result of a long series of painstaking international collaborations we can say that at the moment rhinoviruses are classified in 100 different serotypes (Collaborative report, 1987), on the basis that they have been repeatedly tested and found to be neutralized by homologous sera at a titer 20-fold or more higher than that against other viruses and are not neutralized by similar sera against any other of the known rhinoviruses. This definition, of course, conceals a great deal of variability; for instance, "prime" viruses are often found, i.e., viruses that show one-way cross-neutralization with established serotypes. Furthermore, viruses with less than a 20-fold difference from a prototype virus may yet be sufficiently different to behave differently in nature.

When it was first noticed that there were large numbers of serotypes of virus, it was suggested that further serotypes might be developing all the time. Indeed, when RV2 was propagated in the laboratory in the presence of immune serum, a minor antigenic shift occurred (Acornley *et al.*, 1968) and escape mutants are readily detected with monoclonal antibodies. However, in the most recent publications it has been indicated that new serotypes have not been appearing and that almost all the viruses isolated in recent epidemiologic studies have been allocated to known serotypes (Fox *et al.*, 1985). This suggests that there are some constraints to the appearance of new epitopes. These presumably could not be due to limitations in the possible random variations in RNA sequences of the genome, or to limitations in the immunologic repertoire of humans, though since human sera often contain antibody against several epitopes of a rhinovirus, this might reduce the chance of mutants surviving. It might, however, be due to some limitations imposed by the functions of the viral peptides. Although we concentrate on the antigenic significance of these sequences, it is known that the whole peptides are multifunctional; for example, they have to form part of a virus capsid of the requisite degree of stability, and they may be specifically involved in attachment, uncoating, or other early steps in virus replication. This, in very general terms, might explain why highly neurotropic enteroviruses, i.e., polioviruses, exist in strictly limited serotypes, while those with tropisms for respiratory epithelium show a great antigenic flexibility, which, however, is not unlimited. However, this is an exceedingly generalized suggestion, which will only be validated when proper detailed studies at the molecular level have been completed.

A number of studies have documented the serotypes of viruses isolated from patients with colds in different populations (e.g., Fox *et al.*, 1985). They all show a similar pattern, namely, a variety of different serotypes present in a population at any one time. The prevalent viruses change gradually over the years. Furthermore, there is no conclusive evidence that viruses of certain serotypes are associated with particular types of disease, as there would be, for example, with enteroviruses. Some viruses may be detected relatively often because they are cultivated more easily in tissue cultures, particularly those that grow in monkey kidney cells, the minor receptor groups. But some may genuinely spread more readily. It has, however, been noticed that on repeated isolations of a serotype over the years some evidence of antigenic "drift" may be seen (Stott and Walker, 1969).

We cannot be confident that the separate isolations represent the result of successive cycles of mutation and selection, but the evidence is suggestive and the phenomenon is worth further investigation.

In the natural history of rhinoviruses two factors seem to be particularly important. One is that the virus infects efficiently by the nasal route and that it is shed and dispersed into the environment by sneezing and coughing of respiratory secretions; thus we do not find evidence of frequent asymptomatic infections (as occurs with enteroviruses, which are shed in the feces whether symptoms occur or not) and viruses always seem to be mildly pathogenic. The other factor is the need to evade the immune system of the host; these viruses evoke specific secretory and circulatory antibodies which apparently prevent reinfection or disease (Cate *et al.*, 1964, 1966), yet they exist in many serotypes, and of the volunteers coming to the Unit we find that about two-thirds can be infected and half of these have symptomatic infections. It thus appears that these organisms "balance" the transmission rate so well that in the population at large there are always susceptibles who can be infected when conditions for transmission are appropriate. Slight drifts of antigenic type may be part of the "fine tuning" of this process. It could be worthwhile to use refined immunologic reagents such as monoclonal antibodies to look for the emergence of minor variants in closed epidemics, as has been done with influenza (Oxford *et al.*, 1986). To summarize, it looks as though rhinoviruses as a whole exist and are common pathogens only because they can occur in a variety of serotypes. Drift of serotypes can occur and may assist this process. Infectiousness and pathogenicity are essential properties, and although the genome is so variable, it seems that variability in these biologic properties is not consistent with survival and so is not found in nature.

III. VARIABILITY AND ANTIVIRAL TREATMENT

In recent years two strands of research have led to some powerful antivirus drugs active against rhinoviruses. One strand is that of screening of various source materials for drug inhibitors, which has led, as summarized elsewhere (Al-Nakib and Tyrrell, 1988), to the molecules dichloroflavan (DCF), a chalcone, and the drug R61837, all of which inhibit the uncoating of rhinoviruses, though chemically they are quite different and one of them, R61837, prevents colds when sprayed up the nose of volunteers (Al-Nakib *et al.*, 1989). The other strand began with the recognition by Eggers that rhodanine inhibited uncoating of ECHO virus 12 and then led the Sterling–Winthrop team to synthesize a large series of molecules that had a similar mode of action (Diana *et al.*, 1985; Zeichhardt *et al.*, 1987). Some of them were active against rhinoviruses as well as poliovirus *in vitro* and were active by the oral route against poliovirus in a mouse model, but it is not known whether they are effective in humans. Nevertheless, by an elegant extension of the work of Rossman's group, described earlier, it has been shown that the lead compound, disoxaril, is incorporated into a specific site in the virus particle, namely, a hydrophobic region close to the "canyon" (Smith *et al.*, 1986) (Fig. 4).

In the course of this work, drug-resistant mutants have been selected and are of interest in several ways. In the first place, they have been tested against several different members of this group of drugs and have often been found to be cross-resistant. For example, RV2 selected with chalcone was found to be resistant to dichloroflavan (DCF) (Ninomiya *et al.*, 1984, 1985), while that selected with R61837 was resistant to both DCF and chalcone (C. Dearden, unpublished). On the other hand, RV9 selected with chalcone was not resistant to DCF (Ahmad and Tyrrell, 1986); when tested against disoxaril, it was found that RV2 selected with chalcone and RV9 selected with R61837 showed cross-resistance against this drug. All this suggests that the drug-resistant mutations are due to alterations in similar areas of the virus, and that conversely, all the drugs interact with similar areas of the virus capsid. However, there are detailed discrepancies and the degree of cross-resistance suggests that there are at least two types of mutants (unpublished). Furthermore, these

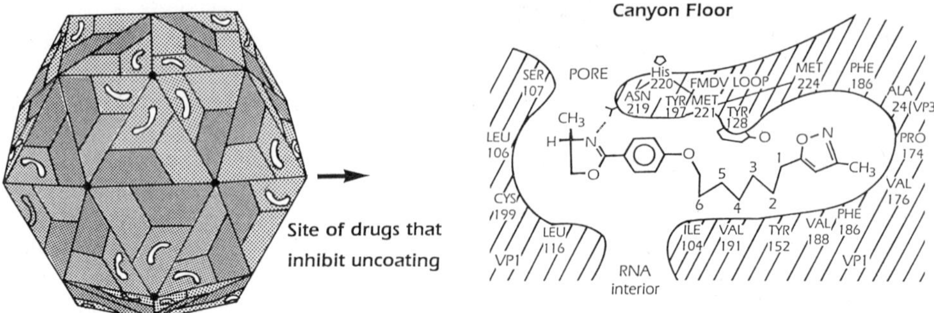

Figure 4. Schematic drawing to show the site of interaction of drugs that inhibit uncoating with the VP1 subunit of the virus capsid. The space beneath the floor of the canyon is shown enlarged with a molecule of disoxaril in place. (Reproduced by permission by M. Rossman.)

compounds vary greatly in their effectiveness against different serotypes. It is clear, therefore, that there are fine differences at the molecular level, and these will have to be resolved by sequencing and structural studies.

The Sterling–Winthrop group and their colleagues have shown that different drugs in their series are inserted into the RV14 crystals in different ways. In complementary studies they have sequenced mutants selected by exposure to disoxaril and find that they display amino acid substitutions in the region around VP1-190. This confirms the crystallographic data on the site of insertion of the molecule into the virus (G. Diana, personal communication).

Studies on drug-resistant mutants are of interest in other ways. For example, these mutants replicate less well than the parental type (Ahmad *et al.*, 1987). Recently we have worked with an RV2 mutant resistant to the chalcone which had had serial passages in culture and grew well *in vitro*. We found that it was less infectious for volunteers and less pathogenic for human volunteers than the wild type, while a drug-dependent mutant derived from it was noninfectious and nonpathogenic, as one might expect (unpublished). However, when some early drug-resistant mutants were propagated in the absence of drug, they recovered sensitivity to the drug. Though we still do not know the nucleotide sequence of either virus, it indicates that changes in the amino acid sequences that would lead to drug resistance are likely to be selected against in clinical practice. However, drug-resistant virus has been isolated from the nasal washings of volunteers infected with RV9 and treated with R61837 (unpublished).

These findings must be carefully considered if these drugs are proposed for clinical use. They suggest that even if a drug-resistant virus appears in the airway of a treated patient, it is unlikely to prevent the drug being clinically effective. On the other hand, if, as a rare event, a mutant that was drug resistant was able to mutate further in such a way that its infectivity and pathogenicity were restored, then a drug-resistant organism might become established in the community and impair the drug's effectiveness.

On the other hand, there is laboratory evidence that there are slight differences at the molecular level in the site of action of the drugs—thus there is a limited amount of true synergy in antiviral activity between different members of this group of molecules (Al-Nakib and Tyrrell, 1987). The effect is, however, much less than the synergy between interferons and between interferons and other drugs (Ahmad and Tyrrell, 1986) and the usual explanation of this is that their site of action is different. Drug synergy might be exploited to enhance the antiviral effects and, in the context of the present discussion, reduce the probability that drug-resistant mutants will emerge. Whatever the practical consequences may be, exploiting the variability and the mutants of rhinoviruses will teach us more about how antivirals act.

IV. VARIABILITY AND DIAGNOSIS

Treatment with drugs and epidemiologic research require sensitive and convenient diagnostic methods. Rhinoviruses have in the past been detected by cultivation from respiratory secretions obtained by means of swabs, nasal washings, and so forth. However, these are slow and labor intensive. Now that molecular reagents have become available for these viruses, alternative direct methods can be considered. For example, ELISA can be used to detect viral antigens in nasal washings. In addition, cDNA has been obtained which hybridizes well with viral RNA. It looks as though the DNA for the 5′ noncoding region, which it was thought might be well conserved, can be used to detect a wide range of serotypes of rhinoviruses. However, in the studies at the Unit the signal obtained equaled that of the homologous pair in only a small proportion of combinations (Al-Nakib *et al.*, 1986)—clearly, the 5′ region is not completely conserved, as sequencing studies already indicate. However, within this region in both rhinoviruses and enteroviruses there are two short sequences that are completely conserved. We have therefore been using a mixture of complementary deoxyoligonucleotides, which appears to detect virus in nasal washings efficiently and specifically and should be equally satisfactory for all serotypes (Bruce *et al.*, 1988)—time will tell, but the method promises to avoid the difficulty presented to the virologist and epidemiologist due to the variability of other regions of the genome.

V. VARIABILITY AND RHINOVIRUS VACCINES

In the light of the antigenic variability of the virus, the early idea that growing the causative virus would quickly lead to producing a vaccine has been abandoned. This has been revived, however, by recent work, in particular the experiments of Bittle *et al.* (1982), which showed that an oligopeptide that mimics an important epitope of foot-and-mouth disease virus can be given by injection as a vaccine and induces antibody that neutralizes viral infectivity and protects animals from disease.

Oligopeptides that mimic the epitopes of rhinoviruses have also been sought, by synthesizing sequences that are likely to appear on the surface of RV2 and that include some of the epitope sequences identified by monoclonal antibody studies. Some stimulate antibodies that bind only to peptide, others to C-type particles as well. One was found to induce antibody that binds to D-type particles and neutralizes virus infectivity (Francis *et al.*, 1987a).

By an independent approach a similar peptide was defined by Blaas and his colleagues (Skern *et al.*, 1987). Antibodies induced by this synthetic peptide are able to block antibodies in human sera, and so presumably the human immune system reacts to the same sequence (W. Barclay, personal communication). We do not yet know whether these peptides can be formulated in such a way as to induce antibodies and confer resistance to infection in humans. If this is achieved, and I see no reason why it should not be, then the problem reduces itself to finding sequences that induce effective antibody against a majority of rhinoviruses, either by defining more of the commonly detected epitopes, or by finding conserved cross-reacting sequences, such as that in the canyon.

This is still a daunting task, but whether or not we defeat the variability that the virus uses as a defense, we shall certainly learn a lot about the biology of rhinoviruses and the human immune system.

REFERENCES

Acornley, J. E., Chapple, P. J., Stott, E. J., and Tyrrell, D. A. J. (1968). *Arch.f.d.ges. Virusforsch.* **23,** 284–287.

Ahmad, A. L. M., and Tyrrell, D. A. J. (1986). *Antiviral Res.* **6,** 241–252.

Ahmad, A. L. M., Dowsett, A. B., and Tyrrell, D. A. J. (1987). *Antiviral Res.* **8,** 27–39.

Al-Nakib, W., and Tyrrell, D. A. J. (1987). *Antiviral Res.* **8,** 179–188.

Al-Nakib, W., and Tyrrell, D. A. J. (1988). In *Clinical Use of Antiviral Drugs* (E. De Clercq, ed.), pp. 241–276, Nijhoff, Boston.

Al-Nakib, W., Stanway, G., Forsyth, M., Hughes, P. J., Almond, J. W., and Tyrrell, D. A. J. (1986). *J. Med. Virol.* **20,** 289–296.

Al-Nakib, W., Higgins, P. G., Barrow, G. I., Tyrrell, D. A. J., Andries, K., Vanden Bussche, G., Taylor, N., and Janssen, P. A. J. (1989). *Antimicrob. Agents Chemother.* **33,** 522–525.

Bittle, J. L., Houghten, R. A., Alexander, H., Shinnick, T. M., Sutcliffe, J. G., Lerner, R. A., Rowlands, D. J., and Brown, F. (1982). *Nature* **298,** 30–33.

Bruce, C. B., Al-Nakib, W., Tyrrell, D. A. J., and Almond, J. W. (1988). *Lancet,* **2,** 53.

Cate, T. R., Couch, R. B., and Johnson, R. M. (1964). *J. Clin. Invest.* **43,** 57–67.

Cate, T. R., Rossen, R. D., Douglas, G. Jr., Butler, W. T., and Couch, R. B. (1966) *Am. J. Epidemiol.* **84,** 352–363.

Collaborative report (1987). *Virology* **159,** 191–192.

Colman, P. M., Laver, W. G., Varghese, J. N., Baker, A. T., Tulloch, P. A., Air, G. M., and Webster, R. G. (1987). *Nature* **326,** 358–363.

Decock, B., and Billiau, A. (1986). *Arch. Virol.* **90,** 337–342.

Diana, G. D., Otto, M. J., and McKinlay, M. A. (1985). *Pharmac. Ther.* **29,** 287–297.

Fox, J. P., Cooney, M. K., Hall, C. E., and Foy, H. M. (1985). *Am. J. Epidemiol.* **122,** 830–846.

Francis, M. J., Hastings, G. Z., Sangar, D. V., Clark, R. P., Syred, A., Clarke, B. E., Rowlands, D. I., and Brown, F. (1987a). *J. Gen. Virol.* **68,** 2687–2691.

Francis, M. J., Hastings, G. Z., Syred, A. D., McGinn, B., Brown, F., and Rowlands, D. J. (1987b). *Nature* **300,** 168–170.

Hughes, P. J., Phillips, A., Minor, P. D., and Stanway, G. (1987). *Arch Virol.* **94,** 141–147.

Lonberg-Holm, K., and Korant, B. D. (1972). *J. Virol.* **9,** 29–40.

MacNaughton, M. R. (1982). *Curr. Topics Microbiol.* **97,** 1–19.

McCray, J., and Werner, G. (1987). *Nature* 329, 736–738.

Mischak, H., Neubauer, C., Kuechler, E., and Blaas, D. (1988). *Virology* **163,** 19–25.

Neubauer, C., Frasel, L., Kuechler, E., and Blaas, D. (1987). *Virology* **158,** 255–258.

Ninomiya, Y., Ohsawa, C., Aoyama, M., Umeda, I., Suhara, Y., and Ishitsuka, H. (1984). *Virology* **134,** 269–276.

Ninomiya, Y., Aoyama, M., Umeda, I., Suhara, Y., and Ishitsuka, H. (1985). *Antimicrob. Agents Chemother.* **27,** 595–599.

Oxford, J. S., Salum, S., Corcoran, T., Smith, A. J., Grilli, E. A., and Schild, G. C. (1986). *J. Gen. Virol.* **67,** 265–274.

Parsons, R., Bynoe, M. L., Pereira, M. S., and Tyrrell, D. A. J. (1960). *Br. Med. J.* **1,** 1776.

Reanney, D. C. (1984). In *The Microbe 1984. 1. Viruses* (B. W. J. Mahy and J. R. Pattison, eds.), pp. 175–196, Cambridge University Press, Cambridge.

Rossmann, M. G., Arnold, E., Erickson, J. W., Frankenberger, E. A., Griffith, J. P., Hecht, H. J., Johnson, J. E., Kamer, G., Luo, M., Mosser, A. G., Rueckert, R. R., Sherry, B., and Vriend, G. (1985). *Nature* **317,** 145–153.

Sherry, B., Mosser, A. G., Colonno, R. J., and Rueckert, R. R. (1986). *J. Virol.* **57,** 246–257.

Skern, T., Sommergruber, W., Blaas, D., Pieler, Ch., and Kuechler, E. (1984). *Virology* **136,** 125–132.

Skern, T., Neubauer, C., Frasl, L., Gründler, P., Sommergruber, W., Zorn, M., Kuechler, E., and Blaas, D. (1987). *J. Gen. Virol.* **68,** 315–323.

Smith, T. J., Kremer, M. J., Luo, M., Vriend, G., Arnold, E., Kamer, G., Rossmann, M. G., McKinlay, M. A., Diana, G. D., and Otto, M. J. (1986). *Science* **233,** 1286–1293.

Stott, E. J., and Walker, M. (1969). *Nature* **224,** 1311–1312.

Tomassini, J. E., and Colonno, R. J. (1986). *J. Virol.* **58,** 290–295.

Tyrrell, D. A. J. (1968). *Monographs in Virology 2,* pp. 67–124, Springer-Verlag, New York.

Wang, M. L., Skehel, J. J., and Wiley, D. C. (1986). *J. Virol.* **57,** 124–128.

Zeichhardt, H., Otto, M. J., McKinlay, M. A., Willingmann, P., and Habermehl, K-O. (1987). *Virology* **160,** 281–285.

14

Genetic Variability and Antigenic Diversity of Foot-and-Mouth Disease Virus

Esteban Domingo, Mauricio G. Mateu, Miguel A. Martínez, Joaquín Dopazo, Andrés Moya, and Francisco Sobrino

I. INTRODUCTION: FOOT-AND-MOUTH DISEASE AND ITS DISTRIBUTION

Foot-and-mouth disease (FMD) is an acute systemic disease of cloven-hooved animals, including cattle, swine, sheep, and goats. Despite mortality rates being generally below 5%, FMD severely decreases livestock productivity and trade. It is considered the economically most important disease of farm animals. Near two thousand million doses of vaccine are used annually to try to control FMD, which, nevertheless, is enzootic in most South American and African countries, parts of Asia, the Middle East, and the south of Europe. The causative agent, foot-and-mouth disease virus (FMDV), is an aphthovirus of the family Picornaviridae, a historically important virus as it was the first recognized viral agent (Loeffler and Frosch, 1898). In this chapter we review briefly the structure of FMDV and the organization and expression of its genome (Section II) and, in more detail, recent results on genetic variability (Section III) and antigenic diversity (Section V), reflected in several serotypes and subtypes of the virus. Such a diversity has implications for vaccine design and disease control, as discussed in Section V. Finally, we propose a model of evolution of FMDV and discuss its implications (Sections VI and VII).

A. The Disease

The infection process may be initiated either by inhalation of as few as one to ten infectious units or by ingestion or penetration, through skin abrasions, of a larger number of particles (Sellers,

Abbreviations used in this chapter: FMDV, foot-and-mouth disease virus; MAb, monoclonal antibody; MAR, monoclonal-antibody-resistant; nMAb, neutralizing MAb; s/s per year, substitutions per nucleotide site per year.

Esteban Domingo and Mauricio G. Mateu • Centro de Biología Molecular, Universidad Autónoma de Madrid, Canto Blanco 28049 Madrid, Spain. *Miguel A. Martínez and Francisco Sobrino* • Departamento de Sanidad Animal, Instituto Nacional de Investigaciones Agrarias, 28012 Madrid, Spain. *Joaquín Dopazo and Andrés Moya* • Laboratorio de Genética, Universidad de Valencia, 46100 Burjassot, Valencia, Spain.

1971; Gibson and Donaldson, 1986; Donaldson *et al.*, 1987). At the site of experimental inoculation, a vesicle is formed (the primary vesicle, or "aphtha"). In the natural transmission of the virus, the most frequent portal of entry appears to be the upper respiratory tract (McVicar and Sutmoller, 1976; Sutmoller and McVicar, 1976; Burrows *et al.*, 1981). An initial viral replication is generally followed by febrile viremia involving a large amplification of virus, which is eventually found in many organs. Among young animals, lesions in the myocardium may lead to high mortality rates; myocarditis may also occur before signs of generalized infection. The usual outcome of viremia is macrophage-mediated transport of virus to the epithelium in the vascular dermis. Individual cells from the stratum spinosum are infected and change to a round morphology. Then, the infection spreads to neighboring cells, that show the same morphologic alterations (Yilma, 1980; Di Girolamo *et al.*, 1985). High virus titers are scored in many tissues, including skin areas where no lesions are apparent (Gailiunas, 1968; Burrows *et al.*, 1981). It has been estimated that an acutely infected animal may contain up to 10^{12} infectious units (Sellers, 1971). This represents a huge viral amplification that takes place in a time span of 3 days to 2 weeks, depending largely on the dose of infecting virus. Detailed descriptions of the disease process can be found in reviews by Shahan (1962), Bachrach (1968), Joubert and Mackowiak (1968), and Timoney *et al.* (1988).

B. Persistent Infection

FMDV may establish an inapparent, persistent infection in ruminants for periods of a few weeks up to several years. In the course of persistence, virus can be isolated from the esophagous and throat fluids of the animals by probang extraction (van Bekkum *et al.*, 1959). This asymptomatic infection may be an outcome of the acute infection (Section I.A) or it may be established either by pharyngeal or nasal exposure to virus, or as a result of immunization with some live-attenuated vaccines (Sutmoller and Gaggero, 1965; Sutmoller *et al.*, 1968). It is a common type of infection in areas where FMD is enzootic (van Bekkum *et al.*, 1959; Sutmoller and Gaggero, 1965; Burrows, 1966; Sutmoller *et al.*, 1968; Augé de Mello *et al.*, 1970), and it has permitted FMDV to be maintained in African buffalo for at least 24 years and through several generations (Condy *et al.*, 1985). There is evidence that carrier animals have been at the origin of FMD outbreaks when brought into contact with susceptible animals (Singh, 1969; Hedger and Stubbins, 1971; Hedger and Condy, 1985). Such transmission, however, appears to be infrequent, perhaps due to the low virus titers [variable, but usually lower than 10^3 plaque-forming units (PFU)/ml] in saliva and throat fluids of carriers. The mechanisms that mediate in this long-term persistence of FMDV are unknown, and they may involve selection of antigenically variant viral subpopulations (Gebauer *et al.*, 1988) (compare Sections III.B and IV.A). Carrier animals cannot be cured of FMDV by vaccination and, moreover, persistence may be established in vaccinated cattle (Sutmoller and Gaggero, 1965). Thus, inapparent infections constitute a major drawback for the control of FMD.

C. Epizootiology

FMDV is stable in a considerable range of environmental conditions, it has several portals of entry into animals hosts (Section I.A) and invades various tissues and organs (Sections I.A and IV.A). As a result, it is an extremely contagious virus that may spread at astonishing rates. Several wild animals (small mammals such as rodents, etc.) are also susceptible to FMDV, and they may contribute to virus spread (Capel-Edwards, 1971; Gomes and Rosenberg, 1984). It may be mechanically transmitted by insects, animals, or farming equipment. The transport of infected animals has contributed to the dissemination of FMDV (Brooksby, 1981a). Long-distance, airborne transmission has also been documented (Sellers, 1981; King *et al.*, 1981; Donaldson, 1987). Important epizootics (such as those in Europe in 1937–1940 and 1951–1952 and in Mexico in 1946–1953) have occurred

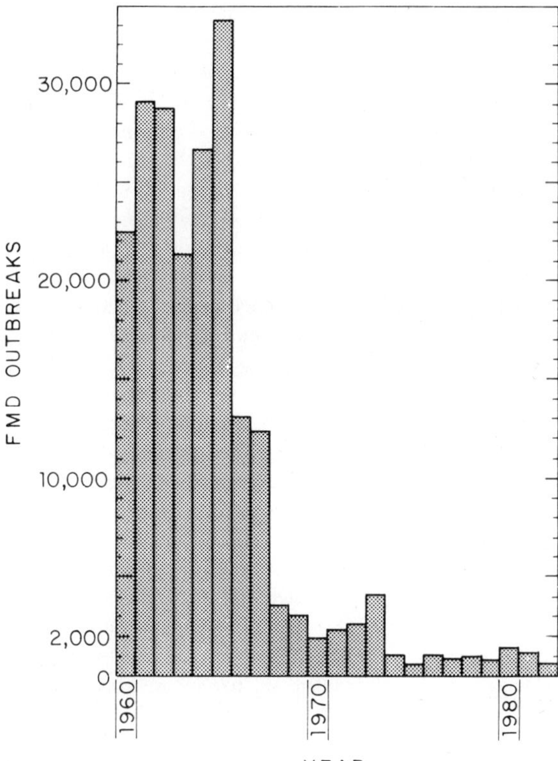

Figure 1. Number of FMD outbreaks officially recorded in Europe from 1960 until 1982. The sharp decrease in the mid-1960s coincides with implementation of compulsory vaccination of cattle in several European countries (data compiled from reports of the sessions of the European Commission for the Control of Foot-and-Mouth Disease of the Food and Agriculture Organization of the United Nations).

where high densities of susceptible animals with poor levels of immunity are found. At present, thousands of outbreaks are recorded each year in South America, and no country in the world is free of the direct or indirect consequences of the disease (Callis, 1979; Callis and McKercher, 1985). Systematic vaccination, updated rapid diagnostic procedures, control of animal movements, and prompt destruction of affected and exposed herds (the "stamping out" procedure) are prerequisites for FMD control. The comparison of the number of outbreaks recorded annually in Europe during the last two decades (Fig. 1) testifies to the benefits of vaccination and other control measures. Considerable international effort is invested in harmonizing policies for FMD control. Among concerned institutions, the Office International des Epizooties, World Health Organization, International Association of Biological Standardization, and Food and Agriculture Organization of the United Nations hold regular meetings and publish reports with up-to-date information on FMD and its distribution. Several in-depth reviews have also covered this subject (Bachrach, 1968; Hyslop, 1970; Brooksby, 1981a; Pereira, 1981; Sellers, 1981).

II. FOOT-AND-MOUTH DISEASE VIRUS

A. The Particle

Pioneer work by H. L. Bachrach, F. Brown, and their colleagues during four decades has greatly contributed to the characterization of FMDV. The properties of the particle and of its genomic RNA are summarized in Table 1. FMDV is an icosahedral virion, about 300 Å in diameter, that conforms to the general architecture of a picornavirus, as revealed by crystallographic analysis (Fox *et al.*, 1987; Acharya *et al.*, 1989). However, FMDV differs from polio, rhino, and Mengo viruses, in that a pit or canyon believed to accommodate the cellular receptor (Luo *et al.*, 1987, 1988; Rossmann and Rueckert, 1987) is absent. Also, the main antigenic site on VP1 of FMDV O_1 BFS (residues 137–156) is an exposed, disordered loop on the virion surface (compare Sections V.B and V.C). Interestingly, this exposed segment includes the highly conserved sequence RGD, thought to be the cell attachment site of the virus (Geysen *et al.*, 1985; Fox *et al.*, 1989) (see also the alignment of VP1 sequences in Section III.A). The three-dimensional structure of FMDV at 2.9 Å resolution has also provided an elegant interpretation of several features of FMDV particles, such as their high density in CsCl (penetration of cesium ions through a hole in the icosahedral fivefold axis) or their lability to acidic media (histidine residues at the pentamer interfaces), among others (Acharya *et al.*, 1989).

Unfortunately, for FMDV one of the main sites relevant to the interaction of virions with antibodies is disordered and could not be accurately positioned on the three-dimensional structures (Acharya *et al.*, 1989). Precise information on the contacts between antigenic sites and antibody molecules is key to the development of new vaccines (Section V). Thus, an understanding of the interaction of FMDV with antibodies will necessitate determining three-dimensional structures of representative FMDV isolates and mutants, as well as of antigen–antibody complexes, and some penetration into the molecular dynamics of the relevant structures. Also, data that are emerging from studies with FMDV and with other viral systems indicate that the single residue makeup of virions may affect their recognition by components of the immune system (see Mateu *et al.*, 1989, for an example concerning FMDV; reviewed by Domingo and Holland, 1988; Domingo, 1989). Thus, in attempting to relate any particular structure with an antigenic behavior (or other biological

Table 1. Properties of FMDV and Its Genomic RNA[a]

Virion
- Sedimentation ≃ 146S
- Density (*CsCl*) ≃ 1.43–1.45 g/cm^3
- Unstable at acidic pH and physiologic temperatures[b]
- Antigenic diversity[c]: seven serotypes (A, O, C, SAT1, SAT2, SAT3, Asia 1), more than 65 subtypes, and many variants
- Proteins:
  - VP1, VP2, VP3 (M.W. = 27,000–30,000); 60 copies of each per particle; phosphorylated
  - VP4 (M.W. = 9,000–10,000); 60 copies per particle; myristilated
  - VP0 (uncleaved VP2, VP4 precursor); 1–2 copies per particle
  - VPg (covalently linked to 5′ end of RNA)

Genomic RNA
- Single stranded, ~8500 nucleotides (~35S)
- Infectious, acts as messenger RNA
- Poly (A) at 3′-end; poly (C) near 5′-end

[a]Data from Bachrach, 1977; Brown, 1979; La Torre *et al.*, 1980; Chow *et al.*, 1987; Fox *et al.*, 1987; Acharya *et al.*, 1989.
[b]Instability varies with strains, isolates, and passage history. Usually FMDV infectivity is rapidly lost at pH ≤ 6.0.
[c]Antigenic diversity is reviewed in Section V and Genetic variability in Section III.

function), it must be remembered that FMDV populations are extremely heterogeneous (Section VI) and that one is not dealing with single, defined genomic species.

FMDV particles include as genome a single-stranded RNA molecule (Table 1), with no ordered nucleotides that could be identified in the crystal structures (Acharya *et al.*, 1989). During its replication, FMDV RNA undergoes remarkable genetic variation (Section III), a salient property of RNA viruses.

B. Genome Organization and Expression

The RNA of FMDV, like that of other picornaviruses, is translated into a polyprotein that is processed to yield the mature proteins (Fig. 2). Removal of the poly (C) tract and of the RNA segment to its 5′ side does not affect translation of the RNA *in vitro* (Sangar *et al.*, 1980). Two internal AUG codons located 84 nucleotides apart are used as start sites, giving rise to two leader (L) proteins at the N-terminus of the polyprotein (Sangar *et al.*, 1987). The smaller L is further processed, late in infection, to yield a third form of L protein (Sangar *et al.*, 1988). L catalyzes its own cleavage from the L/P1 junction (Strebel and Beck, 1986; Vakharia *et al.*, 1987) and initiates cleavage of p220 (a component of the cap binding protein complex) (Devaney *et al.*, 1988), a process involved in the shutoff of host cell protein synthesis. Most of the cleavages of P1, P2, and P3 are catalyzed by 3C, which functions in *trans* (Vakharia *et al.*, 1987; Clarke and Sangar, 1988). The P1/P2 cleavage occurs between 2A/2B in the absence of L and 3C. The activity may be associated with the highly conserved, 16-amino-acid peptide 2A, which shows some homology with residues 119–132 of 2A from encephalomyocarditis virus (Clarke and Sangar, 1988). The differences in protein processing between FMDV and other picornaviruses have been reviewed by Jackson (1989).

Protein 2C is probably involved in viral RNA synthesis, since mutations that confer resistance to the inhibitory action of guanidine (an inhibitor of viral RNA synthesis) map in 2C (Saunders *et al.*, 1985). In poliovirus-infected cells, protein 2C is probably part of the replication complex (Pincus *et al.*, 1987). VPg is encoded by a three-tandem repetition within 3B (Forss and Schaller, 1982), and it is probably involved in initiation of viral RNA synthesis (Semler *et al.*, 1988). Protein 3D is the viral subunit of the RNA replicase, or virus-infection-associated antigen (VIAA) (Polatnick *et al.*, 1967; Polatnick, 1980), diagnostic of virus infection.

According to computer predictions, FMDV RNA depicts a considerable degree of secondary structure, in particular repeated pseudoknotted foldings at its 5′-terminal region (Clarke *et al.*, 1987a). The function of such structures is unknown. The poly (C) tract of FMDV RNA (Brown *et al.*, 1974; Rowlands *et al.*, 1978) in solution is single stranded (Mellor *et al.*, 1985). Its length varies among strains and isolates and it has been a useful marker to reveal genome heterogeneities and fluctuations (Section III.B), but its function remains unknown. FMDV RNA replication takes place via a complementary RNA minus strand. Protein assembly and RNA encapsidation yields

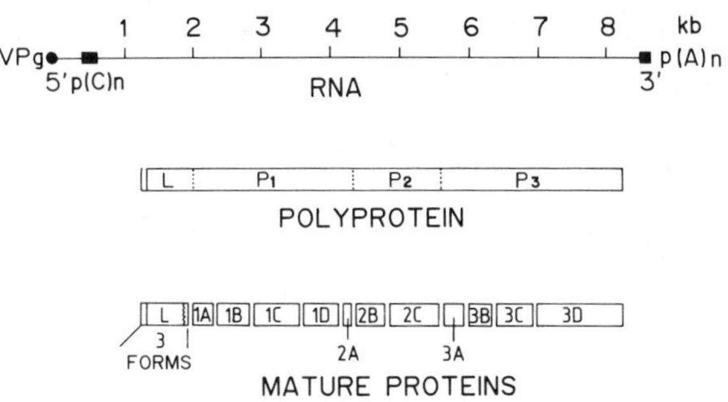

Figure 2. Scheme of the FMDV genome and encoded proteins. Genome length and coding assignments are based on Forss *et al.* (1984), Carroll *et al.* (1984), Robertson *et al.* (1985), Vakharia *et al.* (1987), and Clarke and Sangar (1988). The poly (A) at the 3′-end is extremely heterogeneous within each RNA preparation.

about 10^4–10^5 particles per cell, 0.1–1% of which are infectious. The reason for such low infectivity-to-particle ratio is unclear (Sangar, 1979; Semler *et al.*, 1988), but it does not appear to be due to the presence of defective RNA molecules (Belsham and Bostock, 1988). Progress in FMDV genetics has been hampered by the difficulty in deriving infectious cDNA clones, perhaps due to biologic instability of a double-stranded d(C)d(G) DNA duplex. Recently, an infectious clone representing the genome of FMDV O_1 K has been synthesized by E. Beck and his associates (Zibert *et al.*, 1990) Thus, it will now be possible to construct defined mutants with which to probe the activities of gene products whose function remains obscure.

As documented also for other RNA viruses, it is becoming apparent that FMDV RNA may undergo extensive genetic variation during its replication in animals and in cell culture. We next review recent developments on the study of genetic variability of FMDV.

III. GENETIC VARIABILITY OF FMDV

The concept that high mutability and potential for rapid variation are distinctive properties of some RNA viruses (typically the influenza A viruses) has been gradually substituted by the notion that all RNA viruses, because of their error-prone RNA replication, share a great potential for variation (Holland *et al.*, 1982). The evidence for this view and its biologic implications of RNA genome heterogeneity have been reviewed (Reanney, 1984; Domingo *et al.*, 1985; Steinhauer and Holland, 1987; Smith and Inglis, 1987; Domingo and Holland, 1988; Strauss and Strauss, 1988; Domingo, 1989).

A. Variations among FMDV Types, Subtypes, and Isolates: Alignment of VP1

The early comparisons among FMDV genomes were carried out by nucleic acid competition–hybridization assays. The sequence homology between RNAs of different serotype A, O, or C was estimated in 44–70%. For viruses of different subtype within types A and O the homology was greater than 70% (Dietzshold *et al.*, 1971; Robson *et al.*, 1977). A very early sequence determination revealed little homology among the 40 residues adjacent to the poly A tract of FMDV A-61, O-VI, and C-997 (Porter *et al.*, 1978). By T_1 oligonucleotide mapping, Robson *et al.* (1979) showed differences among the RNAs of isolates of subtypes A_5, A_{22}, and A_{24}. Evidence that natural FMDV populations are genetically heterogeneous was obtained by comparing oligonucleotide fingerprints of RNA of isolates from a single disease outbreak (Domingo *et al.*, 1980). Despite the viruses examined being indistinguishable by classical serologic techniques (complement fixation and serum neutralization tests), each RNA differed from the others in about 0.7–2.2% of their residues. Genetic heterogeneity of FMDV during disease episodes is now well documented (King *et al.*, 1981; Piccone *et al.*, 1988). Moreover, two different nucleotide sequences were found in a single isolate; Domingo *et al.* (1980) proposed a *quasispecies* (or extreme genetic heterogeneity) model to describe FMDV populations. The *quasispecies* concept has proven important to represent RNA viruses and to explain the dynamics of their evolution; it is discussed in Section VI.A.

Application of molecular cloning and rapid nucleic acid sequencing techniques has permitted the derivation of considerable numbers of sequences from FMDV genomes. In particular, attention has been directed to the VP1 coding region, since VP1 includes important determinants for FMDV neutralization (Section V.B) and may be the basis for synthetic vaccine formulations (Section V.D). The VP1 sequence of more than 50 FMDV isolates has been deduced from the corresponding nucleotide sequences (Kurz *et al.*, 1981; Boothroyd *et al.*, 1982; Makoff *et al.*, 1982; Cheung *et al.*, 1983, 1984; Villaneuva *et al.*, 1983; Beck *et al.*, 1983a; Rowlands *et al.*, 1983; Weddell *et al.*, 1985; Onischenko *et al.*, 1986; Sobrino *et al.*, 1986; Beck and Strohmaier, 1987; Martínez *et al.*, 1988; Piccone *et al.*, 1988; de la Torre *et al.*, 1988; Gebauer *et al.*, 1988; Pfaff *et al.*, 1988; Thomas *et al.*, 1988; Marquardt and Adam, 1990; Carrillo *et al.*, 1990; Piccone *et al.*, and Martínez *et al.*, unpublished results).

To analyze the relationship among VP1 genes of FMDVs belonging to different serotypes and subtypes, Dopazo *et al.* (1988) have constructed phylogenetic trees by maximum-parsimony methods. One such tree, relating the VP1 gene of 22 FMDV isolates of serotype A, O, and C, is shown in Fig. 3. The study of Dopazo *et al.* (1988) has documented that the VP1 genes analyzed constitute sets of related, nonidentical sequences, clustered according to serotype. It is a variable segment, with an average of 0.359 substitution per site, a value that is higher for serotype A than for serotypes O or C. This has led to the suggestion that FMDVs of type A may be ancestral to those of types O and C (Dopazo *et al.*, 1988), in line with the greater antigenic diversification (reflected in the number of subtypes scored) of FMDV A. It has also been proposed that copying of the RNA segment that encodes a critical antigenic determinant of VP1 (the 140–160 region, Sections V.B and V.C) may be error prone for FMDV type A because of a looped RNA structure predicted near that site (Weddell *et al.*, 1985). While it is not possible at present to exclude this possibility, other mechanisms must also be in operation since such secondary structures are also predicted in the RNA of FMDV belonging to other serotypes (Villaverde *et al.*, unpublished results). Extremely ramified trees are expected whenever rapid genetic diversification occurs (Fig. 3; compare Sections VI.A and VI.B). Such trees, relating structural and nonstructural protein genes of different picornaviruses, have been derived by Palmenberg (1989).

An alignment of amino acid sequences of VP1 of FMDVs A, O, and C is shown in Figs. 4, 5, and 6, respectively. The identity values (percentage of positions at which the same amino acid is found among the compared proteins) are 73%, 83%, and 83% respectively. Alignment of seven sequences of viruses of different serotypes reduces the identity to 57% (Fig. 7). The variance of the number of amino acid substitutions is eightfold lower between any two VP1s of a different serotype than between VP1s within one serotype. This suggests that considerable variation is allowed around the consensus or prototype sequence that defines one serotype, and more severe restrictions limit jumps between serotypes (Dopazo *et al.*, 1988). There are some highly conserved residues. Among

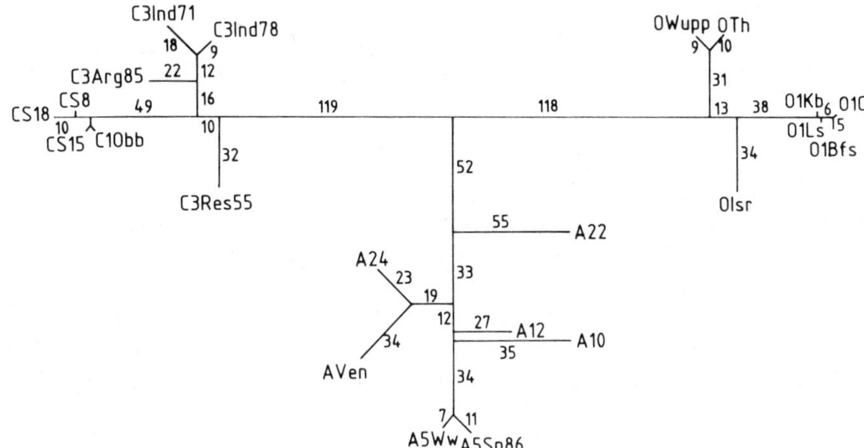

Figure 3. Most-parsimonious tree relating the VP1 gene of 22 isolates of FMDV of serotypes A, O, and C. Genetic distances among closely related viruses are not indicated. Matrices of nucleotide substitutions and other calculations will be supplied upon request. Note that the tree is very similar to that derived by Dopazo *et al.* (1988) using 15 FMDV isolates. The procedures for analysis of sequences and phylogenetic calculations have been described (Dopazo *et al.*, 1988).

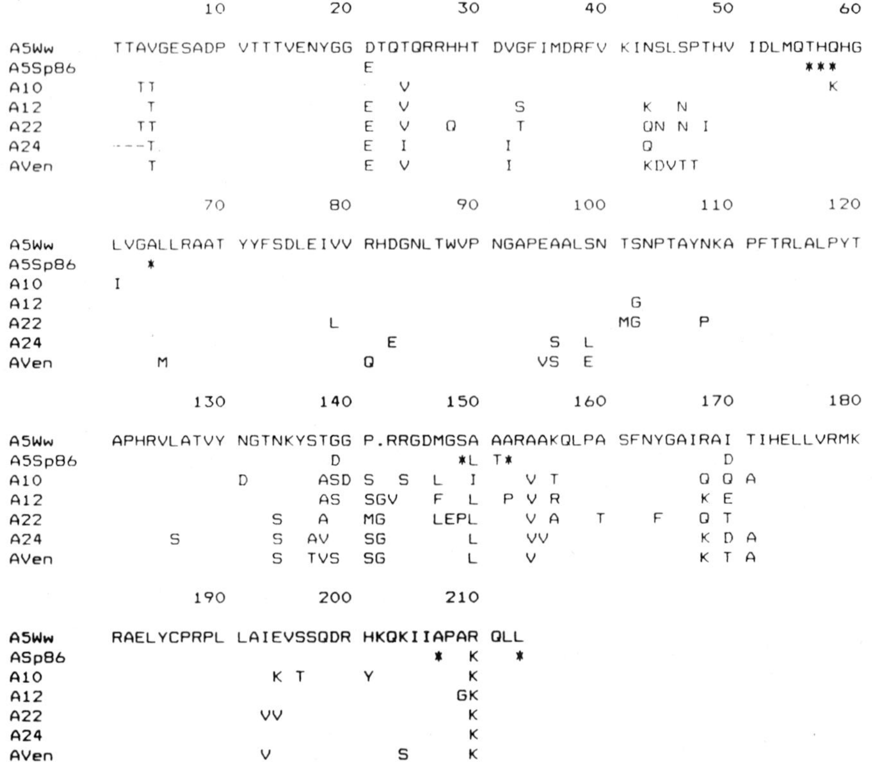

Figure 4. Alignment of the amino acid sequences of capsid protein VP1 of seven FMDVs of serotype A. The sequences have been deduced from the corresponding nucleotide sequences. Asterisk, Undefined amino acid; dash, amino acid not determined; dot (at position 142), deletion introduced to give a better alignment. Only the amino acids that differ from that at the corresponding position of the first row are written. Of a total of 98 substitutions (different for each position), 69 involve amino acids of a different group, as defined by Dayhoff *et al.* (1972). Sequences are from Weddell *et al.* (1985), Beck and Strohmaier (1987), and Carrillo *et al.* (1989).

them, the sequence RGD (positions 146–148 in Fig. 7) is probably part of a cell attachment site (Geysen *et al.*, 1985; Fox *et al.*, 1989).

Sequences for other genomic segments are also available (Boothroyd *et al.*, 1982; Beck *et al.*, 1983b; Forss *et al.*, 1984; Carroll *et al.*, 1984; Martínez-Salas *et al.*, 1985; Robertson *et al.*, 1985; Newton *et al.*, 1985; Thomas *et al.*, 1988; Villaverde *et al.*, 1988; Brown *et al.*, 1989; Sobrino *et al.*, 1989; Dopazo *et al.*, in preparation). These comparisons, and other data summarized in Table 2 and in Sections III.B and III.C, lead to the conclusion that each FMDV genome sequenced to date is genetically unique, and that each population (i.e., each "individual" isolate) is genetically heterogeneous. This includes genome segments regarded as highly invariable, such as the 3D (polymerase). These comparisons acquire new significance when another parameter is taken into consideration: time.

B. Rate of Fixation of Mutations

Rates of fixation of mutations for RNA viral genomes are highly variable and may reach values near 10^{-2} substitutions per nucleotide site (s/s) per year. In contrast, the corresponding values for

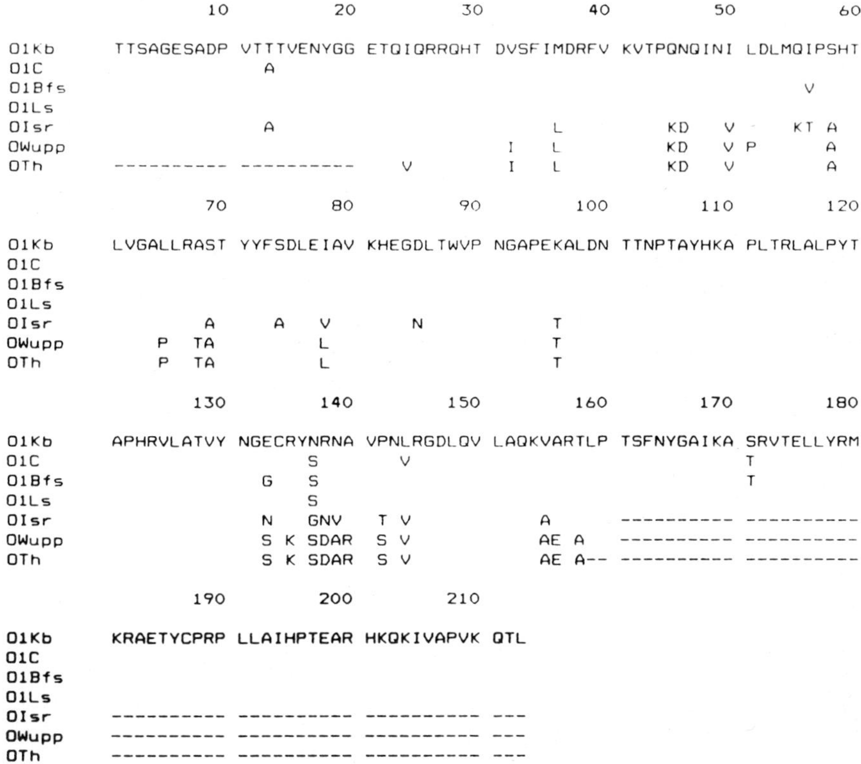

Figure 5. Alignment of amino acid sequences of VP1 of seven FMDVs of serotype O as deduced from the corresponding nucleotide sequences. Symbols are as in Fig. 4. The sequences of O Isr, O Wupp, and OTh have not been completed (Beck and Strohmaier, 1987; Beck, personal communication). Of a total of 47 substitutions (different for each position), 32 involve amino acids of different groups (Dayhoff *et al.*, 1972). Sequences are from Cheung *et al.* (1983), Beck and Strohmaier (1987), and Pfaff *et al.* (1988).

many cellular genes (that have been derived by comparing homologous genes from organisms which diverged at times given by the fossil record) are 10^{-8} to 10^{-9} s/s per year (Britten, 1986; Weissmann and Weber, 1986). Sobrino *et al.* (1986) compared oligonucleotide fingerprints of defined genomic segments from FMDVs obtained in the course of acute disease episodes in Spain from 1970 until 1981. They estimated values that ranged from $<4 \times 10^{-4}$ to 4.5×10^{-2} s/s per year, depending on the time period and the genomic segment considered. Moreover, they observed heterogeneity among three viruses isolated the same day and noted that this introduced an indeterminacy in the evaluation of fixation rates. In a recent study with the same viruses, Villaverde *et al.* (1988) found that the rate of fixation of mutations for the VP1 RNA was 2.1 times higher than for the 3D RNA. However, the average heterogeneity of the VP1-coding region was not significantly different from that of the 3D-coding region among cocirculating viruses. In fact, nine mutations and points of heterogeneity occurred within nucleotide residues 883–1026, which encode an amino acid segment relatively conserved among different RNA viruses (Domier *et al.*, 1987). Thus, conservation of 3D is not due to an intrinsically lower mutability when the viral RNA polymerase copies this segment but, rather, to continuous selection of the same (or closely related) consensus sequence (Villaverde *et al.*, 1988). This study clearly illustrated the potential for long-term genomic conservation, in spite of mutations arising at all times (compare Section VI.A).

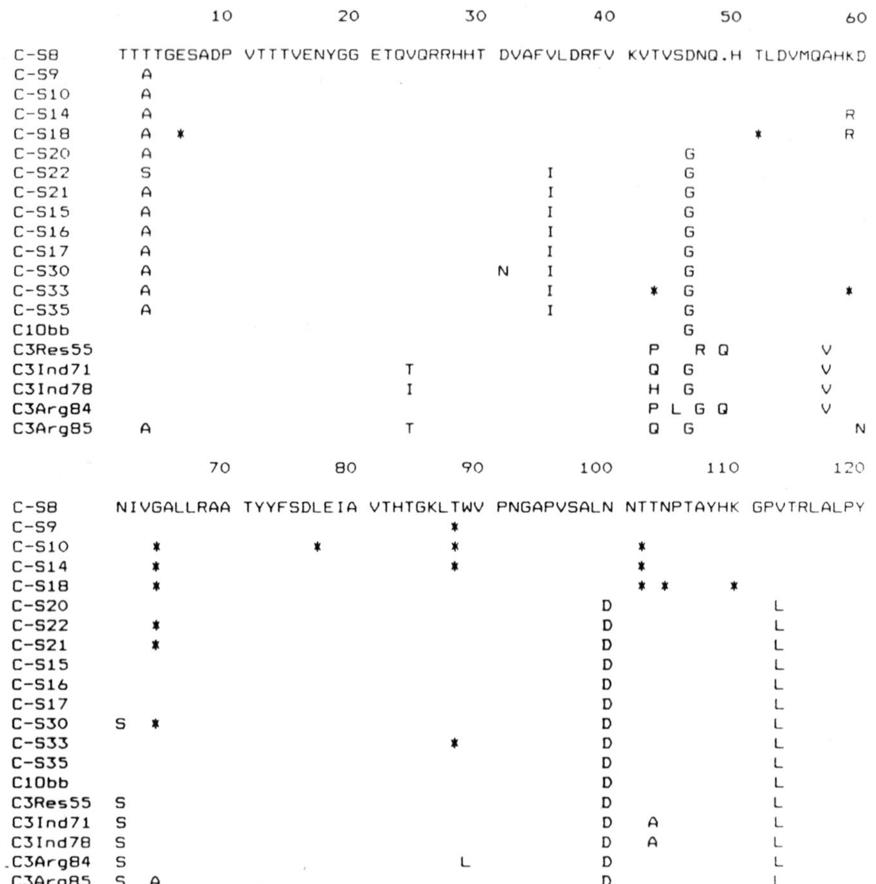

Figure 6. Alignment of amino acid sequences of VP1 of 20 FMDVs of serotype C, as deduced from the corresponding nucleotide sequences. Symbols are as in Fig. 4. Of a total of 58 substitutions (different for each position), 40 involve amino acids of a different group, according to Dayhoff *et al.* (1972). Sequences are from Beck *et al.* (1983a), Sobrino *et al.* (1986), Martínez *et al.* (1988), Piccone *et al.* (1988), and Martínez et al., unpublished results.

An interesting experiment *in vivo* was designed by J. La Torre, P. Augé de Mello, and their colleagues in Argentina and Brazil, which consisted in the establishment of persistent infections in cattle with plaque-purified variants of FMDV C_3 Resende Br- 55 (Costa Giomi *et al.*, 1984, 1988). Virus was recovered for up to 539 days of inapparent, persistent infection (Section I.B). Variations in the length of the genomic poly (C) tract, charge alterations in the capsid proteins, and differences in the nucleotide sequence of the VP1-coding region and in the reactivity of virions with neutralizing monoclonal antibodies (nMAbs) were quantitated (Gebauer *et al.*, 1988). The rates of fixation of substitutions at the VP1 gene of FMDV replicating in one animal were 0.9×10^{-2} to 7.4×10^{-2} s/s per year. In addition, heterogeneity was detected within several of the isolates. Since the infections were initiated with clonal pools of FMDV, the results document the potential for rapid evolution of FMDV. In the course of the infection, antigenically variant viruses were selected (Gebauer *et al.*, 1988; compare Section V.C). This strongly suggests that inapparent infections of ruminants, in addition to being a reservoir of virus, may promote rapid antigenic drift of FMDV.

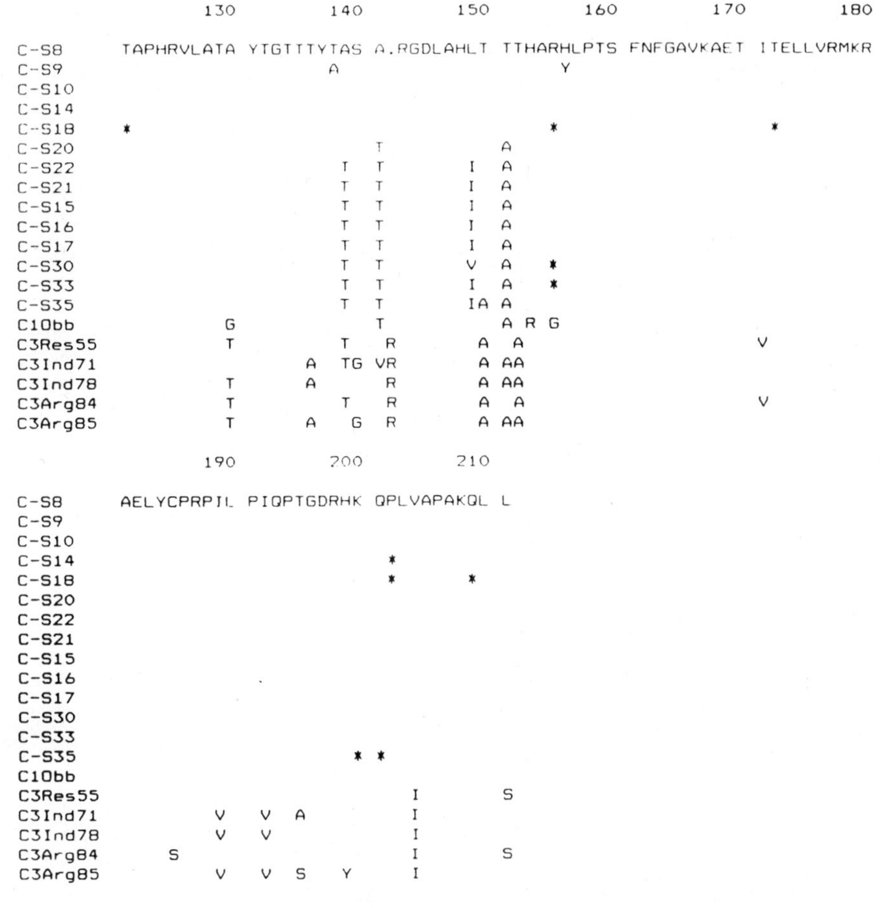

Figure 6. (Continued)

Clearly, this experiment shows that variant FMDVs are generated at high rate, a conclusion expected from previous results of passage of FMDV in cell culture, summarized in the following section.

C. Genetic Heterogeneity in Clonal Pools of FMDV

The serial passage experiment outlined in Fig. 8 permitted an estimate of the genetic divergence among populations and of the heterogeneity within a population, attained by FMDV in a limited number of replication rounds in cell culture. By fingerprinting total RNA from viral populations 1–5 of either FMDV O-S7p28 or C-S8p30 (Fig. 8) (each set derived from a plaque-purified FMDV preparation), Sobrino *et al.* (1983) calculated a range of 14–57 mutations per consensus genome population. Furthermore, and most important, analysis of individual clones from two populations (FS1-22 and ED1-12 in Fig. 8) revealed that each population was heterogeneous, with an average of two to eight substitutions per infectious genome. Very similar quantitations were derived from the analysis of RNA from FMDVs rescued from cells persistently infected in culture (de la Torre *et al.*, 1985, 1988; compare Section IV.B) (Table 2). Recently, Diez *et al.* (1989) have used an *in situ* plaque immunoassay to show that plaque-purified FMDVs passaged in cell culture

```
                    10         20         30         40         50         60

C1Obb       TTTTGESADP VTTTVENYGG ETQVQRRHHT DVAFVLDRFV KVTVSGNQ.H TLDVMQAHKD
C-S15         A                                I
C3Res55                                                   P  DR Q        V
A5Ww          AV                  D  T         G IM      INSLSPT   VI L  T QH
A12           A                                S IM      IKSLNPT   VI L  T QH
O1Kb          SA                     I   Q     S IM        PQNQI N I  L  IPSH
OWupp         SA                     I   Q    IS I         PKDQI N VP L  IPAH

                    70         80         90        100        110        120

C11Obb      NIVGALLRAA TYYFSDLEIA VTHTGKLTWV PNGAPVSALD NTTNPTAYHK GPLTRLALPY
C-S15
C3Res55     S
A5Ww        GL                  V  R D N          EA  S   S     N  A F
A12         GL                  V  R D N          EA  S   G        A F
O1Kb        TL       S             K E D          EK               A
OWupp       TL   P  T          L   K E D          ET               A

                   130        140        150        160        170        180

C1Obb       TAPHRVLATG YTGTTTYTAS ....TRGDLA HLTATRAGHL PTSFNFGAVK AETITELLVR
C-S15                A         T              I   H R
C3Res55              T         T     AR        ATAH R                 V
A5Ww                 V  N  NK STG GP  R   MG SAA RA KQ   A   Y  IR  I  H
A12                  V  N  NK S   GS GV   FG S APRV RQ   A   Y  I      H
O1Kb                 V  N ECR NRN AVPNL    Q V AQKV RT       Y  I   SRV    Y
OWupp                V  N SCK SDA RVSNV    Q V AQKAERA  ---------- ----------

                   190        200        210

C1Obb       MKRAELYCPR PILPIQPT.G DRHKQPLVAP AKQLL
C-S15
C3Res55                                  I       S
A5Ww                    L A EVSSQ      KII    R
A12                     L A EVSSQ      KII   G
O1Kb             T      L A H   E A    KI    V  T
OWupp       ---------- ---------- ---------- -----
```

Figure 7. Alignment of amino acid sequences of VP1 of some representative FMDVs of serotype A, O, and C. Symbols are as in Fig. 4, and the origins of the sequences are indicated in the legends for Figs. 4, 5, and 6. Of a total of 223 amino acid substitutions (different for each position), 160 involve amino acids of a different group, according to Dayhoff *et al.* (1972). Note that for several positions that show substitutions, the same amino acid is shared by viruses of different serotype.

are also antigenically heterogeneous. The conclusion that such clonal FMDV populations consist of a pool of variant genomes is identical to that reached previously by Weissmann and colleagues with phage Qβ (Domingo *et al.*, 1978). Such distributions, in which the genomes are statistically defined but individually indeterminate, are termed *quasispecies* (Eigen, 1971, 1987; Eigen and Schuster, 1979; Eigen and Biebricher, 1988, and Section VI.A). Repeatedly, whether analyzing field isolates or laboratory-adapted viruses, FMDVs appear as complex distributions of genomes rather than single, defined species (compare Sections VI and VII).

D. Genetic Recombination

Molecular recombination occurs at high frequency during the replication of several positive-stranded RNA viruses (reviewed by King, 1988). Direct proof of a covalent linkage between two different parental molecules to yield a recombinant genome was obtained for the first time with FMDV (King *et al.*, 1982). There are many recombination sites on the FMDV genome, perhaps with some preference for the more conserved segments, and with no evidence of site-specific events (King *et al.*, 1985). Crossover zones show a significantly higher potential for secondary structure than recombination-free zones (Wilson *et al.*, 1988). Distantly related FMDVs (i.e., belonging to a different serotype) appear to recombine at a lower frequency than closely related isolates (McCahon

Table 2. Genetic Heterogeneity and Variability of FMDV

Heterogeneity during an FMD outbreak[a]	
Among consensus sequences of different isolates	60–200 subst./genome
Among consensus sequences of contemporary isolates	2–20 subst./genome
Among individual genomes of the isolate	0.6–2 subst./genome
Heterogeneity of clonal populations in cell culture[a]	
Among consensus sequences of independently passaged plaque-purified viruses	14–57 subst./genome
Among individual genomes of clonal, passaged population	2–8 subst./genome
Rate of fixation of mutations[b]	
Acute disease	$<0.04 \times 10^{-2}$ to 4.5×10^{-2} s/s per year
Persistent infection in cattle (VP1 gene)	0.9×10^{-2} to 7.4×10^{-2} s/s per year

[a]Estimates based on data from Domingo *et al.* (1980), Sobrino *et al.* (1986), de la Torre *et al.* (1988), Villaverde *et al.* (1988), and Carrillo *et al.* (1990).

[b]Values are given in substitutions per nucleotide site (s/s) per year. They vary for different genomic segments and are not always proportional to the time intervals between isolations. Based on data from Sobrino *et al.* (1986) and Gebauer *et al.* (1988).

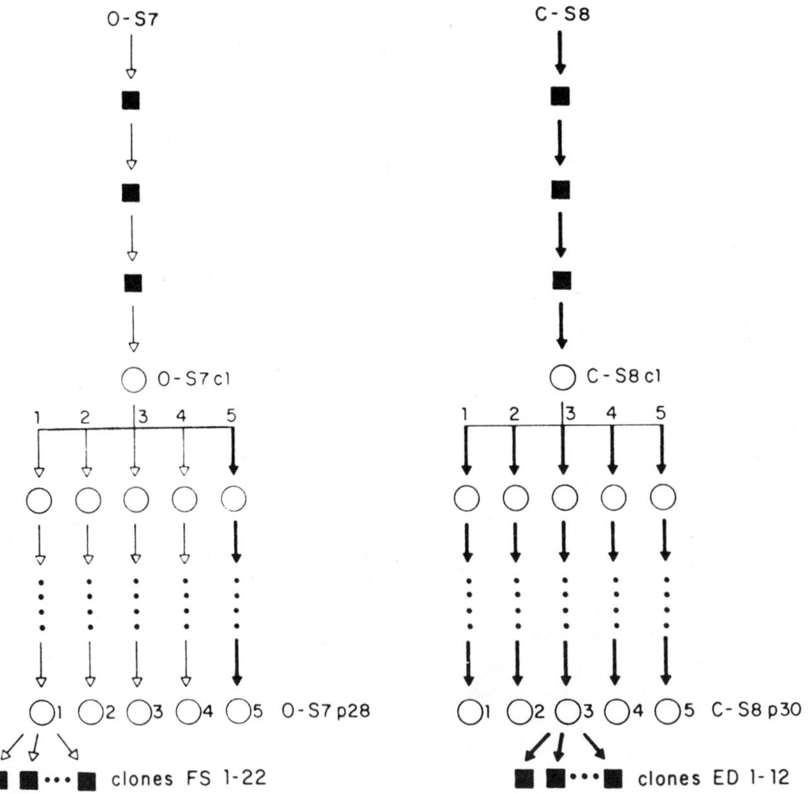

Figure 8. Serial passage of plaque-purified FMDV O-S7 and C-S8 in cell culture. The viruses were isolated from lesions of diseased animals. Each passage was at a multiplicity of infection of about 0.02–0.04 PFU/cell. Methods for plaque purification of FMDV and infection of cell cultures are described in Sobrino *et al.* (1983). ■, Plaque purification; O, uncloned population; filled arrow, infection of IBRS-2 cells; open arrow, infection of BHK-21 cells. [From Sobrino et al. (1983), reprinted from *Virology,* Academic Press, with permission.]

et al., 1985). Evidence of homologous and heterologous recombination has been obtained for a variety of RNA viruses, and it permits long evolutionary jumps (Section VI.A), although other possible functions of RNA recombination have been proposed (King, 1988, and references therein). The contribution of RNA recombination to the evolution of FMDV in nature is largely unexplored.

IV. PHENOTYPIC VARIABILITY OF FMDV

Unless most mutations fixed in the viral genome were selectively neutral, frequent changes in the biologic behavior of FMDV are expected. One of such changes, which complicates FMD diagnosis and control, is antigenic variation, discussed in detail in Section V. Next we review other phenotypic changes observed with FMDV, some of which play a role in pathogenesis.

A. Implication of Phenotypic Variability in Viral Pathogenesis

In the early studies with FMDV, it was shown that several genetic traits, such as thermal or acid sensitivity, and plaque morphology varied among strains and isolates (review in Bachrach, 1968). Reduction in the plaque size of FMDV (and of other viruses) is due to mutations at any of several genes. In some cases, the mutations cause sensitivity of the virions of sulfated polysaccharides present in the agar overlay of tissue culture media; in others, the capsid is modified in such a way that the virus is prone to aggregate in solution (review on small plaque morphology of FMDV in Parry, 1982).

Alterations of FMDV cell and tissue tropisms have been observed. In some FMD outbreaks, a single animal species is preferentially infected, while in others, several species are affected. Likewise, on occasion a strain prone to produce myocarditis circulates and causes high mortality rates among young animals (Section I.A); fulminant and degenerative forms of myocarditis have been described. FMDV can invade nervous tissue, leading to paraplegia, or the pancreas, causing diabetes (reviewed by Joubert and MacKowiak, 1968). Many host influences intervene in leading to a particular viral pathology, but it is likely that viral genetic variation also plays a role. A FMDV type C was adapted either to guinea pigs by 36 intradermal passages or to mice by 60 intramuscular passages. The guinea-pig-adapted virus maintained its epitheliotropic character for guinea pigs, while the mice-adapted strain was both epitheliotropic and myotropic (Ciaccio, 1963).

Those different pathologies speak of the many intricate ways by which FMDV interacts with its host organisms. Some novel and interesting phenotypes, selected in response to unusual environmental pressures, have recently been identified in cell cultures persistently infected with FMDV.

B. Variations during Persistent Infections in Cell Culture

Early studies by Dinter *et al.* (1959) and Seibold *et al.* (1964) suggested that FMDV could persist in cultured cells, and that the virus underwent genetic modification. Because of technical difficulties, however, those cells were not propagated. De la Torre *et al.* (1985) established cell lines persistently infected with FMDV. One such culture, termed C_1-BHK-Rc1, was initiated by infection of cloned BHK-21 cells with plaque-purified FMDV C-S8c1. The cells that survived the initial lytic infection were propagated. Upon serial passage, two distinct stages were observed. During the first 60 to about 100 passages (depending on the particular subline of C_1-BHK-Rc1 cells analyzed) infectious FMDV was detected at decreasing concentrations (about 10^6 PFU/ml in early passages, down to 10^2 PFU/ml at later passages) in the culture medium. Then, the cultures went through a transient stage in which subgenomic FMDV RNAs and antigens were detected, and finally they were cured of any detectable FMDV (de la Torre *et al.*, 1985). Analysis of virus from C_1-BHK-Rc1 cultures at successive cell passages during the productive stage demonstrated gradual changes in genetic composition and phenotypic behavior. It was estimated that by passage 58, the consensus viral RNA population differed in 0.3% of its genomic nucleotides from the parental FMDV C-S8c1 RNA (de la Torre *et al.*, 1985). By repetitive sequencing of the VP1 gene segment of nine viral clones derived from passage 59, heterogeneity was found at the level of 5×10^{-4} substitutions per nucleotide (de la Torre *et al.*, 1988). Thus, again, evolution of FMDV occurred via heterogeneous distributions of genomes (Sections III.C and VI.A).

This genetic variation in persistently infected cultures was accompanied of several phenotypic changes, some of which (*ts* character, small plaque morphology) were expected from studies with other animal viruses (Youngner and Preble, 1980). Other traits, however, were unexpected. Upon infection of BHK-21 cells with virus rescued from C_1-BHK-Rc1 cultures at passages 50–100, the viral replication cycle was completed in a shorter time than in a standard infection with FMDV C-S8c1, the cytopathology was extensive, and the viral yield normal. Thus, viruses had been selected that were hypervirulent for the parental BHK-21 cells. Contrary to our findings, we had considered that virus attenuated for BHK-21 cells would permit a more stable long-term virus and cell survival. The solution to this enigma came with the observation that C_1-BHK-Rc1 cells cured of any detectable FMDV by treatment with the antiviral agent ribavirin (de la Torre *et al.*, 1987) maintained a specific resistance to infection by C-S8c1. A detailed analysis of the course of infection of cured C_1-BHK-Rc1 cultures and of cell clones derived from them indicated that the host cells had evolved to become resistant to FMDV C-S8c1 and the latter, in turn, evolved to overcome the cellular block. In this way, viruses that were increasingly virulent for the parental (nonevolved) cells were gradually selected (de la Torre *et al.*, 1988). More recent results indicate that cell heterogeneity is extensive in C_1-BHK-Rc1 cultures, with individual cell clones showing different degrees of cellular transformation (as judged by cell morphology, growth in liquid culture, and ability to form colonies in semisolid agar) and of resistance to FMDV C-S8c1 (de la Torre *et al.*, 1989b). Cell fusion experiments have indicated that cell resistance is a dominant trait, probably due to one or several *trans*-acting cellular products (de la Torre *et al.*, 1989a). An assay designed to determine the minimum number of cells, isolated from a culture persistently infected with a virus, required to reconstruct the carrier state yielded for C_1-BHK-Rc1 at an early passage a value of 10^3 to 10^4 cells (de la Torre and Domingo, 1988). Thus, the C_1-BHK-Rc1 carrier cells are a complex system that must be described as a dynamic interaction between heterogeneous populations of cells and of FMDV.

It is noteworthy that an unusual and complex FMDV phenotype (namely, hypervirulence for a particular type of cell) was selected in response to an alteration of that host cell. Since persistence was established with plaque-purified virus, the result illustrates the potential for adaptation of FMDV when selective forces are continuously applied. During its replication in animals (whether in acute or in persistent infections, Section I) the virus faces a variable but continuous immune response of its hosts, and this is an important driving force for antigenic variation (Hyslop, 1965; Fagg and Hyslop, 1966).

V. ANTIGENIC DIVERSITY OF FMDV

A. Antigenicity

FMDV induces an immune response that generally results in protection against challenge with homologous virus. Circulating, neutralizing antibodies probably play a major role in the overall response. Unfortunately, the virus is antigenically diverse and heterogeneous. The initial observation was the recognition of several serologic types of FMDV. Types A and O were identified by Vallée and Carré (1922), type C by Waldmann and Trautwein (1926), and, later, SAT1, SAT2, SAT3, and Asia 1 serotypes were diagnosed by Brooksby and colleagues (Brooksby and Rogers, 1957; Brooksby, 1958). Infection or vaccination with virus of one type does not confer protection against virus of another type. By complement fixation and cross-neutralization tests carried out generally with guinea-pig sera raised against reference and field strains, more than 65 subtypes, showing various degrees of serologic relatedness, have been identified (Pereira, 1977). Subtyping has helped in vaccine strain selection, but it has resulted in a considerably confused picture of FMDV strains and isolates. Results do not always agree with those of other techniques—such as reactivity with MAbs and nucleotide sequencing—and with conclusions of cross-protection tests. In addition, internal inconsistencies have been detected in several serologic assays. The problem may be partly due to rapid antigenic variation in the field (Section V.C) and/or the consequence of using nonstandardized reagents in semiquantitative assays, based on a very complex set of antigen–antibody interactions (discussions in Rweyemamu, 1984; Kitching *et al.*, 1988, 1989).

Even selecting adequate strains for vaccine production, the level and duration of the immunity provided by the current whole-virus inactivated vaccines depend on a number of variables. Among them, to mention just a few, are the following: (1) The quality (concentration of 140S antigen) and type of formulation. Oil-based adjuvants provide longer-lasting immunity than many aqueous adjuvants. (2) Age of the animal. In young animals, maternal antibodies may interfere with the response, by unknown mechanisms. (3) The number of vaccinations. FMDV particles prime an immune response, even to synthetic peptides (Section V.D). (4) Variations in response among individual animals, due to genetic (or other, unknown) conditionants. In a survey of 870 cattle, carried out in Germany, even after three vaccinations, 10–15% of the animals had low serum neutralization titers (Gaschutz *et al.*, 1986). Clearly, the antigenicity of FMDV is affected by a wide range of circumstances, including the nature of FMDV particles. Next we review recent results of the antigenic structure of FMDV as deduced mainly from the use of MAbs.

B. Epitopes Involved in Neutralization of FMDV

Early experiments showed that *in situ* trypsin cleavage of VP1 decreased the immunogenicity and cell attachment of FMDV particles (Wild *et al.*, 1969). Isolated VP1 and fragments derived from its carboxy-terminal half, but neither VP2 nor VP3, induced neutralizing antibodies and partial protection against challenge with homologous virus (Laporte *et al.*, 1973; Bachrach *et al.*, 1975, 1979; Kaaden *et al.*, 1977; Adam *et al.*, 1978; Meloen *et al.*, 1979; Strohmaier *et al.*, 1982). Antigenic determinants of FMDV O_1 Kaufbeuren were mapped in fragments including amino acids 138–154 and 200–212 (Strohmaier *et al.*, 1982). Synthetic peptides representing these VP1 segments induced neutralizing antibodies in animals (Bittle *et al.*, 1982; Pfaff *et al.*, 1982). More recently, experiments involving use of nMAbs, synthetic peptides, and MAb-resistant (MAR) mutants have led to extensive recognition of the VP1 140–160 region as a site of interaction of neutralizing antibodies with FMDV of serotypes A, O, and C (Duchesne *et al.*, 1984; Robertson *et al.*, 1984; Geysen *et al.*, 1984, 1985; Parry *et al.*, 1985, 1989a; Pfaff *et al.*, 1985, 1988; Stave *et al.*, 1986, 1988; Mateu *et al.*, 1987, 1989, 1990; McCullough *et al.*, 1987; Xie *et al.*, 1987; Meloen *et al.*, 1987; Thomas *et al.*, 1988; Bolwell *et al.*, 1989b; Baxt *et al.*, 1989a). This segment is a protruding, disordered loop (the "FMDV loop") on the surface of FMDV O_1BFS (Acharya *et al.*, 1989) (compare Section II.A). The region is trypsin-sensitive and includes the conserved sequence RGD—probably part of a cell attachment site (Fox *et al.*, 1989)—flanked by two hypervariable amino acid stretches involved in recognition by neutralizing antibodies (compare Sections II.A and III.A). For FMDV type C, several nonconserved epitopes recognized by nMAbs involve amino acids from both hypervariable regions (Mateu *et al.*, 1989, 1990). Neutralization of FMDV by these antibodies could thus be explained by a blockade of the virion–cell interaction.

The structure of this antigenic determinant has proved very complex, in that (1) it is composed of multiple overlapping and/or nonoverlapping epitopes; (2) some of these epitopes are virion-conformation-dependent and include also residues from the C-terminus of VP1 and/or other residues from VP1, VP2, or VP3; and (3) there may be important differences in this determinant among different FMDVs.

We next review some of the experimental support for those conclusions. At least two epitopes were initially located by Ouldridge *et al.* (1984) within a trypsin-sensitive VP1 site of FMDV O_1BFS. By analyzing the competition exerted by antipeptide sera on the binding of nMAbs to virions, and also through sequencing of MAR mutants, several virion-conformation-dependent epitopes of O_1BFS have been identified, and each of them involves residues from segments 143–150 and 200–213 of VP1 (Parry *et al.*, 1985, 1989a; Barnett *et al.*, 1989). Consistent with these results, the C-terminus of VP1 from one capsid protomer has been positioned very close to the FMDV loop from the adjacent protomer in FMDV O_1BFS (Acharya *et al.*, 1989). Similarly, work with FMDV O_1 Kaufbeuren has shown that the virus displays several epitopes that include residues within the 144–154 VP1 region; some of the epitopes involve also residue 208 of VP1 or possibly some residue(s) from VP2 or VP3 (Xie *et al.*, 1987; McCahon *et al.*, 1989; Kitson *et al.*, 1989). Pfaff *et al.* (1988) have found, in agreement with Xie *et al.* (1987), two overlapping epitopes at the 144–160 VP1 region of the same virus. One of them included residues from segment 144–154 as well as from the C-terminal region of VP1. In contrast to these results, the C-terminus of VP1 of FMDVs of types A and C may constitute a minor antigenic site, independent—at least functionally—of the FMDV loop. Epitopes that involve residues of the C-terminal region of VP1 have been identified in FMDV A_{10}Holland (Meloen *et al.*, 1987, 1988; Thomas *et al.*, 1988), A_{12}119 (Baxt *et al.*, 1989a), and a number of type C viruses, including C-S8 (Mateu *et al.*, 1990). Such epitopes have been identified by sequencing MAR mutants and by analyzing the reactivity of nMAbs with synthetic peptides. No epitope of any of these viruses has unambiguously been found to involve residues from both the FMDV loop and the C-terminus of VP1—though a link has been suggested for epitopes of FMDV A_{12} (Baxt *et al.*, 1989a). On the other hand, several epitopes have been detected within the FMDV loop of viruses of types C (Mateu *et al.*, 1987, 1989, 1990) and A (Thomas *et al.*, 1988; Bolwell *et al.*, 1989b; Baxt *et al.*, 1989a). Most of these epitopes are virion-conformation-independent, in that the corresponding nMAbs reacted with isolated VP1 and/or immobilized synthetic peptides. For FMDV C-S8, at least ten distinguishable overlapping epitopes involved in viral neutralization have been found within the 138–150 VP1 region that, upon alignment, matches very closely the FMDV loop of O_1BFS (Mateu *et al.*, 1990). The observation of multiple epitopes within this short amino acid stretch is consistent with the suggestion that the FMDV loop is highly flexible (Harrison, 1989) and that it may easily adopt a number of immunogenic conformations recognized by antibodies.

Other antigenic sites have been also found on FMDV, though their characterization is far from complete. Multiple sites were detected on FMDVs of serotypes A, O, and C by using the elegant "peptide scan" or "PEPSCAN" procedure (with overlapping peptides as antigens) and amino acid replacement set methods (involving peptides in which the amino acid for each residue position in the parental sequence is substituted by each of the other 19 common amino acids, keeping the other positions invariant) (Geysen *et al.*, 1984, 1985, 1987). Overlapping peptides of VP1 of isolates O_1, A_{10}, and C_1 were reacted with either anti-140S, anti-12S, and anti-VP1 rabbit sera or nMAbs. In addition to a few positions scattered along the molecule, peaks of antigenic activity were located at the N-terminus, segment 40–50, around position 100, segment 140–150, and the C-terminus (Meloen and Barteling, 1986a; Meloen *et al.*, 1987). Peptides of VP2 and VP3 were also found reactive with anti-FMDV nMAbs, and epitopes appeared to be scattered on the surface of the virion (Meloen *et al.*, 1986).

Other approaches, such as MAb-competition experiments and sequencing and reactivity of MAR mutants, have led to the identification of several independent, discrete antigenic sites of FMDV. Neutralizing MAbs raised against O_1BFS reacted with intact particles (even trypsin-treated), but not with isolated capsid proteins (Meloen *et al.*, 1983), thus suggesting for the first time the existence of virion-conformation-dependent epitope(s). Two infectivity-associated and conformation-dependent determinants were identified, in addition to the 140–160 site, in FMDVs of serotype O using MAbs raised against FMDV O_1 Suisse in competition assays (McCullough *et al.*, 1987) and MAR mutants of O_1 Brugge in cross-reactivity assays (Stave *et al.*, 1988). Work with MAR mutants of O_1 Kaufbeuren allowed identification of four antigenic sites, and all of them were to some extent conformation-dependent (Xie *et al.*, 1987; McCahon *et al.*, 1989; Kitson *et al.*, 1989). In addition to the complex site that involves the FMDV loop, three other sites of FMDV O_1 Kaufbeuren have been roughly located on the capsid by sequencing MAR mutants and positioning the mutated residues on the three-dimensional structure of FMDV O_1 BFS. One site involves at least amino acids 43–45 of VP1, a second includes residues within segment 70–77 of VP2 and also amino acid 131 of VP2, and the third contains amino acid 58 of VP3 (Kitson *et al.*, 1989). These four antigenic sites appear to be in positions closely analogous to those reported for poliovirus sites.

For serotypes other than O, antigenic sites not located on the FMDV loop or the C-terminus of VP1 have been less characterized. In this respect, several virion-conformation-dependent antigenic sites and epitopes have been distinguished for FMDV A_{12}119 (Baxt *et al.*, 1984, 1989a; Grubman and Morgan, 1986), A_{10} Holland (Thomas *et al.*, 1988), A_{22}Iraq (Bolwell *et al.*, 1989b), A_5 isolates (Saiz *et al.*, 1989), and C-S8 and other type C viruses from Europe and South America (Mateu *et al.*, 1988, and unpublished results). In addition to the 140–160 and the C-terminus of VP1, some additional sites have been partly located in these viruses. Apparently independent conformational sites of FMDV A_{10} Holland (Thomas *et al.*, 1988) and of A_{12} 119 (Baxt *et al.*, 1989a) involve multiple residues of VP3; such residues are not located within equivalent regions in VP3 of the two viruses. Also, Thomas *et al.* (1988) and Baxt *et al.* (1989a) have found epitopes that include VP1 positions 169 in FMDV A_{10} and 173 in A_{12} and that may define another antigenic site of FMDV type A.

It is not possible at present to draw a conclusive picture of the antigenic structure of FMDV relevant to the neutralization by antibodies. The panels of MAbs available are still limited, and localization and characterization of the sites and epitopes they define are far from complete. Immunodominance of each site and epitope on the different FMDVs is an open question. Also, responses of different species and individual animals to FMDV antigenic determinants may vary both qualitative and quantitatively, and adequate statistical analyses of the results are required.

In spite of our limited knowledge, the emerging picture is that of an antigenically complex virus, with several (perhaps three or four) different, nonoverlapping sites involved in neutralization of infectivity. At least some of these antigenic sites are virion-conformation-dependent, include several epitopes, and/or may be located in positions analogous to some determinants found on other picornaviruses. The FMDV loop (segment 135–160 of VP1) constitutes, or is a part of, a complex immunodominant site composed of multiple overlapping epitopes. Many of the epitopes are subject to variation, and the precise differences in the antigenic structure of distinct FMDVs are largely unknown.

C. Molecular Basis of Antigenic Variation

Classical serologic techniques indicated extensive antigenic variation of FMDV, reflected in many subtypes and variants within subtypes (Arrowsmith, 1977; compare Section V.A). Use of MAbs and nucleotide sequencing techniques has permitted the identification of some antigenic sites and epitopes that conform the antigenic structure of FMDV (Section V.B) and of amino acid differences in structural proteins among viral variants (Section III.A). Such studies have shown that the fixation of amino acid substitutions at epitopes involved in neutralization of FMDV is the basis of an extensive antigenic variation of the virus, detected even within single isolates and which occurs during disease outbreaks.

It has been suggested that because the 140–160 VP1 segment is most heterogeneous among types and subtypes (see Figs. 4–7), it may be a determinant of serotype and subtype specificity, while the 40–60 VP1 region may confer serotype specificity (Beck *et al.*, 1983a; Cheung *et al.*, 1983). That part of the antigenic diversity of FMDV is determined by the sequence and structure of VP1 segment 140–160 is suggested by the specificity of antibodies raised against synthetic peptides representing the sequence of that segment for viruses of different serotypes and subtypes (Bittle *et al.*, 1982; Clarke *et al.*, 1983). However, other regions of the viral capsid must be also involved. For example, some FMDVs within type O (Ouldridge *et al.*, 1986) or within type C (Piccone *et al.*, 1988) are serologically different, yet they depict the same sequences in the 140–160 VP1 site (see also Section V.B).

Studies with well-characterized MAbs and field FMDV isolates indicate extensive variation in the epitopes involved in viral neutralization. This refers both to virion-conformation-independent epitopes located within the VP1 140–160 segment and to more complex, conformational epitopes. Variation has been well documented among FMDV serotypes, subtypes (Grubman and Morgan, 1986; Pfaff *et al.*, 1988; Mateu *et al.*, 1988), different representatives of a single subtype (Stave *et al.*, 1986, 1988; Pfaff *et al.*, 1988; Mateu *et al.*, 1988), and even among viruses isolated during one epizootics (Mateu *et al.*, 1987, 1988, 1989). Mutants with decreased reactivity with nMAbs that define epitopes in the 140–160 VP1 region, and which include substitutions in this segment, were selected in the course of a persistent infection of cattle, established with plaque-purified FMDV C_3 (Gebauer *et al.*, 1988; compare Section III.B). In this case, rapid selection of antigenic variants probably occurred as a result of a local immune response at the site of viral replication. Overall, results with MAbs suggest that each viral isolate may be not only genetically unique, but also antigenically unique (Sections VI.A and VII). Because of the relevance of antigenic variation to practical aspects of FMD control (diagnosis, epidemiologic surveillance, etc.), it would be desirable to organize a data bank of reactivities of FMDV strains with MAbs, available to all scientists. Also, an exchange of standardized reagents and procedures is clearly needed.

Knowledge of the amino acid substitutions responsible for antigenic changes in field isolates of FMDV is also most important to the design of new vaccines. Three variants, differing at VP1 residues 148 and 153, were identified by Rowlands *et al.* (1983) in one isolate $A_{12}119$. Sera induced by each of the variants, or by synthetic peptides representing their repective 141–160 VP1 segments, neutralized more efficiently the homologous than the heterologous virus. Also, chemically conservative substitutions, fixed at positions 138 and 147 at both sides of the RGD sequence (Section V.B) during the course of an epizootic caused by FMDV C_1, affected several epitopes in the FMDV loop (Mateu *et al.*, 1989, 1990). The effects that such substitutions have on the reactivity of viruses with nMAbs have been quantitatively mimicked with synthetic peptides including the relevant replacements (Mateu *et al.*, 1989, and unpublished results). This provides evidence that those replacements—and not others present in the viral quasispecies (see Section VI.A)—are responsible for the modified interactions of the antigens with antibodies. These substitutions are of a type frequently found during FMDV evolution and are likely to be highly represented in a viral quasispecies. In addition, one chemically conservative substitution at position 146 of VP1, fixed in a number of MAR mutants, was sufficient to abolish all ten epitopes found in the FMDV loop of virus C-S8 (Mateu *et al.*, 1990). A substitution fixed in the equivalent position during an experimental persistent infection of cattle with FMDV C_3 Resende affected also most of the tested epitopes of this antigenic region (Gebauer *et al.*, 1988; Mateu *et al.*, 1990). Thus, very modest genetic changes can result in remarkable antigenic drift, perhaps providing an interpretation of the extensive antigenic heterogeneity among natural isolates (Section V.B, Fig. 9).

From these studies, it appears highly unlikely that serologic types and subtypes of FMDV will correspond to well-defined amino acid sequences at specific antigenic sites on the viral capsid. On the contrary, several sequence combinations may yield the same or similar "average" serologic behavior. Some replacements may eliminate one or many epitopes defined by MAbs, while other substitutions may generate new epitopes, or be of no noticeable consequence (Mateu *et al.*, 1990, and unpublished results). Such dynamic antigenic conversions are also strongly suggested by the analyses of 46 field isolates of FMDV type C using MAbs, carried out by Mateu *et al.* (1988) (Fig. 9), and agree with previous conclusions on epitope variability of coxsackievirus B4 reached by A. L. Notkins, B. S. Prabhakar, and their colleagues (Prabhakar *et al.*, 1982, 1985; Webb *et al.*, 1986). To complicate matters even further, each FMDV population, such as a field isolate, is not only genetically heterogeneous (Sections III.A and III.C), but probably also antigenically heterogeneous (Rowlands *et*

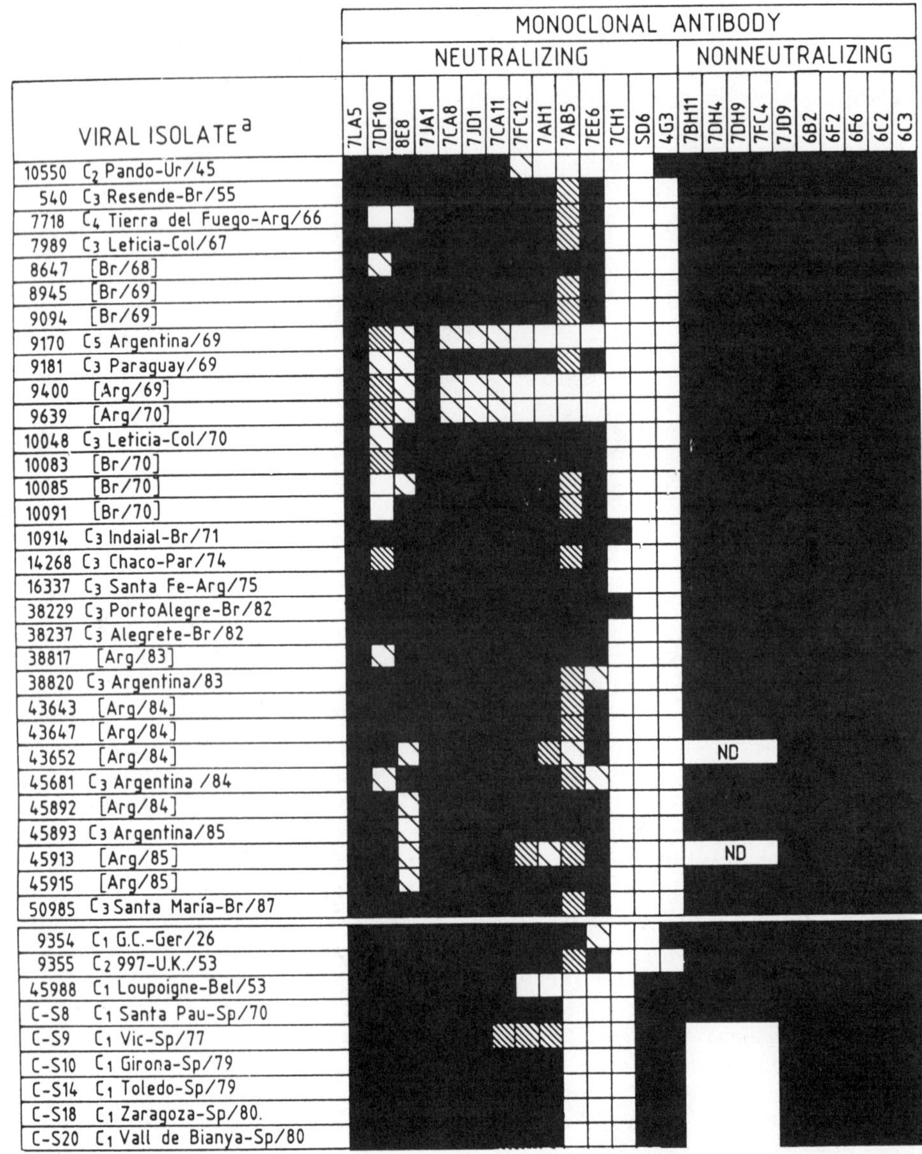

Figure 9. Antigenic heterogeneity of field isolates of FMDV of serotype C from South America (upper section) and from Europe (lower section), as detected with the MAbs indicated at the top row. MAbs 7LA5 to 8E8 (neutralizing) and MAbs 7BH11 to 6B2 (nonneutralizing) are conformation-specific; 7JA1 to 4G3 react with protein VP1 in Western blot assays; 6F2 reacts with VP2, and 6F6 to 6C3 react with VP3. MAbs 7LA5 to 7CH1 and 7BH11 to 7JD9 were raised against FMDV C_3 Indaial Br-71; MAbs SD6, 4G3, and 6F2 to 6C3 against FMDV C_1 Santa Pau Sp-70 (C-S8c1); and 8E8 was raised against FMDV C_1 Vosges (France, 1960). MAbs that did not react with an isolated protein (conformation-specific) were tested in enzyme-linked immunodot assays; those that reacted with one protein were tested in Western blot assays. For each MAb, a relative reactivity (r_i) of any virus i with respect to the homologous virus s (the virus used to raise the MAb) is defined as the ratio of MAb titer (reciprocal of the limit dilution) for virus i to that for virus s. For the relative reactivity: ■, Positive ($r_i > 0.1$); ▨, intermediate ($0.01 < r_i < 0.1$); ⧅, very weak ($0.001 < r_i < 0.01$); □, negative (no signal detected, $r_i < 0.001$). Blocks with ND, not determined. [a]When no designation is given to an isolate, the country and year of isolation are abbreviated in brackets: Ur, Uruguay; Br, Brazil; Arg, Argentina; Col, Colombia; Par, Paraguay. [Reprinted from Mateu *et al.* (1988), with permission from *Virology*, Academic Press.]

al., 1983; Mateu *et al.*, 1989), consisting of an indeterminate spectrum of variants. Díez *et al.* (1989) have used an *in situ* immunoassay—that permits analysis of the reactivity of individual viral plaques with MAbs—to show that upon serial passage of plaque purified FMDV C-S8 in cell culture, antigenic variants were selected in the absence of antibodies. These results and those of Bolwell *et al.* (1989a)—who showed that an antigenic variant of FMDV A_{22} was selected upon adaptation of the virus to suspension spinner cultures—suggest the possibility that shifts in the equilibrium distribution of genomes (Section VI.A) caused by environmental changes may lead to antigenic variation. Thus, it may not be always necessary that the immune system exerts its selective action in order to increase the proportion of antigenically variant viruses. In this view, occurrence of replacements at antigenic domains would precede exposure to the selective activity of the immune system (Díez *et al.*, 1989). Thus, fluctuations in the antigenic behavior of FMDV are expected to be rather unpredictable (Rowlands *et al.*, 1983; Mateu *et al.*, 1988, 1990; Gebauer *et al.*, 1988), as a reflection of heterogeneities and variations at the genetic level (compare Section VI.A).

D. Design of New Vaccines

Vaccination has been the most effective means of controlling FMD (Section I.C). In areas where the number of outbreaks has decreased as a result of regular vaccination, such as in Europe (see Fig. 1), evidence has recently been provided that some FMD outbreaks had a vaccine origin (Beck and Strohmaier, 1987; Carrillo *et al.*, 1990). Thus, it would be very important to develop synthetic FMD vaccines whose manufacture would not require handling or administering entire virus particles (review in Brown, 1988, 1989). The development of recombinant DNA techniques permitted the molecular cloning of cDNA copies of the FMDV genome and the expression of VP1 (the protein that carries major antigenic determinants, Section V.B) in *Escherichia coli* and other cells (Boothroyd *et al.*, 1981; Küpper *et al.*, 1981; Cheung and Küpper, 1984). Kleid *et al.* (1981) succeeded in inducing protection in cattle and swine with VP1 of strain $A_{12}119$, expressed as a protein fused to *trp* L-E. In spite of the enthusiasm generated by the above reports (Brooksby, 1981b), it soon became apparent that the immunity levels induced by engineered VP1 were 100–1000 times lower than those produced by the equivalent amounts of protein in the form of intact virion. Several expression vectors and constructions including VP1 segments (rather than the entire protein) have been used to express FMDV antigens. The systems applied include vaccinia virus recombinants (Newton *et al.*, 1986, 1987), hepatitis B core proteins (Clarke *et al.*, 1987b, 1988), and bacterial fusion proteins (Winther *et al.*, 1986; Broekhuijsen *et al.*, 1986, 1987). Proteins that contain tandem repeats of the main antigenic segment of VP1 evoked in animals a far better response and protection than did their monomeric counterparts. Also, the 142–160 VP1 segment of FMDV O_1 on hepatitis B core particles was 500 times more immunogenic than the free synthetic peptide (see below). This construction gave the best response so far scored for a synthetic immunogen of FMDV (Clarke, *et al.*, 1988). These findings justify renewed optimism in the development of a synthetic anti-FMD vaccine (Brown, 1988, 1989).

An alternative approach has been to chemically synthesize oligopeptides representing the sequences of the relevant antigenic sites (Anderer and Schlumberger, 1965; Pfaff *et al.*, 1982; Bittle *et al.*, 1982; Sutcliffe *et al.*, 1983). A peptide construction with the structure C-C-200-213-P-P-S-141-158-P-C-G (in which the numbers represent the amino acid residues of VP1 of FMDV O_1 Kaufbeuren) induced protection in cattle (Di Marchi *et al.*, 1986), with both VP1 segments being important for the response (Doel *et al.*, 1988). There is increasing reason to believe, however, that some important elements of the immune response against FMDV have yet to be identified. Meloen and Barteling (1986b) have provided evidence that an epitope located at the C-terminal end of VP1 elicited neutralizing activity but poor protection. In a different line of research, McCullough *et al.*

(1986) showed that MAbs against FMDV type O passively protected neonatal mice at dilutions at which the antibodies could not neutralize infectivity *in vitro*. They suggested that protection against FMDV may require opsonization of virions by antibodies and phagocytosis of virus–antibody complexes. These studies indicate that induction of good levels of neutralizing antibodies does not entail protection against infection. Th-cell epitopes probably contribute to the immune response against FMDV. A Th-cell epitope was located within the 140–160 VP1 region of FMDV O_1 Kaufbeuren (Francis *et al.*, 1987a,b). Foreign Th-cell determinants were effective in overcoming a genetically determined nonresponsiveness of mice to synthetic peptides (Francis *et al.*, 1987c, 1988). In mice, a T-cell-independent response (Borca *et al.*, 1986) appears to be followed by a T-cell-dependent stage (Collen *et al.*, 1989). Thus, it is likely that by administering as vaccine entire (or highly structured) virus particles, a variety of epitopes are provided that contribute to a global protective response, humoral and cellular. If this were the case, synthetic peptides or fusion proteins would be lacking in some (or many) of the determinants involved in evoking such global immune stimulation. Not surprisingly, formulations that include activators of B cells and macrophages, as well as Th-cell epitopes, may induce improved immunity levels in laboratory animals (Wiesmüller *et al.*, 1989).

A second important problem for the design of a synthetic, anti-FMD vaccine is virus variation, increasingly recognized as an obstacle for viral vaccine efficacy (Holland *et al.*, 1982; Domingo *et al.*, 1985; Steinhauer and Holland, 1987; Cloyd and Holt, 1987; Domingo and Holland, 1988; Mateu *et al.*, 1987, 1988, 1989, 1990; Martínez *et al.*, 1988; Domingo, 1989). Classical, whole-virus inactivated vaccines prepared with either FMDV C-S8 or C-S15 (two epidemiologically related viruses of type C_1) induced complete protection of swine (the natural host of these viruses) against homologous virus, but only partial protection against heterologous virus (Martínez *et al.*, 1988). VP1 of the two viruses differ in nine amino acids, four of them located within residues 138 to 151 (compare sequences in Fig. 6). The two substitutions at positions 138 and 147 affected several of the epitopes located at this antigenic site on VP1 (Mateu *et al.*, 1989, 1990; see also Section V.C). Since very few additional differences between the two viruses were found in the other structural proteins (Sobrino *et al.*, 1989), it is likely that the changes in VP1 contributed to the discernible difference in vaccine potency.

As documented in Section V.C, each epitope will admit certain substitutions, and the tolerance will depend on the role that each amino acid (and its coding nucleotides in the RNA) plays in virion stability or, more broadly, in the life cycle of the virus. The conservation of epitopes recognized by nonneutralizing MAbs in FMDV C (Mateu *et al.*, 1988 and Fig. 9) is probably due not only to absence of selective pressure for fixation of replacements, but to structural roles played by the relevant sequences. In this view, some replacements fixed at neutralizing epitopes may be antigenically neutral, but others will eliminate previously reactive epitopes or generate new ones (Sections V.B and V.C). Depending on the number and nature of the amino acid substitutions fixed on one infectious particle, that particle may evade an immune response previously directed to its parental virus or to another variant. The replicating ability of the mutant virus in relation to all other mutants present and arising in the same population will determine its fate toward fixation or extinction. Thus, the picture is clearly more complex than that of a number of strains to be cloned and sequenced to provide the desired peptides (Beale, 1982). The precise antigenic composition is indeterminate in the sense that it is only identified as an "average" or "consensus" antigenicity (Sections V.C and VI.A). As expected, the probability of generating MAR mutants of FMDV is high (in the order of 10^{-3} to 10^{-5}, as for other picornaviruses; reviewed by Domingo and Holland, 1988). Viruses resistant to *several* different antibodies directed at independent epitopes will be present in proportions orders of magnitude lower than those resistant to *one* antibody species. As a result of these considerations, the formulation of a synthetic vaccine should contemplate (1) the inclusion of several independent B- and T-cell epitopes. (2) Since some epitopes are complex, and they involve more than one capsid protein (Section V.B), they should be provided in the form of capsid structures or anti-idiotypic antibodies, and these approaches are in progress (Clarke and Sangar, 1988; Baxt *et al.*, 1989b). Capsid structures would also have the advantage of incorporating several antigenic sites. (3) The vaccine should also incorporate a number of constructs (e.g., peptides) representing variant amino acid sequences of the important and highly heterogeneous epitopes. For example, some of the different VP1 140–160 sequences shown in Fig. 6 for the C-S isolates should be considered for a vaccine intended to control type C virus in Spain. In this respect, it may be an advantage that antibodies raised against synthetic peptides showed decreased specificity for the virus and for the homologous peptide, perhaps reflecting lower conformational constraints in a free peptide (Geysen *et al.*, 1985; Parry *et al.*, 1989b). (4) Carriers and adjuvants should be carefully tested since they may have enhancing or suppressive effects on the response.

Sequences recognized by T cells vary among animal species and among individuals of one species, because of the polymorphism of the proteins of the major histocompatibility complex through which antigen-specific T cells recognize antigens. This is an important difficulty for vaccine design (Schwartz, 1986), and it will require either the search for very immunodominant sequences or investigation of the responses of the animal species and populations to be protected. Other factors that influence the immunity against FMDV have been discussed in Section V.A. In keeping with the conclusions of previous paragraphs, the vaccine potency test should ideally include as challenge virus a collection of natural FMDV variants likely to be well represented in the viral populations whose spread is to be controlled. We suspect that a better assessment of vaccine potency could be obtained if the doses of challenging viruses currently used were reduced. Such amounts are unlikely to be encountered in the process of natural infection (Section I).

In summary, the progress toward development of synthetic vaccines is slow, but much has been learned about the nature of the immune response to FMDV. Such knowledge will eventually find application to control other viral diseases as well. Variation of FMDV is greater than anticipated by most researchers, and vaccine formulations with multiple sites and epitopes will probably be required to afford solid protection. Despite the problems outlined in previous paragraphs, the improved responses evoked by several engineered constructs (Clarke *et al.*, 1988) will probably be extended to additional epitopes of several FMDV serotypes. If this were the case, one could consider a gradual replacement of classical by synthetic vaccines, beginning in areas where the incidence of FMD is so low as to be candidates for a nonvaccination policy. Of course, such a proposal would require extensive planning along with severe control measures, and its chances of success are an entirely open question.

VI. THE EVOLUTION OF FMDV

A. Quasispecies and Population Equilibrium

The genetic and antigenic diversification of FMDV is not the result of a steady fixation and accumulation of mutations in the genome over extended time periods. Rather, as documented in Sections III–V, evolutionary rates vary for different genomic segments and are not constant with time. Heterogeneities in the nucleotide sequence of RNA and in the epitopes involved in neutralization of infectivity abound among epidemiologically related viruses, and even within single isolates and clonal pools. Whenever a FMDV population has been examined in some detail either by repetitive nucleotide sequence sampling or by determining the proportion of MAR mutants, the conclusion has been that FMDV consists essentially of *pools of variants*. Current estimates are that in RNA from individual clonal pools derived from FMDV populations one mutated base is expected for each 1000–4000 residues sequenced (Sobrino *et al.*, 1983; de la Torre *et al.*, 1985, 1988), but populations with far larger heterogeneities have also been found (J. Díez, C. Escarmís, M. Medina, and E. Domingo, unpublished results). Mutation rates (defined as the frequency of a misincorporation event during a single replicative round of RNA synthesis) have not been determined for FMDV, as a mutation rate cannot be inferred from a rate of fixation of substitutions or from the genetic heterogeneity in an RNA population (Domingo and Holland, 1988). On average, however, the intrinsic mutability of the FMDV genome does not appear to be significantly different for a highly variable gene such as VP1 and for a relatively conserved one such as 3D (polymerase): Both depict a similar heterogeneity among cocirculating viruses (Villaverde *et al.*, 1988; see also Section III.B).

Irrespective of the mutation rate operating during copying of each genomic residue, the net result is a remarkable genetic and phenotypic heterogeneity that has important biologic implications. Such implications are only apparent if one considers the following parameters. An infected animal may contain about 10^{12} infectious units (Sellers, 1971; Section I.A), although this may well be a minimal estimate because of the thermal instability of FMDV (Section II.A). Given those population numbers and the levels of heterogeneities observed (Table 2), assuming a random distribution of mutations among the RNA molecules, it can be calculated that *all possible* single and double mutants, as well as decreasing proportions of triple, quadruple, etc., mutants, are *potentially* present in most FMDV populations. Of course, the actual distributions will deviate from those predicted theoretically, since the relative fitness of each variant is an important determinant of the frequency with which it will be present in the distribution. If a variant has a strong selective disadvantage in a given environment, its proportion will be so low that it probably will never be found in a sampling of genomes. This point was clearly illustrated by the effect of an extracistronic mutation on the replication of Qβ RNA (Domingo *et al.*, 1976; Batschelet *et al.*, 1976). Even if not lethal from a biochemical standpoint, its effect on fitness was so adverse that the proportion of mutant sequence was rapidly reduced in the populations.

Even mutations that are necessarily lethal (such as those leading to a termination codon within an open reading frame) have been detected in proportions of 10^{-3} to 10^{-4} in RNA from vesicular stomatitis virus (VSV) populations (Steinhauer *et al.*, 1989a). The presence of such molecules doomed to extinction is best explained as the result of a high mutational input during RNA replication. In this scenario, what determines the evolution of RNA genomes? The most adequate theoretical approach to this question has been provided by the studies of M. Eigen and his colleagues on the error-prone replication of informational molecules during the early evolution of life (Eigen, 1971, 1987; Eigen and Schuster, 1979; Eigen and Biebricher, 1988). They showed that molecules that replicate with a limited fidelity will be organized in distributions that they term *quasispecies*. Such distributions include one (or several) *master* sequence(s) and a mutant spectrum. The *master* may, nevertheless, represent a very small proportion of the total number of genomes, estimated in about 10% for phage Qβ (Domingo *et al.*, 1978) and in less than 1% for FMDV (Sobrino *et al.*, 1983). Thus, most molecules belong to the *mutant spectrum*, which constitutes a huge reservoir of genetic and phenotypic variants of a size that can be appreciated, considering an epidemiological situation as the one depicted in Fig. 1 (Section 1.C). There, each of the 10^5 disease outbreaks could involve about 10^2 infected animals with at least 10^{12} genomes only at the peak of each infectious process (Section I.A). The total approaches 10^{19} genomes in two decades and a limited geographic area!

The evolution of a genome distribution is best explained by the *quasispecies* concept in what has been termed the *population equilibrium* model of RNA genomes (Domingo and Holland, 1988). During RNA replication variants are continuously arising, and they are rated against all other variants present and being generated in the same population. The ranking of mutants will be the composite result of their ability not only to replicate RNA, but also of any relevant parameter for that genome: ability to form a capsid, to spread from cell to cell, and so forth. Unless endowed with a selective advantage—and excluding founder effects (see Section VI.B)—a variant will be kept at a low frequency in spite of being generated at a high rate. If the environmental conditions are such that a previously unfavored mutant is now selected, a *shift in the equilibrium distribution* will take place. One virus has a dual potential for either long-term conservation of a consensus sequence under equilibrium conditions or rapid evolution if the equilibrium is ruptured. This has been most elegantly demonstrated by J. Holland and his associates working with VSV. Rapid evolution of the infectious genomes was driven by defective-interfering VSV particles (Holland *et al.*, 1979; Spindler *et al.*, 1982; Horodyski *et al.*, 1983; O'Hara *et al.*, 1984a,b; DePolo *et al.*, 1987). In the absence of the selective pressure exerted by defective-interfering RNAs, the same or a similar sequence was maintained for at least 523 passages. Yet, the RNA of individual VSV clones isolated from this population differed in an estimated 10–40 positions from the consensus sequence (Steinhauer *et al.*, 1989b). This demonstrates that it is possible to ensure long-term conservation of sequences while maintaining an heterogeneous viral population.

In line with those results is the observation of Villaverde *et al.* (1988) that the polymerase (3D) gene is heterogeneous among epidemiologically related FMDVs of type C in spite of its long-term conservation (Section III.B). As documented in Section III, the nucleotide sequence of the RNA of each independent FMDV isolate examined to date differs in some residues from that of previous isolates. In some instances, the sequence determined corresponds to a consensus—i.e., when the RNA is sequenced by primer extension—but in other cases, the sequence is that of one of the genomes that constitute the quasispecies—i.e., when a molecularly cloned cDNA copy of the genome is sequenced. The uniqueness of the consensus sequences examined implies that equilibrium was frequently disturbed during FMDV replication and transmission in the field. Shifts in equilibrium are probably favored by the diversity of animal hosts and of cell types able to support FMDV multiplication (Section I). Adaptation to cell culture entailed fixation of amino acid substitutions in VP1 (Laplace *et al.*, 1987). There is also evidence that the virus produced in the first days after infection of BHK-21 differed in several biologic characteristics from the virus produced late in infection; changes in VP1 and VP3 were detected (Dawe and King, 1983).

There is a constant uncertainty regarding which mutants will arise during any replication event, and thus, an RNA virus genome is statistically defined, but individually indeterminate (Domingo *et al.*, 1978; Domingo and Holland, 1988; Domingo, 1989). This fact is important to the understanding of RNA genetics.

A consequence of the quasispecies nature of RNA genomes is that at any one time an RNA virus encompasses a *range* of phenotypes able to express themselves with some probability in a given environment. For example, the neurotropic and other cell tropism variants of FMDV mentioned in Section IV.A were among the *range* of phenotypes present in the viral quasispecies. It is now clear that one or a few mutations may profoundly alter the phenotype of an RNA virus. Several examples have been reviewed (Domingo *et al.*, 1985; Domingo and Holland, 1988; Domingo, 1989), and additional ones are often reported. Among the relevant properties subject to frequent change, virulence vs. attenuation is worth mentioning. Attenuation is applicable only to one or a group of hosts. A strain may be highly attenuated for one host species and highly virulent for another host. This *relative* attenuation may result from one or a combination of genomic alterations that yield a suboptimal virus. The problem is that a virus may be suboptimal in one environment (i.e., one host) and perfectly fit in another. Reversion to virulence may also be frequent, as documented in the case of polio and other viruses (reviewed in Domingo and Holland, 1988, and Domingo, 1989). Naturally, attenuation due to deletions may also be lost as a result of compensating mutations fixed at other sites. Thus, attenuated strains of FMDV (and probably of other viruses with a wide host range) should not be used as vaccines unless lengthy tests to ensure that the attenuation phenotype is stably maintained in the natural hosts of the virus have been carried out.

Other phenotypic changes cannot be produced by a few mutations and, thus, are unlikely to be among the range of phenotypes of a quasispecies. For a virus to produce an entirely new pathology large evolutionary jumps are likely to be required. Jumps of this kind may be the result of accumulation of many point mutations or of recombination, or both. Perhaps new viral diseases such as hemorrhagic conjunctivitis (Takeda *et al.*, 1984), AIDS, or the fatal disease of seals caused by some morbillivirus-like agent (Mahy *et al.*, 1988; Cosby *et al.*, 1988) are the result of long evolutionary jumps in viruses.

B. Multiple Evolutionary Lineages

If the evolving FMDV population is genetically heterogeneous, diversification is expected to be observed in relatively short time spans. Two clearly distinguishable VP1 genes that provided FMDV with distinct antigenic properties were characterized among the type C isolates in Spain (1970–1982) (Sobrino *et al.*, 1986; Martínez *et al.*, 1988; Mateu *et al.*, 1987, 1988, 1989, 1990). It is likely that these two sublines had a common and recent FMDV C ancestor.

E. L. Palma and his colleagues in Buenos Aires have shown that isolates of FMDV C_3 or A_{24} obtained in Argentina in recent years not only are heterogeneous but can also be grouped in what appear to be evolutionary sublines (Piccone *et al.*, 1988, and manuscript in preparation). This was particularly striking in the comparison of VP1 from isolates of FMDV C_3. One subline was related to FMDV C_3- Resende Br-55 and the other to FMDV C_3 Indaial.* Interestingly, isolate C_3 Argentina-84 belonged to the first one while FMDV C_3 Argentina-85 belonged to the second. The two sublines are distinguished not only by point mutations, but also by a three-base deletion (or insertion) in the VP1 gene, which by virtue of its corresponding to a less frequent event than a base substitution is more definitory of kinship. Several viruses related either to C_3 Resende Br-55 or to C_3 Indaial were isolated in South America, and thus, the studies of Piccone *et al.* (1988) provide evidence of cocirculation of two related but different classes of FMDV C_3 genomes. It is likely that examination of additional genomes would reveal increasing number of sublines, in a complicated network of relationships of the kind described by P. Palese and his associates for type C influenza viruses (review in Smith and Palese, 1988). The shape of phylogenetic trees derived for FMDV (Dopazo *et al.*, 1988; Fig. 3) supports this suggestion.

Genetic diversification is achieved not only by fixation of substitutions (or equilibrium shifts, Section VI.A) but also by transmission bottlenecks. An infected animal will generate variant genomes (Gebauer *et al.*, 1988; Carrillo *et al.*, manuscript in preparation), and each has certain probability of being transmitted to a new host. Since infection is often initiated by one or a few virions (Section I.A), it is not a complex mixture of genomes from the donor animal which is amplified, but rather one or a few representatives of the quasispecies. The likelihood of arriving at consensus sequence identical to that in the donor host is, consequently, decreased. All indications are that transmission bottlenecks are a major driving force in FMDV diversification (Domingo *et al.*, 1980; Domingo, 1989). The latter process occurs by unpredictable alternations of periods of relative stability with others of abrupt change.

VII. CONCLUSION

The quasispecies concept has been timely in providing a theoretical background to interpret the experimental observations on RNA variability and evolution (Domingo *et al.*, 1978, 1980, 1985; Eigen and Biebricher, 1988; Steinhauer and Holland, 1986, 1987; Domingo and Holland, 1988; Domingo, 1989; Meyerhans *et al.*, 1989). Variation of FMDV, like that of other RNA viruses, deals with unpredictable shifts of genomic distributions. As expected, fluctuations affect also phenotypic traits, in particular antigenicity. The situation demands new strategies for vaccine design (Section V.D) and, more broadly, for the control of viral disease (Domingo, 1989). Even for viruses believed to be antigenically stable (rabies, rinderpest, etc.) it may be a matter of time before variants able to evade an immune response are selected. In line with the mechanisms of evolution delineated in Sections VI.A and VI.B, it is unpredictable how many of the amino acid substitutions fixed at the

*The two C_3 Indaial viruses termed Indaial-71 and Indaial-78 are in reality one single natural isolate differing in passage history in the laboratory (personal communication of E. Ouldridge to E. D., July 1988). Since VP1 of the two preparations differ in seven amino acids (Makoff *et al.*, 1982; Cheung *et al.*, 1983), the unpredictability of FMDV variations is clearly illustrated.

several antigenic sites of FMDV (Sections V.B and V.C) will be of any consequence regarding the immune response against the virus. Undoubtedly, this may greatly affect FMDV diagnosis, an important topic that lies beyond the scope of this chapter. In our view, one valid approach is the accumulation of data on nucleotide sequences and reactivities of many isolates and variants with MAbs and polyclonal antibody preparations. Perhaps some generalizations will emerge that will make it possible to design better vaccines and new strategies for FMD control. Both the nature of FMDV populations and their mode of evolution appear to have many features in common with those of other RNA viruses. Efforts to understand FMDV acquire, by this very fact, additional justification.

VIII. SUMMARY

FMD is the economically most important disease of cattle. It is caused by FMDV, an aphthovirus of the family Picornaviridae. The virus may produce either an acute infection to a wide range of animal hosts or an inapparent, persistent infection in ruminants. This chapter has focused on the genetic variability of FMDV and one of its important consequences: antigenic diversity. The FMD virion contains several complex antigenic sites, each composed of several epitopes. Those involved in neutralization of infectivity are remarkably heterogeneous and subject to unpredictable variation. We have emphasized that FMDV populations, whether clonal or not, consist of complex distributions of related but nonidentical genomes, termed quasispecies. This genomic structure is shared by most other RNA viruses and provides them with a dual potential for either relative conservation (conditions of population equilibrium) or rapid evolution (shift in the equilibrium). This mode of evolution has important theoretical and practical implications, some of which have been reviewed here. In particular, we have considered the difficulties encountered in the design of new, synthetic anti-FMD vaccines. We have suggested that a vaccine should include multiple, unrelated epitopes to greatly decrease the probability of selecting variant viruses in the field. The identification of new epitopes involved in neutralization of infectivity and the variations that they are likely to undergo will provide information that may finally be integrated in the formulation of a synthetic vaccine. This work is now in progress.

ACKNOWLEDGMENTS

We are indebted to J. C. de la Torre, J. Díez, A. Villaverde, C. Carrillo, and M. Dávila for unpublished results that have been included in the text. We have benefited from many informative and stimulating discussions with A. Donaldson, A. M. Q. King, R. P. Kitching, N. J. Knowles, E. Beck, E. L. Palma, J. Plana, M. Lombard, D. M. Moore, and J. J. Holland, as well as with the delegates of the research group of the European Commission for the control of foot-and-mouth disease (F.A.O.). Thanks are also given to B. W. J. Mahy, R. Casas Olascoaga, B. Carrillo, E. Beck, J. Callis, and R. Brease for their hospitality during visits to their laboratories at Pirbright (U.K.), Rio de Janeiro (Brazil), Castelar, Buenos Aires (Argentina), Heidelberg (F.R.G.), and Plum Island (U.S.A.), respectively.

REFERENCES

Acharya, R., Fry, E., Stuart, D., Fox, G., Rowlands, D., and Brown, F. (1989). *Nature* **337,** 709–716.
Adam, K. H., Kaaden, V. R., and Strohmaier, K. (1978). *Biochem. Biophys. Res. Commun.* **84,** 677–683.
Anderer, F. A., and Schlumberger, H. D. (1965). *Biochim. Biophys. Acta* **97,** 503–509.

Arrowsmith, A. E. M. (1977). In *Developments in Biological Standardization,* Vol. 35 (C. MacKowiak and R. H. Regamey, eds.), pp. 221–230, Karger, Basel.
Augé de Mello, P., Honigman, M. H., Fernández, M. V., and Gomes, I. (1970). *Bull. Off. Int. Epizoot.* **73,** 489–505.
Bachrach, H. L. (1968). *Annu. Rev. Microbiol.* **22,** 201–244.
Bachrach, H. L. (1977). In *Betsville Symposia in Agricultural Research. I. Virology in Agriculture* (J. A. Romberger, ed.), pp. 3–22, Allanheld, Osmun, Montclair, NJ.
Bachrach, H. L., Moore, D. M., McKercher, P. A., and Polatnick, J. (1975). *J. Immunol.* **115,** 1636–1641.
Bachrach, H. L., Morgan, D. O., and Moore, D. M. (1979). *Intervirology* **12,** 65–72.
Barnett, P. V., Ouldridge, E. J., Rowlands, D. J., Brown, J., and Parry, N. R. (1989). *J. Gen. Virol.* **70,** 1483–1491.
Batschelet, E., Domingo, E., and Weissmann, C. (1976). *Gene* **1,** 27–32.
Baxt, B., Morgan, D. O., Robertson, B. H., and Timpone, C. A. (1984). *J. Virol.* **51,** 298–305.
Baxt, B., Vakharia, V., Moore, D. M., Franke, A. J., and Morgan, D. O. (1989a). *J. Virol.* **63,** 2143–2151.
Baxt, B., Garmendia, A. E., and Morgan, D. O. (1989b). *Viral Immunol.* **2,** 103–113.
Beale, J. (1982). *Nature* **298,** 14–15.
Beck, E., and Strohmaier, K. (1987). *J. Virol.* **61,** 1621–1629.
Beck, E., Feil, G., and Strohmaier, K. (1983a). *EMBO J.* **2,** 555–559.
Beck, E., Forss, S., Strebel, K., Cattaneo, R., and Feil, G. (1983b). *Nucleic Acids Res.* **11,** 7873–7885.
Belsham, G. J., and Bostock, C. J. (1988). *J. Gen. Virol.* **69,** 265–274.
Bittle, J. L., Houghten, R. H., Alexander, H., Shinnick, T. M., Sutcliffe, J. G., Lerner, R. A., Rowlands, D. J., and Brown, F. (1982). *Nature* **298,** 30–34.
Boothroyd, J. C., Highfield, P. E., Cross, G. A. M., Rowlands, D. J., Lowe, P. A., Brown, F., and Harris, T. J. R. (1981). *Nature* **290,** 800–802.
Boothroyd, J. C., Harris, T. J. R., Rowlands, D. J., and Lowe, P. A. (1982). *Gene* **17,** 153–161.
Bolwell, C., Brown, A. L., Barnett, P. V., Campbell, R. O., Clarke, B. E., Parry, N. R., Ouldrige, E. J., Brown, F., and Rowlands, D. J. (1989a). *J. Gen. Virol.* **70,** 45–57.
Bolwell, C., Clarke, B. E., Parry, N. R., Ouldridge, E. J., Brown, F., and Rowlands, D. J. (1989b). *J. Gen. Virol.* **70,** 59–68.
Borca, M. V., Fernández, F. M., Sadir, A. M., Braun, M., and Schudel, A. A. (1986). *Immunology* **59,** 261–267.
Britten, R. J. (1986). *Science* **231,** 1393–1398.
Broekhuijsen, M. P., Blom, T., Van Rijn, J., Pouwels, P. H., Klasen, E. A., Fasbender, M. J., and Enger-Valk, B. E. (1986). *Gene* **49,** 189–197.
Broekhuijsen, M. P., Van Rijn, J. M. M., Blom, A. J. M., Pouwels, P. H., Enger-Valk, B. E., Brown, F., and Francis, M. J. (1987). *J. Gen. Virol.* **68,** 3137–3143.
Brooksby, J. B. (1958). *Adv. Virus Res.* **5,** 1–37.
Brooksby, J. B. (1981a). In *Virus diseases of food animals,* Vol. 1 (E. P. J. Gibbs, ed.), pp. 69–78, Academic Press, London.
Brooksby, J. B. (1981b). *Nature* **289,** 535.
Brooksby, J. B., and Rogers, J. (1957). In *Methods of Typing and Cultivation of Foot-and-Mouth Disease Viruses,* OEEC, Paris.
Brown, A. L., Campbell, R. O., and Clarke, B. E. (1989). *Gene* **75,** 225–234.
Brown, F. (1979). In *The Molecular Biology of Picornaviruses* (R. Perez-Bercoff, ed.), pp. 49–72, Plenum Press, New York.
Brown, F. (1988). *Vaccine* **8,** 180–182.
Brown, F. (1989). In *Molecular Aspects of Picornavirus Infection and Detection* (B. L. Semler and E. Ehrenfeld, eds.), pp. 179–191, ASM, Washington, DC.
Brown, F., Newman, J. F. E., Stott, J., Porter, A., Frisby, D., Newton, C., Carey, N., and Fellner, P. (1974). *Nature* **251,** 342–344.
Burrows, R. (1966). *J. Hyg.* **64,** 81–90.
Burrows, R., Mann, J. A., Garland, A. J. M., Greig, A., and Goodridge, D. (1981). *J. Comp. Pathol.* **91,** 599–609.
Callis, J. J. (1979). Foot-and-mouth disease. A world problem. *Proceedings of the 83rd Annual Meeting of the United States Animal Health Association,* pp. 261–269.
Callis, J. J., and McKercher, P. D. (1985). In *Fiebre aftosa,* Bovis, Vol. 3 (J. M. Sánchez-Vizcaíno, ed.), Luzán S. A. de Ediciones, Madrid.

Capel-Edwards, M. (1971). *J. Comp. Path.* **81,** 433–438.

Carrillo, C., Dopazo, J., Moya, A., González, M., Martínez, M. A., Saiz, J. C., and Sobrino, F. (1990). *Virus Res.* **15,** 45–56.

Carroll, A. R., Rowlands, D. J., and Clarke, B. E. (1984). *Nucleic Acids Res.* **12,** 2461–2472.

Cheung, A. K., and Küpper, H. (1984). *Biotechnol. Genetic Engineering Rev.* **1,** 223–259.

Cheung, A., DeLaMarter, J., Weiss, S., and Küpper, H. (1983). *J. Virol.* **48,** 451–459.

Cheung, A. K., Whitehead, P., Weiss, S., and Küpper, H. (1984). *Gene* **30,** 241–245.

Chow, M., Newman, J. F. E., Felman, D., Hogle, J. M., Rowlands, D. J., and Brown, F. (1987). *Nature* **327,** 482–486.

Ciaccio, G. (1963). *Ann. Inst. Pasteur* **104,** 529–534.

Clarke, B. E., and Sangar, D. V. (1988). *J. Gen. Virol.* **69,** 2313–2325.

Clarke, B. E., Carroll, A. P., Rowlands, D. J., Nicholson, B. H., Houghten, R. A., Lerner, R. A., and Brown, F. (1983). *FEBS Lett.* **157,** 261–264.

Clarke, B. E., Brown, A. L., Currey, R. M., Newton, S. E., Rowlands, D. J., and Carroll, A. R. (1987a). *Nucleic Acids Res.* **15,** 7067–7079.

Clarke, B. E., Newton, S. E., Carroll, A. R., Francis, M. J., Appleyard, G., Syred, A. D., Highfield, P. E., Rowlands, D. J., and Brown, F. (1987b). *Nature* **330,** 381–384.

Clarke, B. E., Carroll, A. R., Francis, M. J., Appleyard, G., Syred, A. D., Highfield, P. E., Rowlands, D. J., and Brown, F. (1988). In *Vaccines 88* (H. Ginsberg, F. Brown, R. A. Lerner, and R. M. Chanock, eds.), pp. 127–131, Cold Spring Harbor Laboratory, New York.

Cloyd, M. W., and Holt, M. J. (1987). *Virology* **161,** 286–292.

Collen, T., Pullen, L., and Doel, T. R. (1989). *J. Gen. Virol.* **70,** 395–403.

Condy, J. B., Hedger, R. S., Hamblin, C., and Barnett, I. T. R. (1985). *Comp. Immunol. Microbiol. Infect. Dis.* **8,** 259–265.

Cosby, S. L., McQuaid, S., Duffy, N., Lyons, C., Rima, B. K., Allan, G. M., McCullough, S. J., Kennedy, S., Smyth, J. A., and McNeilly, F. (1988). *Nature* **36,** 115–116.

Costa Giomi, M. P., Bergmann, I. E., Scodeller, E. A., Augé de Mello, P., Gomes, I., and La Torre, J. L. (1984). *J. Virol.* **51,** 799–805.

Costa Giomi, M. P., Gomes, I., Tiraboschi, B., Augé de Mello, P., Bergmann, I. E., Scodeller, E. A., and La Torre, J. L. (1988). *Virology* **162,** 58–64.

Dawe, P. S., and King, A. M. Q. (1983). *Arch. Virol.* **76,** 117–126.

Dayhoff, M. O., Eck, R. V., and Park, C. M. (1972). In *Atlas of Protein Sequence and Structure,* Vol. 5 (M. O. Dayhoff, ed.), pp. 89–99, National Biomedical Research Foundation, Washington, DC.

de la Torre, J. C., and Domingo, E. (1988). *Microbiol. SEM* **4,** 161–166.

de la Torre, J. C., Dávila, M., Sobrino, F., Ortín, J., and Domingo, E. (1985). *Virology* **145,** 24–35.

de la Torre, J. C., Alarcón, B., Martínez-Salas, E., Carrasco, L., and Domingo, E. (1987). *J. Virol.* **61,** 233–235.

de la Torre, J. C., Martínez-Salas, E., Díez, J., Villaverde, A., Gebauer, F., Rocha, E., Dávila, M., and Domingo, E. (1988). *J. Virol.* **62,** 2050–2058.

de la Torre, J. C., de la Luna, S., Díez, J., and Domingo, E. (1989a). *J. Virol.* **63,** 2385–2387.

de la Torre, J. C., Martínez-Salas, E., Díez, J., and Domingo, E. (1989b). *J. Virol.* **63,** 59–63.

DePolo, N. J., Giachetti, C., and Holland, J. J. (1987). *J. Virol.* **61,** 454–464.

Devaney, M. A., Vakharia, V. N., Lloyd, R. E., Ehrenfeld, E., and Grubman, M. J. (1988). *J. Virol.* **62,** 4407–4409.

Di Girolamo, W., Salas, M., and Laguens, R. P. (1985). *Arch. Virol.* **83,** 331–336.

Di Marchi, R., Brooke, G., Gale, C., Cracknell, V., Doel, T., and Mowat, N. (1986). *Science* **232,** 639–641.

Dietzshold, B., Kaaden, O. R., Tokui, T., and Bohm, H. C. (1971). *J. Gen. Virol.* **13,** 1–7.

Díez, J., Mateu, M. G., and Domingo, E. (1989). *J. Gen. Virol.* **70,** 3281–3289.

Dinter, Z., Philipson, L., and Wesslen, T. (1959). *Virology* **8,** 542–544.

Doel, T. R., Gale, C., Brooke, G., and Di Marchi, R. (1988). *J. Gen. Virol.* **69,** 2403–2406.

Domier, L., Shaw, J. G., and Rhoads, R. E. (1987). *Virology* **158,** 20–27.

Domingo, E. (1989). In *Progress in Drug Research,* Vol. 33 (E. Jucker, ed.), pp. 93–133, Birkhäuser Verlag, Basel.

Domingo, E., and Holland, J. J. (1988). In *RNA Genetics,* Vol. III. *Variability of RNA Genomes* (E. Domingo, J. J. Holland, and P. Ahlquist, eds.), pp. 3–36, CRC Press, Boca Raton, FL.

Domingo, E., Flavell, R. A., and Weissmann, C. (1976). *Gene* **1,** 3–25.

Domingo, E., Sabo, D., Taniguchi, T., and Weissmann, C. (1978). *Cell* **13,** 735–744.

Domingo, E., Dávila, M., and Ortín, J. (1980). *Gene* **11,** 333–346.

Domingo, E., Martínez-Salas, E., Sobrino, F., de la Torre, J. C., Portela, A., Ortín, J., López-Galindez, C., Pérez-Breña, P., Villanueva, N., Nájera, R., VandePol, S., Steinhauer, D., De Polo, N., and Holland, J. J. (1985). *Gene* **40,** 1–8.

Donaldson, A. (1987). *Irish Vet. J.* **41,** 325–327.

Donaldson, A. I., Gibson, C. F., Oliver, R., Hamlin, C., and Kitching, R. P. (1987). *Res. Vet. Sci.* **43,** 339–346.

Dopazo, J., Sobrino, F., Palma, E. L., Domingo, E., and Moya, A. (1988). *Proc. Natl. Acad. Sci. USA* **85,** 6811–6815.

Duchesne, M., Cartwright, T., Crespo, A., Boucher, F., and Fallourd, A. (1984). *J. Gen. Virol.* **65,** 1559–1566.

Eigen, M. (1971). *Naturwissenschaften* **58,** 465–523.

Eigen, M. (1987). *Cold Spring Harbor Symp. Quant. Biol.* **52,** 307–320.

Eigen, M., and Biebricher, C. K. (1988). In *RNA Genetics.* Vol. III. *Variability of RNA Genomes* (E. Domingo, J. J. Holland, and P. Ahlquist, eds.), pp. 211–245, CRC Press, Boca Raton, FL.

Eigen, M., and Schuster, P. (1979). *The Hypercycle,* Springer, Berlin.

Fagg, R. M., and Hyslop, N. S. G. (1966). *J. Hyg.* **64,** 397–404.

Forss, S., and Schaller, H. (1982). *Nucleic Acids Res.* **10,** 6441–6450.

Forss, S., Strebel, K., Beck, E., and Schaller, H. (1984). *Nucleic Acids Res.* **12,** 6587–6601.

Fox, G., Stuart, D., Acharya, K. R., Fry, E., Rowlands, D., and Brown, F. (1987). *J. Mol. Biol.* **196,** 591–597.

Fox, G., Parry, N. R., Barnett, P. V., McGinn, B., Rowlands, D. J., and Brown, F. (1989). *J. Gen. Virol.* **70,** 625–637.

Francis, M. J., Fry, C. M., Rowlands, D. J., Bittle, J. L., Houghten, R. A., Lerner, R. A., and Brown, F. (1987a). *Immunology* **61,** 1–6.

Francis, M. J., Fry, C. M., Clarke, B. E., Rowlands, D. J., Brown, F., Bittle, J. L., Houghten, R. A., and Lerner, R. A. (1987b). In *Vaccines 87* (R. M. Channock, H. Ginsberg, R. A. Lerner, and F. Brown, eds.), pp. 60–67, Cold Spring Harbor Laboratory, New York.

Francis, M. J., Hastings, G. Z., Syred, A. D., McGinn, B., Brown, F., and Rowlands, D. J. (1987c). *Nature* **330,** 168–170.

Francis, M. J., Hastings, G. Z., Syred, A. D., McGinn, B., Brown, F., and Rowlands, D. J. (1988). In *Vaccines 88* (H. Ginsberg, F. Brown, R. A. Lerner, and R. M. Chanock, eds.), pp. 1–7, Cold Spring Harbor Laboratory, New York.

Gailiunas, P. (1968). *Arch. ges. Virusforsch.* **25,** 188–200.

Gaschutz, S., Hess, G., and Wittmann, G. (1986). In *Report of the Session of the Research Group of the Standing Technical Committee of the European Commission for the Control of Foot-and-Mouth Disease,* Madrid, pp. 9–11, FAO, Rome.

Gebauer, F., de la Torre, J. C., Gomes, I., Mateu, M. G., Barahona, H., Tiraboschi, B., Bergmann, I., Augé de Mello, P., and Domingo, E. (1988). *J. Virol.* **62,** 2041–2049.

Geysen, H. M., Meloen, R. H., and Barteling, S. J. (1984). *Proc. Natl. Acad. Sci. USA* **81,** 3998–4002.

Geysen, H. M., Barteling, S. J., and Meloen, R. H. (1985). *Proc. Natl. Acad. Sci. USA* **82,** 178–182.

Geysen, H. M., Rodda, S. J., Mason, T. J., Tribbick, G., and Schoofs, P. G. (1987). *J. Immunol. Methods* **102,** 259–274.

Gibson, C. F., and Donaldson, A. I. (1986). *Res. Vet. Sci.* **41,** 45–49.

Gomes, I., and Rosenberg, F. J. (1984). *Preventive Vet. Med.* **3,** 197–205.

Grubman, M. J., and Morgan, D. O. (1986). *Virus Res.* **6,** 33–43.

Harrison, S. C. (1989). *Nature* **338,** 205–206.

Hedger, R. S., and Condy, J. B. (1985). *Vet. Rec.* **117,** 205.

Hedger, R. S., and Stubbins, A. G. J. (1971). *State Vet. J.* **26,** 45–50.

Holland, J. J., Grabau, E. A., Jones, C. L., and Semler, B. L. (1979). *Cell* **16,** 495–504.

Holland, J. J., Spindler, K., Horodyski, F., Grabau, E., Nichol, S., and VandePol, S. (1982). *Science* **215,** 1577–1585.

Horodyski, F. M., Nichol, S. T., Spindler, K. R., and Holland, J. J. (1983). *Cell* **33,** 801–810.

Hyslop, N. St. G. (1965). *J. Gen. Microbiol.* **41,** 135–142.

Hyslop, N. St. G. (1970). *Adv. Vet. Sci. Comp. Med.* **14,** 261–307.

Jackson, R. J. (1989). In *Molecular Aspects of Picornavirus Infection and Detection* (B. L. Semler and E. Ehrenfeld, eds.), pp. 51–71, ASM, Washington, DC.

Joubert, L., and Mackowiak, C. (1968). *La Fièvre Aphteuse,* Expansion Scientifique Française.

Kaaden, O. R., Adam, K. H., and Strohmaier, K. (1977). *J. Gen. Virol.* **34,** 397–400.

King, A. M. Q. (1988). In *RNA Genetics.* Vol. II. *Retroviruses, Viroids, and RNA Recombination* (E. Domingo, J. J. Holland, and P. Ahlquist, eds.), pp. 149–165, CRC Press, Boca Raton, FL.

King, A. M. Q., Underwood, B. O., McCahon, D., Newman, J. W. I., and Brown, F. (1981). *Nature* **293,** 479–480.

King, A. M. Q., McCahon, D., Slade, W. R., and Newman, J. W. I. (1982). *Cell* **29,** 921–928.

King, A. M. Q., Underwood, B. O., McCahon, D., Newman, J. W. I., and Slade, W. R. (1985). *Virus Res.* **3,** 373–384.

Kitching, R. P., Rendle, R., and Ferris, N. P. (1988). *Vaccine* **6,** 1–6.

Kitching, R. P., Knowles, N. J., Samuel, A. R., and Donaldson, A. I. (1989). *Trop. Animal Health Production* **21,** 153–166.

Kitson, J. D. A., Belsham, G. J., Burke, K. L., and Almond, J. W. (1989). *Abstracts of the 6th Meeting of the European Group of Molecular Biology of Picornaviruses,* Bruges, Belgium, Abstract E12.

Kleid, D. G., Yansura, D., Small, B., Dowbenko, D., Moore, D. M., Grubman, M. J., McKercher, P. D., Morgan, D. O., Robertson, B. H., and Bachrach, H. L. (1981). *Science* **214,** 1125–1129.

Küpper, H., Keller, W., Kurz, C., Forss, S., Schaller, H., Franze, R., Strohmaier, K., Marquardt, O., Zaslavsky, V. G., and Hofschneider, H. (1981). *Nature* **289,** 555–559.

Kurz, C., Forss, S., Küpper, H., Strohmaier, K., and Schaller, H. (1981). *Nucleic Acids Res.* **9,** 1919–1931.

La Torre, J. L., Grubmann, M. J., Baxt, B., and Bachrach, H. L. (1980). *Proc. Natl. Acad. Sci. USA* **77,** 7444–7447.

Laplace, E., Rey-Senelonge, A., Kohen, G., and Rivière, M. (1987). In *Report of the Session of the Research Group of the Standing Technical Committee of the European Commission for the Control of Foot-and-Mouth Disease,* Lyons, pp. 117–119, FAO, Rome.

Laporte, J., Grosclaude, J., Wantyghem, J., Bernard, S., and Rouze, P. (1973). *Co. r. hebdoma. seánces Acad. Sci. Paris* **276,** 3390–3401.

Loeffler, F., and Frosch, P. (1898). *Zentralbl. Bacteriol. Parasitenkunde Infektionskrankh.* **23,** 371–391.

Luo, M., Vriend, G., Kamer, G., Minor, I., Arnold, E., Rossmann, M. G., Boege, U., Scraba, D. G., Duke, G. M., and Palmenberg, A. C. (1987). *Science* **235,** 182–191.

Luo, M., Rossmann, M. G., and Palmenberg, A. C. (1988). *Virology* **166,** 503–514.

Mahy, B. W. J., Barrett, T., Evans, S., Anderson, E. C., and Bostock, C. J. (1988). *Nature* **336,** 115.

Makoff, A. J., Poynter, C. A., Rowlands, D. J., and Boothroyd, J. C. (1982). *Nucleic Acids Res.* **10,** 8285–8295.

Marquardt, O., and Adam, K. H. (1988). *Virus Genes* **2,** 283–291.

Martínez, M. A., Carrillo, C., Plana, J., Mascarella, R., Bergadá, J., Palma, E. L., Domingo, E., and Sobrino, F. (1988). *Gene* **62,** 75–84.

Martínez-Salas, E., Ortín, J., and Domingo, E. (1985). *Gene* **35,** 55–61.

Mateu, M. G., Rocha, E., Vicente, O., Vayreda, F., Navalpotro, C., Andreu, O., Pedroso, E., Giralt, E., Enjuanes, L., and Domingo, E. (1987). *Virus Res.* **8,** 261–274.

Mateu, M. G., da Silva, J. L., Rocha, E., de Brum, D. L., Alonso, A., Enjuanes, L., Domingo, E., and Barahona, H. (1988). *Virology* **167,** 113–124.

Mateu, M. G., Martínez, M. A., Rocha, E., Andreu, D., Parejo, J., Giralt, E., Sobrino, F., and Domingo, E. (1989). *Proc. Natl. Acad. Sci. USA* **86,** 5883–5887.

Mateu, M. G., Martínez, M. A., Capucci, L., Andreu, D., Giralt, E., Sobrino, F., Brocchi, E., and Domingo, E. (1990). *J. Gen. Virol.* (in press).

McCahon, D., King, A. M. Q., Roe, D. S., Slade, W. R., Newman, J. W. I., and Cleary, A. M. (1985). *Virus Res.* **3,** 87–100.

McCahon, D., Crowther, J. R., Belsham, G. J., Kitson, J. D. A., Duchesne, M., Have, P., Meloen, R. H., Morgan, D. O., and De Simone, F. (1989). *J. Gen. Virol.* **70,** 639–645.

McCullough, K. C., Crowther, J. R., Butcher, R. N., Carpenter, W. C., Brocchi, E., Capucci, L., and De Simone, F. (1986). *Immunology* **58,** 421–428.

McCullough, K. C., Crowther, J. R., Carpenter, W. C., Brocchi, E., Capucci, L., De Simone, F., Xie, Q., and McCahon, D. (1987). *Virology* **157,** 516–525.

McVicar, J. W., and Sutmoller, P. (1976). *J. Hyg.* **76,** 467–481.

Mellor, E. J. C., Brown, F., and Harris, T. J. R. (1985). *J. Gen. Virol.* **66,** 1919–1929.

Meloen, R. H., and Barteling, S. J. (1986a). *Virology* **149,** 55–63.

Meloen, R. H., and Barteling, S. J. (1986b). *J. Gen. Virol.* **67,** 289–294.

Meloen, R. H., Rowlands, D. J., and Brown, F. (1979). *J. Gen. Virol.* **45,** 761–763.

Meloen, R. H., Briaire, J., Woortmeyer, R. J., and Van Zaane, D. (1983). *J. Gen. Virol.* **64,** 1193–1198.

Meloen, R. H., Puyk, W. C., Meijer, D. J. A., Lankhof, H., Posthumus, W. P. A., and Schaaper, W. M. M. (1986). *Protides Biol. Fluids Proc. Colloq.* **34,** 103–106.

Meloen, R. H., Puyk, W. C., Meijer, D. J. A., Lankhof, H., Posthumus, W. P. A., and Schaaper, W. M. M. (1987). *J. Gen. Virol.* **68,** 305–314.

Meloen, R. H., Puyk, W. C., Posthumus, W. P. A., Lankhof, H., Thomas, A., and Schaaper, W. M. M. (1988). In *Vaccines 88* (H. Ginsberg, F. Brown, R. A. Lerner, and R. M. Chanock, eds.), pp. 35–40, Cold Spring Harbor Laboratory, New York.

Meyerhans, A., Cheynier, R., Albert, J., Seth, M., Kwok, S., Sninsky, J., Morfeldt-Manson, L., Asjö, B., and Wain-Hobson, S. (1989). *Cell* **58,** 901–910.

Newton, S. E., Carroll, A. R., Campbell, R. D., Clarke, B. E., and Rowlands, D. J. (1985). *Gene* **40,** 331–336.

Newton, S. E., Francis, M. J., Brown, F., Appleyard, G., and Mackett, M. (1986). In *Vaccines 86* (F. Brown, R. M. Chanock, and R. A. Lerner, eds.), pp. 303–309, Cold Spring Harbor Laboratory, New York.

Newton, S. E., Clarke, B. E., Appleyard, G., Francis, M. J., Carroll, A. R., Rowlands, D. J., Skehel, J., and Brown, F. (1987). In *Vaccines 87* (R. M. Chanock, R. A. Lerner, F. Brown and H. Ginsberg, eds.), pp. 12–21. Cold Spring Harbor Laboratory, New York.

O'Hara, P. J., Horodyski, F. M., Nichol, S. J., and Holland, J. J. (1984a). *J. Virol.* **49,** 793–798.

O'Hara, P. J., Nichol, S. T., Horodyski, F. M., and Holland, J. J. (1984b). *Cell* **36,** 915–924.

Onishchenko, A. M., Petrova, N. A., Blinov, V. M., Vassilenko, S. K., Sandakhchiev, L. S., Burdov, A. N., Ivanyushchenkov, V. N., and Perevozchikova, N. A. (1986). *Bioorganicheskóya Khimiya* **12,** 416–419.

Ouldridge, E. J., Barnett, P. V., Parry, N. R., Syred, A. D., Head, M., and Rweyemamu, M. M. (1984). *J. Gen. Virol.* **65,** 203–207.

Palmenberg, A. (1989). In *Molecular Aspects of Picornavirus Infection and Detection* (B. L. Semler and E. Ehrenfeld, eds.), pp. 211–241, ASM, Washington, DC.

Parry, N. R. (1982). Ph.D. thesis. *Studies on plaque variation of foot-and-mouth disease virus,* University of London.

Parry, N. R., Ouldridge, E. J., Barnett, P. V., Rowlands, D. J., Brown, F., Bittle, J. L., Houghten, R. A., and Lerner, R. A. (1985). In *Vaccines 85* (R. A. Lerner, R. M. Chanock, and F. Brown, eds.), pp. 211–216, Cold Spring Harbor Laboratory, New York.

Parry, N. R., Barnett, P. V., Ouldridge, E. J., Rowlands, D. J., and Brown, F. (1989a). *J. Gen. Virol.* **70,** 1493–1503.

Parry, N. R., Ouldridge, E. J., Barnett, P. V., Fox, J. D., Francis, M. J., Rowlands, D. J., and Brown, F. (1989b). *Abstracts of the 6th Meeting of the European Group of Molecular Biology of Picornaviruses,* Bruges, Belgium, Abstract D13.

Pereira, H. G. (1977). In *Developments in Biological Standardization,* Vol. 35 (C. MacKowiak and R. H. Regamey, eds.), pp. 167–174, Karger, Basel.

Pereira, H. G. (1981). In *Virus Diseases of Food Animals,* Vol. 2 (E. P. J. Gibbs, ed.), pp. 333–363, Academic Press, London.

Pfaff, E., Mussgay, M., Böhm, H. O., Schultz, G. E., and Schaller, H. (1982). *EMBO J.* **1,** 869–874.

Pfaff, E., Kuhn, C., Schaller, H., Leban, J., Thiel, H. J. and Böhm, H. O. (1985). In *Vaccines 85* (R. A. Lerner, R. M. Chanock, and F. Brown, eds.), pp. 199–202, Cold Spring Harbor Laboratory, New York.

Pfaff, E., Thiel, H-J., Beck, E., Strohmaier, K., and Schaller, H. (1988). *J. Virol.* **62,** 2033–2040.

Piccone, M. E., Kaplan, G., Giavedoni, L., Domingo, E., and Palma, E. L. (1988). *J. Virol.* **62,** 1469–1473.

Pincus, S. E., Rohl, H., and Wimmer, E. (1987). *Virology* **157,** 83–88.

Polatnick, J. (1980). *J. Virol.* **33,** 774–779.

Polatnick, J., Arlinghaus, R. B., Graves, J. H., and Cowan, K. M. (1967). *Virology* **31,** 609–615.

Porter, A. G., Fellner, P., Black, D. N., Rowlands, D. J., Harris, T. J. R., and Brown, F. (1978). *Nature* **276,** 298–301.

Prabhakar, B. S., Haspel, M. V., McClintock, P. R., and Notkins, A. L. (1982). *Nature* **300,** 374–376.

Prabhakar, B. S., Menegus, M. A., and Notkins, A. L. (1985). *Virology* **146,** 302–306.

Reanney, D. C. (1984). In *The Microbe,* Part I (B. W. J. Mahy and J. R. Pattison, eds.), pp. 175–196, Cambridge University Press, Cambridge, MA.

Robertson, B. H., Morgan, D. O., and Moore, D. M. (1984). *Virus Res.* **1,** 489–500.

Robertson, B. H., Grubman, M. J., Wendell, G. N., Moore, D. H., Welsh, J. D., Fischer, T., Dowbenko, D. J., Yansura, D. G., Small, B., and Kleid, D. G. (1985). *J. Virol.* **54,** 651–660.

Robson, K. J. H., Harris, T. J. R., and Brown, F. (1977). *J. Gen. Virol.* **37,** 271–276.

Robson, K. J. H., Crowther, J. R., King, A. M. Q., and Brown, F. (1979). *J. Gen. Virol.* **45,** 579–590.

Rossmann, M. G., and Rueckert, R. R. (1987). *Microbiol. Sci.* **4,** 206–214.

Rowlands, D. J., Harris, T. J. R., and Brown, F. (1978). *J. Virol.* **26,** 335–343.

Rowlands, D. J., Clarke, B. E., Carroll, A. R., Brown, F., Nicholson, B. H., Bittle, J. L., Houghten, R. A., and Lerner, R. A. (1983). *Nature* **306,** 694–697.

Rweyemamu, M. M. (1984). *J. Biol. Stand.* **12,** 323–337.

Saiz, J. C., González, M. J., Morgan, D. O., Card, J. L., Sobrino, F., and Moore, D. M. (1989). *Virus Res.* **13,** 45–60.

Sangar, D. V. (1979). *J. Gen. Virol.* **45,** 1–13.

Sangar, D. V., Black, D. N., Rowlands, D. J., Harris, T. J. R., and Brown, F. (1980). *J. Virol.* **33,** 59–68.

Sangar, D. V., Newton, S. E., Rowlands, D. J., and Clarke, B. E. (1987). *Nucleic Acids Res.* **15,** 3305–3315.

Sangar, D. V., Clark, R. P., Carroll, A. R., Rowlands, D. J., and Clarke, B. E. (1988). *J. Gen. Virol.* **69,** 2327–2333.

Saunders, K., King, A. M. Q., McCahon, D., Newman, J. W. I., Slade, W. R., and Forss, S. (1985). *J. Virol.* **56,** 921–929.

Schwartz, R. H. (1986). *Curr. Topics Microbiol. Immunol.* **130,** 79–85.

Seibold, H. R., Cottral, G. E., Patty, R. E., and Gailiunas, P. (1964). *Am. J. Vet. Res.* **25,** 806–814.

Sellers, R. F. (1971). *Vet. Bull.* **41,** 431–439.

Sellers, R. F. (1981). In *Virus Diseases of Food Animals,* Vol. 1 (E. P. J. Gibbs, ed.), pp. 19–29, Academic Press, London.

Semler, B. L., Kuhn, R. J., and Wimmer, E. (1988). In *RNA Genetics.* Vol. I. *RNA-Directed Virus Replication* (E. Domingo, J. J. Holland, and P. Ahlquist, eds.), pp. 23–48, CRC Press, Boca Raton, FL.

Shahan, M. S. (1962). *Ann. NY Acad. Sci.* **101,** 445–454.

Singh, B. S. (1969). *NZ Vet. J.* **17,** 173–177.

Smith, D. B., and Inglis, S. C. (1987). *J. Gen. Virol.* **68,** 2729–2740.

Smith, F. I., and Palese, P. (1988). In *RNA Genetics.* Vol. III. *Variability of RNA Genomes* (E. Domingo, J. J. Holland, and P. Ahlquist, eds.), pp. 123–135. CRC Press, Boca Raton, FL.

Sobrino, F., Dávila, M., Ortín, J., and Domingo, E. (1983). *Virology* **128,** 310–318.

Sobrino, F., Palma, E. L., Beck, E., Dávila, M., de la Torre, J. C., Negro, P., Villanueva, N., Ortín, J., and Domingo, E. (1986). *Gene* **50,** 149–159.

Sobrino, F., Martínez, M. A., Carrillo, C., and Beck, E. (1989). *Virus Res.* **14,** 273–280.

Spindler, K. R., Horodyski, F. M., and Holland, J. J. (1982). *Virology* **119,** 96–108.

Stave, J. W., Card, J. L., and Morgan, D. O. (1986). *J. Gen. Virol.* **67,** 2083–2092.

Stave, J. W., Card, J. L., Morgan, D. O., and Vakharia, V. N. (1988). *Virology* **162,** 21–29.

Steinhauer, D., and Holland, J. J. (1986). *J. Virol.* **57,** 219–228.

Steinhauer, D., and Holland, J. J. (1987). *Annu. Rev. Microbiol.* **41,** 409–433.

Steinhauer, D., de la Torre, J. C., and Holland, J. J. (1989a). *J. Virol.* **63,** 2063–2071.

Steinhauer, D., de la Torre, J. C., Meier, E., and Holland, J. J. (1989b). *J. Virol.* **63,** 2072–2080.

Strauss, J. H., and Strauss, E. G. (1988). *Annu. Rev. Microbiol.* **42,** 657–668.

Strebel, K., and Beck, E. (1986). *J. Virol.* **58,** 893–899.

Strohmaier, K., Franze, R., and Adam, K. H. (1982). *J. Gen. Virol.* **59,** 295–306.

Sutcliffe, J. G., Shinnick, T. M., Green, N., and Lerner, R. (1983). *Science* **219,** 660–666.

Sutmoller, P., and Gaggero, A. (1965). *Vet. Rec.* **77,** 968–969.

Sutmoller, P., and McVicar, J. W. (1976). *J. Hyg.* **77,** 235–243.

Sutmoller, P., McVicar, J. W., and Cottral, G. E. (1968). *Arch. Ges. Virusforsch.* **23,** 227–235.

Takeda, N., Miyamura, K., Ogino, T., Natori, K., Yamazaki, S., Sakwiai, N., Nakazono, N., Ishii, K., and Kono, R. (1984). *Virology* **134,** 375–388.

Thomas, A. A. M., Woortmeijer, R. J., Puijk, W., and Barteling, S. J. (1988). *J. Virol.* **62,** 2782–2789.

Timoney, J. F., Gillespie, J. H., Scott, F. W., and Barlough, J. E. (1988). *Hagan and Bruner's Microbiology and Infectious Diseases of Domestic Animals,* pp. 647–667, Comstock Publishing Associates, Cornell University Press, Ithaca, London.

Vakharia, V. N., Devaney, M. A., Moore, D. M., Dunn, J. J., and Grubman, M. J. (1987). *J. Virol.* **61,** 3199–3207.

Vallée, H., and Carré, H. (1922). *C. r. hebdom. séances Acad. Sci. Paris* **174,** 207–208.
van Bekkum, J. G., Frenkel, H. S., Frederiks, H. H. J., and Frenkel, S. (1959). *Tijdschr. Diergeneeskd.* **84,** 1159–1164.
Villanueva, N., Dávila, M., Ortín, J., and Domingo, E. (1983). *Gene* **23,** 185–194.
Villaverde, A., Martínez-Salas, E., and Domingo, E. (1988). *J. Mol. Biol.* **204,** 771–776.
Waldmann, O., and Trautwein, K. (1926). *Berl. Munch. tierarztliche Wochenschr.* **42,** 569–570.
Webb, S. R., Kearse, K. P., Foulke, C. L., Hartig, P. C., and Prabhakar, B. S. (1986). *J. Med. Virol.* **12,** 9–15.
Weddell, G. N., Yansura, D. G., Dowbenko, D. J., Hoatlin, M. E., Grubman, M. J., Moore, D. M., and Kleid, D. G. (1985). *Proc. Natl. Acad. Sci. USA* **82,** 2618–2622.
Weissmann, C., and Weber, H. (1986). *Prog. Nucleic Acid Res. Mol. Biol.* **33,** 251–300.
Wiesmüller, K. H., Jung, G., and Hess, G. (1989). *Vaccine* **7,** 29–33.
Wild, T. F., Burroughs, J. N., and Brown, F. (1969). *J. Gen. Virol.* **4,** 313–320.
Wilson, V., Taylor, P., and Desselberger, U. (1988). *Arch. Virol.* **102,** 131–139.
Winther, M. D., Allen, G., Bomford, R. H., and Brown, F. (1986). *J. Immunol.* **136,** 1835–1840.
Xie, Q. C., McCahon, D., Crowther, J. R., Belsham, G. J., and McCullough, K. C. (1987). *J. Gen. Virol.* **68,** 1637–1647.
Yilma, T. (1980). *Am. J. Vet. Res.* **41,** 1537–1542.
Youngner, J., and Preble, O. T. (1980). In *Comprehensive Virology,* Vol. 16 (H. Fraenkel-Conrat and R. R. Wagner, eds.), pp. 73–135, Plenum Press, New York.
Zibert, A., Maas, G., Strebel, K., Falk, M. M., and Beck, E. (1990). *J. Virol.*, in press.

15

Analysis of Rotavirus Proteins by Gene Cloning, Mutagenesis, and Expression

G. W. Both, S. C. Stirzaker, C. C. Bergmann, M. E. Andrew, D. B. Boyle, and A. R. Bellamy

I. INTRODUCTION

Rotaviruses have been isolated from most mammalian species and in many cases are a cause of diarrhea in the young (Kapikian and Chanock, 1985). In domestic animals such as cattle and pigs, the economic loss is often significant (Holmes, 1983). Rotaviruses were discovered in humans in 1973 (Bishop *et al.*, 1973) and in recent times have been intensively studied as their importance as animal and human pathogens has become apparent. At present no approved human vaccine is available, although animal rotavirus isolates and reassortants derived from them are undergoing clinical trials (Vesikari *et al.*, 1988; Midthun *et al.*, 1985). Such live heterologous or reassortant rotaviruses may provide a first-generation rotavirus vaccine. The aims of our research are to characterize rotaviruses in molecular terms, in order to develop an understanding of the structure and function of key viral proteins in the belief that a better understanding of the molecular features of these proteins might lead to the development of second-generation recombinant vaccines. Although the general feasibility of developing recombinant rotavirus vaccines has yet to be established, this might occur through the development of engineered forms of subunit viral proteins; alternatively, procedures by which novel, engineered viral genes could be incorporated by reassortment into live viral vaccines might be developed. Progress in either of these areas requires that we develop a better understanding of the molecular biology of both rotaviruses in general and rotavirus antigens in particular.

II. VIRUS STRUCTURE

Rotaviruses are members of the family Reoviridae and contain 11 segments of double-stranded RNA. These segments code for the six structural and five nonstructural polypeptides summarized in Table 1. The nomenclature was recently revised to take into account the discovery of the minor

---

G. W. Both and S. C. Stirzaker • CSIRO Division of Biotechnology, Laboratory for Molecular Biology, North Ryde, NSW 2113, Australia. *C. C. Bergmann and A. R. Bellamy* • Department of Cellular and Molecular Biology, University of Auckland, Auckland, New Zealand. *M. E. Andrew and D. B. Boyle* • CSIRO Australian Animal Health Laboratory, Geelong, Victoria 3220, Australia.

Table 1. Coding Assignments of SA11 Rotavirus Genome Segments

Gene segment	Polypeptide	Location[a]
1	VP1	IC
2	VP2	IC
3	VP3	IC
4	VP4	OC
5	NS53	NS
6	VP6	IC
7	NS34	NS
8	NS35	NS
9	VP7 (g)[b]	OC
10	NS28 (g)	NS
11	NS29	NS?

[a]IC, Inner core structure; OC, outer capsid; NS, nonstructural.

[b](g), Glycosylated with N-linked, high-mannose carbohydrate.

structural protein VP3 (Liu *et al.*, 1988). As a consequence, the viral hemagglutinin, formerly known as VP3, has now been designated as VP4. To date no proteins translated from alternate reading frames have been described for rotaviruses, although the possibility remains that such a product is produced from gene segment 11 (Mitchell and Both, 1988).

Rotaviruses possess an inner corelike structure (the so-called single-shelled particles) surrounded by an outer capsid layer (forming the double-shelled particles) (Holmes, 1983). In this respect rotaviruses share structural homology with reoviruses (Joklik, 1983), although unlike reovirus, infection of cells with rotavirus leads to the accumulation in the cell of both cores and whole virions (Holmes, 1983). Unfortunately, the resolution generally afforded by the dehydrating techniques of electron microscopy have not provided much insight into the fine structure of rotavirus particles. Metcalf (1979) and Roseto *et al.* (1979) used freeze etch techniques to examine the rotavirus core, which exhibits obvious surface projections (probably VP6) arranged in a T = 13L structure. However, the more immunologically relevant whole virion possesses a surface that is more difficult to examine, having been described by Roseto *et al.* (1979) as a smooth surface sphere with small holes. Recently Prasad *et al.* (1988) determined the three-dimensional structure of rotavirus to a resolution of 40°C Å by image processing of unstained, unfixed particles embedded in vitreous ice. We have also applied this technique to rotavirus (Fig. 1A) and observed the 60 surface projections not previously observed in negatively stained specimens. The spikes are also evident on the virus when freeze etching and shadowing is used to examine the surface of the particle (Fig. 1B). It seems likely, based on the relative proportions of VP4 and VP7 in double-shelled particles, that VP4 comprises the spikes and VP7 the bulk of the outer capsid. However, an unequivocal assignment of viral proteins to the spike structures requires further investigation.

III. SPECIFIC APPROACHES TO THE STUDY OF ROTAVIRUS GLYCOPROTEINS

We have carried out work aimed at characterizing the features of the two rotavirus glycoproteins which direct them to the ER and cause them to be retained in that organelle. Apart from the

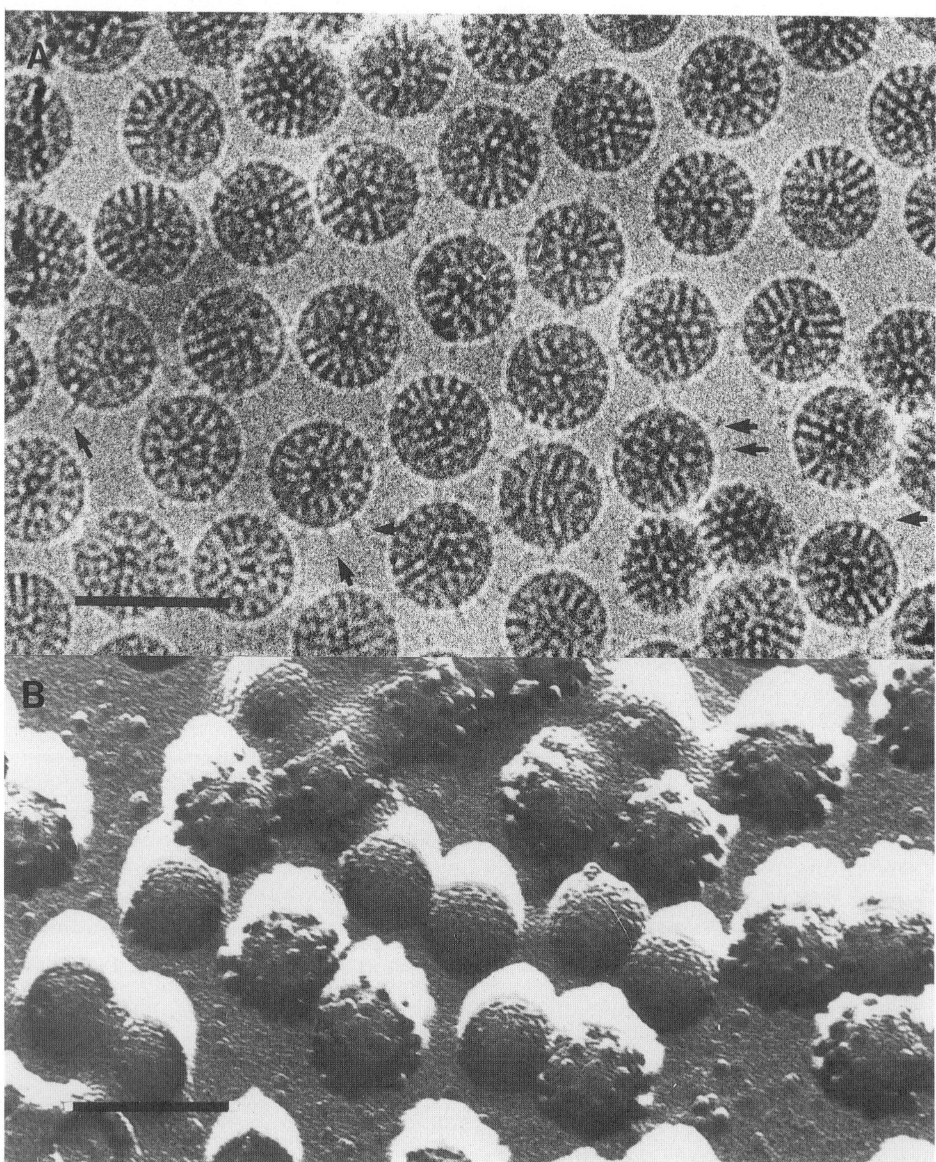

Figure 1. Images of SA11 rotavirus prepared by two different cryoelectron microscopy techniques. The double-shelled virus particles were purified by banding on a CsCl gradient at 1.36 g/ml and dialyzed into a low-ionic-strength buffer. In A, a thin layer of virus suspension on holey carbon grids was quick frozen in liquid ethane. The sample was maintained at −174°C on a Gatan cold stage, and low-dose images, underfocused by 6 μm, were taken with a Philips CM12 at 80 kV by John White and John Berryman of Auckland's Structural Biology Group. B shows a deep-etched rotary shadowed replica of a virus suspension (C. I. Bailey, M.Sc. thesis, 1988). The smooth "golf ball"-like particles are rotavirus cores, while the intact viruses show large projections, or knobs, which correspond to the spikes (arrowed) in the frozen-hydrated image. Scale bar, 100 nm.

intrinsic scientific interest in characterizing these two novel glycoproteins, this work also impacts on our understanding of the antigenic properties of VP7 for the following reasons.

Current knowledge of the events that occur when a specific immune response is mounted against a foreign protein indicates that peptide fragments of the antigen (derived from intracellular processing) are bound to the major histocompatibility complex (MHC), probably at a single site. The recently derived crystalline structure of the human class I MHC molecule localizes the peptide binding site to a cleft on the outer surface of the complex (Bjorkman *et al.*, 1987) and class II complexes are thought to contain a similar region. Thus, for both the "T-cell help" function in B-cell production of antibodies, as well as the cytotoxic T-cell response, the intracellular processing of antigen represents an important step in the mounting of the immune response. For a protein such as VP7, which is sequestered in the ER, engineered modifications that could make the protein more accessible either to the processing pathway or to the other cells involved clearly would be of interest, particularly in view of the success recently achieved in improving on the immunogenicity of the S antigen of the malarial parasite (Langford *et al.*, 1986).

For the nonstructural glycoprotein NS28, which is not a component of the virion, our aim has been to identify the domain of the protein involved in membrane insertion and anchoring, and to identify the region of the molecule that is displayed on the cytoplasmic side of the membrane as the receptor for the core. Although this interaction is intracellular and therefore not accessible to antibody during the course of the viral replication cycle, it is an interaction that might be accessible to interference by antiviral compounds provided that compounds able to cross the membrane could be identified. Recent success in obtaining crystals of the rotavirus core (B. Harris, S. Harrison, I. Anthony, and A. R. Bellamy, unpublished observations, 1988) indicates that the molecular details of the interaction between NS28 and the core will be susceptible to structural analysis: An understanding of the details of this interaction could provide the opportunity for novel antiviral strategies to be developed.

The ability to mutate and express rotavirus glycoprotein genes in several systems has been crucial to our analysis of NS28 and VP7. For example, RNA transcripts produced *in vitro* can be translated in rabbit reticulocyte lysates in the presence or absence of microsomes to investigate whether the protein can be glycosylated, processed, and translocated across the membrane. Transfection of COS cells with SV40 vectors that carry modified forms of the gene(s) permits the cellular location of an expressed protein to be determined. Immunogenicity of viral proteins may be assessed by incorporating the genes into appropriate recombinant vaccinia viruses and determining the antibody response in the whole animal. Alternatively, the genes can be expressed in insect cells using recombinant baculoviruses or in *Escherichia coli* using appropriate expression plasmids. These proteins then may be evaluated for immunologic activity in appropriate animal systems.

IV. MORPHOGENESIS OF ROTAVIRUSES

A partial understanding of the process by which rotaviruses are assembled has been gleaned both from electron micrographs (Estes *et al.*, 1983; Holmes, 1983; Poruchynsky *et al.*, 1985) and from immune fluorescent microscopy (Petrie *et al.*, 1984; Poruchynsky *et al.*, 1985; Kabcenell *et al.*, 1988). Replication occurs entirely in the cytoplasm, in electron-dense bodies known as viroplasmic inclusions. The minor inner viral capsid proteins VP1, 2, and 3 and the major inner component of the core, VP6, probably condense with RNA segments to form corelike structures and these can sometimes be seen in sections taken through the inclusion bodies (Kabcenell *et al.*, 1988). Collections of subviral particles also are observed at the interface between the inclusion body and the membrane of the endoplasmic reticulum (ER). Immature particles reach the interior of the ER by a budding process which results in the cores becoming transiently enveloped in a membrane vesicle, and treatment with tumicamycin causes these enveloped forms to accumulate in the cells (Estes *et al.*, 1983). A mutant SA11 rotavirus, however, has been isolated in which VP7 is not glycosylated. Therefore, the above observation implies that glycosylation of NS28 is required for the membrane to be removed and the budding process itself is not dependent on glycosylation.

Since the mature virus is located in the lumen, all viral structural proteins must at some stage cross the membrane. For VP7 crossing the membrane poses no particular conceptual difficulties because it is directed to the ER and is retained there as a membrane-associated protein oriented to the luminal side (see Section VIII). However, VP4 appears to lack a signal peptide and is nonglycosylated (Both, 1988): It seems reasonable to infer that it must in some way localize to the cytoplasmic face of the ER membrane and be acquired by the virus as the core particle buds through the membrane. There is now good evidence that the nonstructural glycoprotein NS28 plays a crucial role in this budding process by acting as the receptor for the core particle (see Section VI). Whether NS28 is involved in interactions with VP7 and VP4 as they cross the membrane is not known. Nor is it known whether the transient membrane envelope is removed from the virion before or after the outer shell proteins are assembled on the surface.

V. STRUCTURE AND FUNCTION OF VIRAL GLYCOPROTEINS

Over recent years we and others have cloned and sequenced most of the rotavirus gene segments (Both, 1988) and the availability of these clones has opened the way for gene mutagenesis and expression studies. This information, together with biochemical analyses of the corresponding viral proteins, has yielded important clues as to both their location in the cell and their possible function in virus assembly. For example, the nucleotide sequence of gene segment 10, the segment that codes for the nonstructural glycoprotein NS28, showed that the two predicted N-linked glycosylation sites were located at residues 8 and 18 of the molecule, indicating that the signal peptide directing the protein to the ER membrane was not cleaved (Both *et al.*, 1983). Translation of NS28 *in vitro* in the presence of microsomes produced a protein that was glycosylated, yet still sensitive to digestion with proteases, showing that it had not been completely translocated across the ER membrane (Ericson *et al.*, 1983; Chan *et al.*, 1988; Bergmann *et al.*, 1989). Analysis of the attached carbohydrate revealed that the N-linked core glycosylation, Glc3Man9GlcNac2, had been trimmed only to the mannose-9 derivative (Man9GlcNac2) (Both *et al.*, 1983; Kabcenell and Atkinson, 1985). Since further trimming of the carbohydrate occurs in more distal regions of the transport pathway, particularly in the Golgi apparatus of the cell (Farquhar, 1985), the presence of the Man9GlcNac2 form indicated that the protein had not been transported out of the ER but, rather, that it was simply inserted "head-first" through the membrane. Collectively, the data suggested that the exposed cytoplasmic domain of this protein might be involved in interacting with the core during budding, a prediction now confirmed by the recent work of Au *et al.* (1989).

Similarly, pulse labeling of SA11-infected cells revealed that VP7 was glycosylated and possessed a cleavable signal peptide (Ericson *et al.*, 1983). Sequence analysis of the gene showed the location of the single glycosylation site and the probable location of the signal peptide (Both *et al.*, 1983). Analysis of the carbohydrate on VP7 showed that, in the main, trimming of the N-linked core carbohydrate had not occurred beyond the mannose-6 derivative, i.e., Man6GlcNac2 (Kabcenell and Atkinson, 1985), indicating that VP7 also was not transported out of the ER to the Golgi of the cell. Thus, VP7 and NS28 are unlike many other viral glycoproteins in that they are targeted to and retained in the ER, consistent with the evidence from electron microscopy and immune and fluorescent microscopy that this organelle is the site of virus maturation.

VI. ORIENTATION OF NS28 IN THE MEMBRANE AND THE ROLE OF NS28 IN VIRUS BUDDING

Recently Au *et al.* (1989) have demonstrated that the NS28 protein is able to act as a receptor for the rotavirus *in vitro* when membranes from infected cells that contain this protein are mixed with radiolabeled core particles. These workers also demonstrated that membranes from insect cells expressing only the NS28 protein were able to act as receptors for the core, demonstrating conclusively that it is the NS28 gene that confers receptor function.

We have confirmed the receptor activity of membranes that contain NS28 using an *in vitro* assay based on the binding of ^{125}I-labeled core particles. We have also confirmed the identity of the gene product involved using recombinant vaccinia virus to deliver the NS28 gene to MA104 cells. A typical binding assay is shown in Fig. 2A, where the receptor activity present in membranes from SA11-infected cells is compared with that present in uninfected cells. This assay has enabled us to investigate the basic features of the interaction of SA11 cores with the receptor: This work has revealed that the dissociation constant is of the order of 5×10^{-11} M and that binding is stimulated by divalent ions such as Ca^{2+} and Mg^{2+}. Furthermore, the binding of SA11 cores is competitively inhibited by preincubation of the membranes with unlabeled homologous (SA11, subgroup 1) or heterologous (Wa, subgroup 2) cores. The more distantly related reovirus cores do not compete

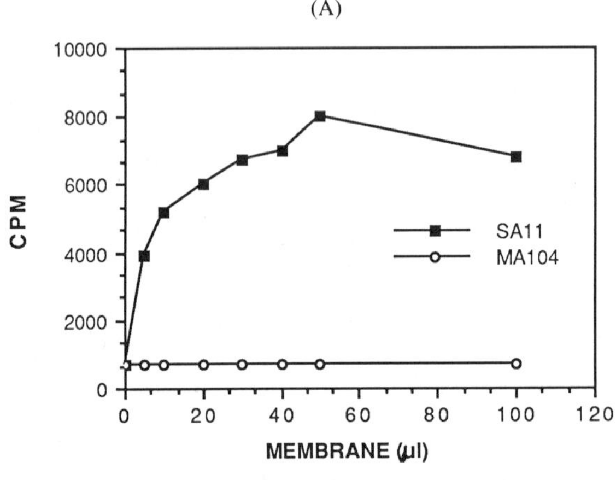

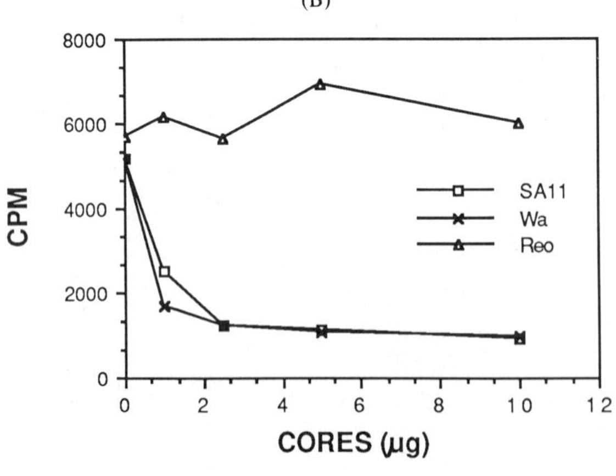

Figure 2. (A) Binding of ^{125}I-labeled rotavirus cores to membranes prepared from rotavirus-infected MA104 cells; 50 ng of ^{125}I-rotavirus cores were incubated with a crude membrane preparation for 90 min at room temperature. Membrane-bound cores were then pelleted by centrifugation and the bound radioactivity measured; 1 μl of the membrane preparation is equivalent to 10^4 cells. (B) Competitive inhibition of binding of ^{125}I-SA11 cores to membranes from SA11-infected cells by nonradioactive SA11, Wa, and reovirus cores. Membranes from SA11-infected cells were preincubated with increasing amounts of nonradioactive SA11, Wa, or reovirus cores for 1 hr, then processed as in A.

(Fig. 2B). Thus, the interaction between receptor and ligand is specific for rotavirus cores and the binding domain is conserved between subgroup 1 and 2 rotaviruses.

Of central importance to our understanding of this novel interaction is the extent of the cytoplasmic domain of NS28 that is involved in receptor : ligand interaction. We have used the approach of gene mutation and expression as a first step toward clarifying the domains of NS28 that are involved in its interaction with the ER membrane. The inferred amino acid sequence of the bovine NS28 gene is presented in Fig. 3. In common with all other rotavirus NS28 proteins, the bovine gene is 175 amino acids in length with three in-frame AUGs and three putative hydrophobic sequences which could be involved in membrane anchoring. The two potential glycosylation sites, which are filled (Ericson *et al.*, 1983), are located in the first of these hydrophobic domains. Since both sites are glycosylated, the NH_2 terminus of the protein must extend into the lumen of the ER to an extent sufficient for the two sites to receive their carbohydrate from the luminally located glycosylation machinery.

To clarify which of the hydrophobic domain(s) of the protein constitutes the transmembrane region, we have constructed a series of site-directed mutants of the protein and then examined the functional abilities of these mutants in an *in vitro* system supplemented with canine pancreatic microsomes. This work has revealed that the second hydrophobic domain (aa30–54) is the membrane-spanning domain and that it acts as a combined signal peptide and anchor sequence (Bergmann *et al.*, 1989).

The availability of the deletion mutants used in this work has also enabled us to determine the point at which the polypeptide chain emerges from the membrane. Figure 4A shows a typical result where a range of proteolytic enzymes was used to investigate the susceptibility of the cytoplasmic domain of NS28 to proteolytic cleavage. These types of experiments have enabled us to localize the point at which the polypeptide emerges from the membrane with some precision. For example, the protease-resistant product of 12.5 kDa, which is derived by cleavage of the protein in the region between aa44 and aa48 (Fig. 4A, upper panel), is absent from a mutant (δ40–48) that lacks these proteolytic sites (Fig. 4A, lower panel). These and other data (Bergmann *et al.*, 1989) suggest that the polypeptide chain emerges from the membrane at aa44 and leads to the conclusion that approximately 131 residues of the protein are potentially available in the cytoplasm for receptor : ligand interaction.

This conclusion varies somewhat from that reached by Chan *et al.* (1988), who identified the third (aa70–84) rather than the second hydrophobic (H2) region (aa29–46) as the transmembrane anchor. We are unable to account for these differences, but they may relate to the protection of certain protease sites near residue 84, which Chan *et al.* attributed to the transmembrane status of this region but which we consider to be due to factors not directly related to the integral membrane status of this protein. A model for the orientation of NS28 in the membrane (Fig. 4B) summarizes our current view of the manner in which the receptor protein is displayed in the endoplasmic reticulum.

```
                                          *********************                           ***********
NCDV     MEKLTDLNYT SSVITLMNST LHTILEDPGM AYFPYIASVL TVLFTLHKAS IPTMKIALKT SKCSYKVVKY CIVTIFNTLL KLAGYKEQIT
UK       .......... .......... .......... .......... .......... .......... .......... .......... ..........
WA       .D..A..... .....S..D. ..S.IQ.... ...L...... .......... .......... .......I.. .....I.... ........V.
SA11     .......... .......... .......... .......... ....A.N... .......... .......... .......... ..........
                 10         20         30         40         50         60         70         80         90

NCDV     TKDEIEKQMD RVVKEMRRQL EMIDKLTTRE IEQVELLKRI HDKLMIRAVD EIDMTKEINQ KNVRTLEEWE NGKNPYEPKE VTAAM
UK       .......... .......... .......... .......... Y......STG .......... .......... .......... .....
WA       ......Q... .I........ .......... .......... ..N.IT.PV. V...S..F.. ..IK..D... ........S. ...S.
SA11     .......... .......... .......... .......... Y...T.QTTG .......... .......... ........R. .....
                100        110        120        130        140        150        160        170        180
```

Figure 3. Comparison of NS28 amino acid sequences from four rotavirus isolates. NCDV is used as the reference. ***, Hydrophobic regions; ______, glycoprotein sites. Other sequences are taken from Both (1988).

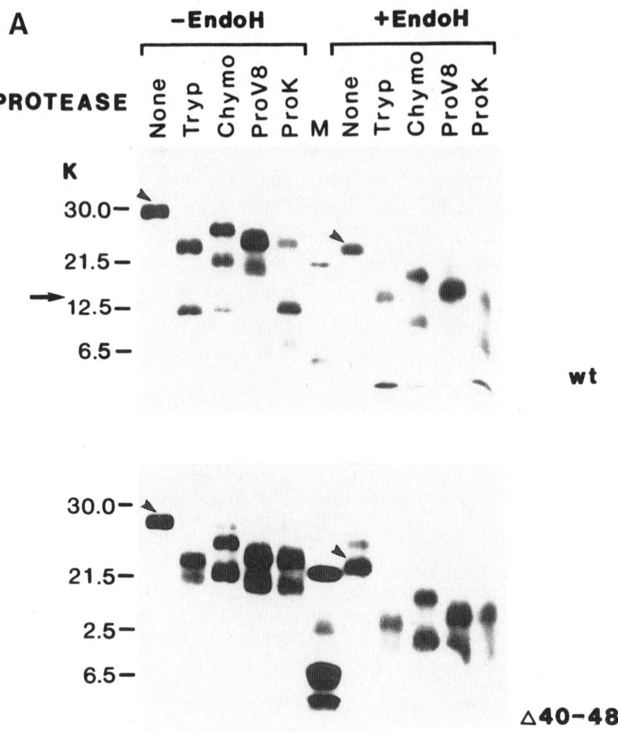
A
−EndoH
+EndoH
PROTEASE
None
Tryp
Chymo
ProV8
ProK
M
None
Tryp
Chymo
ProV8
ProK
K
30.0
21.5
12.5
6.5
wt
30.0
21.5
2.5
6.5
△40-48

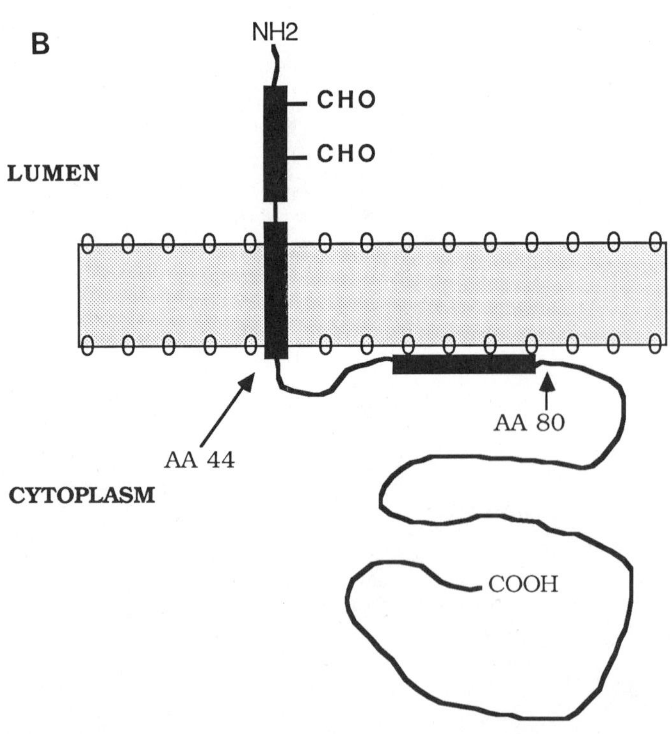
B
NH2
CHO
CHO
LUMEN
AA 44
AA 80
CYTOPLASM
COOH

```
                          H1                                   H2
                ******************               ********************↓            ▼
CONSENSUS  MYGIEYTTIL .FLISIILLN YILKSITRIM DYIIYRFLLI IVVLSP..RA QNYGINLPIT GSMD.AYANS TQEE.FLTST LCLYYPTEAS
STTHOMAS   ........V. FY...FV.VS ....T.IK.. ......ITFV ..V..VLSN. .......... ....T..... ..DNN..F.. ......S..P
WA         .......... I......... .....V.... .......... T.A.FALT.. ....L..... ....AV.T.. ....V..... ..........
S2         .......... TI........ ....T..NT. ....F....L .ALI..FV.T ....MY.... ..L.AV.T.. .SG.S..... ......A..K
SA11       ........V. T......... .....L.... .F......F. ......FL.. .......... ....I..... ....T..... .........A
NCDV       .......... I..T..T... .M........ .......... V...ATIIN. ....V..... ....T...D. ..S.P..... ......V...
                   10         20         30         40         50         60         70         80         90

CONSENSUS  TEI.DTEWKD TLSQLFLTKG WPTGSVYFKE YSNIA.FSVD PQLYCDYNVV LMKYD.TLEL DMSELADLIL NEWLCNPMDI TLYYYQQT.E
STTHOMAS   .Q.S...... .......... ........N. ...VLE..I. .K........ .IRFVFGE.. .I........ .......... ........G.
WA         .Q.N.GD... S...M..... .......... ....VD.... ........L. .....QS... .......... .........V .......SG.
S2         N..S.D..EN .......... .........D .ND.TT..MN .......... ..R..N.S.. .V........ .......... S......NS.
SA11       ...N.NS... .......... .......... .T...S.... .......... .....A..Q. .......... .......... ........D.
NCDV       N..A...... .......... .......L.. .AD..A...E ........L. .....S.Q.. .......... .......... ........D.
                  100        110        120        130        140        150        160        170        180

CONSENSUS  ANKWISMGSS CTVKVCPLNT QTLGIGC.TT NVDTFETVAT SEKLVITDVV DGVNHKLN.T T.TCTIRNCK KLGPRENVAV IQVGGSNILD
STTHOMAS   .......... .......... .......Q.. .TA......D ....A.I... .S......I. ST.......N .........I ..........
WA         S......... .......... .......Q.. ...S..MI.E N...A.V... ..I...I.L. .T........ .......... .......V..
S2         S.......TD .......... .......K.. D.....I..S .......... N.....I.IS IS.......N .........I .....P.A..
SA11       .......... ..I....... .......L.. DAT...E... A......... .......DV. .A........ .......... .......D...
NCDV       ......T... .......... .......LI. .P........ T......... ........V. .A........ .......... .....A....
                  190        200        210        220        230        240        250        260        270

CONSENSUS  ITADPTT.PQ TERMMRVNWK KWWQVFYTVV DYINQIIQVM SKRSRSLNSA AFYYRV
STTHOMAS   .......S.. .......... .......... ......V... .......D.S S.....
WA         .......N.. .......... .........I. ......V... .......... ......
S2         .......V.. VQ.I...... .......... .......... .......DT. .....I
SA11       .......A.. ......I... .......... ..VD...... .......... ......
NCDV       .......T.. ......I... .......... ..V.....T. ......F..S ......
                  280        290        300        310        320
```

Figure 5. Amino acid sequences of serotypes I (Wa), II (S2), III (SA11), IV (St. Thomas), and VI (NCDV). The St. Thomas sequence was determined by Reddy *et al.* (1988). Other sequences are from Both (1988). The location of the H1 and H2 hydrophobic domains, the single glycosylation site (▼), and the cleavage site (↓) in SA11 are shown.

VII. CONSERVATION AND VARIATION IN VP7 SEQUENCES

Comparison of VP7 proteins from different serotypes reveals that some portions of the molecule vary whereas some are highly conserved (Fig. 5). Considerable variation between serotypes 1, 2, 3, 4, and 6 is evident through amino acids 90–100, 143–152, and 208–221, in particular (Fig. 5) (Gunn *et al.*, 1985), and antigenic variants selected with neutralizing monoclonal antibodies against VP7 map to these regions (Dyall-Smith *et al.*, 1986; Taniguchi *et al.*, 1988). In fact, it is now clear that the major antigenic site of VP7 is conformational (Dyall-Smith *et al.*, 1986) and that a detailed understanding of the major epitopes will need to await information on the three-dimensional structure of the protein on the surface of the virion.

Other features of the neutralizing gene and protein are of interest with respect to targeting of the protein to the ER. For example, the nucleotide sequence of SA11 VP7 reveals two potential initiation codons preceding two regions of hydrophobic amino acids (H1 and H2) at the N-terminus of the protein (Fig. 5) (Both *et al.*, 1983; Both, 1988). While numerous amino acid changes have

←

Figure 4. (A) Proteolytic cleavage sites upstream of residues 70–84 are accessible. Proteolysis of membrane-associated wt NS28 (upper panel) and a deletion mutant Δ40–48 (lower panel) was carried out at 30°C using trypsin, α-chymotrypsin, protease K, and protease V8. After addition of protease inhibitors, membrane-associated fragments were sedimented through a high salt sucrose cushion. Half of the membrane fraction was treated with endo H (reducing the apparent molecular weight of fragments with carbohydrate attached by approximately 8 K) and samples were separated by 14% SDS-PAGE. Lanes 1 and 7 (numbered from the left) show intact forms of the proteins (indicated by arrowheads) before and after endo H treatment, respectively. The proteolytic fragments obtained from the indicated proteases are shown in lanes 2–5 (− endo H) and 6–11 (+ endo H). Lane 7 contains molecular weight markers. (B) A model for the topology of NS28 in the ER membrane. The blocks represent the putative hydrophobic domains, (CHO)-carbohydrate sites at residues 8 and 18.

been observed in the various VP7 sequences that have been determined (Fig. 5), the hydrophobic character of these regions is conserved and the initiation codons are inviolate, suggesting that there are intrinsic features of the amino-terminal region of the protein that are biologically important. However, it should be noted that the first AUG codon occurs in a weak context for initiation within a sequence highly conserved at the genetic level (Both, 1988), possibly reflecting some structural requirement in the RNA sequence, rather than in the protein sequence. Nevertheless, both hydrophobic regions have the characteristics of signal peptides which could direct VP7 into the ER. Therefore, the specific questions concerning VP7 processing, targeting, and transport which we sought to answer in our work were (1) where is the signal peptide in VP7 and (2) at which site is it cleaved? (3) What is the signal in VP7 that targets the protein to the ER and (4) how is targeting achieved?

VIII. LOCATION OF THE VP7 SIGNAL PEPTIDE(S) AND CLEAVAGE SITE

To determine the location of the signal peptide in VP7, we constructed mutants in which one or the other, or both, of the H1 and H2 hydrophobic domains were retained in the protein. Signal peptide function was assayed indirectly by determining whether the mutant protein expressed in COS cells could be glycosylated. The data indicated that signal peptide function was present upstream of methionine 63 of the open reading frame and that either of the two hydrophobic domains was capable of directing VP7 into the ER (Whitfeld *et al.*, 1987). This observation suggested that the second initiation codon and hydrophobic domain were sufficient to produce authentic VP7. However, it is still not clear whether the first initiation codon is used *in vivo*, or whether the upstream hydrophobic domain plays any role in the production of VP7.

The blocked N-terminus on viral VP7 prevented the use of classical protein sequencing techniques to determine the site of signal peptide cleavage (Arias *et al.*, 1984). Therefore, gene mutagenesis was used to determine the location of the cleavage site. Standard techniques were used to construct mutants for which initiation of translation began specifically at Met1, Met30, or Met63 of the open reading frame. A further mutation (Asn to Asp) was also introduced at Asn69 to eliminate the single glycosylation site in SA11 VP7. This enabled processing of these proteins to be examined *in vitro* in the absence of compensating mobility changes due to glycosylation. The protein translated from Met63 was not processed by microsomes; i.e., it lacked signal peptide function. However, the proteins translated in rabbit reticulocyte lysates from either of the first two initiation codons were processed in the presence of microsomes to yield products of the same apparent size, indicating that they were probably cleaved at a common site.

Empirical rules formulated by von Heijne (1986) suggested that the most probable cleavage site lay between Ala50 and Gln51 and that cleavage at this site could be prevented by mutating Ala50 to Val. This mutation was introduced into proteins initiated at either AUG and its effect on cleavage by microsomes was determined. In one case, cleavage was not prevented, but nevertheless occurred two residues downstream at an unpredicted site. In the second case, the mutation completely inhibited processing, providing indirect evidence that the hypothetical cleavage site was probably used *in vivo*. Cleavage after Ala50 would yield mature VP7 with an N-terminal glutamine residue and this suggested that the viral protein could be blocked by the formation of pyroglutamic acid. Indeed, when we attempted to obtain partial N-terminal sequence data for radiolabeled VP7 synthesized *in vitro* or purified directly from virus, we were unsuccessful unless the proteins were first digested with pyroglutamate aminopeptidase, an enzyme that removed the putative blocked N-terminal glutamine (Stirzaker *et al.*, 1987). Following such treatment, radioactivity was released from ^{35}S-methionine labeled proteins at cycle 12 of Edman degradation, the expected result if glutamine51 was the original N-terminal residue (Stirzaker *et al.*, 1987). Comparison of VP7 sequences from other serotypes using von Heijne's (1986) rules suggests that a similar cleavage site is conserved in each VP7 protein (Fig. 5).

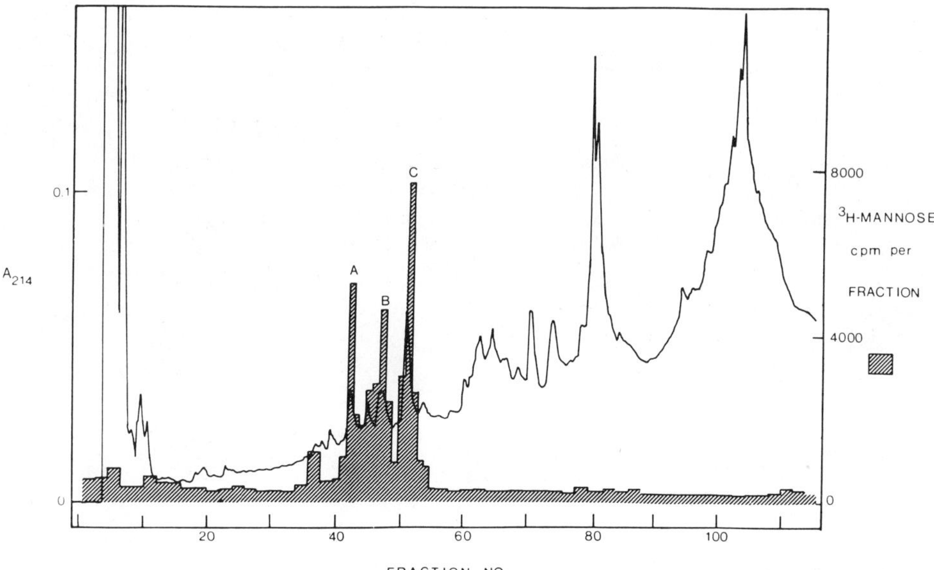

Figure 6. Resolution of VP7 V8 protease peptides by high-pressure liquid chromatography (HPLC). ^{3}H-mannose-labeled VP7 was purified by gel electrophoresis, digested with V8 protease, and loaded onto an HPLC column. Aliquots of each fraction were counted to determine the location of ^{3}H-mannose-carrying peptides. Solid line, A214 nm; hatched areas, ^{3}H cpm. The peptides were prepared and analyzed by D. Christie.

Table 2. Amino Acid Compositions of ^{3}H-Mannose-Containing Peptides Produced by Digestion of VP7 with S. aureus, V8 Protease

Amino acid	Amino acid compositions[b] (residues/mol. peptide)			
	Peak A(#43)	Peak B(#48)	Peak C(#52)	Expected
(D) asp[a]	3.0	2.6	4.1	4
(E) glu[a]	3.0	2.8	4.3	4
(S) ser	1.5	1.4	2.0	2
(G) gly	1.5	2.7	2.9	2
(H) his	—	—	—	—
(R) arg	—	0.5	0.3	—
(T) thr	2.4	2.1	2.0	3
(A) ala	2.0	2.0	2.0	2
(P) pro	0.9	1.3	1.5	1
(Y) tyr	0.9	1.3	1.9	2
(V) val	—	—	—	—
(M) met	—	—	0.3	1
(I) ile	1.8	1.5	2.8	2
(L) leu	1.7	0.7	0.9	1
(F) phe	—	—	—	—
(K) lys	—	—	—	—

[a]Includes Asn(N) and Gln(Q).
[b]All compositions are based on 24-hr hydrolysis values.

To further confirm that cleavage occurred between Ala50 and Gln51, the amino-terminal glycopeptide of VP7 that had been isolated from purified virus was released by digestion with *Staphylococcus aureus* V8 protease and isolated by HPLC. Digestion of viral VP7 with this protease was expected to produce an NH_2-terminal peptide corresponding to residues Gln51–Glu74 (Fig. 5), and to simplify its identification, VP7 was biosynthetically labeled with ^{3}H-mannose. Three peaks (designated A, B, and C) were obtained by HPLC (Fig. 6) and subjected to amino acid analysis. The composition of peak C most closely agreed with that expected for the amino-terminal peptide (Table 2). The elution pattern and amino acid compositions of peaks A and B suggested that these were shorter, related peptides, although their origin is unclear. Collectively, the data clearly indicate that the mature form of VP7 contains neither of the two hydrophobic N-terminal regions of VP7; these clearly were removed from the protein during translocation across the membrane, implying that sequences other than these must be involved in the retention of VP7 in the ER.

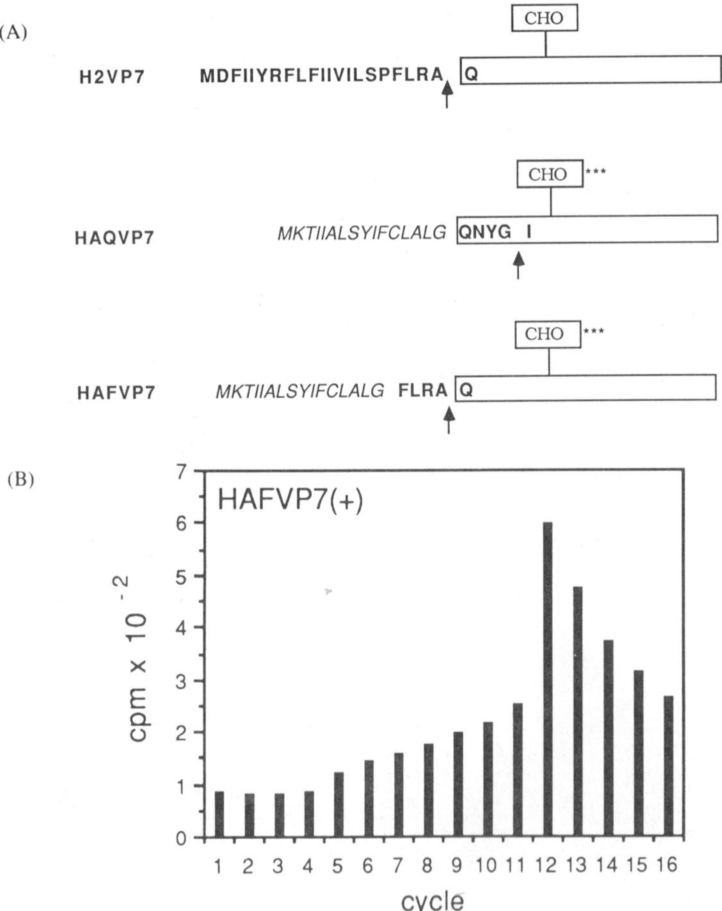

Figure 7. (A) Construction of VP7 genes with altered signal peptides. Bold type and lines indicate VP7 sequences; other sequences are derived from influenza hemagglutinin. CHO, Endo H-sensitive carbohydrate; CHO***, endo H-resistant carbohydrate. Arrows indicate cleavage sites. (B) Partial N-terminal sequence of ^{35}S-methionine-labeled HAFVP7 after treatment of the protein with pyroglutamate aminopeptidase (+). Radioactivity released after each cycle of Edman degradation was determined by liquid scintillation counting.

IX. ROLE OF THE H2 SIGNAL PEPTIDE IN VP7 TARGETING

At this point we reviewed some earlier data which showed that secreted variants of VP7 could be produced by making short, internal deletions in the protein at the C-terminal end of the H2 region (Poruchynsky *et al.*, 1985). Specifically, deletion of residues 51–61 had little effect on the ability of VP7 to remain in the ER. However, deletion of a further four residues in mutant 47–61 caused rapid secretion of the protein. From the above data it is now clear that the difference between the two proteins was that the former retained an intact signal peptide while the latter, secreted variant did not. This raised the intriguing possibility that the H2 signal peptide was not only a "passport" for entry of VP7 into the ER, but that it also participated in an active way in targeting the polypeptide to the correct ER location.

To examine the possible role of the H2 signal peptide we constructed chimeric VP7 molecules in which the rotavirus protein was fused to the leader sequence of the influenza virus hemagglutinin (HA). Influenza HA is normally transported to the cell surface. When the signal peptide of HA was spliced to Gln51 of VP7 to create HAQVP7 (Fig. 7A) [conserving the natural cleavage site of the HA signal peptide (Ward and Dopheide, 1980)], the protein was unexpectedly cleaved four residues downstream, making it difficult to interpret the effect of the changed signal peptide. The HA sequences were therefore fused to Phe47 of VP7 to create HAFVP7 (Fig. 7A). This protein was

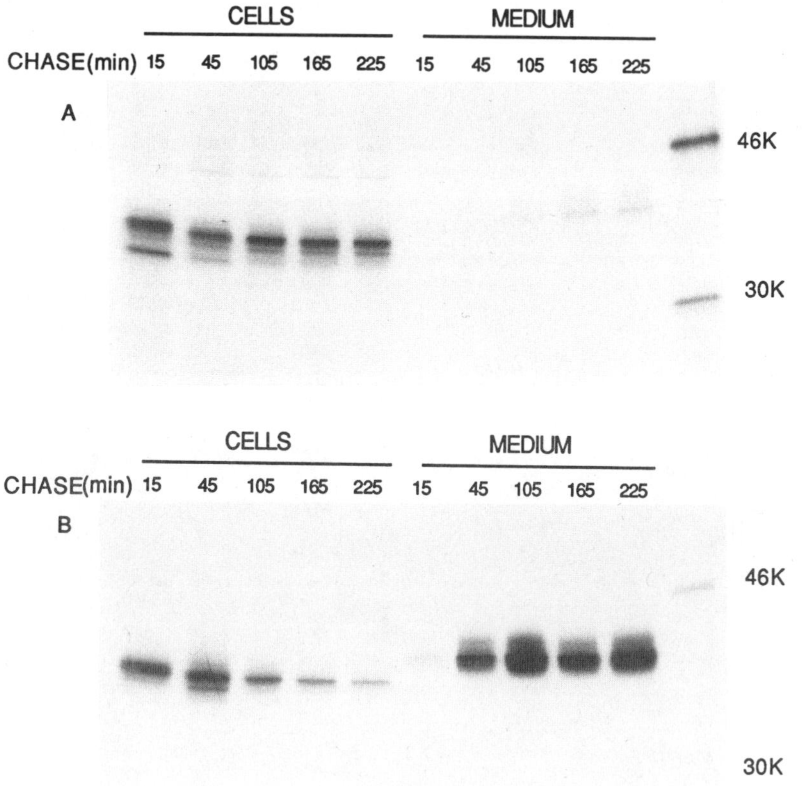

Figure 8. Cellular location and transport of VP7 produced from precursor H2VP7 (A) or HAFVP7 (B). Transfected COS cells were pulse-labeled for 15 min with ^{35}S-methionine and chased for the times indicated. VP7 was recovered by immunoprecipitation.

correctly cleaved by microsomes *in vitro* (Fig. 7B). Thus, VP7 with the same, correct N-terminus was generated from two precursors which differed only in their signal peptides.

The effect of the changed signal peptide on VP7 transport was then examined by expressing the hybrid protein using appropriate SV40 vectors in COS cells that were pulse-labeled. Whereas wild-type protein remained in the cells during a long chase, the same protein produced from the hybrid precursor was rapidly secreted from the cells (Fig. 8). Clearly, the H2 signal peptide contained information that was fundamentally important to the final destination of the protein. To our knowledge, such a function has not previously been reported for a signal peptide. However, we have now determined that not all the targeting information is present in the H2 signal peptide. When the H2 signal peptide was spliced to a secreted S antigen of malaria in place of its own signal peptide, the S protein was still secreted (Fig. 9); i.e., the H2 signal peptide is necessary, but not sufficient for localization in the ER.

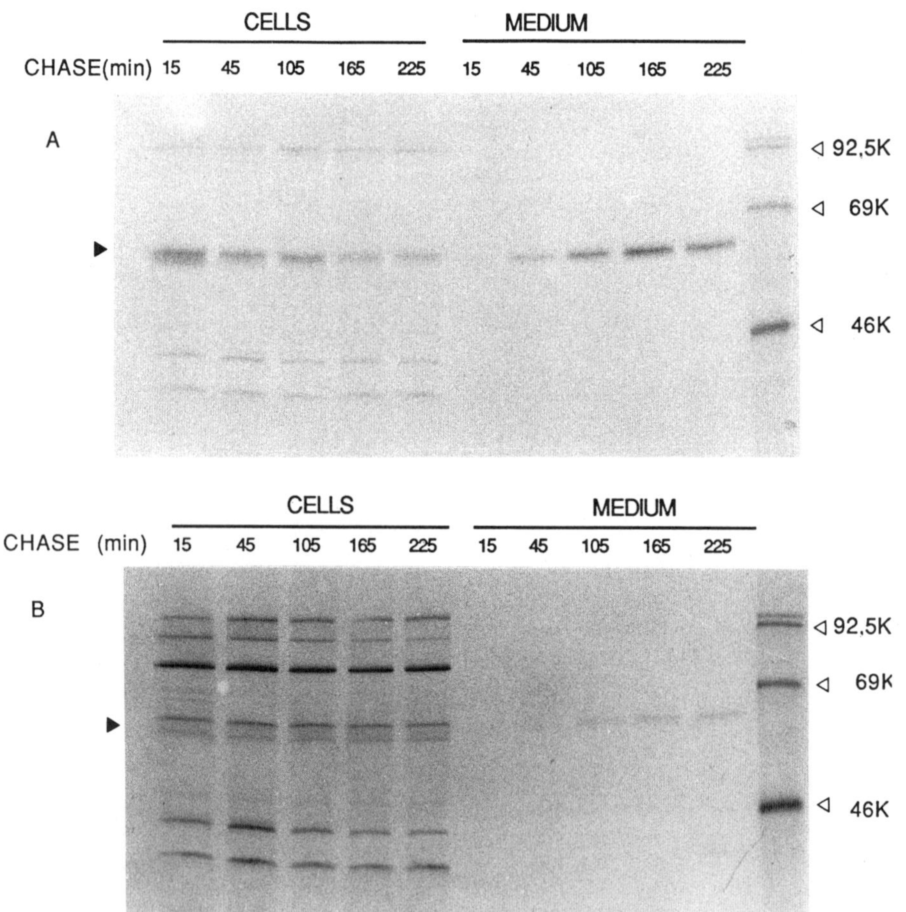

Figure 9. The rotavirus H2 signal peptide alone does not retain proteins in the ER. COS ceils transfected with (A) S- or (B) H2 S-antigen genes were pulse-labeled with ^{35}[S]methionine for 15 min, then chased for the times indicated. Cells and medium were harvested and the proteins recovered by immunoprecipitation using antiserum against S-antigen.

X. RECOMBINANT VP7 EXPRESSED AS A VACCINE ANTIGEN

Rotaviruses are icosahedral particles made up of an inner, corelike structure and an outer capsid layer (Holmes, 1983). Interactions between VP6, the major protein of the core, and VP4 and VP7, the two polypeptides that make up the outer capsid, stabilize the double-shelled particle (Prasad *et al.*, 1988). Therefore, it is likely that the shape of these proteins could be influenced by their interactions with neighboring subunits. A precedent for such interactions was provided by the proteins comprising the picornavirus capsid, where the antigenic shape of VP1 in the whole virus differed greatly from that of the isolated protein VP1 (Bachrach *et al.*, 1975). In contrast with this situation, it has been shown that three subunits of the influenza HA interact with *each other* to form a trimeric spike on the surface of the virus (Wilson *et al.*, 1981). The HA expressed from a recombinant gene is also able to trimerize (Gething *et al.*, 1986) to form a fully functional spike embedded in the cell surface membrane (Gething and Sambrook, 1981). Therefore, we asked whether the shape of VP7 expressed from a recombinant vector was antigenically similar to the shape of the protein in double-shelled virus particles.

This was examined using three different expression systems. Parts of VP7 were expressed in *E. coli* using appropriate plasmid vectors. Second, the gene was inserted into baculoviruses and expressed at high levels in insect cells. Third, recombinant vaccinia viruses were constructed so that the immune response to the antigen in the whole animal could be assessed.

A. Expression in E. coli

Expression of rotavirus proteins in *E. coli* or some other prokaryotic organism that can colonize the Peyer's patch cells of the gut offers the potential to prime an effective antirotavirus immune response. However, this depends on whether some or all of VP7 can be expressed satisfactorily in a form that stimulates the formation of neutralizing antibodies or induces a protective immune response.

Several different gene constructions were tried, but only one is discussed here in any detail because the results were similar in each case. A VP7 fragment N-X that extended from an NcoI site at Met63 to a XhoI site beyond the stop codon was incorporated in-frame into the expression plasmid pEVvrf downstream of the powerful bacteriophage pL promoter (Crowl *et al.*, 1985). Expression was achieved in the *E. coli* when the temperature-sensitive pL repressor was inactivated by raising the temperature of the culture to 42°C.

Unfortunately, in common with many other eukaryotic proteins expressed in *E. coli*, expression of N-X VP7 yielded insoluble aggregates of viral protein, which we were unable to maintain in a soluble form, despite the use of powerful protein denaturants such as guanidine hydrochloride and urea. The protein aggregates were, however, relatively easily purified to about 90% homogeneity to yield two bands upon SDS polyacrylamide gel electrophoresis, which we attributed to the presence of differing conformational variants of the same protein. The two species of VP7 were interconvertible, depending on the particular method adopted for denaturing the protein prior to analysis on the gel.

The partially purified recombinant protein was used to prime rabbits, the animals having been screened by ELISA to exclude those already reactive to rotaviruses. This latter precaution was essential since rotaviruses are ubiquitous among laboratory animals and preexposure to the virus invalidates the assessment of the immunogenicity of a protein preparation. All preparations of recombinant protein expressed in *E. coli* induced antibody that reacted strongly by Western analysis, yielding a single band of 37 kDa when reacted with SA11 viral proteins. No reactivity was observed for an antiserum derived from a mock-inoculated, sentinel rabbit, indicating that it was unlikely that the antibody observed originated from a rotavirus infection among the animals during the priming and boosting period.

Table 3. Serotype Specificity of Antibody Induced in Rabbits by VP7 Protein (Serotype III) Produced in E. coli or in Baculovirus-Infected Insect Cells[a]

Immunization	Rabbit	A_{490} (sample)-A_{490} (blank)			
		Serotype I	Serotype II	Serotype III	Serotype IV
E. coli/N-X VP7	*A26*	*0.781*	*0.538*	*1.10*	—
	D02	*1.016*	*0.429*	*1.453*	—
SDS-VP7	*A11*	*1.170*	*0.433*	*1.438*	—
E. coli/N-X VP7	A26	0.080	0.012	0.089	0.069
	D02	0.116	0.003	0.266	0.046
SDS-VP7	A11	0.178	0.006	0.284	0.160
Baculovirus-SA11 VP7	A_3 (prebleed)	0.054	0.054	0.032	0.053
	(post 2nd boost)	0.033	0.012	0.361	0.026
	(post 3rd boost)	0.017	0.012	0.424	0.021
VV-SA11 VP7	R5 (post 2nd boost)	0.009	0.005	0.389	0.006
Monoclonals	2C9 (serotype I)	0.565	0.017	0.016	0.011
	2F1 (serotype II)	0.012	0.387	0.145	0.160
	RV3:1 (serotype III)	0.000	0.008	0.197	0.011
Polyclonal (serotype III)	H26	0.725	0.736	1.029	0.767

[a]Roman text—assay performed with intact virus that was captured with goat-antirotavirus serum. Italic text (top three lines)—target virus was bound directly to the dish.

The antisera raised against N-X VP7 were assessed for their ability to recognize VP7 in whole virus of each serotype both by a capture ELISA and by binding virus directly to the ELISA plate. If virus was first captured by goat-antirotavirus serum, the N-X anti-VP7 sera recognized the virus poorly and with little specificity, a result similar to that obtained with an antiserum prepared against SDS-denatured VP7 purified from a gel (Table 3). In contrast, target virus bound directly to the ELISA plate was recognized with far greater efficiency by these same antisera, but with even less specificity (Table 3, data in italics). This observation suggested that the majority of antibodies induced by the recombinant protein were targeted at linear epitopes that became exposed only when the virus was partially denatured during attachment to the ELISA plate.

The absence of neutralizing activity in these antisera was consistent with these data. Similarly, when the recombinant protein was used to prime mice, no animals developed titers higher than 2000 when these were determined using "captured" virus as target. When protection was measured in the mouse EDIM model (Sheridan *et al.*, 1984), no evidence of protection was obtained. Thus, the antibody induced by the recombinant protein, although clearly specific for VP7, in our hands did not recognize virus particularly well and exhibited little or no neutralizing activity.

Presumably the failure to induce antibody that correctly recognized the appropriate (conformational?) epitopes reflects the unfolded and aggregated state of the protein produced in the prokaryotic system. Our experience with VP7 expressed in *E. coli* therefore differs from that of others who have reported the priming of a specific neutralizing response. However, as Matsui *et al.* (1989) have pointed out, it is often difficult to determine whether the expressed protein is eliciting an anamnestic or a primary immune response.

B. Expression in Baculovirus-Infected Cells

Recently a number of eukaryotic proteins have been expressed in *Spodoptera frugiperda* cells under the control of the powerful baculovirus polyhedrin promoter (Summers and Smith, 1987;

Matsuura *et al.*, 1987: Kuroda *et al.*, 1986) and rotavirus VP6 protein expressed in this system was able to fold correctly and assemble into tubes that are indistinguishable from those found in clinical samples (Estes *et al.*, 1987). In view of the difficulty encountered in expressing correctly folded VP7 in *E. coli,* this gene was also expressed using two different baculovirus vectors that were found by others to yield large amounts of foreign protein in insect cells (Summers and Smith, 1987; Matsuura *et al.*, 1987). The complete SA11 gene VP7 that retained both hydrophobic leader sequences was expressed after appropriate modification. Both recombinant baculoviruses induced the synthesis of VP7 in insect cells but the yield in both cases was low, despite the use of the improved vector pAcRP23, which retains the A of the initiating ATG of the polyhedrin gene, which for other genes is claimed to enhance levels of expression (Matsuura *et al.*, 1987).

Each product reacted as a single band by Western analysis and comigrated with viral VP7, suggesting it had been correctly processed in the insect cells. Therefore, this material was used to prime rabbits and the specificity of the antisera was investigated by ELISA using the capture assay outlined in Section X.A. Antisera prepared in rabbits vaccinated with a vaccinia recombinant (see Section X.C) were included for comparison. While the titer induced by the baculovirus-derived VP7 was low (Table 3), the antibody response showed serotype specificity similar to that achieved when VP7 was expressed by a recombinant vaccinia virus (Andrew *et al.*, 1987). Thus, the preliminary data indicated that expression of VP7 in *Spodoptera* cells yielded a product that was able to induce serotype-specific antibody, but yields of the protein were disappointing. Further work is required to clarify the nature of the product obtained and particularly to clarify the nature of the carbohydrate that the insect cells add to the molecule.

C. Expression Using Recombinant Vaccinia Virus

A recombinant vaccinia virus carrying the wild-type VP7 gene was constructed and used to infect cells in tissue culture to confirm that the protein located correctly in the ER. This virus was then used to vaccinate prescreened rabbits. The serum from these animals was assayed for the presence of antibodies that recognized VP7 in whole virus in solution using a capture ELISA and a plaque neutralization reduction assay. Preimmune sera showed no significant binding activity by either test and failed to precipitate radiolabeled rotavirus proteins from SA11-infected cells. In contrast, both assays indicated that significant levels of antibodies that recognized whole virus were induced in rabbits when VP7 was expressed during the vaccinia virus infection (Andrew *et al.*, 1987). When the ability of these rabbit antisera to react with *denatured* VP7 antigens was examined by Western transfer, reactivity with serotypes 1, 3, and 4 was obtained (Andrew *et al.*, 1987). However, when the serotype specificity of these antibodies was examined using *native* virus, it was found that they reacted well with the homologous serotype 3 virus but poorly, if at all, with viruses from serotypes 1, 2, or 4. Together these results suggested that some epitopes that were not exposed on whole virus were nevertheless accessible on VP7 located in the ER membrane. Perhaps, more importantly, the epitopes defining serotype specificity were presented in a native configuration during the vaccinia infection.

XI. RELOCATION OF VP7 TO THE CELL SURFACE

Although the results concerning the antigenicity of VP7 expressed by recombinant baculo and vaccinia viruses were encouraging, we reasoned that it might be possible to improve the immune response to the antigen if it could be expressed at the surface of the cell, rather than as an internal protein. This was based on ideas prevailing at the time which indicated that the cellular immune system recognized antigen presented at the cell surface in conjunction with histocompatibility antigens. It now seems probable that helper T cells do not see the whole protein, but rather a peptide(s) derived by processing (Townsend *et al.*, 1986; Maryanski *et al.*, 1986), which is presented in the cleft of a class II histocompatibility antigen (Bjorkman *et al.*, 1987). In the light of this information it perhaps is now an open question as to whether cell-surface expression would be expected to enhance the antigenicity of VP7. However, the site of antigen processing in the cell is presently unknown and changing the location of VP7 in the cell could conceivably alter its ability to be processed. Alternatively, antibody-producing B cells may more efficiently recognize a greater concentration of antigen presented in a "localized" area on a cell surface membrane, compared with antigen released by cell lysis. Indeed, while our work was in progress, encouraging results were obtained by Langford *et al.* (1986), who added a C-terminal transmembrane anchor domain to a secreted antigen of *Plasmodium* and obtained a significant improvement in its antigenicity.

Several different gene constructions were made in our attempts to relocate VP7 to the cell surface. The location of VP7 in the cell was ascertained by the endo H-sensitivity of carbohydrate attached to the protein, since VP7 possesses endo H-sensitive carbohydrate while secreted VP7 is modified *en route* with complex, endo H-resistant carbohydrate (Poruchynsky *et al.*, 1985). We first fused the signal peptide of influenza HA to methionine 63 of VP7 in place of the normal VP7 signal peptides (Fig. 10). This protein was secreted into the medium. Next the C-terminal transmembrane anchor and cytoplasmic domain of HA (amino acids 186–221 of HA2 of strain A/NT/60/68/29C) were also spliced to VP7 at the penultimate amino acid residue (HAVP7A, Fig. 10). However, to our surprise, this molecule failed to leave the ER. In alternative constructions, the N-terminal signal peptide/transmembrane anchor domain of influenza neuraminidase (residues 1–36 of strain A/PR/8/34) or that region plus the residues comprising the neuraminidase stalk (residues 1–77)

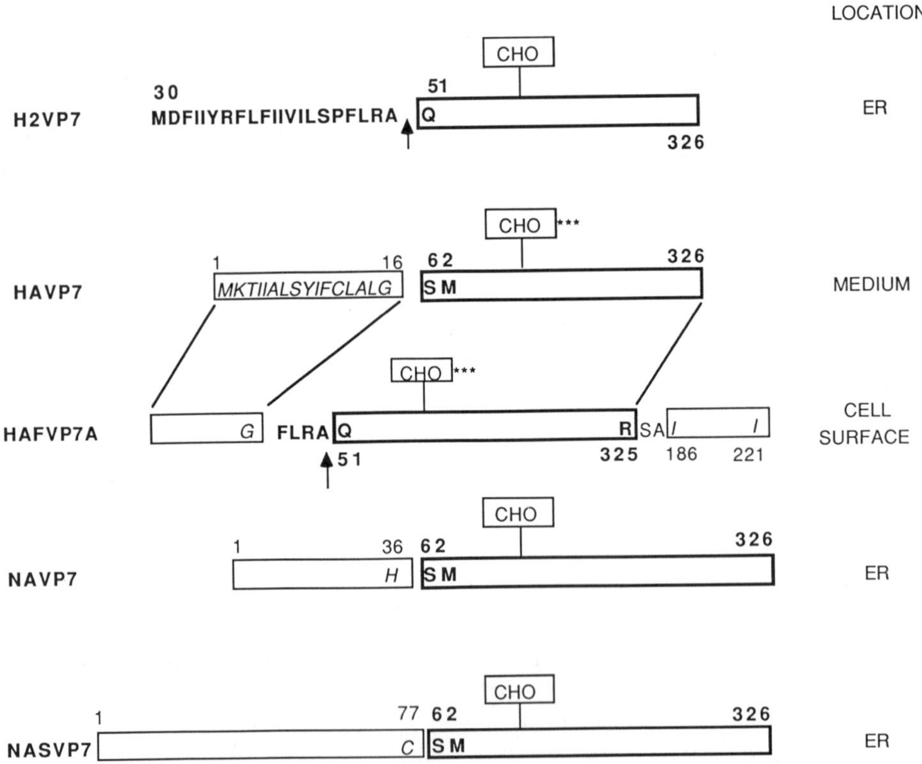

Figure 10. Modification of VP7 with signal peptide and anchor domains from influenza hemagglutinin or neuraminidase. Annotations are described in Fig. 7.

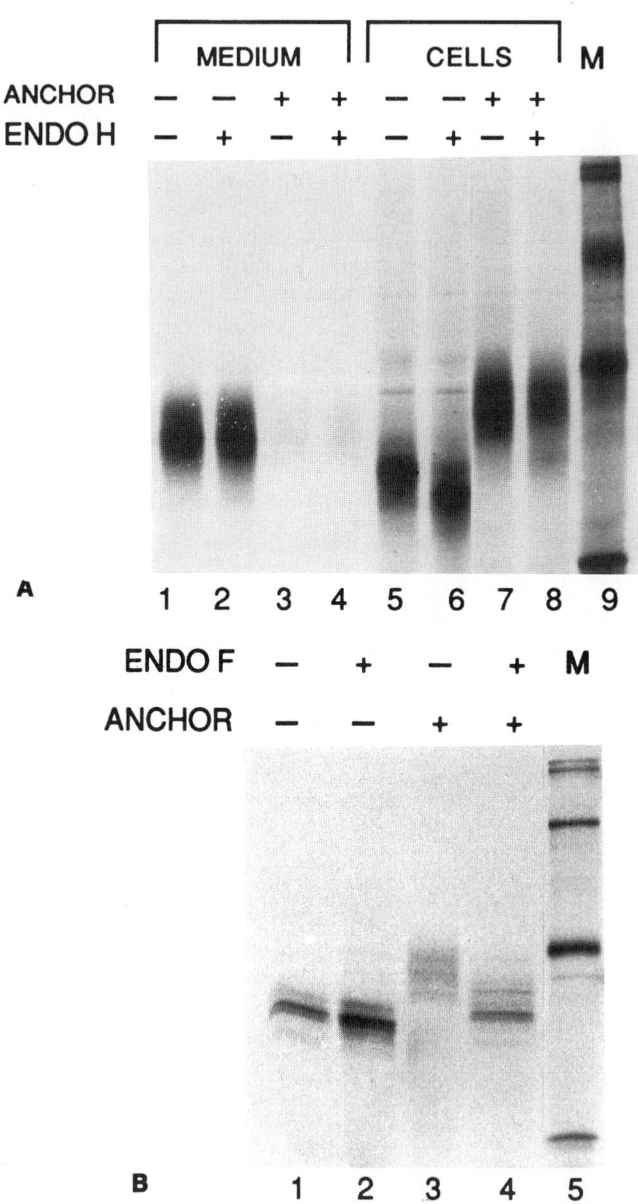

Figure 11. (A) Cellular location and transport of secreted variant HAFVP7 (tracks 1, 2, 5, 6) or C-terminally anchored variant HAFVP7A (tracks 3, 4, 7, 8). VP7 recovered from transfected COS cells by immunoprecipitation was digested with endo H as indicated (+). Track 9 contains standard marker proteins of 30, 46, 69, and 92.5 kDa. (B) Sensitivity of proteins expressed in COS cells to digestion with endo glycosidase F. COS cells were transfected with HAFVP7 (tracks 1, 2) or HAFVP7A (tracks 3, 4) and intact cells were digested with endo F.

(Varghese *et al.*, 1983) were spliced to VP7 at Ser62, yielding constructs NAVP7 and NASVP7, respectively (Fig. 10). However, neither of these hybrid molecules left the ER, since their carbohydrate remained sensitive to digestion with the endo H. The reason why these proteins were not transported out of the ER is unclear but may be related to the way in which they were folded. It appears that cells have a mechanism to prevent transport of aberrantly folded polypeptides (Pfeffer and Rothman, 1987).

With the availability of the correctly cleaved, secreted VP7 derived from HAFVP7, we attempted once more to obtain cell-surface expression of the antigen. The HA signal peptide was fused to phe47 of VP7 and the C-terminal anchor domain of the HA was attached in place of the last residue in VP7 (HAFVP7A, Fig. 10). Expression of this gene in COS cells resulted in the synthesis of a protein that was larger, both because of the presence of the C-terminal anchor domain and the endo H-resistant carbohydrate (compare Fig. 11A, tracks 5, 6 and 7, 8). The protein was not secreted into the medium (Fig. 11A, tracks 3, 4). Incubation of whole cells with endo F, an enzyme that removes complex carbohydrate from proteins, showed that intracellular VP7 was resistant to the enzyme (Fig. 11B, tracks 1, 2); i.e., the cells were intact, but that polypeptide HAFVP7A was

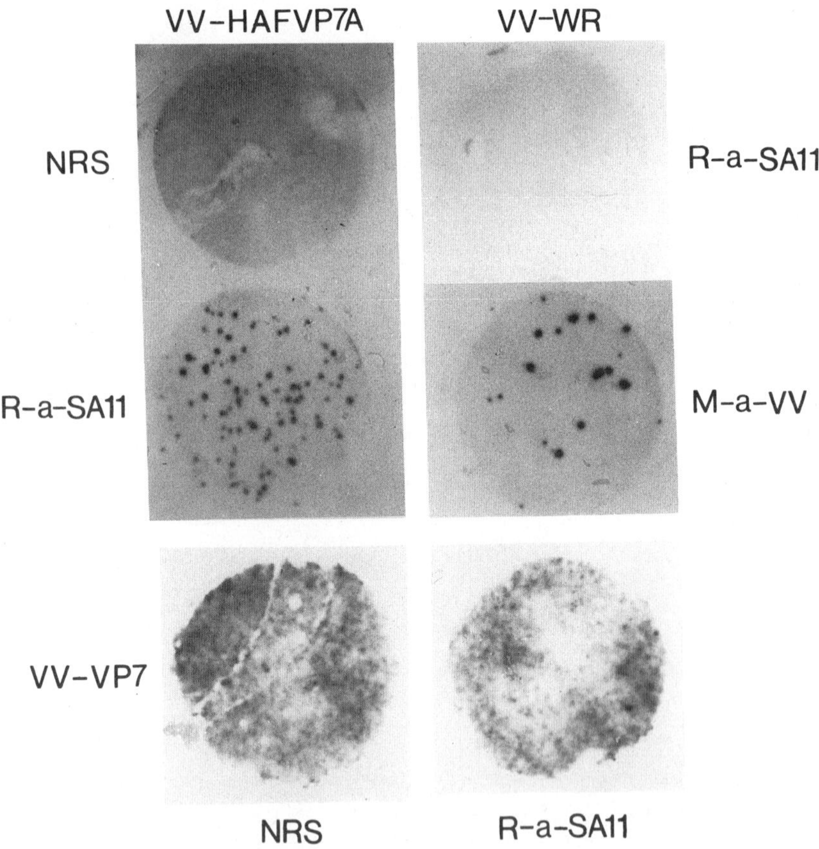

Figure 12. Screening of vaccinia virus plaques. Cells were infected with vaccinia virus strain WR or recombinants VV-VP7 or VV-HAFVP7A (cell-surface variant). Plaques were incubated with normal rabbit serum (NRS), mouse anti-vaccinia (M-a-VV), or rabbit anti-SA11 (R-a-SA11) serum labeled with iodine 125-labeled protein A.

Table 4. Immunization of Rabbits with Vaccinia Virus Recombinants Carrying Genes for Wild Type (wt) or Cell-Surface (sc) Expressed VP7

	vv-VP7 WT		vv-VP7 cell surface	
	Rabbit 2	Rabbit 4	Rabbit 1	Rabbit 3
Day	Antibody titer			
0[a]	50[b]	50	50	50
14	150	450	4,050	36,450
21[a]	50	450	4,050	36,450
31	1,350	1,350	12,150	>100,000
48	1,350	1,350	36,450	36,450
112[a]	150	1,350	4,050	12,150
119	12,150	12,150	12,150	36,450

[a]Animals vaccinated with 10^7 pfu at days 0, 21, and 112.

susceptible to digestion with the enzyme (Fig. 11B, tracks 3, 4). These results indicate that the HAFVP7A had been successfully relocated to the surface of the cells.

A recombinant vaccinia virus carrying the gene for the cell-surface-expressed VP7 protein was then constructed. Recombinant virus plaques expressing either cell-surface VP7 or the native antigen were screened using a radiolabeled antibody. Plaques expressing the modified gene gave a strong signal compared with those carrying the wild-type gene, which gave a weak signal (Fig. 12). This was consistent with the surface location of the modified antigen. The recombinant virus expressing surface located VP7 was then used to vaccinate rabbits. Sera from the small number of animals tested so far indicate that by capture ELISA the antigenicity of the modified antigen is improved over the wild-type protein (Table 4). Clearly, further tests, including plaque neutralization, need to be performed to confirm the improved antigenicity, but the improvement obtained is of the same order as that obtained for the cell-surface display of the malarial S antigen (Langford *et al.*, 1986).

XII. CONCLUSIONS AND FUTURE DIRECTIONS

Since the discovery of the clinical significance of human rotaviruses (Holmes, 1983), a great deal of information has accumulated concerning the identity of the antigenically important proteins and the mechanisms by which these proteins are synthesized and processed ready for assembly into complete virions. However, what is not yet understood is how these proteins fold to form the major antigenic determinants that are recognized by the immune system. Development of such a truly three-dimensional view of rotavirus proteins will require X-ray crystallographic information of the type already available for the surface proteins of influenza virus (Wilson *et al.*, 1981) and for human poliovirus (Hogle *et al.*, 1985) and rhinovirus (Rossman *et al.*, 1985). It is therefore encouraging that the rotavirus single-shelled (core) particle has proved to be amenable to crystallization (unpublished results). Should crystallization of the more immunologically relevant whole virion also be feasible, structural analysis would enable the linear amino acid sequence information for VP7 and VP4 to be more readily interpreted.

During processing of VP7 for its assembly into mature virus particles, the NH_2-terminal hydrophobic regions of the protein are removed. Despite its rapid cleavage, however, the H2 signal peptide is involved in targeting VP7 to the ER and appears to play a role in retaining it at that intracellular site. The mechanism by which this occurs is likely to be novel and remains to be determined. Also unknown is the way in which VP4, the other major surface protein, crosses the ER membrane and assembles on the surface of the virus. This protein appears to lack membrane-spanning or insertion sequences and may therefore cross the membrane by interacting with the core prior to budding. Our *in vitro* results imply that an interaction between the rotavirus core and the cytoplasmic receptor domain of NS28 precedes budding and that conserved binding domains are involved. Indeed, the receptor : ligand interaction between NS28 and the core may in future provide a valuable model system for the study of processes involved in the transfer of viral components across membranes.

The large body of work on rotaviruses undertaken in a comparatively short time has been stimulated by the importance of these viruses in pediatric disease and the hope that progress might be made toward the development of an effective rotavirus vaccine. That goal may be within reach. Rhesus rotaviruses carrying human serotype genes (Midthun *et al.*, 1985) are currently in clinical trials and our studies on the VP7 antigen show that it is possible to express this protein in a form that induces neutralizing antibodies. What is needed now is an appropriate delivery system for these antigens. A live, attenuated, recombinant virus that could be given orally would seem to be the most appropriate delivery system for a second-generation rotavirus vaccine. Unfortunately, systems available at present must necessarily be heterologous because it is not yet possible to reintroduce the cloned, modified genes into an infectious rotavirus. Irrespective of the outcome of the various vaccine strategies that in future may be adopted, it is clear that rotaviruses will continue to be of interest to both cell biologists and molecular biologists because of their value in providing fascinating model systems for the study of important biologic processes.

ACKNOWLEDGMENTS

We thank our many collaborators, in particular D. Christie, B. Coupar, D. Elliott, L. Lockett, J. Meyer, D. Reddy, J. Sheridan, and P. L. Whitfeld, who have participated in this work over recent years. This work was supported by grants from the World Health Organization, the N. Z. Medical Research Council, and the N. Z. National Childrens Health Research Foundation. C. C. B. is a postdoctoral fellow of the N. Z. Universities Research Grants Committee. S. C. S. was supported by an Australian Wool Corporation Postgraduate Scholarship. Drs. Robert Crowl, R. Possee, and M. Summers kindly made available the pEVvrf and baculovirus expression vectors used in this work. We thank A. McGill for typing the manuscript.

REFERENCES

Andrew, M., Boyle, D. B., Coupar, B. E. H., Whitfeld, P. L., Both, G. W., and Bellamy, A. R. (1987). *J. Virol.* **61,** 1054–1060.

Arias, C. F., Lopez, S., Bell, J. R., and Strauss, J. H. (1984). *J. Virol.* **50,** 657–661.

Au, K-S., Chan, W-K., and Estes, M. K. (1989). In *UCLA Symposium on Cell Biology of Viral Entry, Replication and Pathogenesis* (R. Compans, A. Helenius, M. Oldstone, eds.), pp. 257–267, Alan Liss, New York.

Bachrach, H. L., Moore, D. M., McKercher, P. D., and Polatnick, J. (1975). *J. Immunol.* **115,** 1636–1641.

Bergmann, C. C., Maass, D., Poruchynsky, M. S., Atkinson, P. H., and Bellamy, A. R. (1989). *EMBO J.* **8,** 1695–1703.

Bishop, R. F., Davidson, G. P., Holmes, I. H., and Ruck, B. J. (1973). *Lancet,* **2,** 1281–1283.

Bjorkman, P. J., Saper, M. A., Samraoui, B., Bennett, W. S., Strominger, J. L., and Wiley, D. C. (1987). *Nature* **329,** 506–512.

Both, G. W. (1988). In *RNA Genetics* (E. Domingo, J. J. Holland, and P. Ahlquist, eds.), pp. 171–193, CRC Press, Boca Raton, FL.

Both, G. W., Siegman, L. J., Bellamy, A. R., and Atkinson, P. H. (1983). *J. Virol.* **48,** 335–339.

Chan, W-K., Au, K-S., and Estes, M. K. (1988). *Virology* **164,** 435–442.

Crowl, R., Seamans, C., Lomedico, P., and McAndrew, S. (1985). *Gene* **38,** 31–38.

Dyall-Smith, M. L., Lazdins, I., Tregear, G. W., and Holmes, I. H. (1986). *Proc. Natl. Acad. Sci. USA* **83,** 3465–3468.

Ericson, B. L., Graham, D. Y., Mason, B. B., Hansenn, H., and Estes, M. K. (1983). *Virology* **127,** 320–332.

Estes, M. K., Palmer, E. L., and Obijeski, J. F. (1983). *Curr. Topics Microbiol. Immunol.* **105,** 123–184.

Estes, M. K., Crawford, S. E., Penaranda, M. E., Petrie, B. L., Burns, J. W., Chan, W., Ericson, B., Smith, G. W., and Summers, M. D. (1987). *J. Virol.* **61,** 1488–1494.

Farquhar, M. G. (1985). *Annu. Rev. Cell Biol.* **1,** 447–488.

Gething, M-J., and Sambrook, J. (1981). *Nature* **293,** 620–625.

Gething, M-J., McCammon, K., and Sambrook, J. (1986). *Cell* **46,** 939–950.

Gunn, P. G., Sato, F., Powell, K. F. H., Bellamy, A. R., Napier, J. R., Harding, D. R. K., Hancock, W. S. Siegman, L. J., and Both, G. W. (1985). *J. Virol.* **54,** 791–797.

Hogle, J. M., Chow, M., and Filman, D. J., (1985). *Science* **229,** 1358–1365.

Holmes, I. H. (1983). In *The Reoviridae* (W. K. Joklik, ed.), pp. 359–423, Plenum Press, New York.

Kabcenell, A. K., and Atkinson, P. A. (1985). *J. Cell. Biol.* **101,** 1270–1280.

Kabcenell, A. K., Poruchynsky, M. S., Bellamy, A. R., Greenberg, H. B., and Atkinson, P. A. (1988). *J. Virol.* **62,** 2929–2941.

Kapikian, A. Z., and Chanock, R. M. (1985). In *Virology* (B. N. Fields, ed.), pp. 863–906, Raven Press, New York.

Kuroda, K., Hauser, C., Rott, R., Klenk, H., and Doerfler, W. (1986). *EMBO J.* **5,** 1359–1365.

Joklik, W. K. (1983). In *The Reoviridae* (W. K. Joklik, ed.), Chapter 1, Plenum Press, New York.

Langford, C. J., Edwards, S. J., Smith, G. L., Mitchell, G. F., Moss, B., Kemp, D. J., and Anders, R. F. (1986). *Mol. Cell. Biol.* **6,** 3191–3199.

Liu, M., Offit, P. A., and Estes, M. K. (1988). *Virology* **163,** 26–32.

Maryanski, J. L., Pala, P., Corradin, G., Jordan, B. R., and Cerottini, J-C. (1986). *Nature* **324,** 578–579.

Matsui, S. M., Mackow, E., and Greenberg, H. B. (1989). *In Advances in Virus Research* (K. Maramorosch, F. A. Murphy, and A. J. Shatkin, eds.), pp. 181–214, Academic Press, San Diego, CA.

Matsuura, Y., Possee, R. D., Overton, H. A., and Bishop, D. H. L. (1987). *J. Gen. Virol.* **68,** 1233–1250.

Metcalf, P. (1979). Ph.D. thesis, University of Auckland, New Zealand.

Midthun, K., Greenberg, H. B., Hoshino, Y., Kapikian, A. Z., Wyatt, R. G., and Chanock, R. M. (1985). *J. Virol.* **53,** 949–954.

Mitchell, D. B., and Both, G. W. (1988). *Nucl. Acids Res.* **16,** 6244–6244.

Petrie, B. L., Greenberg, H. B., Graham, D. Y., and Estes, M. K. (1984). *Virus Res.* **1,** 133–152.

Pfeffer, S. R., and Rothman, J. E. (1987). *Annu. Rev. Biochem.* **56,** 829–852.

Poruchynsky, M. S., Tyndall, C., Both, G. W., Sato, F., Bellamy, A. R., and Atkinson, P. A. (1985). *J. Cell Biol.* **101,** 2199–2209.

Prasad, B. V. V., Wang, G. J., Clerx, J. P. M., and Chiu, W. (1988). *J. Mol. Biol.* **199,** 269–275.

Reddy, D. A., Greenberg, H. B., and Bellamy, A. R. (1989). *Nucl. Acids Res.* **17,** 449.

Roseto, A., Escaig, J., Delain, E., Cohen, J., and Scherrer, R. (1979). *Virology* **98,** 471–475.

Rossman, M. G., Arnold, E., Erikson, J. W., Frankenberger, E. A., Griffith, J. P., Hecht, H. J., Johnson, J. E., Karner, G., Luo, M., Mosser, A. G., Reuckert, R. R., Herry, B., and Vriend, G. (1985) *Nature* **317,** 145–153.

Sheridan, J. F., Smith, C. C., Manak, M. M., and Aurelian, L. (1984). *J. Infect. Dis.* **149,** 434–438.

Stirzaker, S. C., Whitfeld, P. L., Christie, D. L., Bellamy, A. R., and Both, G. W. (1987). *J. Cell Biol.* **105,** 2897–2903.

Summers, M. D., and Smith, G. E. (1987). *Tex. Agric. Exp. Stn. (Bull.)* **1555,** 1–48 appendix.

Taniguchi, K., Hoshino, Y., Nishikawa, K., Green, K. Y., Maloy, W. L., Kapikian, A. Z., Chanock, R. M., and Gorziglia, M. (1988). *J. Virol.* **62,** 1870–1874.

Townsend, A. R. M., Rothbard, J., Gotch, F. M., Bahadur, G., Wraith, D., and McMichael, A. J. (1986). *Cell* **44,** 959–968.

Varghese, J. N., Laver, W. G., and Colman, P. M. (1983). *Nature* **303,** 35–40.

Vesikari, T., Isolauri, E., Ruuska, T., Delem, A., and Andre, F. E. (1988). In *Applied Virology Research,* vol. 1

(E. Kurstak, R. G. Marusyk, F. A. Murphy, and M. H. V. Van Regenmortel, eds.), pp. 55–62, Plenum Press, New York.

von Heijne, G. (1986). *Nucl. Acids Res.* **14,** 4683–4690.

Ward, C. W., and Dopheide, T. A. (1980). *Virology* **103,** 37–53.

Whitfeld, P. L., Tyndall, C., Stirzaker, S. C., Bellamy, A. R., and Both, G. W. (1987). *Mol. Cell. Biol.* **7,** 2491–2497.

Wilson, I. A., Skehel, J. J., and Wiley, D. C. (1981). *Nature* **289,** 366–373.

IV

Virus Hemorrhagic Fevers

Diversity and Epidemiology

16

The Molecular Epidemiology of Dengue Viruses

Genetic Variation and Microevolution

Dennis W. Trent, Charles L. Manske, George E. Fox, May C. Chu, Srisakul C. Kliks, and Thomas P. Monath

I. INTRODUCTION

Dengue (DEN) fever is a mosquito-transmitted flavivirus disease of humans which has affected untold millions of people over the world during the past two centuries (reviewed by Schlesinger, 1979; Gubler, 1988). The DEN viruses occur as four distinct serotypes, which can be serologically (Westaway *et al.*, 1985; De Madrid and Porterfield, 1974) and biochemically differentiated (Vezza *et al.*, 1980; Blok, 1985; Blok *et al.*, 1984). The viruses are most frequently transmitted from viremic humans to susceptible humans by the bite of *Aedes aegypti* mosquitoes (reviewed by Gubler, 1988). Subsequent studies in the Philippines, Indonesia, and the Pacific Islands showed that *Ae. albopictus* and *Ae. polynesiensis* are efficient vectors of DEN viruses. In Malaysia, Vietnam, and Africa there is evidence that a forest maintenance cycle for DEN virus exists in which the virus is maintained in a cycle involving canopy-dwelling *Aedes* spp. and wild monkeys (Siler *et al.*, 1926; Rudnick, 1965). Potential vector *Aedes* spp. are found throughout the tropics, where more than half the world population live in conditions that frequently expose them to mosquito bites (Halstead, 1980, 1988). Since the end of World War II, the incidence of DEN fever has increased dramatically with the increase in air travel, the introduction of multiple serotypes of virus to many parts of the world, urbanization of the tropics, the breakdown of effective mosquito control programs, and the deterioration of public health programs due to economic and social problems in many areas of the world (Halstead, 1988; Gubler, 1987, 1988). Because of these conditions, DEN fever is currently the most important arthropod-borne viral disease of humans in terms of both morbidity and mortality (Halstead, 1988). Coincident with the increase in number of cases has been the appearance and spread of a severe, sometimes fatal form of the infection designated DEN hemorrhagic fever and DEN shock syndrome (DHF/DSS; Technical Advisory Group on Dengue Hemorrhagic Fever/

Dennis W. Trent and May C. Chu • Division of Vector-Borne Infectious Diseases, Center for Infectious Diseases, Centers for Disease Control, Public Health Service, U.S. Department of Health and Human Services, Fort Collins, Colorado 80522. *Charles L. Manske and George E. Fox* • University of Houston, Department of Biochemical and Biophysical Sciences, Houston, Texas 77004. *Srisakul C. Kliks* • University of California, School of Public Health, Department of Biomedical and Environmental Health Sciences, Berkeley, California 94705. *Thomas P. Monath* • SGRD-UIV, Virology Division, USAMRIID, Fort Derrick, Frederick, Maryland 21701–5011

Dengue Shock Syndrome, 1986). Since the early 1960s, DHF/DSS has been the major cause of death among children in Southeast Asia (WHO, 1986a,b) and, in recent years, of increasing concern in the Pacific Islands and the Caribbean (Guzman *et al.*, 1984a,b).

The importance of DEN as a public health problem with serious medical, economic, and political implications has stimulated increased research on the disease. Major advancements have been made in understanding the molecular biology of the flaviviruses and the immunology of their proteins (Rice *et al.*, 1986). The host and virus factors involved in the pathogenesis of severe DHF/DSS remain an important research focus. The purpose of this review is to present current knowledge on the molecular epidemiology of DEN viruses throughout the world and to discuss the impact of this new information on prevention, control, and understanding the role of virus variants in human disease.

II. THE FLAVIVIRUSES

Dengue virus is a member of the family Flaviviridae, comprising more than 60 agents that are serologically closely related and share a unique structure, morphogenesis, genome organization, and replication strategy (Westaway *et al.*, 1985). The mosquito-borne flaviviruses have been organized on the basis of virus antibody neutralization into six distinct subgroups: DEN, yellow fever (YF), Japanese encephalitis (JE), Uganda S, Ntaya, and Spondweni (Porterfield, 1980; De Madrid and Porterfield, 1974). Four serologically distinct DEN viruses are currently recognized (Porterfield, 1980). The virus particles are 60–70 nm in diameter and are composed of a lipid-bilayer outer envelope in which are embedded the glycosylated envelope (E) and membrane proteins (M) (Fig. 1; Brinton, 1986). The positive-sense, single-stranded RNA is surrounded by a nucleocapsid composed of a single virus protein. The flavivirus genome is approximately 11,000 nucleotides long, has a type 1 cap at the 5′ end of the RNA, and does not have a polyadenylic acid tract at the 3′ end (Fig. 2; Rice *et al.*, 1986). Nucleotide sequences of the entire genomes of DEN type 2 (Hahn *et al.*, 1988; Deubel *et al.*, 1988) and DEN 4 (Zhao *et al.*, 1986) have been determined. The structural genes of DEN 1 (Mason *et al.*, 1987; Chu *et al.*, 1988) and the New Guinea strain of DEN 2 (Gruenberg *et al.*, 1988) have been cloned and sequenced. For the DEN viruses, and other members of the Flaviviridae, the order of genes on the genomic RNA is 5′ -C-prM(M)-E-NS1-ns2a-ns2b-

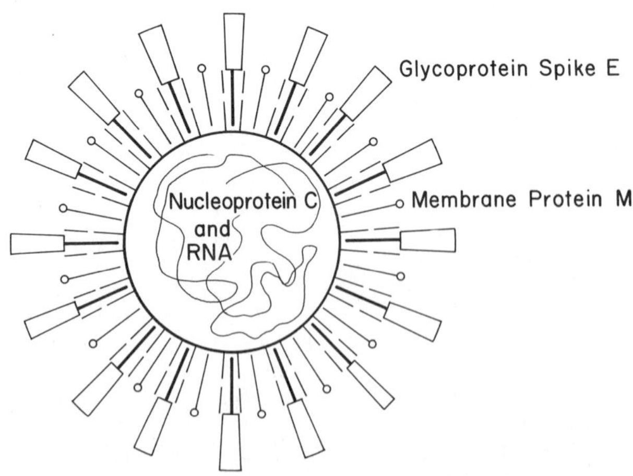

Figure 1. Theoretical model of the flavivirus particle.

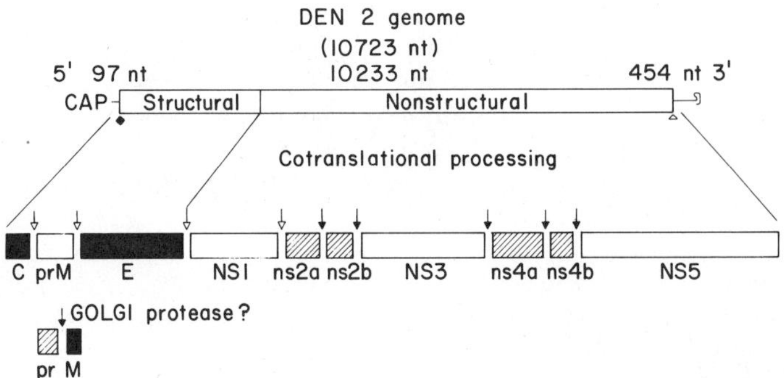

Figure 2. DEN 2 virus genome organization and expression. Untranslated regions are shown as single lines flanking the translated region in the open box. A solid diamond marks the first AUG codon initiating the long open reading frame; the open triangle indicates the termination codon UGA. The protein nomenclature is that of Rice *et al.* (1986). Signalase-like proteolytic cleavage sites are indicated by open arrows.

NS3-ns4a-ns4b-NS5-3′, with the proteins encoded in a single polyprotein that is cleaved by both viral and cellular proteases to form the viral polypeptides (Fig. 2; Rice *et al.*, 1985; Castle *et al.*, 1986; Coia *et al.*, 1988; Deubel *et al.*, 1988).

Comparison of flavivirus genome sequences has shown that there is conservation of the amino acid sequence among the DEN 1, DEN 2, and DEN 4 viruses as well as other members of the genus. The amino acid sequence of the DEN 2 virus is similar to that of West Nile virus and distinct from that of yellow fever virus (Deubel *et al.*, 1986, 1988). Of the nonstructural proteins ns2a, ns2b, ns4a, and ns4b are the least conserved and NS3 and NS5 are the most conserved (Fig. 3). Each flavivirus protein exhibits regions of high similarity in sequence and physical character. Domains in DEN, yellow fever, and West Nile virus proteins share more than a 50% amino acid identity. In addition to these sequence similarities, cysteine residues are conserved in the NS1, prM(M), and E proteins with two glycosylation sites on the NS1 proteins of all flaviviruses (Deubel *et al.*, 1988).

III. MOLECULAR EPIDEMIOLOGY OF DENGUE VIRUSES

Fingerprinting of RNAse T1-resistant oligonucleotides has been used to characterize and distinguish members of the mosquito-borne flaviviruses (Vezza *et al.*, 1980; Trent *et al.*, 1981, 1983; Monath *et al.*, 1983; Repik *et al.*, 1983; Hori *et al.*, 1986b; Walker *et al.*, 1988). DEN viruses that are antigenically distinct have unique oligonucleotide fingerprint images and cDNA restriction enzyme patterns (Vezza *et al.*, 1980; Blok *et al.*, 1984; Blok, 1985). Viruses that are serologically

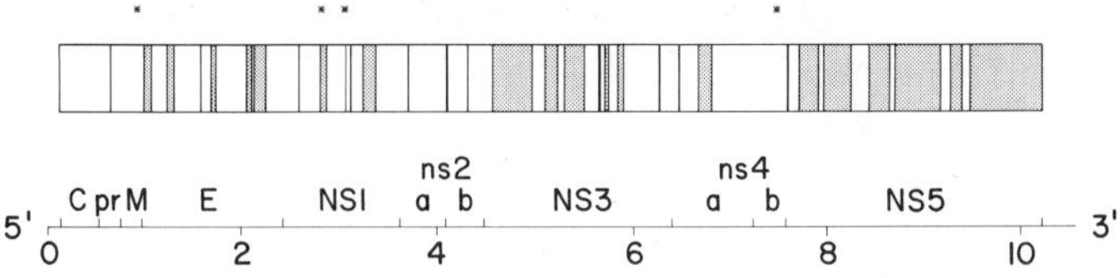

Figure 3. Schematic representation of the flavivirus polyprotein showing the highly conserved regions (shaded areas). Domains are indicated by windows of 20 amino acids and are shaded when there is greater than 55% sequence homology (Deubel *et al.*, 1988).

indistinguishable may have distinct oligonucleotide fingerprints and still retain 50–80% identity in their large oligonucleotides (Trent *et al.*, 1981, 1983). Within the serotype, DEN viruses isolated from different parts of the world have been further classified to a topotype (Trent *et al.*, 1981) if 70% or greater similarity exists in their large resolved oligonucleotides. When sample sizes are greater (Trent *et al.*, 1988), a topotype may be extended to virus strains that have 50% similarity. Nucleotide sequence similarity can be estimated from oligonucleotide fingerprint similarity. When the nucleotide sequence of RNA species differs by 1, 5, or 10%, viruses share approximately 85, 50, or 25% of their large oligonucleotides, respectively (Aaronson *et al.*, 1982). Comparing virus RNAse T1 oligonucleotides is valid, therefore, only when the overall sequence of the nucleic acids being compared is similar. For example, the total nucleotide sequence of the Puerto Rican and Jamaican DEN 2 virus genomes are 96.8% similar (Table 1; Hahn *et al.*, 1988; Deubel *et al.*, 1988). However, the oligonucleotide maps of the two viruses are very distinct and share only 20% of their large resolved oligonucleotides (Fig. 4). Nucleotide variation between the Jamaican and Puerto Rican DEN 2 viruses is random in the genome, with divergence ranging from 8.5 to 12.4% in different genes within the translated portion of the genome. Only 13.3% of the nucleotide differences result in amino acid substitutions (Deubel *et al.*, 1988). Similarly, although the nonstructural gene regions of the Jamaican and New Guinea C DEN 2 viruses have 90.3% identity in their nucleotide sequences, the fingerprints of the two DEN 2 viruses are distinct and each represents a unique topotype (Yaegashi *et al.*, 1986; Trent *et al.*, 1983). Therefore, although the nucleotide sequences of these geographically distinct DEN 2 strains are highly conserved, their oligonucleotide fingerprints are distinct, thus permitting classification into different topotypes. Thus, for epidemiologic purposes, comparing the oligonucleotide fingerprint images remains a very sensitive tool for determining the origin of DEN virus strains.

Does the RNAse T1 oligonucleotide fingerprint image equally represent all regions of the DEN virus genome? The large oligonucleotides of the Jamaican DEN 2 virus have been sequenced and their location determined on the genome of the virus (Fig. 5). The number of large oligonucleotides, representing each of the virus genes in the fingerprint, is directly proportional to the size of the individual gene and they are located randomly throughout the fingerprint. Although oligonucleotides representing genes of different virus strains will change because of nucleotide sequence differences, comparison of viruses of the same genetic topotype permits comparison of total RNA genome similarity.

To examine the importance of genetic variation in the evolution and epidemiology of DEN

Table 1. Similarity (in Percentage) between the Aligned Nucleotide (NT) and Amino Acid (AA) Sequences of Dengue Viruses[a]

	NT				
AA	DEN2JAM[b]	DEN2S1[c]	DEN2NGC[d]	DEN4	DEN1[b]
DEN2JAM	—	90.7	94.4	66.2	66.2
DEN2S1	96.8	—	93.1	66.1	66.5
DEN2NGC[e]	97.7	97.6	—	ND	ND
DEN4	68.2	67.9	ND[e]	—	64.4
DEN1[e]	68.8	68.5	ND	62.9	—

[a]Deubel *et al.*, 1988.
[b]DEN 2 Jamaica.
[c]DEN 2 Puerto Rico.
[d]DEN 2 New Guinea C.
[e]Only partial sequences are compared.

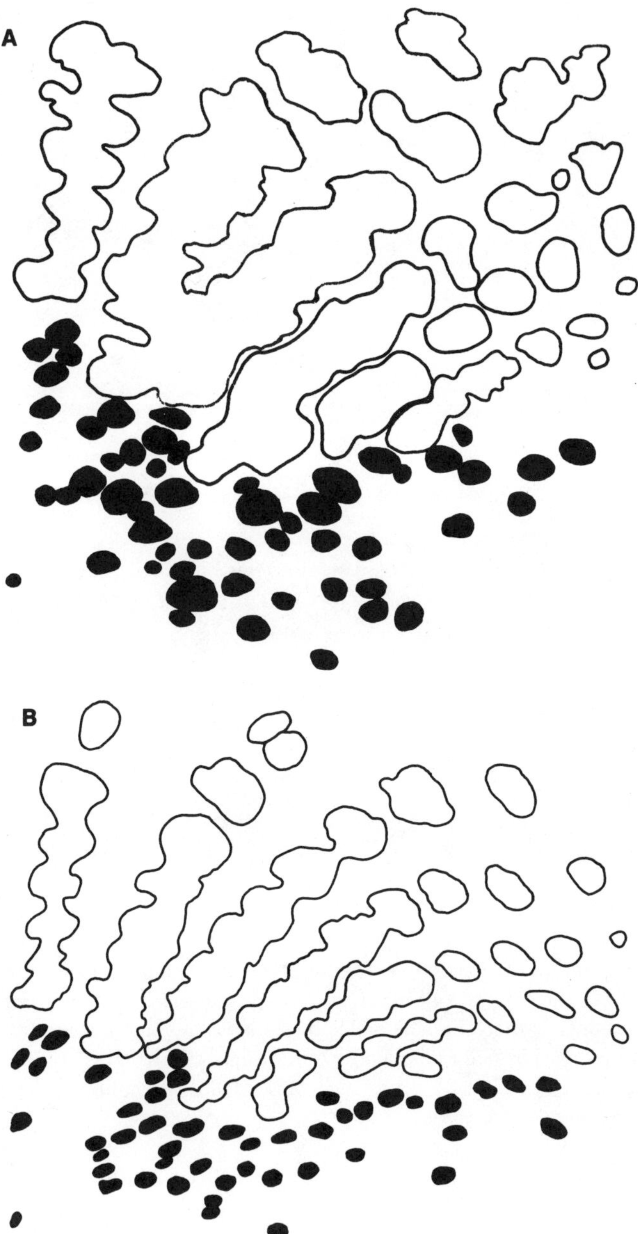

Figure 4. Schematics of oligonucleotide fingerprint images of DEN 2 prototypic topotypes: (A) Jamaican (872-2) and (B) Puerto Rican (PR-159).

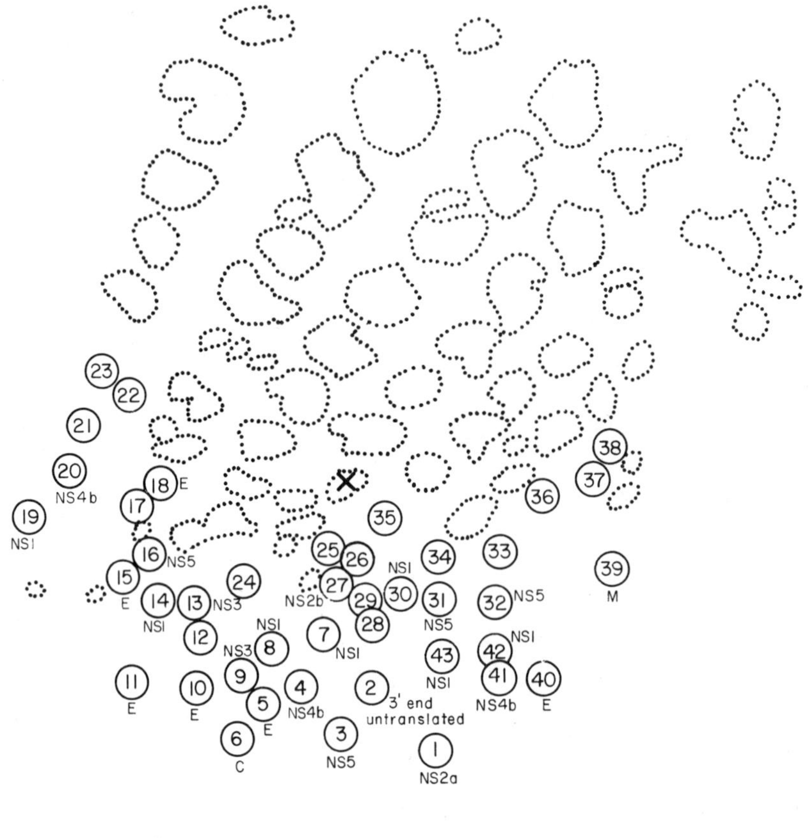

Figure 5. Schematic showing location of oligonucleotides representing different DEN 2 virus genes in the oligonucleotide fingerprint of DEN 2 Jamaican virus ARAC81110827. Numbers indicate isolated RNAse T1-resistant oligonucleotides present in the fingerprint pattern.

virus, we made fingerprints of viruses of all four serotypes from patients, mosquitoes, and monkeys throughout the world. Fingerprints of viral RNA were prepared and visually compared with other fingerprints of the same serotype by geographic origin and time of isolation (Trent *et al.*, 1983). Virus strains were grouped by percent similarity of the oligonucleotide fingerprint images (Trent *et al.*, 1981, 1983). For each topotype a representative of each different oligonucleotide pattern was selected and used as the comparison standard.

A. Dengue 2

By analyzing oligonucleotide fingerprint images of DEN 2 virus isolates from throughout the world, we grouped the viruses into ten different topotypes (Fig. 6; Table 2). As stated earlier, in the analysis, a topotype is designated when eight or more viruses are found to share at least 70% of their large oligonucleotides. This permits us to determine the extent of variation within the virus population and to establish confidence in the unique identity of each topotype.

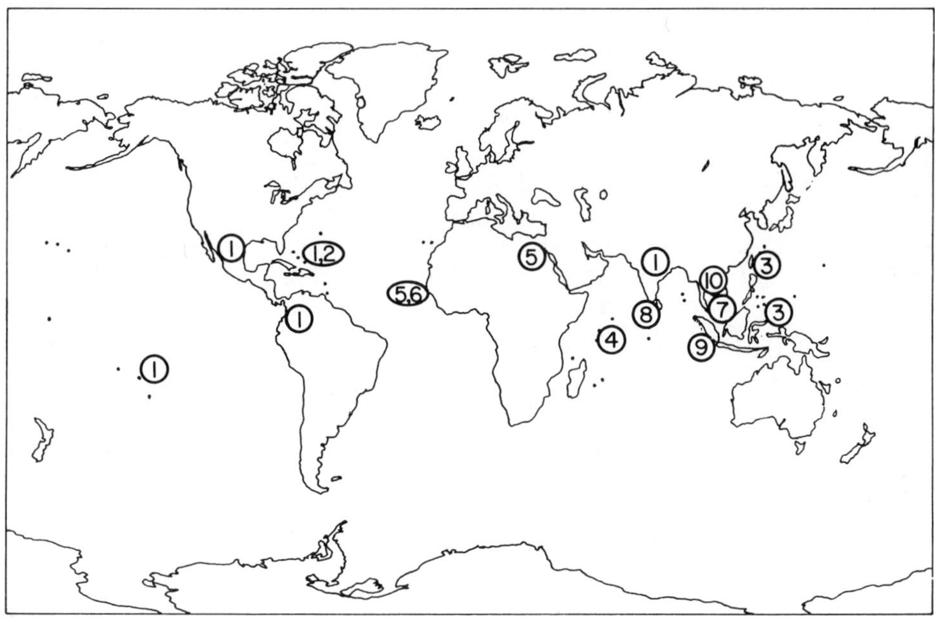

Figure 6. Worldwide distribution of DEN 2 virus topotypes. Individual topotypes or genetic varieties are indicated by numbers located on the map where the DEN 2 genotype strains were isolated. 1, Americas/India/South Pacific; 2, Caribbean; 3, Philippines/Taiwan; 4, Seychelles; 5, Africa (endemic); 6, Africa (enzootic); 7, Malaysia; 8, Sri Lanka; 9, Indonesia; 10, Thailand.

1. The Americas

Fingerprints of the Puerto Rican and Jamaican topotypes (Fig. 4) represent the two distinct DEN 2 virus genotypes in the Western Hemisphere. The Puerto Rican topotype of DEN 2 virus was introduced into the Caribbean in 1969, where it caused an epidemic of DEN fever without complications (Likosky *et al.*, 1973). This same genetic variety of DEN 2 virus subsequently spread

Table 2. Geographic Distribution of Dengue Virus Gene and Topotypes

DEN 1	DEN 2	DEN 3	DEN 4
1. Philippines[a]	1. Americas/India/ South Pacific	1. Thailand	1. The Americas
2. Malaysia		2. Indonesia	2. Philippines
3. Thailand	2. Caribbean	3. Caribbean	3. Thailand
4. Africa/Sri Lanka	3. Philippines/Taiwan	4. Sri Lanka	4. Sri Lanka
5. The Americas	4. Seychelles	5. Philippines	5. Africa
6. India	5. Africa (endemic)		
7. Indonesia	6. Africa (enzootic)		
8. Pacific Islands	7. Malaysia		
	8. Sri Lanka		
	9. Indonesia		
	10. Thailand		

[a]Topotypes are designated by arabic numbers for each serotype.

throughout Mesoamerica and northern South America, where it caused numerous epidemics (Trent *et al.*, 1983; Gubler, 1988). Analysis of DEN 2 viruses isolated in the South Pacific and India revealed that the topotype spread from India to the South Pacific and then into the Caribbean (Trent *et al.*, 1983; Trent, unpublished observations). In 1981, an epidemic of severe DEN 2 virus infection with large numbers of DHF/DSS in children occurred in Cuba (Guzman *et al.*, 1984a,b). Although fingerprints of the viruses isolated during the Cuban epidemic were never made, a DEN 2 virus with a distinct fingerprint image was isolated in 1983 from a patient in Jamaica (Jamaican topotype; Trent *et al.*, 1983). Viruses of the Jamaican topotype were subsequently isolated from patients in Puerto Rico, Haiti, and Dominican Republic. Currently both the Jamaican and Puerto Rican topotypes of DEN 2 virus are being transmitted in the Western Hemisphere. This circumstance provides a unique setting in which DEN 2 viruses of different topotypes could recombine and give rise to a new DEN 2 virus variant.

2. Africa

Since 1927, there has been serologic evidence of DEN in Africa (Kokernot *et al.*, 1956). In the last two decades, DEN 2 virus-associated illness has been documented in Nigeria (Carey *et al.*, 1971), Senegal (Robin *et al.*, 1980), the Seychelles (Calisher *et al.*, 1981), Reunion Island (Coulanges *et al.*, 1979), Kenya (Johnson *et al.*, 1982), Somalia (Saleh *et al.*, 1985), and the Sudan (Hyams *et al.*, 1986). DEN 2 virus remains endemic in forest mosquitoes in many of these countries (Cornet *et al.*, 1984). Fingerprints of seven DEN 2 virus strains isolated from either humans or *Ae. aegypti* mosquitoes in Africa were found to be distinct from fingerprints of isolates from the Seychelles islands (Kerschner *et al.*, 1986). Fingerprints of the isolates from humans and those from forest mosquitos, however, are distinct and form two different topotypes. The oligonucleotide fingerprints of isolates from monkeys and forest mosquitoes in Ivory Coast and Burkina Faso are genetically distinct; although they can be transmitted by *Ae. aegypti* mosquitoes, they are not usually involved in the human-to-human transmission cycle (reviewed by Monath, 1984).

3. Asia

Oligonucleotide fingerprints of DEN 2 isolates from outbreaks of disease in Indonesia, Malaysia, the Philippines/Taiwan, Sri Lanka, and Thailand are all distinct and represent unique genetic variants of the virus. Epidemics of DEN 2 have repeatedly occurred in Asia since 1956, with outbreaks in the Philippines (1956–1977), Thailand (1958–1985), Indonesia (1973–1983), and Burma (1970–1983) (Halstead, 1980; Bang and Sanyakorn, 1984; Setyorogo, 1984). The oligonucleotide fingerprints of the DEN 2 virus isolates from Burma and Thailand are similar and represent a single topotype (Trent *et al.*, 1983).

B. Dengue 1

DEN 1 virus was first isolated during World War II in northern Australia, Japan, New Guinea, Fiji, Hawaii, and the Society Islands (Bres, 1979). The virus is thought to have come from Southeast Asia since there was no evidence of endemic transmission of DEN in the South Pacific region. Since World War II, DEN 1 virus has been responsible for major outbreaks of disease in India, Burma, Thailand, Vietnam, Cambodia, Australia, Malaysia, Singapore, the Philippines, China, Taiwan, the Caribbean Islands, Brazil, and Mesoamerica (Gubler, 1988; Halstead, 1988).

Fingerprints of virion RNA from 12 geographically distinct DEN 1 viruses were shown by Repik *et al.* (1983) to be unique. Similarity among isolates, based on the percent of shared oligonucleotides, was then used to establish three geographic groups. Topotypes established from these studies included the Caribbean, the Pacific Islands/Southeast Asia, and Africa/Sri Lanka.

Furthermore, the similarity of the fingerprint images led Repik *et al.* (1983) to suggest that the DEN 1 virus strain involved in the 1969 Sri Lankan and 1977 Caribbean DEN 1 epidemics originated in Africa.

We have analyzed the oligonucleotide fingerprint images of 142 DEN 1 virus strains isolated from six geographic regions from 1973 through 1988 and placed them into eight topotypes (Fig. 7; Table 2). Virus isolates from Fiji, India, and the Americas have unique fingerprint images with 85–90% similarity within each topotype. The genetic complexity observed in the fingerprints of isolates from Thailand, Malaysia, the Philippines, and Sri Lanka suggests that extensive genetic diversity exists in virus populations from these regions. Comparison of 14 DEN 1 virus strains from Sri Lanka (1982–1985) with more than 50 strains from Mexico, Puerto Rico, Brazil, and Aruba failed to confirm a genetic relationship (Chu, unpublished observations). Since the DEN 1 virus was introduced into the Americas in 1977, a single genetic variant has spread from Kingston, Jamaica, to nearly all parts of the Caribbean and central and northern South America, causing large outbreaks wherever it is introduced. There is no indication from our observations that the strains of DEN 1 isolated in the Americas are related to DEN 1 virus strains from other parts of the world.

C. Dengue 3 and 4

We examined the oligonucleotide fingerprints of DEN 3 and 4 virus strains from geographic regions where the viruses are endemic. There are five different topotypes of DEN 3 virus, each a single genetic variety from the geographic region of its origin (Fig. 8; Table 2). DEN 3 virus is not presently transmitted in the Americas. However, the virus is actively transmitted in many parts of the world and, therefore, could be introduced into the Western Hemisphere. Fingerprints of the Thai

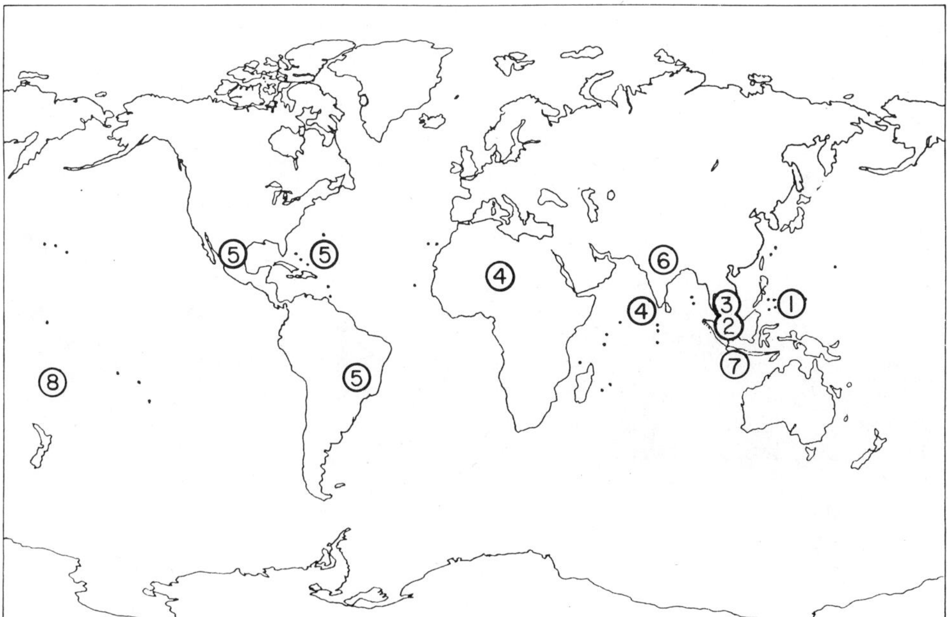

Figure 7. Worldwide distribution of DEN 1 virus topotypes. Individual topotypes or genetic varieties are indicated by numbers located on the map where the DEN 1 genotype strains were isolated. 1, Philippines; 2, Malaysia; 3, Thailand; 4, Africa/Sri Lanka; 5, the Americas; 6, India; 7, Indonesia; 8, Pacific Islands.

Figure 8. Worldwide distribution of DEN 3 virus toptotypes. Individual topotypes or genetic varieties are indicated by numbers located on the map where the DEN 3 genotype strains were isolated. 1, Thailand; 2, Indonesia; 3, the Americas; 4, Sri Lanka; 5, Philippines.

DEN 3 virus isolates from the 1960s are unique, but they contain many oligonucleotides similar in position to the 1980s Thai DEN 3 viruses. We have examined the DEN 3 viruses from Thailand over a 20-year period and can see microevolution in the virus genome similar to that observed with the DEN 2 viruses in Thailand (Trent *et al.*, 1988).

The oligonucleotide fingerprint images of the DEN 4 viruses from the Americas, Africa, the Philippines, Thailand, Indonesia, and the South Pacific are distinct (Fig. 9; Table 2). Fingerprint images of DEN 4 viruses isolated in Thailand during the 1970s are related to the 1980 virus isolates. The reason for these changes in the virus genome is yet unclear since epidemic patterns of transmission throughout this 10-year period have not significantly changed. DEN 4 virus infection occurred in 1979 in French Polynesia (Parc *et al.*, 1981a,b) and in New Caledonia (Le Gonidec *et al.*, 1982). From there, the virus spread to other islands, causing a widespread epidemic in the South Pacific. More than 5600 cases were reported in the South Pacific Islands during 1980 (WHO, 1986c). Attack rates were very high in some islands, such as Niue, where 20% of the population was infected. The fingerprints of DEN 4 viruses from other islands in the Pacific are similar, indicating that the same topotype was involved in the epidemic.

IV. DENGUE VIRUS VARIATION AND MICROEVOLUTION

New technologies enable virologists to study small genetic and antigenic differences between field isolates of the RNA-containing animal viruses at a high level of discrimination. The molecular techniques used include monoclonal antibodies, peptide mapping, fingerprinting of whole RNA virus genomes, RNA : RNA hybridization, restriction enzyme analysis of cDNA, direct sequencing

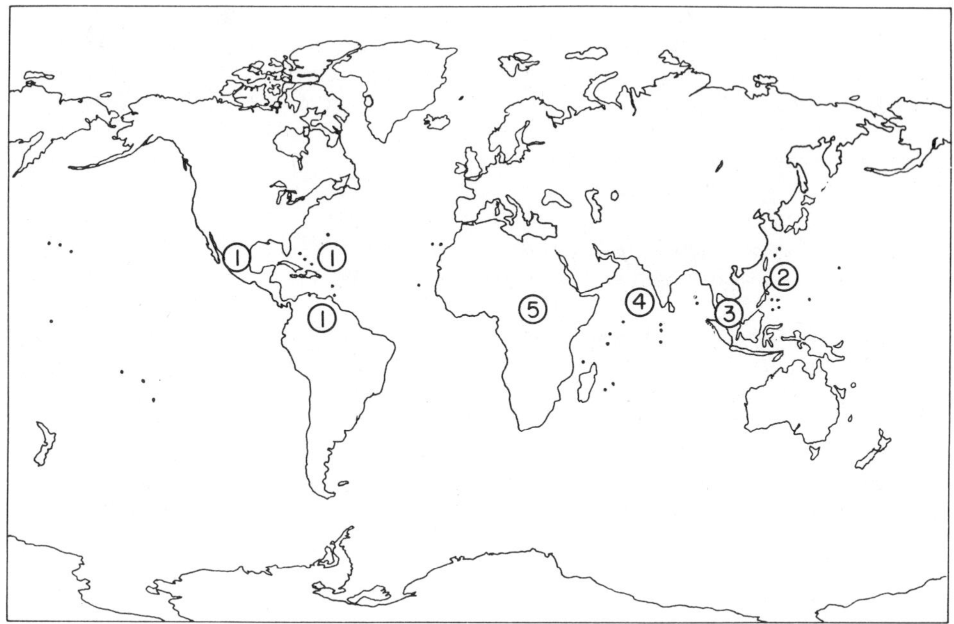

Figure 9. Worldwide distribution of DEN 4 virus toptotypes. Individual topotypes or genetic varieties are indicated by numbers located on the map where the DEN 4 genotype strains were isolated. 1, The Americas; 2, Philippines; 3, Thailand; 4, Sri Lanka; 5, Africa.

of the RNA, and molecular cloning (reviewed by Palese and Roizman, 1980). Understanding genetic and phenotypic heterogeneity among viruses is of importance to understanding disease epidemiology. The molecular characteristics of DEN viruses have been useful in determining the origin of virus strains, antigenic characteristics of specific virus populations, and differentiating between vaccine and wild-type virus strains.

A. Oligonucleotide Fingerprint Comparisons

Molecular techniques have been used to determine the biologic and genetic diversity of DEN viruses transmitted in both endemic and epidemic situations. We investigated the genetics of DEN virus populations in Thailand over a 24-year period to determine whether genetic variants are responsible for serious disease, how much change occurs in the genome over time, and whether this variation affects the antigenic structure of the virus.

Genetic mutations may occur at a high frequency in the genomes of RNA viruses (Holland *et al.* 1982; Steinhauer and Holland, 1986). The poliovirus genome evolves at a rate of about two base substitutions per week during epidemic transmission, so that oligonucleotide fingerprinting only recognizes relationships between isolates separated from the ancestral infection by no more than 3–5 years (Nottay *et al.*, 1981). Mutants could emerge during a single infection and, given a selective advantage, they could predominate in the progeny virus. Studies with the St. Louis encephalitis viruses isolated in the Mississippi River basin over a 30-year period indicate that flaviviruses do not undergo rapid genetic changes (Trent *et al.*, 1981). Oligonucleotide fingerprint analysis of Japanese encephalitis (Hori *et al.*, 1986b) and Murray Valley encephalitis viruses (Coelen and MacKenzie, 1988) indicated that the flavivirus genome changes very little with passage in mosquitoes, in natural hosts, or in cell cultures (Hori *et al.*, 1986a,b).

To investigate the microevolution of endemic DEN 2 virus, we characterized the RNA of the virus by oligonucleotide fingerprinting and analyzed the images with computer-assisted technology (Trent *et al.*, 1988). The RNAse T1 oligonucleotide fingerprint images of 57 strains of DEN 2 virus isolated from DHF/DSS patients in Thailand were digitized and the coordinate systems of each image were transformed to a single strain (D79-069) as a frame of reference; this allowed the comparison of oligonucleotides between any two images (Fig. 10). The analysis area we selected encompassed 38 oligonucleotides in the image of D79-069 (filled circles) and 59 additional ones not in its image but present in the images of other strains (open circles). The number of large oligonucleotides in individual isolates was remarkably constant (39 ± 2.7 oligonucleotides per image), which suggests constancy of size and of base composition of the RNA.

The conserved oligonucleotides identified in the images were used to construct consensus fingerprints of all oligonucleotides that occur in more than 50% of the images. Figure 11 shows the consensus fingerprints for all 57 DEN 2 viruses analyzed and presents three subsets divided by the decade in which the viruses were isolated. By including all oligonucleotides in the analysis, it is possible to calculate an association coefficient (Sab) for each pair of images (Fox and Stackebrandt, 1972). A phylogenetic tree was constructed using the reference ratio method (Fig. 12; Manske and Chapman, 1987), which is a modified form of the simple average clustering method (formally known as UPGMA; Sneath and Skoal, 1973). Using this method, the branching patterns of the first

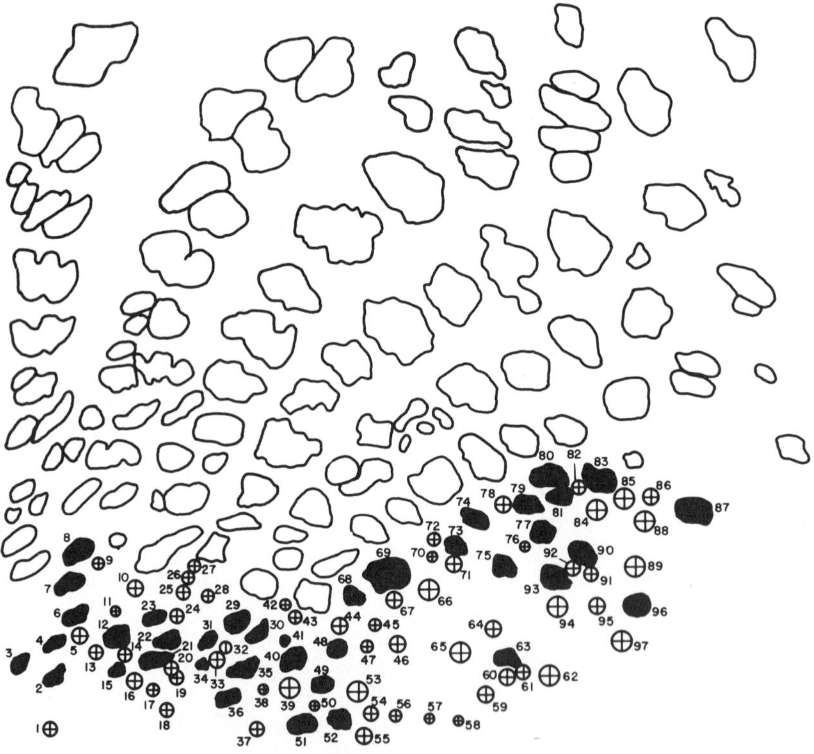

Figure 10. Composite oligonucleotide fingerprint of the Thailand topotype of DEN 2 viruses isolated between 1962 and 1986. RNAse T1-resistant oligonucleotides in the map are numbered in the diagram for identification: oligonucleotides in the fingerprint of strain D79-069 (●) and oligonucleotides in the fingerprints of the other 56 strains (⊗) (Trent *et al.*, 1989).

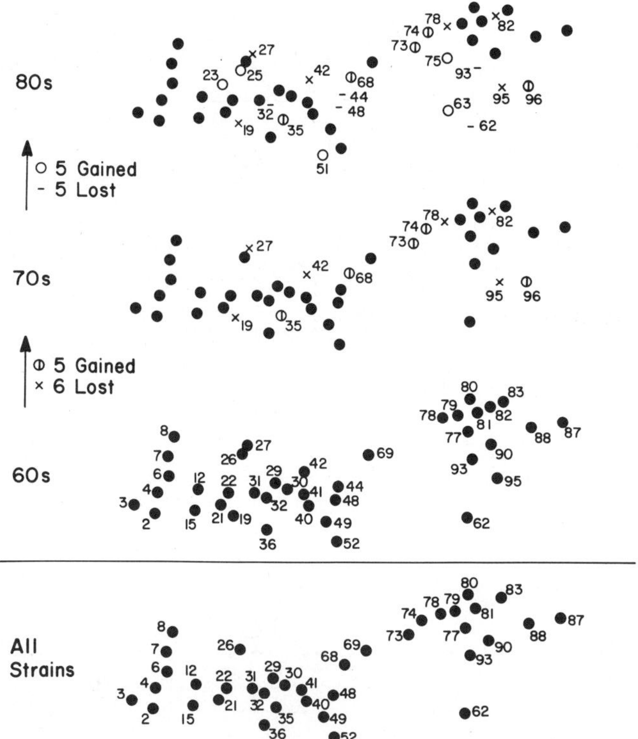

Figure 11. Consensus fingerprints showing oligonucleotides that appear in 50% of the DEN 2 virus strains from Thailand in each decade from 1960 through 1980. RNAse T1-resistant oligonucleotides in the map of each decade are numbered in this diagram as previously identified in Fig. 10 (Trent *et al.*, 1989).

95 strains of Fig. 12 were adjusted to compensate for nonuniform rates of evolution by considering their similarities to the deepest branch in the tree, reference strains 1635-63 and 3227/63. Based on this analysis, the DEN 2 viruses from Thailand were divided into four genetically related clusters (labelled I–IV in Fig. 12). The dendrogram establishes a time-dependent variation in the consensus fingerprint over the 24-year period. The database was divided into three groups according to the decade of isolation, and the groups were examined for similarity. Examining changes in the consensus fingerprints over time reveals that the original 1960s consensus fingerprint changes at a remarkably uniform rate (Fig. 11). The similarity of the fingerprint images decreases by 1.4% (0.014 Sab unit) per year over the 24-year period, with an average of one oligonucleotide image change per year. A true phylogenetic tree of these viruses should have their end-points staggered to reflect different times of isolation (a hypothetical example is shown in Fig. 13A), which are essentially extinctions of the strain. In order to estimate this effect, separate dendrograms were constructed for each set of viruses grouped according to the decade of isolation. These data were combined to form a composite tree (Fig. 13B).

To examine whether such classification would have relevance to biologic characteristics of the viruses, we further noted that eight of the virus strains studied were isolated from *Ae. aegypti* mosquitoes from 1962–1966. The fingerprints of these isolates were similar to those of viruses isolated from DHF/DSS patients during this same period. Thus, the genotype of viruses causing severe disease was not different from that of viruses in the mosquitoes, and there was not a unique genotype causing severe disease.

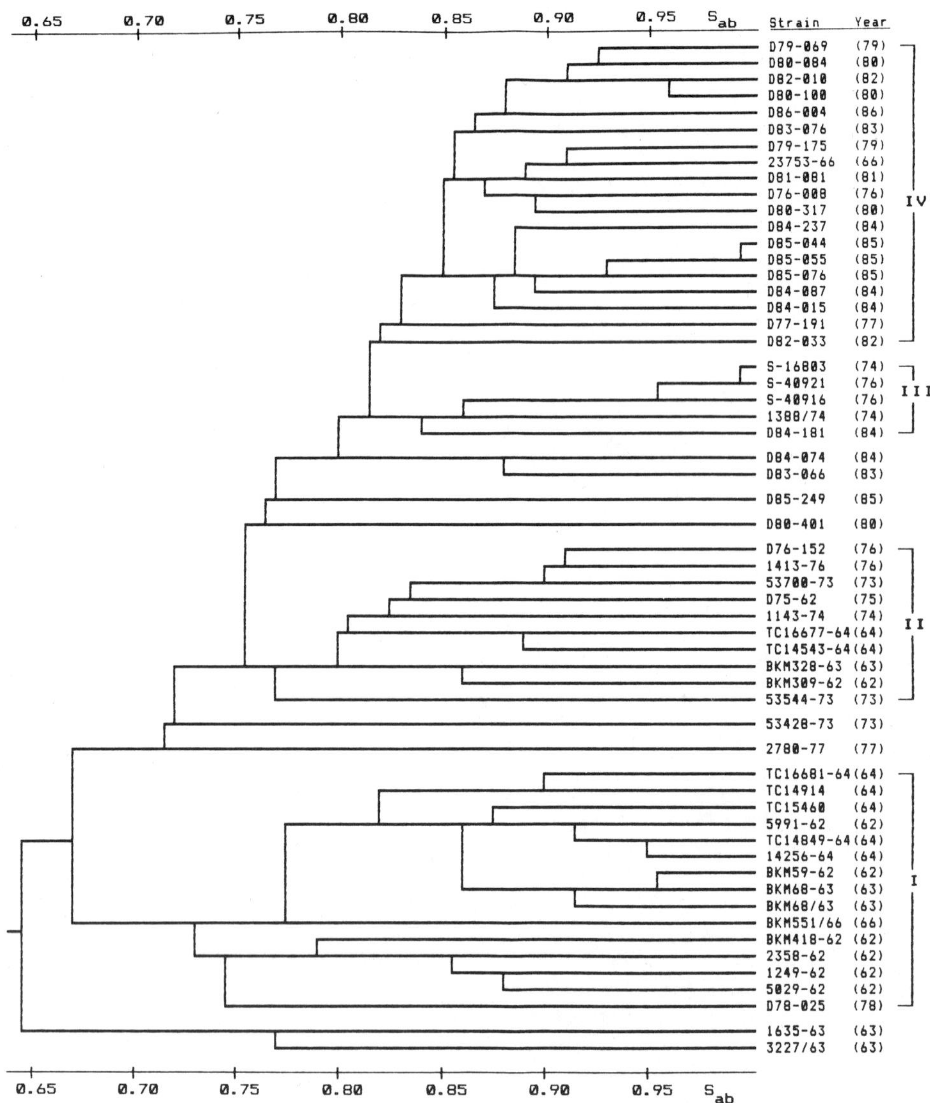

Figure 12. Dendrogram of genetic relationships among Thailand DEN 2 strains as determined by the reference ratio method (Manske and Chapman, 1987).

B. Nucleotide Sequence Comparisons

Comparing nucleotide sequences of the structural gene regions of DEN and other flaviviruses has shown that the structural genes of viruses within a topotype are conserved and that within the nucleotide sequence there are hypervariable regions. To understand the relationship between unique DEN virus fingerprints of the whole genome and of the nucleotide sequence, we cloned and analyzed the structural genome regions of three different DEN 1 virus strains from Thailand (AHF82-80), the Philippines (836-1), and the Caribbean (CV1636/77) (Chu *et al.*, 1988). The fingerprints of the viruses are shown in Fig. 14. Each fingerprint is a topotype representative of

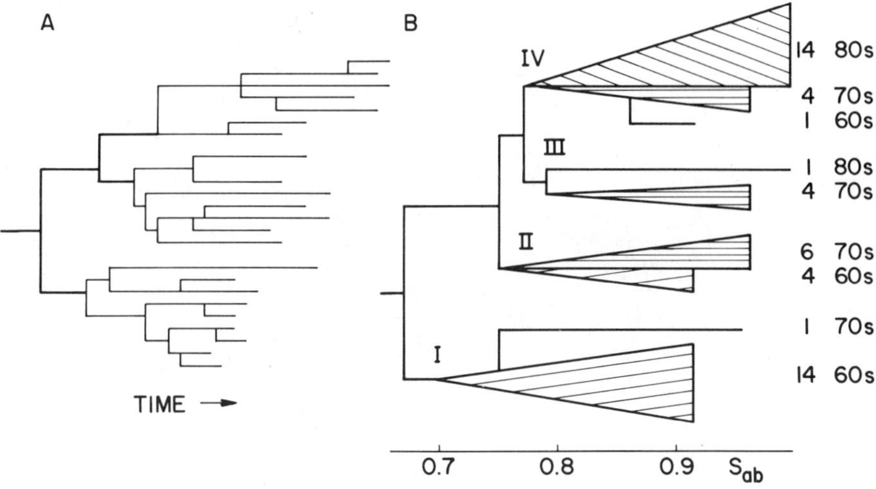

Figure 13. Dendrograms illustrating the evolution of the DEN 2 virus genome. (A) Hypothetical dendrogram illustrating microevolution when the range of isolation times is considered significant in proportion to the rate of genetic change. (B) Composite phylogeny of DEN 2 viruses showing the genetic differences relative to time of isolation, grouped by decade.

other DEN 1 virus strains from their respective region and has less than 50% identity with the others.

The nucleotide sequences encompassing the structural genome regions of these viruses are 97% similar to each other (Fig. 15). Variation between the viruses was most pronounced in the capsid gene, with an overall similarity of 75%. Five amino acid differences in the AHF82-80 DEN 1 virus capsid protein gives a sequence similar to the DEN 2 Jamaican and Puerto Rican topotype sequences (Deubel *et al.*, 1988; Hahn *et al.*, 1988). Divergence in the sequence coding for the premembrane protein of the different DEN 1 strains represented an overall 3% change. Amino acid changes in the premembrane protein at positions 56–60 denote a possible change in the secondary structure of the Thai and Philippine viruses from that of the Caribbean isolate. The only amino acid change in the premembrane protein was located at the same position for all three viruses. The nucleotide sequence of the envelope glycoproteins revealed that nucleotide sequence changes were primarily located in the first one-third regions of the envelope protein gene (Fig. 15). There were 23 nucleotide differences between the Thai and Caribbean viruses (20 transitions and three transversions), resulting in five amino acid differences. Most changes were clustered in the envelope gene nucleotide sequence between nucleotides 217 and 448. In contrast, Philippine strain 836-1 contains unique nucleotide changes before nucleotide 250 in the envelope genome, thus making this strain clearly different from the other two DEN-1 strains.

Signature analysis of DEN 2 viruses using monoclonal antibodies reacting with the envelope protein has permitted separation of the topotypes (Monath *et al.*, 1986). Nucleotides 289–333 of the DEN 1 envelope genome code for an amino acid region that is conserved in all flaviviruses. This suggests that there are variable regions prior to and just after the highly conserved antigenic domain in the envelope protein (Nowak and Wengler, 1987) which play an important role as antigenic determinants of the topotype. Sequence analysis of the envelope protein of Murray Valley encephalitis viruses isolated in Australia and Papua New Guinea (PNG) revealed a 1.7% divergence among the Australian isolates, a 6.8% divergence between Australian and PNG isolates, and a 9–10% divergence between the prototype Australian and PNG viruses (Lobigs *et al.*, 1988).

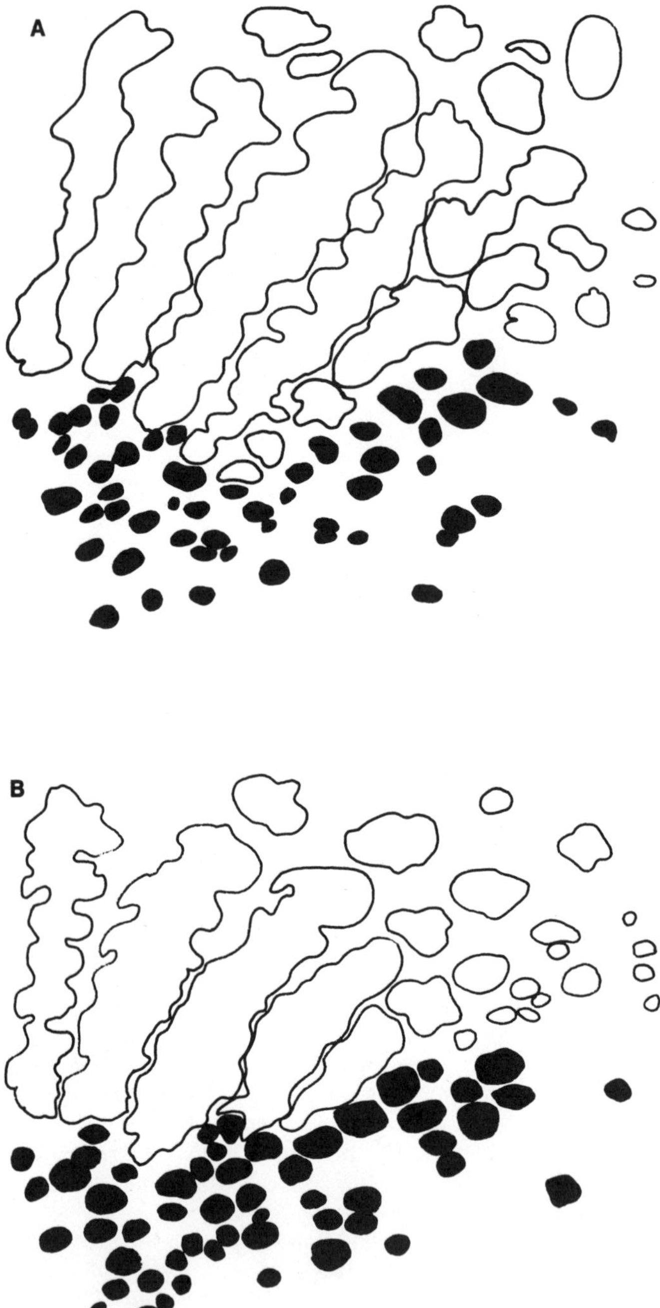

Figure 14. Oligonucleotide fingerprint images of DEN 1 virus strains representing different topotypes: (A) Thailand (AHF 82-80), (B) Philippines (836-1), and (C) Jamaica (CV1636/77).

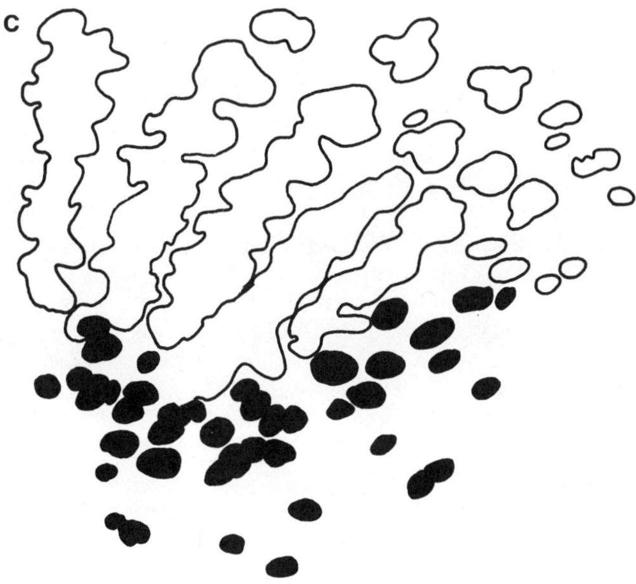

Figure 14. (*Continued*)

V. MOLECULAR CORRELATES OF SEVERE DENGUE VIRUS DISEASE

Two hypotheses have been advanced to explain the pathogenesis of DHF/DSS. The sequential infection hypothesis suggests that DHF/DSS is produced by immunologic mechanisms elicited by a second, heterotypic DEN virus infection, which usually occurs within 4–6 years after the first infection (Halstead *et al.*, 1970). Severe disease may also occur after a primary infection in infants

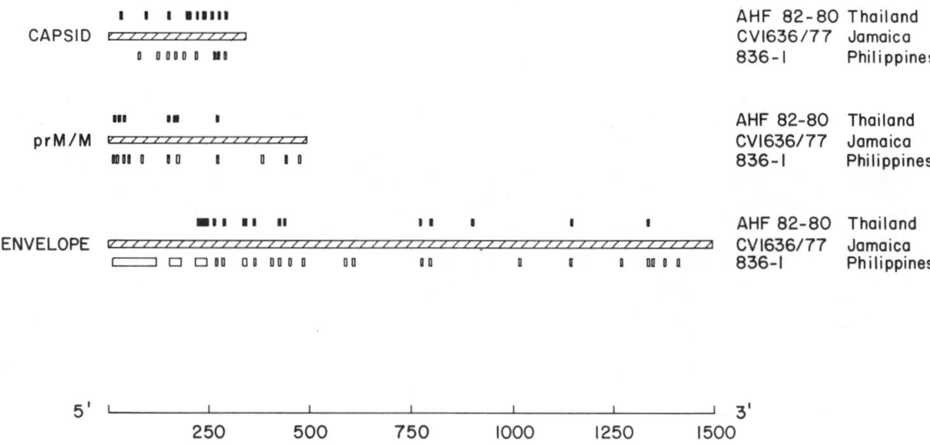

Figure 15. Sequence comparisons of DEN 1 virus through the structural gene regions using CV1636/77 (Jamaica) as the reference sequence. Bars above and below the reference strain indicate regions in the genomes of the AHF82-80 (Thailand) (▮) and 836-1 (Philippines) (▯) strains where nucleotide sequence differences are located.

who have a critical level of circulating anti-DEN antibody from their mothers. In this model, antibody-dependent enhancement of DEN virus replication in mononuclear phagocytes plays a major role in the immunopathology of severe diseases (Halstead *et al.*, 1977; Kliks *et al.*, 1988). This hypothesis suggests that the destruction of monocytes and macrophages during virus replication releases biologic mediators that cause the hemodynamic disorders observed in patients with DHF/DSS (Halstead, 1982). DHF/DSS with secondary infection in Thailand (Burke *et al.*, 1988), Tahiti (Moreau *et al.*, 1973), and Cuba (Guzman *et al.*, 1984,b) supports the sequential infection hypothesis. DHF/DSS has been observed, however, in patients without previous history of DEN virus infection (Barnes and Rosen, 1974). The absence of DHF/DSS with secondary infection in the Tonga Islands (Gubler *et al.*, 1978) and in the Caribbean before the Cuban outbreak raises questions about whether preexisting antibody must be present to cause severe DEN virus disease (Bres, 1979).

The second hypothesis suggests that a virulent strain is responsible for severe DEN virus infection in primary DEN cases (Rosen, 1977; Gubler *et al.*, 1985). Although genetic variants of each serotype are known (Repik *et al.*, 1983; Trent *et al.*, 1983, unpublished data), there is no direct virologic or genetic evidence that supports the second hypothesis.

Epidemiologic data have not ruled out either hypothesis. Epidemiologic and virologic analysis of a DEN virus outbreak in Bangkok, Thailand, indicated that though both DEN 1 and DEN 2 viruses cocirculated in the population, approximately 80% of the DHF/DSS cases were caused by

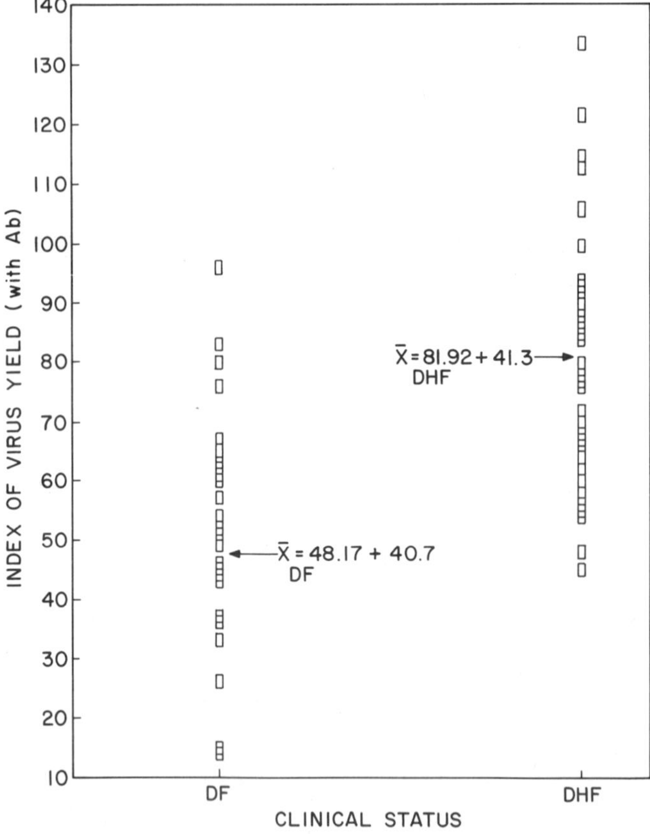

Figure 16. Differences in monocyte infectivity between DEN 2 viruses isolated from patients with dengue fever (DF) and dengue hemorrhagic fever/dengue shock syndrome (DHF/DSS).

DEN 2 virus infection in association with preexisting DEN 1 antibodies (Burke *et al.*, 1988), while in persons without preexisting DEN 2 antibodies, DEN 1 virus infections generally were mild. However, it was also noted that during the mid-1980s in Thailand, DHF/DSS has been associated with primary and secondary DEN 3 infections, suggesting that both hypotheses may be important to understanding severe DEN virus infection.

Replication of DEN viruses can be enhanced by low concentrations of antibody in human, primate, and mouse mononuclear cell cultures (Brandt *et al.*, 1982; Halstead *et al.*, 1973, 1977, 1983). The marked difference in virus replication between DEN 2 virus strain PR159 and its S1 vaccine derivative led to the suggestion that growth of DEN viruses in mononuclear phagocytes *in vitro* could provide a rational basis for measuring virulence (Halstead *et al.*, 1981). More recently,

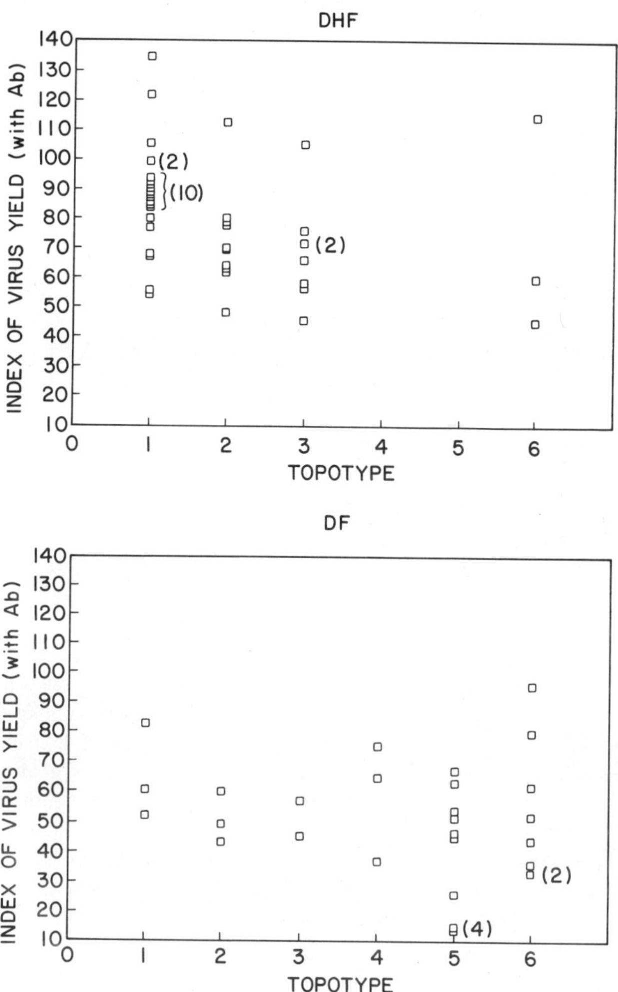

Figure 17. Distribution of virus isolates from DF and DHF/DSS patients among the different DEN 2 topotypes: 1, Thailand/Burma; 2, the Philippines; 3, Indonesia; 4, Sri Lanka; 5, Puerto Rico; 6, Jamaica (Trent *et al.*, 1983). Closed squares represent viruses from DF patients, and open squares represent virus strains from patients with DHF/DSS.

monocyte infectivity of DEN 2 virus strains isolated from patients with DHF/DSS, pyrexia of unknown origin (PUO), or classical DEN fever (DF) in the same epidemic has been studied (Kliks, unpublished observations). When the DEN 2 isolates were ranked according to the amount of virus produced in the presence of enhancing antibodies, all of the isolates associated with DHF yielded more than 5 $\log_{10}$ of virus. The virus yield in the cultures infected with the reference strain (DEN 2 16681) was compared with the yield of other strains and standardized to give a yield ratio or infectivity index. Infectivity indices for the 39 DEN 2 viruses isolated from patients with DHF/DSS were significantly different from those isolated from patients with PUO or DF. The mean indices from DHF/DSS-related strains ranged from 45 to 135, with a mean of 81.92 ($\pm$2 SD 41.32) as compared to a mean of 48.17 ($\pm$2 SD 40.70) in the DF-related strains (Fig. 16). Thus the analysis of 72 DEN 2 virus strains in the monocyte infectivity assay suggests that virus isolates obtained from DHF/DSS patients were correlated with (1) virus production in the presence of enhancing antibodies, (2) increased virus yield in cultures with DHF/DSS-associated strains over strains obtained from PUO/DF patients, and (3) distinct patterns of fluorescent antibody staining of monocytes infected with DHF/DSS-associated isolates as compared to the PUO/DF group. These findings demonstrate that antibody-enhanced monocyte infectivity assay can be a reliable *in vitro* correlate for defining virulence of DEN 2 strains isolated from all grades of DHF cases. In addition, this model suggests that DHF/DSS may be caused by the interaction of specific virus strains, the avidity of the enhancing antibodies, the sensitivity of the host monocytes to infection, and the host's physiologic response to metabolites of DHF/DSS.

The monocyte-infectivity indices of the 72 DEN 2 virus strains were correlated with the topotype of the virus as described by Trent *et al.* (1983). The 39 DEN 2 strains associated with severe disease not only had a higher monocyte-infectivity index, but also belonged to certain fingerprinting topotypes originating from Thailand, the Philippines, Indonesia, and Jamaica (topotypes 1, 2, 3, and 6, respectively; Trent *et al.*, 1983). In contrast, the 33 strains from patients with mild disease were randomly distributed among all of the topotypes (Fig. 17).

VI. CONCLUSIONS

The correlation between genetic variation and biologic characteristics of the virus has permitted the identification of virus topotypes, the understanding of genetic changes that may influence antigenic characteristics, and an insight into virus virulence. The pattern of genetic change in the DEN virus genome over time suggests that new genetic varieties arise through directional selection, random genetic drift. At any point in time the genome that defines the species is in reality a population of more or less similar genomes. As a consequence, the typical genome continues to change so that, at any point, the virus population contains minor genetic components that either are lost or flourish depending on their selective advantage. Fingerprinting of the whole viral RNA and nucleotide sequencing of selected regions of the genome have been used to define DEN virus genetic species and to elucidate changes within antigenically important portions of the envelope protein.

Divergence of DEN 2 viruses from different regions of the world (e.g., Jamaica and Thailand [Trent *et al.*, 1983]) requires that strains be assigned to different topotypes. To establish a topotype, consensus fingerprints of multiple isolates obtained over a period of time from each geographic region must be constructed and used to evaluate genetic diversity. By studying genome change and evolution over a 24-year period in Thailand, we have documented genetic change over time and have established similarities that determine a topotype to include new strains isolated within the same region that share only 50% similarity of their large oligonucleotides (Walker *et al.*, 1988).

Topotypes of all four serotypes of DEN virus have been established, and the geographic limits of their known distribution determined. By using computers to analyze the database, we have established a foundation for continued analysis of DEN virus molecular epidemiology.

Nucleotide sequence data for the entire genomes of DEN 4 (Mackow *et al.*, 1987) and DEN 2 viruses (Hahn *et al.*, 1988; Deubel *et al.*, 1988) and for the genome structural regions of DEN 1 (Mason *et al.*, 1987; Chu *et al.*, 1988) have confirmed that regions of the genome are highly conserved. Although the antigenic structure of the envelope is variable, the critical neutralization site appears to be stable (Monath *et al.*, 1986; Roehrig *et al.*, 1983). In the nonstructural region, NS3 and NS5 are highly conserved (Deubel *et al.*, 1988). Thus, the conservation of the surface glycoprotein and of the putative polymerase genes NS3 and NS5 is believed to maintain integrity of the functions necessary for virus replication in both mosquito vectors and mammalian hosts. An examination of nucleotide sequence variability within different topotypes of a single serotype has revealed remarkable conservation in the nucleotide sequence (Hahn *et al.*, 1988; Deubel *et al.*, 1988; Yaegashi *et al.*, 1986). Although the rate of mutation in the DEN virus genome is unknown, sequence and oligonucleotide fingerprint information predict a 0.43% change per year. Changes in the genome of DEN 1 strains observed by sequence analysis in the envelope protein were less than 0.3% per year over an 8-year period, suggesting that biologically acceptable mutations are infrequent.

The relationship of virus virulence and genetic variability remains an open question. The degree of severe DEN illness is likely the interaction of both the presence of preexisting DEN virus antibodies and the ability of the virus to replicate in its host macrophages (Barnes and Rosen, 1974; Halstead, 1988). The apparent association of DEN 2 isolates from DHF/DSS endemic regions with increased replication in antibody-mediated monocyte cell cultures provides an *in vitro* correlate to study virus virulence and genetic variation. With the knowledge that biologically viable changes in genomic RNA are infrequent, and that the nucleotide sequences of DEN strains representing differing disease designations (Hahn *et al*, 1988; Chu, unpublished observations) are highly conserved, we can begin to associate specific genomic sequences with changes in the biologic characteristics of the virus.

ACKNOWLEDGMENTS

The authors thank Drs. D. Burke, S. B. Halstead, B. Innis, D. Gubler, and L. Rosen for virus strains used in this study and B. J. Blandford for preparation of the manuscript. Dr. A. V. Vorndam, J. A. Grant, R. Tsuchiya, and S. Sviat provided technical support. The work was supported by U.S. Army Medical Research and Development Command contracts PP-3809 and DAHD 17-83C-3167.

REFERENCES

Aaronson, R. P., Young, J. F., and Palese, P. (1982). *Nucleic Acids Res.* **10,** 237–246.

Bang, Y. H., and Sanyakorn, C. K. (1984). *Dengue Newsl.* **10,** 1–7.

Barnes, W. J. S., and Rosen, L. (1974). *Am. J. Trop. Med. Hyg.* **23,** 495–506.

Blok, J. (1985). *J. Gen. Virol.* **66,** 1323–132.

Blok, J., Henchal, E. A., and Gorman, B. M. (1984). *J. Gen. Virol.* **65,** 2173–2181.

Brandt, W. E., McCown, J. E., Gentry, M. K., and Russell, P. K. (1982). *Infect. Immun.* **36,** 1036–1041.

Bres, P. (1979). In *Dengue in the Caribbean, 1977,* pp. 4–10, PAHO Scientific Publication No. 375, Pan American Health Organization, Washington, D.C.

Brinton, M. A. (1986). In *The Togaviridae and Flaviridae* (S. Schlesinger and M. Schlesinger, eds.), pp. 327–374, Plenum Press, New York.

Burke, D. S., Nisalak, A., Johnson, D. E., and Scott, R. McN. (1988). *Am. J. Trop. Med. Hyg.* **38,** 172–180.

Calisher, C. H., Mati, M., Lazuick, J. S., Ferrari, J. D. M., and Kappus, K. D. (1981). *Bull. WHO* **59,** 619–622.

Carey, D. E., Causey, O. R., Reddy, S., and Cooke, A. R. (1971). *Lancet* **1,** 105–106.

Castle, E., Leidner, U., Nowak, T., Wengler, G., and Wengler, G. (1986). *Virology* **149,** 10–26.

Chu, M. C., O'Rourke, E. J., and Trent, D. W. (1989). *J. Gen. Virol.* **70,** 1701–1712.

Coelen, R. J., and MacKenzie, J. S. (1988). *J. Gen. Virol.* **69,** 1903–1912.

Coia, G., Parker, M., Speight, G., Byrnk, M. E., and Westaway, E. G. (1988). *J. Gen. Virol.* **69,** 1–21.

Cornet, M., Saluzzo, J. F., Digoutte, J. P., Germain, M., Chauvancy, M. F., Eyraud, M., Ferrara, L., Heme, G., and Legros, F. (1984). *Cashiers Ser. Entomol. Med. Parasit.* **12,** 313–323.

Coulanges, P., Clerc, Y., Fousse, F. X., Rodhain, F., and Hannoun, C. (1979). *Bull. Soc. Path. Exot. Filiales* **72,** 205–209.

De Madrid, A. T., and Porterfield, J. S. (1974). *J. Gen. Virol.* **23,** 91–96.

Deubel, V., Kinney, R. M., and Trent, D. W. (1986). *Virology* **155,** 365–377.

Deubel, V., Kinney, R. M., and Trent, D. W. (1988). *Virology* **165,** 234–244.

Fox, G. E., and Stackebrandt, E. (1972). *Methods Microbiol.* **19,** 405–458.

Gruenberg, A., Woo, W. S., Biedrazycka, A., and Wright, P. J. (1988). *J. Gen. Virol.* **69,** 1391–1398.

Gubler, D. J. (1987). In *Current Topics in Vector Research,* Vol. III (K. F. Harris, ed.), pp. 37–50, Springer-Verlag, New York.

Gubler, D. J. (1988). In *The Arboviruses: Biology and Epidemiology* (T. P. Monath, ed.), pp. 223–258, CRC Press, Boca Raton, FL.

Gubler, D. J., Reed, D., Rosen, L., and Hitchock, J. C., Jr. (1978). *Am. J. Trop. Med. Hyg.* **27,** 581–589.

Gubler, D. J., Kuno, G., Sather, G. E., and Waterman, H. S. H. (1985). *Am. J. Trop. Med. Hyg.* **78,** 235–238.

Guzman, M. G., Kouri, G. P., Bravo, J., Calunga, M., Soler, M., Vasquez, S., and Venero, C. (1984a). *Trans. R. Soc. Trop. Med. Hyg.* **78,** 235–238.

Guzman, M. G., Kouri, G. P., Bravo, J., Calunga, M., Soler, M., Vasquez, S., Santos, M., Villaescuso, R., Basanta, P., Indan, G., and Ballester, J. M. (1984b). *Trans. R. Soc. Trop. Med. Hyg.* **78,** 239–241.

Hahn, Y. S., Galler, R., Hunkapiller, T., Dalrymple, J. M., Strauss, J. H., and Strauss, E. G. (1988). *Virology* **162,** 167–180.

Halstead, S. B. (1980). *Bull. WHO* **58,** 1–21.

Halstead, S. B. (1982). *Prog. Allergy* **31,** 301–364.

Halstead, S. B. (1988). *Science* **239,** 476–481.

Halstead, S. B., Nimmanitya, S., and Cohen S. N. (1970). *Yale J. Biol. Med.* **42,** 311–328.

Halstead, S. B., Chow, J. S., and Marchette, N. J. (1973). *Nature* **243,** 24–26.

Halstead, S. B., O'Rourke, E. J., and Allison, A. C. (1977). *J. Exp. Biol. Med.* **146,** 218–229.

Halstead, S. B., Tom, M. C., and Elm, Jr., J. L. (1981). *Infect. Immun.* **31,** 102–106.

Halstead, S. B., Larsen, K., Kliks, S., Peiris, J. S. M., Cardosa, J., and Porterfield, J. S. (1983). *Am. J. Trop. Med. Hyg.* **32,** 157–163.

Holland, J., Spindler, K., Harodyski, F., Grabau, E., Nochol, S., and Vanderpal, S. (1982). *Science* **215,** 1577–1585.

Hori, H., Igarashi, A., Yoshida, I., and Takagi, M. (1986a). *Acta Virol.* **30,** 428–431.

Hori, H., Morita, K., and Igarashi, A. (1986b). *Acta Virol.* **30,** 353–359.

Hyams, K. C., Oldfield, E. C., Scott, R. McN., Bourgeois, A. L., Garadiner, H., Pazzaglla, G., Mooussa, M., Saleh, A. S., Dawi, O. E., and Daniell, F. D. (1986). *Am. J. Trop. Med. Hyg.* **35,** 860–865.

Johnson, B. K., Musoke, S., Ocheng, D., Gichogo, A., and Rees, P. H. (1982). *Lancet* **1,** 208–209.

Kerschner, J. H., Vorndam, A. V., Monath, T. P., and Trent, D. W. (1986). *J. Gen. Virol.* **67,** 571–586.

Kliks, S. C., Nimmanitya, S., Nisalak, A., and Burke, D. S. (1988). *Am. J. Trop. Med. Hyg.* **38,** 411–419.

Kokernot, R. H., Smithburn, K. C., and Weinbren, M. P. (1956). *J. Immunol.* **77,** 313–323.

Le Gonidec, P. G., Queue, J. P., and Fauran, P. (1982). *Bull. Soc. Pathol. Exot. Filiales* **75,** 141–150.

Likosky, W. H., Calisher, C. H., Michelson, A. L., Correa-Coronas, R., Henderson, B. E., and Feldman, R. A. (1973). *Am. J. Epidemiol.* **97,** 264–275.

Lobigs, M., Marshall, I. D., Weir, R. C., and Dalgarno, L. (1988). *Virology* **165,** 245–255.

Mackow, W., Makino, Y., Zhao, B., Zhang, Y. M., Markoff, L., Buckler-White, A., Guiler, M., Chanock, R., and Lai, C. J. (1987). *Virology* **159,** 217–228.

Manske, C. L., and Chapman, D. J. (1987). *J. Mol. Evol.* **26,** 226–251.

Mason, P. W., McAda, P. C., Mason, T. L., and Fornier, M. J. (1987). *Virology* **161,** 262–267.

Monath, T. P. (1984). In *Applied Virology* (E. Kurstak, ed.), pp. 377–400, Academic Press, New York.

Monath, T. P., Kinney, R. M., and Schlesinger, J. J. (1983). *J. Gen. Virol.* **64,** 627–637.

Monath, T. P., Wands, J. A., Hill, L. J., Brown, N. V., Marciniak, R. A., Wong, M. A., Gentry, M. K., Burke, D. S., Grant, J. A., and Trent, D. W. (1986). *Virology* **154,** 313–324.

Moreau, J. P., Rosen, L., Saugrain, J., and Lagraulet, J. (1973). *Am. J. Trop. Med. Hyg.* **22,** 237–241.

Nottay, B. K., Kew, O. M., Hatch, M. H., Heyward, J. T., and Obijeski, J. F. (1981). *Virology* **108,** 405–423.
Nowak, T., and Wengler, G. (1987). *Virology* **156,** 127–137.
Palese, P., and Roizman, B., eds. (1980). *Ann. NY Acad. Sci.* **354,** 1–507.
Parc, F., Pichon, G., and Tetaria, C. (1981a). *Med. Trop. (Mars.)* **41,** 93–96.
Parc, F., Pichon, G., and Tetaria, C. (1981b). *Trop. Med. (Mars.)* **41,** 97–101.
Porterfield, J. S. (1980). In *The Togaviruses* (R. W. Schlesinger, ed.), pp. 13–16, Plenum Press, New York.
Repik, P. M., Dalrymple, J. M., Brandt, W. E., McCown, J. M., and Russell, P. K. (1983). *Am. J. Trop. Med. Hyg.* **32,** 577–589.
Rice, C. M., Lences, E. M., Eddy, S. R., Shin, S. J., Sheets, R. S., and Strauss, J. H. (1985). *Science* **229,** 726–733.
Rice, C. M., Strauss, E. G., and Strauss, J. H. (1986). In *The Togaviridae and Flaviviridae* (S. Schlesinger and M. Schlesinger, eds.), pp. 279–326, Plenum Press, New York.
Robin, Y., Cornet, M., Heme, G., and LeGonidec, G. (1980). *Ann. Virol.* **131,** 149–154.
Roehrig, J. T., Mathews, J. H., and Trent, D. W. (1983). *Virology* **128,** 118–126.
Rosen, L. (1977). *Am. J. Trop. Med. Hyg.* **26,** 337–343.
Rudnick, A. (1965). *J. Med. Entomol.* **2,** 203–208.
Saleh, A. S., Hassan, A., Scott, R. McN., Mellick, P. W., Oldfield III, E. C., and Podgore, J. K. (1985). *Lancet* **1,** 211–212.
Schlesinger, R. W. (1979). *Dengue Viruses,* Virology Monographs Vol. 16, Springer-Verlag, New York.
Setyorogo, S. (1984). *Dengue Newsl.* **10,** 8–19.
Siler, J. F., Hall, M. W., and Hitchens, A. P. (1926). *Philip. J. Sci.* **29,** 1–302.
Sneath, P. H. A., and Skoal, R. R. (1973). *Numerical Taxonomy,* W. H. Freeman, San Francisco.
Steinhauer, D. A., and Holland, J. J. (1986). *J. Virol.* **57,** 219–228.
Technical Advisory Group on Dengue Hemorrhagic Fever/Dengue Shock Syndrome (1986). *Dengue Hemorrhagic Fever: Diagnosis, Treatment and Control,* World Health Organization, Geneva, Switzerland.
Trent, D. W., Grant, J. A., Vorndam, A. V., and Monath, T. P. (1981). *Virology* **114,** 319–332.
Trent, D. W., Grant, J. A., Rosen, L., and Monath, T. P. (1983). *Virology* **128,** 271–284.
Trent, D. W., Grant, J. A., Monath, T. P., Manske, C. L., Corina, M., and Fox, G. E. (1989). *Virology* **172,** 523–535.
Vezza, A., Rosen, L., Repik, P., Dalrymple, J. M., and Russell, P. K. (1980). *Am. J. Trop. Med. Hyg.* **29,** 643–652.
Walker, P. J., Henchal, E. A., Blok, J., Repik, P. M., Henchal, L. S., Burke, D. S., Robbins, S. J., and Gorman, B. M. (1988). *J. Gen. Virol.* **69,** 591–602.
Westaway, E. G., Brinton, M. A., Gaidamovich, S., Horzinek, M. C., Igarashi, A., Kaarainen, L., Lvov, D. K., Porterfield, J. S., Russell, P. K., and Trent, D. W. (1985). *Intervirology* **24,** 183–192.
World Health Organization (1986a). *Weekly Epidemiol. Rec.* **61,** 20–21.
World Health Organization (1986b). *Weekly Epidemiol. Rec.* **61,** 205–208.
World Health Organization (1986c). *Weekly Epidemiol. Rec.* **61,** 306–307.
Yaegashi, T., Vakharia, V. N., Page, K., Sasaguri, Y., Feighny, R., and Padmanaban, R. (1986). *Gene* **46,** 257–267.
Zhao, B., Mackow, E., Buckwalter-White, A., Markoff, L., Chanock, R. M., Lai, C. J., and Makino, Y. (1986). *Virology* **155,** 77–88.

17

Hantavirus Variation and Disease Distribution

Guido van der Groen

I. INTRODUCTION

At present, the term "hantavirus" is not well known to most clinicians or even to many virologists. Hantavirus is a new genus of the Bunyaviridae family of viruses (Executive Committee, 1987) and the causative agent of a disease called hemorrhagic fever with renal syndrome (HFRS) (World Health Organization, 1983).

We still do not know the true prevalence of human disease caused by these viruses, but it is clear that they are distributed much more widely than Korea, where they were discovered originally. Hantaviruses cause chronic, silent infections of various rodents and some other animal species in all continents of the world, and HFRS may be recognized in those areas of the world where infected rodents come into close contact with humans. Transmission occurs mainly through the aerogenic route.

Since the discovery of the prototype virus Hantaan by Lee and Lee (1976), a large amount of new data concerning hantaviruses and HFRS became available. Much of these data were reviewed recently (Yanagihara and Gajdusek, 1987; LeDuc, 1987; Gajdusek *et al.*, 1987).

Different types of hantaviruses have been isolated and studied. They share antigenic, genetic, epidemiologic, and ecologic characteristics. In this chapter I summarize the virologic and serologic characteristics of hantaviruses and the clinical disease pattern they may induce in humans.

II. HANTAVIRUS INFECTION IN RODENTS

For a better understanding of the natural evolution of hantavirus infection in humans, it is important to mention very briefly the critical role played in rodents in virus maintenance and transmission. Hantavirus cause chronic, apparently asymptomatic infection in their host. Wild rodents as well as laboratory rodents are the most important hosts. The majority of rodent species harboring hantaviruses are probably nonpathogenic for humans, except under unusual circumstances. Rats clearly serve as the primary reservoirs of hantaviruses in urban centers around the world. Mild forms of HFRS is expected in areas where hantavirus-infected bankvoles (*Clethrionomys glareolus*) are present, and a more severe form of HFRS is expected in areas where infected *Apodemus agrarius* is the principal reservoir. In Eastern Europe, where both *Apodemus* and

---

Guido van der Groen • Institute of Tropical Medicine, 2000 Antwerp, Belgium.

Clethrionomys species served as reservoirs, both mild and severe forms of HFRS were observed. Other small mammals, including the peridomestic musk shrew (*Suncus murinus*) (Tang *et al.*, 1985; Tsai *et al.*, 1984), the domestic cat (*Felis catus*) (Hung *et al.*, 1985), and a hamster (Lee, 1988b), have been found to be infected with hantaviruses. The polyhostal nature of hantaviruses was extensively reviewed by van der Groen *et al.* (1986) and Yanagihara and Gajdusek (1987).

Susceptible rodents are viremic for about a week. Hantaviral antigen is dessiminated throughout the body (Kurata *et al.*, 1987) and usually present abundantly in the host lungs, spleen, and kidneys. Perhaps antigen expression persists for the total lifetime of the rodents and is not diminished when indirect immunofluorescent assay (IFA) and neutralization-type antibodies are formed. The virus is shed for long periods in urine, feces, or saliva. Transmission to humans, as well as horizontal passage of the virus between rodents, is thought to occur primarily via aerosolized urine or feces. Hantaan and Seoul viruses are pathogenic to suckling mice, where the virus is inoculated intracerebrally (Tsai *et al.*, 1984). It was shown that after inoculation of suckling mice, differences in virulence of hantaviruses in China could be demonstrated (Dashi *et al.*, 1987). To what extent these differences in virulence also have pathogenic consequences in the human being is not known. The role of rodents in the spread of hantaviruses was covered in an excellent review by LeDuc *et al.* (1986) and LeDuc (1987); the mode of transmission to humans was reviewed by Tsai (1987) and Yanagihara and Gajdusek (1987).

III. HANTAVIRUS VARIATION

In order to study the correlation between the clinical outcome in humans of hantavirus infection and the virologic and serologic characteristics of the hantavirus, one should isolate the virus either from the patient or from infected animals living in the vicinity of clinical documented cases.

In 1976, a viral agent designated Hantaan virus was isolated (Lee and Lee, 1976) from the Korean striped field mouse (*Apodemus agrarius corea*) captured in the endemic area where a severe form of HFRS, also called Korean hemorrhagic fever (KHF), was described. Since French *et al.* (1981) and McCormick *et al.* (1984) were able to adapt and assay hantavirus strain 76-118 in continuous cell lines such as A 549 (human lung carcinoma) and the E6 clone of Vero E6 cells, the number of hantavirus strains being isolated from humans and animals has increased enormously.

A. Hantavirus Isolation from Humans

Hantaviruses antigenically very close to the Hantaan 76-118 prototype virus have been isolated from patients with a severe form of HFRS in Korea (Lee and Lee, 1977; Lee *et al.*, 1978), in China (Yu-tu, 1983; Men Ruhe *et al.*, 1983; Song *et al.*, 1984; Yen and Qiu, 1984), where the disease is also called epidemic hemorrhagic fever (EHF), in the far eastern part of the U.S.S.R. (Tkachenko *et al.*, 1983, 1984; Dantas *et al.*, 1987), and in Greece (Antoniadis *et al.*, 1987a,b).

Hantaviruses have also been isolated from patients with a mild form of HFRS in the western part of the U.S.S.R. (Tkachenko, personal communication).

B. Hantavirus Isolation from Animals

1. Apodemus Mice Strains

Hantaviruses have been isolated from wild *Apodemus* mice in endemic areas with a severe form of HFRS in Korea (Lee and Lee, 1977), China (Song *et al.*, 1982), the far eastern part of the U.S.S.R. (Tkachenko *et al.*, 1988), and the southern and northern part of Yugoslavia (Gligic *et al.*, 1986; Avsic-Zupanc *et al.*, 1988).

2. Urban Rat Strains

Rattus-derived hantaviruses, some of which have been associated with clinical disease in Korea (Lee *et al.*, 1982) and China (Song *et al.*, 1983), have been isolated from wild rats in Korea (Lee *et al.*, 1982), Japan (Sugiyama *et al.*, 1984; Arikawa *et al.*, 1985b), the People's Republic of China (Song *et al.*, 1982, 1983) Brazil (LeDuc *et al.*, 1985), the United States (LeDuc *et al.*, 1984; Tsai *et al.*, 1985; Childs *et al.*, 1987, 1988), Egypt, Thailand, Hong Kong, Singapore, and recently Italy (Lee, 1988b; Nuti *et al.*, 1987, 1988).

3. Laboratory Rat Strains

Hantaviruses have also been isolated from laboratory rats in Japan and Korea (Kitamura *et al.*, 1983) as well as from rat tumors in Japan (Yamanishi *et al.*, 1983) and England (Lloyd and Jones, 1986).

4. Clethrionomys Mice Strains (Bankvoles)

Hantaviruses have been isolated in areas with a mild form of HFRS in Finland (Lee, 1988b), Sweden (Niklasson and LeDuc, 1984; Yanagihara *et al.*, 1984), the European part of the U.S.S.R. (Tkachenko *et al.*, 1983, 1984; Gavrilovskaya *et al.*, 1983; Chumakov *et al.*, 1981), Yugoslavia (Gligic *et al.*, 1986; Avsic-Zupanc *et al.*, 1988a), and Belgium (van der Groen et al., 1983a,b).

5. Microtus Mice Strain

In the United States, Prospect Hill virus was isolated from meadow voles (*Microtus pennsylvanicus*) (Lee *et al.*, 1985a) and isolates Nr 5303 and 5181 from *Microtus fortis* in the far eastern part of the U.S.S.R. (Vladivostok) (Tkachenko *et al.*, 1988).

6. Mus musculus strain

Leaky virus was isolated from *Mus musculus* in the United States (Baek *et al.*, 1987).

7. Hamster Strain

Recently a hantavirus was isolated from a hamster in Korea (Lee, 1988b).

8. Domestic Cat Strain

Hantaviruses have been isolated from domestic cats in endemic areas of EHF in China (Zhao Zhuang *et al.*, 1987).

9. Bandicota indica Strain

This strain was isolated in Thailand (Lee, 1988b).

C. Antigenic Relationships between Hantaviruses

A hantavirus type is defined on the basis of its immunologic distinctiveness as determined by quantitative neutralization of animal or human antisera. A hantavirus type either has no cross-reaction with others or shows a homologous-to-heterologous titer ratio of greater than 16 in both directions. If neutralization shows a certain degree of cross-reaction between two viruses in either or both directions (homologous-to-heterologous titer ratio of 8 to 16), we define the virus as a subtype.

Plaque reduction neutralization (PRN) tests with different hantaviruses have been performed (Schmaljohn *et al.*, 1985; Arikawa *et al.*, 1985a; Goldgaber *et al.*, 1984; Lee *et al.*, 1985b; Elliott *et al.*, 1985; Takenaka *et al.*, 1985; Tanishita *et al.*, 1984; Yamanishi *et al.*, 1984; Lee and Lee, 1987a; LeDuc *et al.*, 1986).

Four antigenically distinct groups or serotypes have been suggested, based on the prototype virus of each major rodent genus: Hantaan for *Apodemus*-derived isolates, Seoul virus for *Rattus*-derived isolates, Puumala virus for *Clethrionomys*-derived isolates, and Prospect Hill for *Microtus*-derived isolates. Probably other serotypes and subtypes do exist, as shown by recent isolations. Lee *et al.* (1985b) suggested the existence of yet other types of hantaviruses in Eastern Europe, specifically in Yugoslavia where some sera from HFRS patients did not conform to any of the four known serotypes described. This was recently confirmed by the plaque reduction assays with rat antisera prepared against the Yugoslavian-isolated Fojnica and Plitvice strains, both isolated from *Apodemus flavicollis* captured in endemic areas (Gligic *et al.*, 1986, 1987), which showed that these viruses were antigenically more related to but not identical with Hantaan 76-118. A clear difference with the Puumala type was observed. So they probably form a subtype in the Hantaan type. Additional evidence was also obtained by P. W. Lee (unpublished data), who showed that FB 79-3 virus, isolated from *Apodemus sp.* in Korea, reacted equally well in plaque reduction assays with rat antiserum against FB 79-3 virus and Hantaan 76-118, but much higher neutralization titers on the FB 79-3 virus were obtained with sera of HFRS patients from Listice ad Bangaluka collected in Yugoslavia as well as with some sera of HFRS patients in Albania. No antigenic differences, however, could be detected between Hantaan 76-118, Plitvice, Fojnica, and FB 79-3 strain when six monoclonal antibodies were tested in the IFA (see Table 1).

The Porogia virus strain, isolated from an HFRS patient in Greece, living in an area where seropositive *A. flavicollis* was also documented, showed slightly higher (less than 16) plaque reduction neutralization antibody titers with Greek patients' sera than on Hantaan virus. This is evidence that Porogia is a subtype of the Hantaan virus type (Antoniadis *et al.*, 1987a). Recently it was shown that the ultrastructural characteristics of the Porogia and Fojnica virus were very similar (White et al., 1988). Another Hantaan virus subtype is the Maaji virus, recently isolated from *A. agrarius* in Korea (P. W. Lee, unpublished data), which was of particular interest because it reacted in PRN very well with some Yugoslavian HFRS patients' sera.

Leaky virus, isolated from *Mus musculus* in the United States (Baek *et al.*, 1987), was inoculated in rats. The rat immune serum revealed, by serum-dilution, PRN tests, no cross-neutralization with prototype Hantaan, Seoul, Puumala, or Prospect Hill viruses. These data indicate that yet another serotype of hantavirus is enzootic among wild rodents in the United States. Whether this leaky virus is pathogenic for humans is not known.

Minor antigenic differences were observed among hantavirus strains belonging to one serotype when monoclonal antibody reaction patterns in the IFA were compared. Among Hantaan-type or *Apodemus* sp.–derived strains, two different monoclonal reaction patterns were observed, whereas among the Seoul type or *Rattus* sp.–derived strains, four different reaction patterns were observed when 36 hantavirus strains of Chinese origin were analyzed by the method cited (Xu *et al.*, 1986). Additional evidence for strain variability among one hantavirus type was also obtained by H. W. Lee (1988b), where the IFA reactivity pattern of nine different anti-Hantaan monoclonal antibodies in IFA on 19 different hantavirus strains from widely different geographic areas were compared. Minor differences in reaction pattern were observed among Hantaan-type viruses isolated from patients with the severe form of HFRS in Korea. At least three different monoclonal antibody reaction patterns were observed with Hantaan-type virus isolated from *A. agrarius* sp..

Using a panel of six monoclonals (four anti-Hantaan 76-118, two anti-B-1 strains) in IFA on 32 different hantavirus strains isolated in nine different countries from rodents, infected human patients, cat and rat (tumors), six different reaction patterns were observed. Four of the reaction

Table 1. Reactivity of Monoclonal Antibodies Elaborated from Hantaan and B-1 Virus Strains with Various Hantaviruses in the Indirect Immunofluorescent Antibody Assay[a]

			Monoclonal antibodies[b]						
No.	Virus strain	Source	80-A	133-E	141-D	B-8	B-61	19-B	Type
1	A5	*A. agr.*, China	+	+	+	−	−	+	
2	A9	*A. agr.*, China	+	+	+	−	−	+	
3	A96	*A. agr.*, China	+	+	+	−	−	+	
4	Hantaan	*A. agr.*, S. Korea	++	++	++	−/−	−/−	++	
5	HA1018	*A. agr.*, China	+	+	+	−	−	+	
6	Chuan-76	*A. agr.*, China	+	+	+	−	−	+	
7	R27	*R. nor.*, China	+	+	+	−	−	+	I
8	Chen	Patient, China	+	+	+	−	−	+	
9	C4	Cat, China	+	+	+	−	−	+	
10	CG 3883	*C. gl.*, USSR	+	+	+	−	−	+	
11	Fojnica 2508	*A. flav.*, Yugoslavia	+	+	+	−	−	+	
12	Plitvice 2829	*A. flav.*, Yugoslavia	+	+	+	−	−	+	
13	FB 79-3	*A. sp.*, Korea	+	+	+	−	−	+	
14	R22	*R. nor.*, USA	+	+	+	+	+	−	
15	GOU-1	*R. nor.*, China	+	+	+	+	+	−	
16	Tchoupitoulas	*R. nor.*, USA	++	++	++	++	++	−/−	
17	Girard-Point	*R. nor.*, China	++	++	++	++	++	−/−	
18	L99	*R. los.*, China	+	+	+	+	+	−	II
19	SR-11	Rat, Japan	++	++	++	++	++	−/−	
20	Urban rat	S. Korea	+	+	+	+	+	−	
21	Seoul	S. Korea	+	+	+	+	+	−	
22	A1	*A. agr.*, China	+	+	+	+	+	−	
23	A3	*A. agr.*, China	+	+	+	+	+	−	
24	Hällnäs	*C. gla.*, Belgium	++	++	++	−/−	−/−	−/−	
25	CG 13891	*C. gla.*, Belgium	++	++	++	−/−	−/−	−/−	
26	CG 14444	*C. gla.*, Belgium	+	+	+	−	−	−	III
27	CG 14445	*C. gla.*, Belgium	+	+	+	−	−	−	
28	CG 18–20	*C. gla.*, USSR	+	+	+	−	−	−	
29	Vranica	*C. gla.*, Yugoslavia	+	+	+	−	−	−	
30	PH	*M. pen.*, USA	+	+	−	−	−	−	IV
31	GB	Immunocytomas, GB	++	++	++	++	−/−	−/−	V
32	Dobrava 3970	*A. fla.*, Yugoslavia	+	+	−	+	−	−	VI

[a]*A. fla., Apodemus flavocillis; A. agr., Apodemus agrarius; C. gla., Clethrionomys glareolus; M. pen., Microtus pennsylvaticus; R. non., Rattus norvegicus; R. los., Rattus losea.* +, Positive immunofluorescent antibody test. Each monoclonal was titrated twice. −, Negative result means no characteristic fluorescence was observed at a 1:16 dilution. ++, Positive results obtained in the Microbiology Depeartment, Institute of Tropical Medicine Antwerp, Belgium. These were confirmed at the Institute of Virology, Chinese Academy of Preventive Medicine, Beijing, China, by Dr. Song and Dr. Xing. −/−, Negative results were confirmed in the above-mentioned institutes.

[b]Monoclonal antibodies were kindly supplied by Dr. Yamanishi (Osaka, Japan). 80-A, 133-E, 141-D, and 19-B are mouse ascitic fluids obtained from BALB/C mice inoculated with the prototype Hantavirus strain 76-118 whereas B-8 and B-6 were prepared using the B-1 hantavirus strain as previously described (Yamanishi *et al.*, 1984).

patterns (patterns I–IV in Table 1) coincided well with the grouping of the hantavirus type according to the main host from which they were isolated and were a confirmation of the neutralization antibody assay classification (van der Groen *et al.*, 1988). Surprisingly, virus strains belonging to the Seoul type from different geographic areas, such as China, the United States, Japan, and Korea, all reacted in a similar way with these six monoclonals, indicating a close antigenic similarity.

Two hantavirus strains, strain GB-B and the Dobrava strain, clearly reacted in a different way compared to the other 30 strains tested. GB-B was isolated from a rat immunocytoma (Lloyd and Jones, 1986) in England. In addition, five other isolates from rat immunocytomas in England all reacted similarly to the GB-B strain (Lloyd *et al.*, personal communication). Dobrava was isolated from *A. flavicollis* in Slovenia, in northern Yugoslavia (Avsic-Zupanc *et al.*, 1988). Neutralization experiments with reference antisera should be performed on these strains, to determine whether they form a subtype among the existing serotypes or whether they are really new hantavirus types.

Hantavirus strains isolated from patients with HFRS in a certain area possess a similar antigenic specificity to the strains isolated from the principal host animal in the same area (Tkachenko *et al.*, 1987).

Hantaviruses contain four virion proteins, a virus nucleocapsid protein (N) of 50–53 kDa, and two envelope glycoproteins (G1 and G2) with molecular weights of 64–74 K and 55–60 K, respectively. A minor large protein of approximately 200 kDa was also reported. G1 and G2 are highly glycosylated. Molecular characteristics of hantaviruses were recently extensively reviewed by Yanagihara and Gajdusek, 1987). The major cross-reactive antigen between the hantavirus types is the N protein. The G2 protein exhibited weak cross-reactivity while the G1 protein appeared to be the least cross-reactive. This was recently confirmed by Lukashevich and Tkachenko (1987). Convalescent sera of patients with a severe form of HFRS in the far eastern part of the U.S.S.R. precipitated only the nucleic protein (45–50 kDa) and the G2 glycoprotein (55–57 kDa) of the CG 18-20 strain in radioimmunoprecipitation assay (RIPA). The G1 protein was not precipitated. The CG 18-20 was a Hantavirus isolated from bankvoles captured in endemic areas in the European part of the U.S.S.R. where a mild form of HFRS was observed. However, all three proteins, G1, G2, and the N protein, were precipitated by convalescent sera of patients with a mild form of HFRS. This antigenic grouping matches the clinical grouping of the HFRS, that is, a severe vs. a milder form of the disease. However, it remains to be determined how these antigenic differences are related to the pathogenic differences between these viruses.

IV. CLINICAL FEATURES OF HANTAVIRUS INFECTION IN HUMANS

Hantavirus infection is difficult to diagnose on clinical grounds alone, because of the large variety of symptoms. HFRS can appear in a subclinical, mild, moderate, or severe form. The severe form of HFRS is best known and studied in Korea, China, and the far eastern part of the U.S.S.R., where it is also known as KHF, EHF, and nephroso nephritis. The mild form of HFRS was initially well documented in Scandinavia and known as nephropathia epidemica (NE).

A. Hantaan Virus and Hantaan-Related Virus Infection

Hantaan-type viruses have been isolated from HFRS patients as well as from *Apodemus* mice in Korea, China, and the far eastern part of the U.S.S.R. Table 1 shows that there are slight antigenic differences among the Hantaan-type viruses. To what extent these differences are reflected in pathogenic differences for humans is not well known. Generally, Hantaan virus–type infection results more often in a more severe form of HFRS, with mortality rates in the range of 5–20% (Tsai, 1987).

1. Korea

In Korea, the major manifestations of the disease are fever, prostration, vomiting, proteinuria, hemorrhagic phenomena, shock, and renal failure (Lee, 1988b). The incubation period is generally 2–3 weeks. Approximately 30% of patients show a mild clinical course, about 50% exhibit a moderate course, and 20% have severe forms of the disease. Criteria for estimating severity of the various phases of HFRS in Korea have been defined (Lee, 1988b). The severe forms can be divided into five phases (febrile, hypotensive, oliguric, diuretic, convalescent). A detailed description of each of these phases according to characteristic clinical, laboratory, and pathophysiologic features may be found in Lee (1988b).

Prognostic factors for a severe course of KHF are the degree of facial flushing, fever, and conjunctival hemorrhage, the duration of fever, the number and degree of petechiae, and the presence of facial petechiae and small amounts of albumin in the urine during the febrile phase (Lee, 1988a). In China, despite the relatively small number of HFRS patients studied (9/95), a number of variables could be associated with fatal outcome. The most important of these on presentation were low serum total protein, serum calcium, and plasma plasminogen and high serum aspartate transaminase (AST). Multivariate analysis revealed that serum total protein and AST were independent predictors of mortality. Serum total protein was the most important of the two (Cosgriff *et al.*, 1988). Identification of patients at high risk of dying early in the course of the disease makes it possible to target these patients for new types of therapy.

Clinical features of Hantaan virus infection in Korea are summarized and compared to other forms of HFRS in Table 2.

The signs and syptoms of HFRS are not pathogenic. A history of exposure to infected rodents in an endemic area and the appearance of fever, abdominal pain and severe retching, marked

Table 2. Hantavirus Types: Preliminary Listing

Hantavirus type	Main host and distribution	Human disease[a]
I. Hantaan type		Severe HFRS
Prototype: Hantaan 76–118	*Apodemus agrarius* (striped field mouse—Asia and central and southern Europe)	KHF ENF
Subtypes?		
Dobrava	*Apodemus flavicollis* (yellow-necked field	Not identified
Fojnica	mouse—Balkan, Yugoslavia)	Severe HFRS?
Plitvice	*A. flavicollis* (Yugoslavia)	Severe HFRS?
Porogia	*A. flavicollis* (Greece)	Severe HFRS?
Maaji	*Apodemus* sp. (Korea)	Not identified
II. Seoul type		
Prototype: Seoul virus	*Rattus norvegicus* (Asia, U.S.A., Japan, Korea, port cities in the world)	Moderate HFRS
III. Puumala type		
Prototype: Puumala virus	*Clethrionomys glareolus* (bankvole; north and western Europe, Balkan western U.S.S.R.	Mild HFRS NE HN
IV. Prospect Hill type		
Prototype: Prospect Hill	*Microtus pennsylvanicus* (meadow vole, U.S.A.)	Not identified
V. Leaky type		
Prototype: Leaky virus	*Mus musculus* (house mouse, U.S.A.)	Not identified

[a]KHF, Korean hemorrhagic fever; EHF, epidemic hemorrhagic fever; NE, nephropathia epidemica; HN, hantavirus nephropathy.

proteinuria, flushing, shock, hemorrhagic diathesis, pulmonary edema, associated with leukocytosis, thrombocytopenia, hemoconcentration, and azotemia, are virtually diagnostic. In a study of 114 clinically diagnosed HFRS cases, 98% were confirmed by serology (Lee, 1988b).

2. HFRS Caused by Hantaan-Related Viruses in Central and Southeastern Europe

Seroepidemiologic studies have demonstrated the presence of antibodies against Hantaan-type virus in either rodent or human sera in Albania* (Eltari *et al.*, 1987), Bulgaria* (Gavrilovskaya *et al.*, 1984), Czechoslovakia* (Gresikova *et al.*, 1988), Greece* (Antoniadis *et al.*, 1987a,b), Hungary* (Yanagihara and Gajdusek, 1987), Italy (Nuti *et al.*, 1988), Poland, and Yugoslavia* (Gligic *et al.*, 1987; Avsic-Zupanc *et al.*, 1988a).

Concerning Bulgaria, not all patients mentioned in Table 3 were seroconfirmed. Only sera of 67 HFRS patients were examined by Chumakov and co-workers (results reported during a WHO meeting in October 1985, Moscow), of which 42 had specific antibodies to the Puumala type and five had antibodies to the Hantaan type.

A correlation between immune response against hantaviruses (Hantaan 76-118 and Puumala) by IFA and the clinical picture of 20 hospitalized cases in Yugoslavia was observed. The clinical picture varied from moderate (fever > 38°C, headache, abdominal pain, myalgia, acute renal insufficiency, weakness, epistaxis, vomiting, hyperemia of face, hemorrhagic risk) in seven patients to severe in 13 patients (all mentioned symptoms plus hemorrhagic syndrome, hematemesis, melena, diarrhea, hiccup, and central nervous system involvement). Of the severe cases, 93% were dialyzed; in 77% of the severe cases the IFA titers were higher with Hantaan antigen, whereas in four of seven moderate cases, IFA titer was higher with Puumala antigen (Gligic *et al.*, 1988).

Some sera from a rather severe form of HFRS in Slovenia, the northern part of Yugoslavia, did show higher titers in IFA on Fojnica strains than on Puumala-type hantaviruses (Avsic-Zupanc *et al.*, 1988a).

The clinical features of the HFRS patients in Greece are summarized in Table 3. In the region where these patients were located, IFA antibodies to Hantaan virus have been found in *Apodemus flavicollis* mice, as well as in some (2/15) *Rattus rattus alexandrikus* captured in a slaughterhouse (Antoniadis *et al.*, 1987b).

The disease in Greece is more severe than NE in Scandinavia and HFRS in Western Europe. Of 23 patients studied, 12 (52%) were severely ill, eight (35%) required renal dialysis, and three (14%) died. The disease is more like the Asian form of HFRS caused by Hantaan virus.

In conclusion, a severe form of HFRS is endemic in central and southern Europe and the clinical manifestations of the disease resemble the Hantaan type. These data suggest that the more severe form of HFRS in Yugoslavia, Czechoslovakia, Bulgaria, and Greece is probably correlated with a new subtype of Hantaan-type viruses, with *A. flavicollis* as the main host. *A. flavicollis* is found throughout the Balkan Peninsula and coincides well with the severe form of HFRS found in Albania, Yugoslavia, Bulgaria, Hungary, and Czechoslovakia (Yanagihara and Gajdusek, 1987) and northern Greece.

B. Seoul Virus Infection

Urban cases of HFRS in Korea, China, Japan, and Southeast Asia and laboratory infections in Korea and Japan are caused by hantaviruses of the Seoul type (Morimoto *et al.*, 1985; Buyn *et al.*, 1986.) Clinical disease with an occasional fatality has been reported among investigators working with laboratory rats (Lloyd and Jones, 1986; Umenai *et al.*, 1979; Desmyter *et al.*, 1983; Lee *et al.*,

*Clinical evidence of disease in humans was noted in addition to serologic evidence. The clinical features, as far as they could be deduced from the literature, are summarized in Table 3.

Table 3. Comparison of Clinical Features of HFRS in Various Countries of Europe and Asia

	Hantaan virus infection (%)			Hantaanlike virus infection (%)			Seoul virus infection (%)	Puumala virus infection (%)		
	Korea[a]	China[a]	Greece[b]	Albania[c]	Yugoslavia[d]	Bulgaria[e]	Japan[a] and Korea[a]	Sweden[f]	Finland[g]	W. Europe[h]
Fever	100	100	100	+	100	98	100	94.1	100	100
Anorexia	–[i]	–	–	–	–	95	96	–	–	–
Chills	92	–	100	–	–	67	70	–	–	–
Nausea	82	72	–	–	75	–	61	–	78	24
Vomiting	63	58	83	+	75	57	45	33	70	–
Myalgia	78	69	91	–	–	98	52	37	20	73
Headache	86	83	100	+	87	–	42	44.5	90	83
Abdominal pain	23	25	96	+	–	55	65	48.1	67	26
Backache	95	–	78	+	87	96	80	66.4	82	–
Constipation	60	–	–	–	–	–	37	–	34	–
Diarrhea	11	37	17	–	19	–	24	11.8	12	7
Dizziness and vertigo	100	41	50	–	–	–	7	–	12	–
Ophthalmalgia	–	34	–	–	–	–	15	–	0	–
Blurred vision	–	18	ND	–	31	–	–	6.5	12	–
Conjunctival injection	64	23	65	+	44	80	79	–	18	17
Pharyngeal or palatal injection	55	64	–	–	–	–	79	–	36	–
Petechiae on body	32	56	15	+	31	–	31	–	12	–
Hemorrhages (epitaxis, melena, hematemesis, etc.)	72	31	5	–	–	–	26	3	10	–
Hepatomegaly	0	–	–	–	–	–	32	–	9	17
Splenomegaly	7	–	–	–	–	–	15	–	0	6
Lymphadenopathy	38	3	–	–	–	–	15	–	15	–
Preorbital edema	9	17	–	–	–	–	4	56	–	–

(*continued*)

Table 3. (Continued)

	Hantaan virus infection (%)			Hantaanlike virus infection (%)			Seoul virus infection (%)	Puumala virus infection (%)		
	Korea[a]	China[a]	Greece[b]	Albania[c]	Yugoslavia[d]	Bulgaria[e]	Japan[a] and Korea[a]	Sweden[f]	Finland[g]	W. Europe[h]
Proteinuria	96	–	–	+	100	94	94	88.7	100	94
Haematuria	85	86	–	+	94	92	73	73.2	74	70
Oliguria ≤ 500 ml	67	59	–	+	63	80	37	–	54	68
Polyuria > 2000 ml	92	87	–	–	–	–	63	40.8	97	–
Leukocytosis 10,000/mm[c]	91	92	–	–	–	80	41	–	57	–
Thrombocytopenia > 100,000/mm[c]	54	78	–	–	–	–	70	–	87	58
Increased ESR > 20 mm/hr	72	–	–	–	–	–	7	–	90	91
Bun > 20 or serum Cr > 2 mg/dl	94	100	–	+	–	–	50	30.8	70	–
Hypotension (<90/60 mm Hg)	80	42	65	–	19	–	22	–	40	14
Purpura	–	–	0	–	–	–	–	–	–	–
Cough	–	–	13	–	–	–	–	–	–	–
Dyspnea	–	–	10	–	–	–	–	–	–	–
Pneumonic infiltrations	–	–	13	–	–	–	–	–	–	–

Pulmonary edema	–	–	8.5	–	–	–	–	–	–	–
Shock	–	–	21	+	–	–	–	–	–	–
Convulsions	–	–	0	–	–	–	–	–	–	–
Renal failure	–	–	–	+	–	90	–	–	–	–
Malaise	–		100	–	–	98	–	–	–	–
Arthralgia	–	–	83	–	–	–	–	–	–	–
Flushing over the head and neck	–	–	65	–	–	–	–	–	–	–
White cell count 5.6 × 10^9/liter	–	–	–	+	–	–	–	–	–	–
Gastrointestinal tract involvement	–	–	5	–	13	–	–	–	–	–
Acute onset	–	–	–	–	–	78	–	–	–	–
Mortality rate	–	–	13	–	6.8	–	–	–	–	–

[a]Lee, 1988b.
[b]Antoniadis A. *et al.*, 1987b.
[c]Eltari *et al.*, 1987.
[d]Gligic *et al.*, 1987.
[e]World Health Organization: Tick-borne encephalitis and Haemorrhagic fever with renal syndrome in Europe. Euro reports and Studies 104, Baden, Oct. 3–5, 1983—Copenhagen WHO Regional office for Europe 1986; 53–64. These data were from persons *not* seroconfirmed.
[f]Settergren *et al.*, 1988.
[g]Lähdevirta, 1971.
[h]van Ypersele de Strihou, and Mery, 1988.
[i]–, data not available.

1986). These viruses, however, were antigenically slightly different from the Seoul viruses (see Table 2).

Urban commensal rats (*Rattus norvegicus* and *Rattus rattus*) and laboratory rats are the main reservoir hosts and transmit the disease to humans. Some cases are severe, but many cases are milder than Hantaan virus infection but more severe than Puumala virus infection.

The clinical characteristics of the disease are high fever, fatigue, anorexia, vomiting, backache, myalgia, abdominal pain, conjunctival infection, petechiae on soft palate, hepatomegaly, proteinuria, microscopic hematuria, lymphocytosis, thrombocytopenia, increase of GOT, GPT, and transient glysuria, based on observations of 56 cases of Seoul virus infection in Korea and Japan (Lee, 1988b).

The most characteristic manifestations are abdominal symptoms, hepatomegaly and hepatic dysfunction, and mild renal dysfunction.

C. Puumala Virus Infection

The prototype of Puumala virus was isolated from bankvoles (*Cl. glareolus*) captured in areas where a mild form of HFRS was described. This mild form is also known as *NE* in Scandinavia (Settergren *et al.*, 1988) and hantavirus nephropathy in western Europe (van Ypersele de Strihou and Mery, 1988). Seroconfirmed, clinically documented cases of the mild form of HFRS have been documented in Belgium (Clement and van der Groen, 1987; Degrez and Colson, 1987; van Ypersele de Strihou and Mery, 1988), Denmark (Yanagihara and Gajdusek, 1987), the Federal Republic of Germany (Pilaski *et al.*, 1987; Gärtner *et al.* 1988), Finland (Lähdevirta, 1971; Brummer-Korvenkontio *et al.*, 1980), France (van Ypersele de Strihou and Mery, 1988; Dournon *et al.*, 1987), England (Lloyd *et al.*, 1984), Norway (Sommer *et al.*, 1988), Sweden (Settergren *et al.*, 1988; Niklasson and Leduc, 1987; Niklasson *et al.*, 1987), and the western part of the U.S.S.R. (Tkachenko and Drozdov, 1988).

The mild form of HFRS was also reported in Yugoslavia, where the Vranica strain has been isolated from *Cl. glareolus,* which is a Puumala-type of virus (see Table 1) (Gligic *et al.*, 1987; Avsic-Zupanc *et al.*, 1988a). Based on seroepidemiologic data obtained in wild rodents in western Europe as well as human serosurveys, silent forms of the Puumala type of infection do exist. The mild form of HFRS in Europe coincides well with the geographic distribution of *Cl. glareolus* in Europe (Yanagihara and Gajdusek, 1987).

A patient was considered to have a clinical disease compatible with NE if four or more of the following criteria were fulfilled: acute onset, headache, fever, increased serum creatinine, proteinuria or hematuria, polyuria (>2000 ml/2 hr). These criteria were successfully applied by Settergren and co-workers (1988) on patients with a mild form of HFRS in Sweden. Antibodies were present in 368 patients, 355 of whom also fulfilled four or more of the criteria mentioned. These criteria may also be used to diagnose the mild form of HFRS as it was described in Belgium, France, Germany, and Yugoslavia (see Table 3). The total number of cases varied per year, which coincided with the fluctuations in small rodent populations. This was observed in Sweden (Settergren *et al.*, 1988), Norway (Sommer *et al.*, 1988), and Belgium (Verhagen *et al.*, 1988). Though cases of melena and hematemesis and disseminated intravascular coagulation (Settergren *et al.* 1988) have been reported for NE in Sweden, such findings were not reported in the large clinical series published from Finland (Lähdevirta, 1971) and western Europe (van Ypersele de Strihou and Mery, 1988); the majority of patients for whom platelet counts were available had thrombocytopenia, which supports a relationship between the mild European form of HFRS (NE) and the severe form of Asia (KHF). In the acute stage, NE can mimic other diseases, such as diabetes mellitus and acute glomerulnephritis and leptospirosis; it is therefore of importance in differential diangosis in endemic regions, and laboratory diagnosis is necessary.

D. Not Yet Identified Hantavirus Infection

Many hantaviruses have been isolated from wild rodents, but to what extent they can cause disease in humans is not yet known (Lee, 1988a). Some viruses, such as PH and Leaky virus (see Table 1), clearly represent a new hantavirus type. However, no one knows to what extent they are pathogenic for humans. Further investigations are necessary.

V. LABORATORY DIAGNOSIS

Although epidemiologic and clinical data may be highly indicative for HFRS, the diagnosis has to be confirmed by serologic and/or virologic diagnostic tests.

Specific antibodies against hantaviruses can be detected by the following methods: IFA using Vero E6 cells infected with Hantaan 76-118 or other hantaviruses capable of propagation in cell culture (Yanagihara and Gajdusek, 1987); IFA using frozen lung sections of wild or laboratory-bred rodents or laboratory rats harboring the hantaviral antigens (van der Groen and Beelaert, 1985); neutralization test in cell culture or laboratory rats (Schmaljohn *et al.*, 1985; Arikawa *et al.*, 1985b; Goldgaber et al., 1984; Lee *et al.*, 1985a; Elliott *et al.*, 1985; Takenaka *et al.*, 1985; Tanishita *et al.*, 1984); enzyme immunoassay (ELISA) using hantavirus-infected E6 cells as antigen (Groen et al., 1989); hemagglutination, inhibition complement fixation, immune adherence hemagglutination, and solid-phase radioimmunoassay [recently reviewed by Yanagihara and Gajdusek (1987)]; ELISA using *Autographa californica* nuclear polyhidrosis virus–expressed nucleocapsid protein of Hantaan virus as antigen (Schmaljohn et al., 1988); ELISA for IgM-type antibodies (Chen Li-Li *et al.*, 1987; Ivanov *et al.*, 1988); Western blot (Avsic-Zupanc *et al.* 1988b; Lee and Lee, 1987b).

Since IgG antibodies occur very early, absence of specific IgG helps to exclude the HFRS diangosis. Presence of hantavirus-specific IgG supports the diagnosis but in an area with a high antibody prevalence, positive serology may be caused by an earlier hantavirus infection and thus not be significant for the acute disease. A fourfold rise in IgG antibody titer and the presence of IgM-type antibodies are the criteria for an acute infection. However, paired sera from patients with NE often failed to show a significant titer rise when tested by IgG ELISA or IgG IFA (Niklasson, 1988). Since in this study all acute and early convalescent sera were positive by IgM ELISA, this test could become an important tool for early diagnosis in acute human NE infections. HFRS antigen or virus can be detected by the following methods: IFA staining of tissue sections (either acetone-fixed frozen sections or trypsinized formalin-fixed sections) (Lee and Lee, 1976; Kurata *et al.*, 1983); antigen-capturing ELISA (Xu *et al.*, 1987; Verhagen *et al.*, 1986; Chen Bo-quan, 1987); propagation of infectious hantaviruses in cell culture (Vero E6, A 549 cells) or suitable rodents. These methods were all recently reviewed by Yanagihara and Gajdusek (1987). Hantaviral genome detection was recently developed by Schmaljohn *et al.* (1988). Radio-labeled RNA probes generated from cDNA clones of the M and S genome segments of Hantaan virus readily detected Hantaan virus and two isolates from KHF patients but were less effective in detecting four other hantaviruses. Better-defined nucleic acid probes are needed.

VI. CONCLUSIONS

1. It is apparent from recent reports that HFRS is being recognized in an increasing number of countries as facilities for hantavirus diagnosis become available.

2. Generally speaking, the disease in Asia and the far eastern part of the U.S.S.R. is more severe, with more pronounced hemorrhagic phenomena, than observed in most of the European

countries. However, in central Europe (Balkan Peninsula) and Greece, a more severe form of HFRS was observed, compared to the mild form in Scandinavia and western Europe.

3. Hantaan-type infection correlates with the severe form of HFRS as well as with the geographic distribution of the infected host, *A. agrarius.*

4. Hantaan-related type of virus, for example Porogia, probably a new subtype in the Hantaan type, does correlate with a more severe form of HFRS observed in central Europe and Greece and coincides well with the geographic distribution of *A. flavicollis.*

5. Many hantaviruses have been isolated from wild rodents, some of them representing a new serotype (Leaky) whose pathogenicity for humans has yet to be elucidated.

6. Hantavirus infection in humans is not always easy to diagnose. A person with a febrile flulike syndrome, headache, lumbar and/or abdominal pain, increased creatinine, proteinuria, hematuria, or polyuria should be suspected to have HFRS and serologic diagnosis should be requested.

7. Differential diagnosis should be considered for the following diseases in the endemic areas of HFRS: acute renal failure, hemorrhagic scarlet fever, acute abdomen, leptospirosis, scrub typhus, murine typhus, spotted fevers, non-A, non-B hepatitis, Colorado tick fever, septicemia, dengue, heatstroke, and disseminated intravascular coagulation (Lee, 1986).

REFERENCES

Antoniadis, A., Grekas, D., Rossi, C. A., and LeDuc, J. W. (1987a). *J. Infect. Dis.* **156,** 1010–1013.

Antoniadis, A., LeDuc, J. W., and Daniel-Alexiou, S. (1987b). *Eur. J. Epidemiol.* **3,** 295–301.

Arikawa, J., Takashima, I., and Hashimoto, N. (1985a). *Arch. Virol.* **86,** 303–313.

Arikawa, J., Takashima, I., Morita, C., Sugiyama, K., Matsuura, Y., Shiga, S., and Kitamura, T. (1985b). *Acta. Virol.* (Praha) **29,** 66–72.

Avsic-Zupanc, T., Hoofd, G., van der Groen, G., Gligic, A. (1988a). Abstract, First International Symposium on Hantaviruses and Crimean Congo Hemorrhagic Fever viruses, Porto Carras, Halkidiki, Greece, Sept. 26–30.

Avsic-Zupanc, T., Cizman, B., Gligic, A., Hoofd, G., and van der Groen, G. (1989). *Acta Virol.* **33,** 327–337.

Baek, L. J., Gligic, A., Yanagihara, R., Gibbs, C. J., Jr., and Gajdusek, D. C. (1987). In Abstracts of the 16th Pacific Science Congress, Seoul, Korea, 33.

Brummer-Korvenkontio, M., Vaheri, A., Hovi., T., von Bonsdorff, C. H., Vuorimies, J., Manni, T., Pentinnen, K., Oker-Blom, N., and Lähdevirta, J. (1980). *J. Infect. Dis.* **141,** 131–134.

Buyn, K. S., Soe, J. B., and Kim, Yh. (1986). *Korean J. Infect. Dis.* **18,** 11–18.

Chen Bo-quan (1987). Abstract, 29th International Colloquium, Hantaviruses, Antwerp, Dec. 10–11, 38.

Chen Li-Li, Wang Hong-Xi, Gu Xian-Shi, Chen Zhang-Zhi, Quin Guan-Ming, Xu Feng-qin, Li Cheng-Ming, and Yang Shi-quan (1987). *Chin. Med. J.* **100,** 402–405.

Childs, J. E., Korch, G. W., Glass, G. E., LeDuc, J. W., and Shah, K. V. (1987). *Am. J. Epidemiol.* **126,** 55–68.

Childs, J. E., Glass, G. E., Korch, G. W., Arthur, R. R., Shah, K. V., Glasser, D., Rossi, C., and LeDuc, J. W. (1988). *Am. J. Epidemiol.* **127,** 875–878.

Chumakov, M. P., Gavrilovskaya, I. N., Bioko, V. A., Zakharova, M. A., Myasnikov, Y. A., Bahskirev, T. A., Apekina, N. S., Safiullin, R. S., and Potapov, V. S. (1981). *Arch. Virol.* **69,** 259–300.

Clement, J., and van der Groen, G. (1987). *Adv. Exp. Med. Biol.* **212,** 251–263.

Cosgriff, T. M., Hsiang, C. M., Huggins, J. W., Guang, M. Y., Smith, I., Wu, Z. O., LeDuc, J. W., Zheng, Z. M., Meegan, J. M., Wang, C. N., Gibbs, P. H., Gui, X. E., Yuan, G. W., and Zhang, T. M. (1988). Abstract, First International Symposium on Hantaviruses and Crimean Congo hemorrhagic fever viruses, Halkidiki, Greece, Sept. 26–30, 27.

Dantas, J. R., Okuno, Y., Tanishita, O., Takahashi, Y., Takahashi, M., Kurata, T., Lee, H. W., and Yamanishi, K. (1987). *Intervirology* **27,** 161–165.

Dashi, N., Hao, H., Ning, Z., Yongzheng, L., and Chao Yang, D. (1987). Abstract, 29th International Colloquium. Hantaviruses, Antwerp, Dec. 10–11, 36.

Degrez, Th., and Colson, P. (1987). *Acta Clin. Belg.* **42,** 316–322.
Desmyter, J., LeDuc., Johnson, K. M., Brasteur, F., Deckers, C., and van Ypersele de Strihou, C. (1983). *Lancet* **2,** 1445–1448.
Dournon, E., Rollin, P., Assous, M., and Sureau, P. (1987). Abstract, 29th International Colloquium, Hantaviruses, Antwerp, Dec. 10–11, 14.
Elliott, L. H., Sauches, A., and McCormick, J. B. (1985). Presented at 85th annual meeting of the American Society for microbiology 1985, Las Vegas, NV, March 3–7, Poster T60.
Eltari, E., Nuti, Hasko, I., and Gina, A. (1987). *Lancet* **2,** 1211.
Executive Committee of the International Committee on Taxonomy of Viruses (1987). Minutes of the 17th meeting, pp. 1–14.
French, G. R., Foulke, R. S., Brand, O. A., Eddy, O. A., Lee, H. W., and Lee, P. W. (1981). *Science* **211,** 1046–1048.
Gajdusek, D., Goldfarb, L. G., and Goldgaber, D. (1987). *Bibliography of Hemorrhagic Fever with Renal Syndrome,* 2nd ed., NIH Publication No. 88-2603, U.S. Department of Health and Human Services, National Institutes of Health, Bethesda, MD.
Gärtner L., Emmerich, P., and Schmitz, H. (1988). *Dtsch. Med. Wochenschr.* **113,** 937–940.
Gavrilovskaya, I. N., Apekina, N. S., Miasnikov, Y. A., Bernstein, A. D., Ryltseva, E. V., Gorbachkova, E. A., and Chumakov, M. P. (1983). *Arch. Virol.* **75,** 313–316.
Gavrilovskaya, I. N., Vasilenko, S., Chumakov, M. P., Chindarov, L., and Katstarov, G. (1984). *Epidemiol. Mikrobiol. Infect. Bol.* **21,** 17–22.
Gligic, A., Obradovic, M., Stojanovic, R., Frusic, M., Yanagihara, R., Gibbs, C. J., and Gajdusek, D. C. (1986). In Abstracts of the XIth International Congress of Infectious and Parasitic Diseases, Munich, 810.
Gligic, A., Yanagihara, R., and Gibbs, C. J. (1987). Abstract, 35th Annual Meeting of the American Society of Tropical Medicine and Hygiene, Denver, CO, Dec. 8–11.
Gligic, A., Obradovic, M., Stajonovic, R., Diglisic, G., and Nastic, D. (1988). Abstract, First International Symposium on Hantaviruses and Crimean Congo Hemorrhagic Fever Viruses, Halkidiki, Greece, Sept. 26–30, 63.
Goldgaber, D., Yanagihara, R., Lee, P. W., Gibbs, C. H., Jr., and Gajdusek, D. C. (1984). Abstract, Sixth International Congress of Virology, Sendai, Japan, Sept. 17, p 27–19.
Gresikova, M., Kozuch, O., Sekeyova, M., Tkachenko, E. A., Rezapkin, G. V., and Lysy, J. (1988). *Acta. Virol.* **32,** 164–167.
Groen, J., van der Groen, G., Hoofd, G., and Osterhaus, A. (1989). *J. Virol. Meth.* **23,** 195–203.
Hung, T., Chou, Z. Y., Zhao, T. X., Xia, S., and Hang, C. S. (1985). *Intervirology* **23,** 97–108.
Ivanov, A. P., Tkachenko, E. A., Petrov, V. A., Pashkov, A. J., Dzagurova, T. K., Vladimirova, T. P. Voronkova, G. M., and van der Groen G. (1988). *Arch. Virol.* **100,** 1–7.
Kitamura, T., Morita, C., Komatsu, T., Sugiyama, K., Arikawa, J., Shiga, S., Takeda, H., Akao, Y., Imaizumi, K., Oya, A., Hashimoto, M., and Urasawa, S. (1983). *Jpn. J. Med. Sci. Biol.* **36,** 17–25.
Kurata, T., Sata, T., Yamanishi, K., Domar, K., Yamanouchi, T., Tsai, T. F., McCormick, J. B., Wear, D. J., and Lähdevirta, J. (1987). Abstract, 29th International Colloquium. Hantaviruses, Antwerp, Dec. 10–11, 8.
Lähdevirta, J. (1971). *Ann. Clin. Res.* **3,** 1–154.
LeDuc, J. W. (1987). *Lab. Anim. Sci.* 413–418.
LeDuc, J. W., Smith, G. A., and Johnson, K. M. (1984). *Am. J. Trop. Med. Hyg.* **33,** 992–998.
LeDuc, J. W., Smith, G. A., Pinheiro, F. P., Vasconcelos, P. F. C., Rosa, E. S. T., and Maiztegui, J. I. (1985). *Am. J. Trop. Med. Hyg.* **34,** 810–815.
LeDuc, J. W., Smith, A. A., Childs, J. E., Pinheiro, F. P., Maiztegui, J. I., Niklasson, B., Antoniadis, A., Robinson, D. M., Khin, M., Shortridge, K. F., Wooster, M. T., Elwell, M. R., Ilberty, P. L. T., Koech, D., Rosa, E. S. T., and Rosen, L. (1986). *Bull WHO* **64,** 139–144.
Lee, H. W. (1986). *J. Korean Soc. Virol.* **16,** 1–5.
Lee, H. W. (1988a). *J. Korean Med. Assoc.* **31,** 581–593.
Lee, H. W. (1988b). *Nephrology* **2,** 816–831.
Lee, H. W., and Lee, P. W. (1976). *Korean J. Intern. Med.* **19,** 371–383.
Lee, H. W., and Lee, P. W. (1977). *J. Korean Soc. Virol.* **7,** 1–9.
Lee, H. W., Lee, P. W., and Johnson, K. M. (1978). *J. Infect. Dis.* **137,** 298–308.
Lee, H. W., Baek, L. J., and Johnson, K. M. (1982). *J. Infect. Dis.* **146,** 638–644.
Lee, H. W., Baek, L. J., and Kim, H. D. (1986). *J. Korean Soc. Virol.* **16,** 113–120.

Lee, P. W., and Lee, H. W. (1987a). *JE HFRS Bull.* **2,** 65–68.

Lee, P. W., and Lee, H. W. (1987b). Abstract, Pacific Science Association 16th Congress, Seoul, Korea, Aug. 20–30, 150.

Lee, P. W., Amyx, H. L., Yanagihara, R., Gajdusek, D. C., Goldgaber, D., and Gibbs, C. J. (1985a). *J. Infect. Dis.* **152,** 826–829.

Lee, P. W., Gibbs, C. J., Gajdusek, D. C., and Yanagihara, R. (1985b). *J. Clin. Microbiol.* **22,** 940–944.

Lloyd, G., and Jones, N. (1986). *J. Infect.* **12,** 2117–125.

Lloyd, G., Bowen, E. T. W., and Jones, N. (1984). *Lancet* **1,** 1175–1176.

Lukashevich, I., and Tkachenko. E. (1987). Abstract, Hantaviruses, 29th International Colloquium, Antwerp, Belgium, Dec. 10–11, 9.

McCormick, J. B., Sasso, D. R., and Palmer, E. L. (1984). *Lancet* **1,** 765–768.

Men Ruhe, K. J., Yuehuai, G., and Xianlu, A., (1983). Abstract, WHO Ad Hoc Meeting on Surveillance of Viral Hemorrhagic Fevers, Antwerp, Sept. 28–30.

Morimoto, Y., Kishimoto, S., and Yamanouchi, T. (1985). *J. Jpn. Infect. Dis.* **59,** 439–458.

Niklasson, B. (1988). Abstract, First International Symposium on Hantaviruses and Crimean Congo Hemorrhagic Fever Viruses, Halkidiki, Greece, Sept. 26–30, 31.

Niklasson, B., and LeDuc, J. W. (1984). *Lancet* **1,** 1012–1013.

Niklasson, B., and LeDuc, J. (1987). *J. Infect. Dis.* **155,** 269–276.

Niklasson, B., LeDuc, J. W., and Nyström, K. (1987). *Epidemiol. Infect.* **99,** 559–562.

Nuti, M., Awaddeo, D., and Costa, M. (1987). Abstract, 29th International Colloquium, Hantaviruses, Antwerp, Dec. 10–11, 43.

Nuti, M;, Amaddeo, D., Costa, M., Cristaldi, M., Leradi, L. A., Gibss, C. J., and Gajdusek, D. C. (1988). In *XII International Congress of Tropical Medicine and Malaria,* (P. A. Kager and A. M. Polderman, eds.), Excerpta medica. Amsterdam, Sept. 19–23 (TUP 10–6), p. 164.

Pilaski, J., Peceny, R., Gorchewskio, O., Zöller, L., Zeier, M., Kraft, V., and Lee, H. W. (1987). Abstract, 29th International Colloquium—Hantaviruses, Antwerp, Dec. 10–11, 25.

Schmaljohn, C. S., Hasty, S. E., Dalrymple, J. M. LeDuc, J. W., Lee, H. W., von Bonsdorff, C. H., Brummer-Korvenkontio, M., Vaheri, A., Tsai, T. F., Regnery, H. L., Goldgaber, D. L., and Lee, P. A. (1985). *Science* **227,** 1041–1044.

Schmaljohn, C. S., Sugiyama, K., Schmaljohn, A. L. and Bishop, D. H. L. (1988). *J. Gen. Virol.* **69,** 777–786.

Settergren, B., Juto, P., Wadell, G., Trollfors, B., and Norrby, S. R. (1988). *Am. J. Epidemiol.* **127,** 801–807.

Sommer, A. I., Traavik, T., Mehl, R., Berdal, B. P., and Dalrymple, J. (1988). *Scand. J. Infect. Dis.* **20,** 267–274.

Song, G., Qiu, X., Ni, D., Zhao, J., and Kong, B. (1982). *Acta Acad. Med. Sinicae* **4,** 73–77.

Song, G., Hang, C. S., Qui, X. Z., Ni, D., Liao, H., Gao, G., Du, Y., Xu, J., Wu, Y., Zhao, J., Kong, B., Wang, Z., Shang, Z., Shen, H., and Zhou, N. (1983). *J. Infect. Dis.* **147,** 654–659.

Song, G., Hang, C. S., Liao, H. X., and Fu, J. L., (1984). In, Abstracts Sixth International Congress of Virology, Sendai, Japan, Sept. 1–7, 267.

Sugiyama, K., Morita, C., Matsuura, Y., Shiga, S., Komatsu, T., Morikawa, S., and Kitamura, T. (1984). *J. Infect. Dis.* **149,** 472.

Takenaka, A., Gibbs, C. J., Jr, and Gajdusek, D. C. (1985). *Arch. Virol.* **84,** 197–206.

Tang, Y. W., Xu, Z. Y., Zhu, Z. Y., and Tsai, T. F. (1985). *Lancet* **1,** 513–514.

Tanishita, O., Takahashi, Y., Okuno, Y., Yamanishi, K., and Takanashi, M. (1984). *J. Clin. Microbiol.* **20,** 1213–1215.

Tkachenko, E. A., and Drozdov, S. G. (1988). Abstract, First International Symposium on Hantaviruses and Crimean Congo Hemorrhagic Fever Viruses, Porto Carras, Halkidiki, Greece, Sept. 26–30, 59.

Tkachenko, E. A., Ivanov, A. P., Donets, M. A., Miasnikov, Y. A., Ryltseva, E. V., Gaponova, L. K., Bashkirtsev, V. N., Okulova, N. M., Drozdov, S. G., Slonova, R. A., Somov, G. P., and Astakova, T. I. (1983). *Ann. Soc. Belg Med. Trop.* **63,** 267–269.

Tkachenko, E. A., Bashkirtsev, V. N., van der Groen, G., Dzagurova, T. K., and Ivanov, A. P. (1984). *Ann. Soc. Belg Med. Trop.* **64,** 425–426.

Tkachenko, E. A., Ryltseva, E. V., Myasnikov, A., Ivanov, A. P. Rezapkin, G. V., Tatiyanchenko, L. A., and Pashkov, A. (1987). *Probl. Virol.* **6,** 709–715.

Tsai, T. F. (1987). *Lab. Animal Sci.* **37,** 428–430.

Tsai, T. F., Tang, Y. W., Hu, S. L., Ye, K. L., Chen, G. L., and Xu, Z. Y. (1984). *J. Infect. Dis.* **150,** 895–898.

Tsai, T. F., Bauer, S. P., Sasso, D. R., Whitefield, S. G., McCormick, J. B., Caraway, T. C., McFarland, L., Bradford, H., and Kurata, T. (1985). *J. Infect. Dis.* **152,** 126–136.

Umenai, T., Lee, H. W., Lee, P. W., Saito, T., Toyoda, T., Hongo, M., Yoshinaga, K., Nobunaga, T., Horiuchi, T., and Ishida, N. (1979). *Lancet* **1,** 1314–1316.

van der Groen, G., and Beelaert, G. (1985). *J. Virol. Meth.* **10,** 53–58.

van der Groen, G., Piot, P., Desmyter, J., Colaert, J. Muylle, L., Tkachenko, E. A. Ivanov, A. P., Verhagen, R., and van Ypersele de Strihou, C. (1983a). *Lancet* **2,** 1493–1494.

van der Groen, G., Tkachenko, E. A., Ivanov, A. P. and Verhagen R. (1983b). *Lancet* **2,** 110–111.

van der Groen, G., Leirs H., and Verhagen, R. (1986). In *Proceedings of the Second Symposium on Recent Advances in Rodent Control, Kuwait,* pp. 197–207.

van der Groen, G., Goyvaerts, D., Hoofd, G., Xing, Z., Song, G., Leirs, H., Verhagen, R., and Yamanishi, K. (1988). *JE HFRS Bull.* (in press).

van Ypersele de Strihou, C., and Mery, J. P. (1988). In *Nephrology Proceedings of the Xth International Congress of Nephrology* (A. M. Davison, ed.), Baillière Tindall, London, Philadelphia, Toronto, Sydney, Tokyo, pp. 802–831.

Verhagen, R., Leirs, H., Tkachenko, E. A., and van der Groen, G. (1986). *Arch. Virol.* **91,** 193–205.

Verhagen, R., Van Ypersele de Strihou, C., Leirs, H., Clement, J., Hoofd, G., and van der Groen, G. (1988). Abstract, XIIth International Congress for Tropical Medicine and Malaria, Amsterdam, The Netherlands, p228.

White, J. P., Geisbert, T. W., Rossi, C. A., and LeDuc, J. W. (1988). Abstract, First International Symposium on Hantaviruses and Congo Crimea, Hemorrhagic Fever Viruses, Halkidiki, Greece, Sept. 26–30, 45.

World Health Organization (1983). *Bull. WHO* **61,** 269–275.

Xu, Z. K., An, X. L., and Wang, M. X. (1986). *J. Hyg. Camb.* **97,** 366–375.

Xu, Z. K., Wang, M. X., and Jiang, S. (1987). *J. Med. Coll. PLA* **2,** 46–48.

Yamanishi, K., Dantas, J. R., Takahashi, M., Yamanouchi, T., Domae, J., and Kurata, T. (1983). *Biken J.* **26,** 155–160.

Yamanishi, K., Dantas, J. R., Takahashi, M., Yamanouchi, T., Domae, K., Takahashi, Y., and Tanishita, O. (1984). *J. Virol.* **52,** 231–237.

Yanagihara, R., and Gajdusek, C. (1987). In *Medical Virology,* Vol. VI (LM de la Maza and EM Peterson, eds.), pp. 171–214, Elsevier Science Publishers, New York.

Yanagihara, R., Goldgaber, D., Lee, P. W., Amyx, H. L., Gajdusek, D. C., and Gibbs, C. J., Jr (1984). *Lancet* **1,** 1013.

Yen, Y.-C., and Qiu, F.-x. (1984). In Abstracts XIth International Congress for Tropical Medicine and Malaria. Calgary, Canada, Sept. 16–22, 225.

Yu-tu, J. (1983). *Chin. Med. J.* **96,** 265–268.

Zhao-Zhuang, L., Yi-Yin, W., and Y-Xin, T. (1987). Abstract, 29th International Colloquium, Hantaviruses, Antwerp, Dec. 10–11, 37.

18

Nairoviruses

Characteristics and Disease Distribution

Stephen M. Eley, Lynn G. Bruce, Julian I. Delic, Robert M. Henstridge, and Norman F. Moore

I. INTRODUCTION

The family Bunyaviridae comprises over 200 viruses (Bishop *et al.*, 1980), which share the following characteristics: The particles are spherical, 80–110 nm in diameter, and have a lipid envelope from which protrude 5- to 11-nm polypeptide spikes. The envelope encloses three helical nucleocapsids; the genome comprises three strands of single-stranded RNA (L, large; M, medium; S, small); each segment is complexed with the nucleocapsid protein (N) and linked by 3′-5′ end sequence hydrogen bonding into a closed circle. In addition to N, the viral polypeptides include two glycoproteins, designated G1 and G2, and an RNA-dependent RNA polymerase. The genome is in negative sense; replication involves transcription of the RNA into positive sense. Although coding assignments have been completed for other bunyavirus genera, this is not the case for the nairoviruses (Bishop, 1985). Several bunyaviruses are capable of heterologous virus genome segment reassortment, thus forming recombinant viruses; Bishop (1985) asserts that this is possible only between closely related members of the same serogroup. The site of maturation for bunyaviruses is mainly at smooth membranes, the virus accumulating in or near the Golgi vesicles. Several members of the family can be transmitted transovarially, venereally, and/or transstadially in arthropod vectors.

The five genera (*Bunyavirus, Nairovirus, Phlebovirus, Uukuvirus,* and *Hantavirus*) are distinguished on the basis of serologic and molecular analysis (Clerx *et a,* 1981). At least three genera, *Bunyavirus, Nairovirus,* and *Phlebovirus* (phlebotomus fever serogroup), are further subdivided into serogroups; single serogroups are formed by the *Hantavirus* and *Uukuvirus* (Uukuniemi virus serogroup) genera. The *Nairovirus* genus, whose type member is Nairobi sheep disease virus (NSD), comprises the serogroups Crimean–Congo hemorrhagic fever (CCHF), NSD, Qalyub (QYB), Dera Ghazi Khan (DGK), Hughes (HUG), and Sakhalin (SAK). These viruses, in addition

Abbreviations used in this chapter: CFT, complement fixation test; ELISA, enzyme-linked immunosorbent assay; HAI, hemagglutination inhibition; IF, immunofluorescence; IFAT, indirect immunofluorescence test; NT, neutralization test; MAF, mouse (hyperimmune) ascitic fluid; CPE, cytopathic effect; i.c., intracerebral inoculation; i.p., intraperitoneal inoculation; s.c., subcutaneous inoculation.

Stephen M. Eley, Lynn G. Bruce, Julian I. Delic, Robert M. Henstridge, and Norman F. Moore • Chemical Defence Establishment, Porton Down, Salisbury SP4 0JQ, England.

to their serologic identity as a serogroup, differ in that the N protein has a greater molecular weight (49–54 kDa) than is the case with other bunyaviruses (20–30 kDa).

Representative members of the serogroups have been investigated and slight size variations in the RNA species have been found (Clerx *et al.*, 1981). Most noticeable variation occurs in the L segments and there is marked similarity between two members of the same serogroup [QYB and Bandia (BDA)]. The 3′ terminal ends of the RNA species in several nairoviruses have been characterized (Clerx-van Haaster *et al.*, 1982). Generally the Bunyaviridae genera have conserved sequences at the 3′ termini of the three RNA species, AGAGUUCU being the sequence for the nairoviruses. The conserved 3′ ends have been proposed to represent enzyme recognition sites for transcription and/or RNA replication (Clerx-van Haaster *et al.*, 1982).

The three major viral polypeptides have been characterized for several of the nairoviruses, and viral-associated intracellular glycosylated polypeptides have been found in virus-infected cells by immune precipitation (Clerx *et al.*, 1981). In this review the serologic, physicochemical, and pathogenetic data available concerning the *Nairovirus* genus are summarized, highlighting points of generic distinction and similarity.

II. THE CONGO–CRIMEAN HEMORRHAGIC FEVER SEROGROUP

This group is comprised of the causative agents of Congo fever (CON), Crimean hemorrhagic fever (CHF), Hazara virus (HAZ), and Khasan virus (KHA) (Clerx *et al.*, 1981; Smirnova, 1979; Lvov *et al.*, 1978). CON virus and CHF virus are serologically indistinguishable (Casals, 1969; Tignor *et al.*, 1980; Casals *et al.*, 1970), and in spanning three zoogeographic zones, the Palearctic, Oriental, and Ethiopic, this complex has one of the widest distributions among arthropod-borne viruses. This geographic range, coupled with the variety of potential hosts and vectors (it has been isolated from at least 25 species or subspecies of ticks, representing eight genera), renders the lack of serologic distinction among CCHF isolates extremely unexpected (Tignor *et al.*, 1980; Hoogstraal, 1979).

The first records of CHF were made by Chumakov in 1944, when it was demonstrated that a febrile disease with high mortality could be reproduced in human volunteers by inoculation with a filtrate of the patient's blood (Chumakov, 1974). CON virus was isolated independently, in 1956, by Simpson *et al.* (1967), from viremic patients displaying symptoms ranging from a brief febrile period, through varying degrees of headache, joint pains, and fever, to mortality following a massive hematemesis. Simpson *et al.* (1967) also noted a marked incidence of laboratory-acquired infection. U.S.S.R. workers differentiate, on a clinical basis, between "Crimean type hemorrhagic fever" (CTHF) and "Central Asian hemorrhagic fever" (CAHF); it appears that the causative agents are serologically indistinguishable. Mortality is markedly higher in CAHF (32/61) than in CTHF (7/35); the incidence of hemorrhagic symptoms appears to be the same (Casals *et al.*, 1970). Both diseases are marked by the sudden onset of fever, headache, and myalgia, lasting on average for 8 days. Leukopenia and thrombocytopenia are common symptoms, with bleeding from the nose, gums, buccal mucosa, stomach, and so forth, beginning on the 4th day. Death is usually due to shock secondary to blood loss. Simpson *et al.* (1967), in describing the original series of cases of CON, described a similar pattern of slowly resolving pyrexia and myalgia. The single mortality in this study was due to shock following hematemesis, but hemorrhage was not evident in any of the other, nonfatal cases. This contrasts with other, later, reports of CCHF virus isolations in Africa, where direct or indirect association of the virus with jaundice and hemorrhage was made in Ethiopia (Wood *et al.*, 1978) and the Union of South Africa (Swanepoel *et al.*, 1983). In separate studies, Casals (1969) and Chumakov (1974) have shown CON virus strains and CHF virus strains to be serologically identical, using both laboratory-prepared sera and sera from convalescent patients, in demonstrating cross-neutralization and cross-protection. Evidence of the presence of CCHF viruses, by direct isolation or seroepidemiology, has been demonstrated across the Asian land mass, including India (Shanmugam *et al.*, 1976), Egypt (Darwish *et al.*, 1978), Iraq (Tantawi *et al.*, 1980; Ellis *et al.*, 1981), Iran (Saidi *et al.*, 1975), Pakistan (Begum *et al.*, 1970a), and China (Yu-Chen *et al.*, 1985), and in southern Europe (Antoniadis and Casals, 1982; Filipe *et al.*, 1985) (Fig. 1).

The tick species that have been shown to be CCHF-associated are primarily from the family Ixodidae. The majority of these species fall into the "original ixodid pattern" as three-host ticks (Hoogstraal, 1979); three different vertebrates act as host to the larval, nymphal, and adult stages of each tick. The most important vectors of CCHF, however, fall into the "two dissimilar host" category, e.g., *Hyalomma marginatum marginatum.* The efficiency of this species as a vector of CCHF is enhanced by the readiness of the adults to attach to a variety of vertebrates, including humans, their domestic animals, and the lagomorphs and rodents that colonize agricultural land. The first outbreak of CCHF followed the dereliction of Crimean agricultural land, and neglect of the local sport of hare shooting, during enemy occupation in World War II. The resulting explosion in the hare population, the concomitant rise in local concentrations of *H. m. marginatum,* and the heavy infestation of cattle with adult ticks led to the 1944–1945 epizootic of CCHF. *H. m. marginatum* will also parasitize avian species, including rooks and migratory passerines. There is no evidence that CCHF will cause a viremia in birds, but they will transport infected ticks, both locally and during migration. Several of the other tick species involved with CCHF frequently parasitize birds in Eurasia and Africa, including the widespread cattle tick, *Ixodes ricinus;* the apparent prime vector in S. Africa, *H. m. rufipes;* and the prime vector in the Crimea, *H. m. marginatum.* Possible confusion can arise over the classification of the vector in the Crimea by Russian workers of *H. plumbeum plumbeum* (Panzer), as Hoogstraal (1979) believed this to be a misidentification of *H. m. marginatum* (Koch).

The CCHF serogroup also includes Khasan, isolated, in 1971, from a pool of ticks, *Haemaphysalis longicornis,* parasitic on spotted deer, in the Khasansk region of the U.S.S.R. by Lvov *et al.* (1978). This virus was shown, by complement fixation tests (CFT) and neutralization tests (NT), to

Figure 1. Geographic distribution of isolates that are included in the Nairovirus genus. ●, Sakhalin; ○, CCHF; ■, Hughes; ▲, NSD; ▼, Dera Ghazi Khan; and △, Qalyub.

be related to CHF, as reported in the International Catalogue of Arboviruses (1985), though no detailed data are available on the degree of relatedness.

The fourth member of the CCHF group is HAZ, isolated from *Ixodes redikorzevi* collected in West Pakistan in 1964 by Begum *et al.* (1970b). This virus has been widely investigated as it appears to be safer to work with (4/150 seropositive humans in the area of origin, with no overt clinical disease reports) and grows to tenfold higher titers than CCHF.

The serologic relationship between CCHF and CON virus has been extensively examined, and the two have been shown to be indistinguishable, with a high degree of antigenic conservation even among isolates causing various degrees of clinical symptoms (Tignor *et al.*, 1980). HAZ has been shown to have a weak cross-neutralization and CFT relationship with the CCHF viruses and is thus included in this serogroup (Buckley, 1974; Begum *et al.*, 1970c).

III. NAIROBI SHEEP DISEASE SEROGROUP

There are three members of this serogroup; NSD, Ganjam (GAN), and Dugbe (DUG).

A. Nairobi Sheep Disease

NSD was first described in 1917 and was shown to have a viral etiology (Montgomery, 1917). The symptoms of NSD include fever and gastroenteritis with a duration of 4–5 days. The mortality rate of NSD in sheep is over 70%. Less severe symptoms are observed in goats (Montgomery, 1917). There have been numerous isolations of the virus throughout East Africa (mainly in Kenya and Uganda) from the blood and organs of infected sheep and goats. The highest titers of virus were found in the liver and spleen (Montgomery, 1917; Daubney and Hudson, 1934; Weinbren *et al.*, 1958; Davies *et al.*, 1978). Antibodies have been found in humans and are thought to be due to a nosocomial subclinical infection (Munz *et al.*, 1984a). The most important vector of NSD is the hard tick, *Rhipicephalus appendiculatus* (Montgomery, 1917; Daubney and Hudson, 1934; Lewis, 1946; Davies and Mwakima, 1982). NSD can remain silently enzootic in an area for long periods as the virus can survive in *Rh. appendiculatus* for periods of at least 871 days (Lewis, 1946). It has been shown that NSD virus can be transmitted transstadially, vertically, and transovarially in *Rh. appendiculatus* (Daubney and Hudson, 1934; Davies and Mwakima, 1982). Vectors of NSD include the tick *Amblyomma variegatum* (Daubney and Hudson, 1934) and the midge *Culicoides tororensis* (Davies *et al.*, 1979).

Cattle, buffalo, horses, mules, donkeys, pigs, dogs, rabbits, and guinea pigs showed no clinical symptoms after being injected with virulent blood (Montgomery, 1917). NSD caused death in suckling mice by the intracerebral (i.c.) and intraperitoneal (i.p.) routes although only the i.c. route was pathogenic in adult mice. Encephalitis, degeneration of neurons, and vascular dilation in the brains of mice were seen after NSD inoculation (Weinbren *et al.*, 1958). The wild field rat, *Arvicanthis abyssinicus nairobiae,* was also susceptible and may be a reservoir for NSD (Daubney and Hudson, 1934). Attempts to grow virus in the chorioallantoic membrane of 12-day-embryonated eggs were only occasionally successful (Weinbren *et al.*, 1958). NSD replicated in BHK-21 and sheep kidney cells but did not always produce a cytopathic effect (CPE) (Davies *et al.*, 1979; Munz *et al.*, 1984b). A tick cell line has also been used for virus propagation (Munz *et al.*, 1984b). NSD virus withstood lyophilization (Daubney and Hudson, 1931). Virulent blood kept in sealed containers remained infective after incubation at 37°C for 42 hr. Montgomery showed in 1917 that infected blood incubated at 50°C for 2 hr caused an attenuated version of the disease whereas blood incubated at 60°C for 5 min was apathogenic (Montgomery, 1917).

Serologic evidence of the presence of NSD stretches across Ethiopia to Somalia and Tanzania. By CFT, NSD is very closely related, and may be identical, to GAN. NSD showed some relationship to DUG and to a few strains of CON by CFT. Indirect fluorescent antibody test (IFAT) also showed NSD to have a close relationship to GAN and some identity with DUG and HAZ but very little with CON. In the indirect hemagglutination test (IHA) there was a close relationship between NSD and GAN, and a weak one with HAZ and CON. From these data, NSD, GAN, and DUG have been formed into a distinct serogroup distantly related to CON (Davies *et al.*, 1978). Indirect immunoperoxidase technique (IPT) (Munz *et al.*, 1984b) and enzyme-linked immunosorbent assays (ELISA) (Munz *et al.*, 1984a) have been developed to detect NSD antigen and antibodies.

B. Ganjam

GAN virus was first isolated in 1954 from a pool of 19 adult ticks, *Haemaphysalis intermedia* (Warburton and Nuttall, 1909), collected from healthy goats in Bhanjanagar, Orissa State, India (Dandawate and Shah, 1969). Since then 30 strains of GAN have been isolated and the viruses have been characterized by CFT and NT. Since 1954, 30 strains of GAN have been isolated, including, in 1956, a strain isolated from a 12-year-old boy with a febrile illness (which lasted 3 days) at Vellore, Madras State, India. Neutralizing antibodies were still present in the patient's serum 9 years after the infection (Dandawate *et al.*, 1969a). GAN has been isolated from a pool of 100 female mosquitoes of the *Culex vishnui* complex (*C. tritaeniorhynchus, C. pseudovishnui,* and *C. vishnui*) collected from Sathuperi, near Vellore, Madras State, India, in 1956 (Dandawate *et al.*, 1969b). Eighteen strains were isolated in 1961 from *H. intermedia* ticks collected from sheep and goats in Shimoga District, Mysore State, India (Boshell *et al.*, 1970). In 1967/1968 two strains of the virus were isolated from ticks, *Haemaphysalis wellingtoni* (Nuttall and Warburton, 1967), again in the Shimoga District (Rajagopalan *et al.*, 1970). One strain was found in a pool of 55 nymphs collected from red spurfowl (*Galloperdix spadiceal*) and the other strain was isolated from a pool of 132 nymphs collected from the forest floor. This showed the possible involvement of birds in the natural cycle of infection. Two isolates were recovered from febrile sheep in Andhira Pradesh, India (Ghalsas *et al.*, 1981).

In 1977, five strains of GAN were recovered from five members of staff at the National Institute of Virology (N.I.V.), Pune, India, who were suffering from a febrile illness with arthralgia, lasting 3–5 days. The virus was isolated from acute-phase serum samples (Mohan Rao *et al.*, 1981). Other laboratory infections with GAN have been reported but no virus has been recovered. In 1956 a laboratory infection occurred at Vellore associated with the 12-year-old boy mentioned earlier. The symptoms developed 3 days after a laboratory worker had handled infectious material (Dandawate *et al.*, 1969a). At the N.I.V., six cases of infection with GAN were reported; these were presumed to be due to aerosols, as there was no history of injury or bites by vectors among the patients (Banerjee *et al.*, 1979).

Antibodies to GAN have been found in beetal and nomadic goats from Bhanjanagar and Patha, Orissa State, and also in sheep at Zaba, Kashmir (Dandawate and Shah, 1969). Human antibodies were recovered at Vellore, Ootacamund (Madras State), Kashmir and North East Frontier Agency (Dandawate *et al.*, 1969a). GAN is pathogenic to several laboratory animals. The original isolate caused a 50% mortality rate in suckling mice after 8 days; in the second passage the mortality rate was 100% after 4 days. The average survival time decreased to 2 days by the 20th passage (Dandawate and Shah, 1969). GAN is lethal by i.c. and i.p. routes in suckling mice and i.c. and occasionally i.p. in adult mice (the symptoms include hyperactivity, ruffled coat, and occasionally paralysis). Chick embryos have a survival time of 3–5 days when inoculated by the yolk sac route. GAN is not pathogenic to guinea pigs but neutralizing antibodies are initiated (Dandawate and Shah, 1969). Viremia occurred in a Langur monkey after subcutaneous (s.c.) inoculation with the virus (International Catalogue of Arboviruses, 1985). In hamsters and *Macaca radiata,* CFT and NT antibodies were initiated but there was no viremia (Dandawate, 1977). GAN causes a febrile illness in sheep with viremia.

Several cell lines are susceptible to GAN. The virus grows in *Aedes albopictus* cells without a

CPE but not in *Ae. aegypti* (Singh and Paul, 1968). GAN grows in a few vertebrate cell lines with a CPE, such as BHK-21 and Vero, but not LLC-MK2 cells (Dandawate and Shah, 1969; Stim, 1969). The CPE in Vero cells is a mixture of two types, cytolytic and polykaryocytotic (Paul and Dandawate, 1970).

The estimated size of the virion is between 100 and 220 nm according to passage through filters (Casals, 1968). GAN is sensitive to deoxycholic acid (Paul and Dandawate, 1970), chloroform, and sodium deoxycholate (1 : 1000) (Dandawate and Shah, 1969). The virus has been shown to multiply in ticks and mosquitoes (*Ae. albopictus* and *Ae. aegypti*) by intrathoracic inoculation (International Catalogue of Arboviruses, 1985; Paul and Dandawate, 1970) and in *Haemaphysalis spinigera* and *H. turtuns* after parenteral inoculation (International Catalogue of Arboviruses, 1985; Singh and Paul, 1968). No hemagglutination or hemadsorption has been reported (Dandawate and Shah, 1969; Paul and Dandawate, 1970). GAN is closely related to NSD and distantly related to DUG by IFAT and CFT (International Catalogue of Arboviruses, 1985).

C. Dugbe

DUG was first isolated in 1964 at the Virus Research Laboratories, University of Ibadan, Nigeria, from *Amblyomma variegatum* ticks collected from white Fulani cattle at a market (Causey, 1970). Since the initial isolation, strains of the virus have been isolated many times from Ixodid ticks, cattle, and humans. The virus is found to be widely distributed in ecologic sites varying from dry *Acacia* savannnah to lakeside forest (Wood *et al.*, 1978). Although DUG has been isolated from a child with a febrile illness (Moore *et al.*, 1975), the occurrence in humans is relatively rare. Of 251 human sera tested from people living in the Ibadan region, only nine were found to have significant titers of NT antibodies as measured using suckling mouse lethality (David-West *et al.*, 1975). Furthermore, it was also reported that only four isolates of DUG had been made from humans in Nigeria over a 10-year period up to 1975.

DUG has been reported to be sensitive to detergents and chloroform (Causey, 1970; David-West and Porterfield, 1974). It has also been shown to be temperature sensitive. The virus is completely inactivated by storage at −20°C or 4°C for 2 months or heating for 12 hr at 37°C (David-West and Porterfield, 1974). DUG was also found to be unstable at pH 3.0, but infectivity was retained at pH 9.0 (David-West and Porterfield, 1974). DUG can be routinely passaged in suckling mouse brain (Causey, 1970; David-West, 1970). In studies using suckling mice David-West (1970) found that both infectious virus and complement-fixing (CF) antigens could be obtained from the brain, in which histopathologic lesions could also be observed. Furthermore, the appearance of virus was relatively quick (within 2 days) and the titer high in this organ. In contrast, although infectious virus was also observed in the liver of the same mice, the appearance of this was slower (and the titer much lower), with no CF antigens extractable and no histopathologic changes. A number of cell lines have been found to be permissive for DUG, including BHK-21, BS-C-1, CER, LLC-MK2, PS, VERO, and XTC-2 cells (Clerx *et al.*, 1981; David-West and Porterfield, 1974; David-West, 1974; Cash, 1985). David-West and Porterfield (1974) found that DUG could be passaged in PS cells with only a slight CPE, whereas Cash (1985) reported that a CPE was evident in BS-C-1 cells although a background of "uninfected" cells persisted until regrowth of the monolayer occurred. PS cells have also been observed to act as persistently infected carriers of DUG (David-West and Porterfield, 1974). Following infection with DUG the cells could be passaged and cell-released virus continually detected in the culture medium, although the titer diminished as the passage number increased. Similar observations have been made for BS-C-1 cells (Cash, 1985). It was also observed in PS cells that the production of DUG was much reduced in chronically infected cells rechallenged with fresh virus (David-West and Porterfield, 1974). This interference phenomenon was not observed in the case of representatives of the alpha- and flaviviruses, but an intermediate effect was observed with NSD virus. In a similar way, cells chronically infected with GAN are less permissive for DUG.

The expression of DUG-virus-induced proteins in cell lines has also been investigated. David-West (1974) observed a single major protein of about 50kDa in DUG-infected PS and BS-C-1 cells labeled with [^{35}S]-methionine. Subsequently Clerx *et al.* (1981) observed an immune-precipitated intracellular protein of about 49kDa from DUG-infected cell extracts. On examination of purified virions, Clerx *et al.* (1981) were able to make estimates of the sizes of the three RNA species and the structural proteins of DUG. A more extensive study was carried out by Cash (1985). Using [^{3}H]-leucine he observed three virus-induced proteins, of molecular weights 92, 82, and 48kDa in virus-infected BS-C-1 cell extracts. The 48-kDa protein first appeared by 24 hr postinoculation (p.i.), peaked at 27 hr, and became undetectable by 48 hr. A similar time course was observed for expression of the 92-kDa protein. In contrast, the 82-kDa protein was first detected at 27 hr and was still present at 48 hr. A 52-kDa protein was intermittently detected. Cash (1985) observed four proteins of 34, 48, 52, and 77kDa in purified virions, of which only the 77-kDa protein could be labeled with [^{3}H]-glucosamine, indicating that it was a glycoprotein. However, the 82-kDa and 92-kDa proteins observed in cell extracts were not obtained from purified virus.

Hemagglutination (HA) antigens for goose erythrocytes have been extracted from suckling mouse brain (SMB) inoculated with a DUG isolate (Causey, 1970). DUG virus antigen has been reported to cross-react with hyperimmune mouse ascitic fluid (MAF) raised against NSD in CFT (Davies *et al.*, 1978). In the same study, MAF raised against DUG was found to cross-react in CFT, IFA, and hemagglutination inhibition (HAI) tests with NSD antigens. In the CF test at the Yale Arbovirus Research Unit (YARU), DUG antigen gave negative results with sera raised against a range of viruses, including polyvalents to groups A and B arboviruses, California group, Thogoto, HUG, UUK, Kemerovo, LCM, and Congo (Causey, 1970). By HAI DUG antiserum was observed to be related to CCHF, HAZ, and Abu Mina (AM) (Casals and Tignor, 1980). Conversely, hemagglutination by DUG antigen was inhibited by sera raised to the following viruses: CCHF, HAZ, GAN, DGK, Abu Hammad (AH), AM, Kao Shuan (KS), Pretoria (PRE), BDA, QYB, HUG, Farallon (FAR), and Soldado (SOL). Furthermore, using NT, DUG antiserum was found to cross-react with CCHF, QYB, and AM. Antiserum raised against DUG has also been found to precipitate the N proteins of HAZ, QYB, and BDA, and the G1 glycoprotein of DGK, AM, and QYB from infected cell extracts (Clerx *et al.*, 1981). Therefore, by serologic tests DUG was related to other nairoviruses, although the strongest cross-reactions were with NSD and GAN.

IV. QALYUB SEROGROUP

The Qalyub serogroup consists of three members; Qalyub (QYB), Bandia (BDA), and Omo (OMO) (Clerx *et al.*, 1981; Bishop, 1985). The type virus was first isolated in 1952 from a pool of *Ornithodoros erraticus* ticks collected from a neglected rodent nest in Qalyub, Egypt (Taylor, 1970). The status of QYB as an arbovirus is still questioned (e.g., International Catalogue of Arboviruses, 1985). However, further isolates of QYB have been obtained from pools of *O. erraticus* (Abdel Wahab *et al.*, 1970), and Miller *et al.* (1985) have demonstrated experimentally that such ticks are capable of infection by, and transmission of, this virus. Both these lines of evidence, therefore, indicate that QYB is an arbovirus. Current evidence indicates that *O. erraticus* is the only arthropod vector. The main vertebrate hosts are likely to be rodents, with the Nile grass rat (*Arvicanthis niloticus niloticus*) predominant. However, antibodies to QYB have been identified in a number of other species ranging from dogs to camels and also humans, although there is no evidence of clinical disease.

BDA virus was isolated in 1965 in the Bandia Forest, Senegal, from a moribund young male of an unidentified species of the multimammate mouse *Mastomys natalensis* (Bres *et al.*, 1967). The virus was also isolated by the same workers from *O. erraticus sonrai,* Sautet and Witwoski, ticks in the same forest. Although antibodies to BDA have been found in humans living in the area, there is, again, no evidence of disease induced by this virus.

OMO was isolated from the kidney and spleen of an adult *Mastomys erythroleucus,* in 1971, caught in a baited trap at Shangura, Omo Valley, Gamo-Gofa province, Ethiopia (Rodhain *et al.*, 1985). No further isolates have been reported.

QYB virus can be routinely passaged in suckling mice (Clerx and Bishop, 1981; Taylor, 1970; Miller *et al.*, 1985), with a peak viremia (titer 10^6 pfu/ml) reported at 48 hr (Miller *et al.*, 1985). No infection was observed in older weaned mice inoculated by the i.c. route (Taylor, 1970). Embryonated hens' eggs can also be used to passage QYB virus. Miller *et al.* (1985) were able to demonstrate experimentally that *O. erraticus* ticks of varying developmental stages could be infected with virus when they fed on viremic mice. Propagation of QYB virus has been attempted in a number of cell lines. Attia and Williams (1970) reported only a marginal CPE in rabbit kidney cells inoculated with mouse-grown virus. However, Clerx and Bishop (1981) and Clerx *et al.* (1981) were able to obtain reasonable titers of QYB in both BHK-21 (between 10^7 and 10^8 pfu/ml) and Vero cells (approximately 10^6). Electron micrographs of cell-released virus (from both BHK-21 and Vero) revealed that the outer fringe of characteristic envelope spikes possessed electron-lucent spherical structures which are unique to this virus (Clerx and Bishop, 1981). It has been suggested that these may be an artifact produced during the preparation for electron microscopy (Clerx *et al.*, 1981). QYB is reported to be ether sensitive (Taylor, 1970). More detailed biochemical analyses of cell-grown virus have produced estimates of the sizes of the RNA segments and structural polypeptides (Clerx *et al.*, 1981). Studies of immune-precipitated intracellular virus-coded proteins have indicated that the two glycoproteins may be derived from the cleavage of two larger precursor proteins of molecular weights 115kDa and 85kDa (Clerx and Bishop, 1981; Clerx *et al.*, 1981).

OMO has been passaged in suckling mice (Rodhain *et al.*, 1985). BDA virus has been passaged in suckling mice (Clerx *et al.*, 1981; Bres *et al.*, 1967); it induced a lethal infection in guinea pigs but was not pathogenic for adult mice or rabbits (Bres *et al.*, 1967). Both BHK-21 and Vero cells were permissive for BDA with titers (at 48 hr) of 10^8 and 10^6 pfu/ml, respectively, obtainable. Unlike QYB, BDA virus did not possess envelope-associated electron-lucent spherical structures (Clerx *et al.*, 1981). BDA and OMO have shown a sensitivity to organic solvents similar to that of QYB (Bres *et al.*, 1967; Rodhain *et al.*, 1985). Biochemical analyses of cell-grown virus have produced estimates of RNA and protein sizes (Clerx *et al.*, 1981). Oligonucleotide fingerprint analyses of the RNA of both QYB and BDA revealed that each RNA species contained unique T1-resistant sites (Clerx and Bishop, 1981; Clerx *et al.*, 1981).

Serologic analysis demonstrated that the only significant cross-reaction in CFT was with BDA (Taylor, 1970). As with QYB, hyperimmune fluids raised against BDA gave negative results in CFT with representative antigens of alpha- and flaviviruses as well as members of the bunyavirus genus (Bres *et al.*, 1967). More recently, Casals and Tignor (1980), using mouse grown virus, reported a partial comparison of serologic cross-reactions of QYB and BDA with other nairoviruses. HAI tests indicated that both QYB and BDA were related to CCHF and HAS of the CCHF serogroup and DUG of the NSD group. These relationships appeared to be stronger for QYB than for BDA. Immune serum to QYB also demonstrated a weak HAI with AM (DGK serogroup) antigen. However, an indirect immunofluorescent test failed to demonstrate any cross-reaction between QYB and BDA immune sera and antigens of CCHF, HAZ, DUG, AH, or AM. In a separate study using cell-grown virus and antiserum raised against QYB, Clerx *et al.* (1981) were able to immune-precipitate N and G1 structural proteins from cells inoculated with DGK and AM as well as BDA. Hyperimmune serum to BDA would only precipitate proteins from cells infected with QYB or BDA. Conversely, viral N and G1 proteins from cells infected with QYB or BDA could be precipitated with heterologous immune sera raised against a number of other nairoviruses (Clerx *et al.*, 1981), but not against other members of the Bunyaviridae. The overall pattern therefore appears to be that QYB and BDA are serologically related to other nairoviruses, but not other Bunyaviridae genera, to which they are morphologically and biochemically similar.

V. DERI GHAZI KHAN SEROGROUP

The DGK serogroup is composed of six members, which are, including the type member, AH, AM, PRE, KS, and Pathum Thani (PTH). The DGK prototype strain was first isolated in 1966 from a pool of ticks, *Hyalomma dromedarii,* collected from camels in the Sakhi Sarwar semidesert region of the DGK district, Pakistan (Begum *et al.* 1970a,b). AH was isolated in 1971 from *Argas hermani* nymphs taken from pigeon houses on a farm at Abu Hammad, Shargiya, Egypt. Two additional isolates of AH were obtained from argasid ticks collected in Egypt (Converse *et al.*, 1974). AM was reported to be isolated from ticks, *Argas (Persicargas) streptopelia,* Kaiser, Hoogstraal, and Horner in Egypt (Converse *et al.*, 1975a). PRE was first isolated in 1975 from ticks, *Argas (Argas) africolumbae,* collected near a nest of the African rock pigeon (*Columba guinea phaeonata,* Gray) in a house at Derdepoort, Pretoria, Republic of South Africa. KS virus was isolated initially in 1970 from *Argas robertsi* nymphs found in the bark of *Acacia* trees, which served as a rookery for night herons (*Nycticorax nycticorax*), in the Kao Shuan district of Taiwan. Further strains of KS have been isolated in Australia, Thailand, and Java. PTH was first isolated in 1970 from *Argas robertsi* nymphs taken from a colony of open-billed stork (*Anastomus ascitans*) at Pathum Thani, Thailand, and a further isolation from the same species of tick has been made in Sri Lanka (Doherty *et al.*, 1976).

The biologic and biochemical characteristics of the DGK group have not been studied to the same extent as other nairoviruses; PRE, PTH, and KS are all reported to be sensitive to detergents, though to varying degrees (International Catalogue of Arboviruses, 1985), and DGK is sensitive to both detergent and ether (Begum *et al.*, 1970b). All the viruses can be passaged in suckling mice. DGK was found to have no adverse effects in guinea pigs or adult hamsters inoculated i.p., or baby hamsters inoculated i.c. (Begum *et al.*, 1970b). PRE has been reported to be lethal for suckling mice by the i.c. and i.p. routes, but only by the i.c. route in suckling hamsters. No effect of PRE was observed in adult mice, hamsters, guinea pigs, or rabbits when inoculated i.c. (Converse *et al.*, 1975b). KS virus was also found to cause death in suckling mice when inoculated i.c. (after 5 days) or i.p. (after 9–13 days). Attempts to propagate DGK viruses in cell lines have been limited. DGK failed to grow in BHK-21 cells (Begum *et al.*, 1970b), and Clerx *et al.* (1981) were unable to develop a plaque assay for DGK. However, in the same study, DGK structural proteins were immune-precipitated, by homologous antiserum, from pelleted extracellular material (the cell line used was not indicated) and size estimates of G1 (75kDa) and N (50kDa) obtained. Clerx *et al.* (1981) also reported that AM virus could be assayed using Vero or CER cells. Using purified virions, estimates of the size of the RNA species and the structural proteins of AM were also made. PRE virus does not grow in chick embryo fibroblasts, Vero, or BHK-21 cells (Converse *et al.*, 1974). The Australian isolates of KS produced no CPE in PS-EK cells, but CF antigen was detected in the culture fluid after two 14-day passages (Doherty *et al.*, 1976).

Immune serum raised against the prototype strain of DGK did not inhibit HA by representative antigens from groups A, B, and C arboviruses (i.e., alpha-, flavi-, and bunyaviruses) (Begum *et al.*, 1970a,b). The same antiserum did not cross-react in CFT with antigens of the bunyavirus genus, CON, DUG, FAR, GAN, HUG, Kemerovo, Sandfly fever, SOL, Tacaribe, or UUK (Begum *et al.*, 1970b). Reciprocal CFT with antigens and sera from the DGK group demonstrated that they were related, but may fall into two serologic subgroups; one African (PRE, AH, AM) and one Asian (KS, PTH) with DGK between the two (Converse *et al.*, 1975a). Using the HAI test, Casals and Tignor

(1980) demonstrated that all of the DGK viruses, except PTH, were related to CCHF and DUG. Furthermore, using NT and IFAT, Casals and Tignor demonstrated that the members of the DGK group were related.

VI. HUGHES SEROGROUP

The Hughes serogroup contains the following viruses: HUG, Punta Salinas (PS), SOL, and Zirqa (ZIR) (Hoogstraal and Feare, 1984). There are six other proposed members of the serogroup: Great Saltee, Farallon, Fraser Point, Puffin Island, Raza, and Sapphire. Members of the Hughes serogroup have been isolated from argasid ticks associated with birds; there are two exceptions, *Amblyomma (Adenopleura) loculosum* Neumann (Hoogstraal and Feare, 1984) and *Ixodes (Ceratixodes) uriae* White (Nuttall *et al.*, 1986). The isolates have been found mainly at coastal locations in seabird ticks (the exceptions are PS from *Argas arboreus* Kaiser in Tanzania (Converse *et al.*, 1975b, 1981) and Sapphire from *Argas cooleyi* Kohls and Hoogstraal in Texas (Yunker *et al.*, 1972).

Several of the viruses are associated with human disease. Humans bitten by infected ticks have had symptoms ranging from severe pruritus (SOL) (Hoogstraal and Feare, 1984; Converse *et al.*, 1975b) through pyrexia (SOL) (Chastel *et al.*, 1981a) to pyrexia with allergic reactions (ZIR) (Varma *et al.*, 1973).

A. Hughes

The prototype for the group (HUG) was first isolated in 1962 from a pool of 12 adult ticks of the *Ornithodoros capensis* complex (Hughes *et al.*, 1964; Philip, 1965) (later identified as *O. denmarki* Kohls (Yunker, 1975)). The ticks were collected from Bush Key in the Dry Tortugas, a group of coral reef islands off the west coast of Florida. Bush Key is a nesting site for brown noddy terns (*Anous stolidus stolidus*) and sooty terns (*Sterna fuscata fuscata*). The ticks were collected from nests and litter in the nesting area (Hughes *et al.*, 1964). The virus was shown to grow and form plaques in Vero, LLC-MK2, and XTC-2 cells (Stim, 1969; Gould *et al.*, 1983). HUG formed persistent infections in Vero cells (Gould *et al.*, 1983) and killed a high percentage of newly hatched chicks inoculated i.c. but was apathogenic when inoculated s.c. Although RAZ was originally classified as an isolate of HUG (Clifford *et al.*, 1968), it is now believed to be a different strain (Matthews, 1982). RAZ was isolated from a pool of ticks *O. denmarki* Kohls (eight nymphs and 16 adults) from Raza Island, Gulf of California, Mexico (Matthews, 1982). The tick extract was inoculated i.c. into suckling mice, which died 15–17 days later. After the initial passage, i.c. inoculation of suckling mice caused death within 4–5 days (Clifford *et al.*, 1968). RAZ did not generally kill newly hatched chicks by any route of inoculation and did not hemagglutinate. It was shown by complement fixation that RAZ and HUG shared complete identity, but variations were seen in cross-neutralization tests (Clifford *et al.*, 1968).

B. Soldado

SOL virus was first isolated from a pool of nymphal ticks (*Ornithodoros capensis* Neumann and/or *Ornithodoros denmarki* Kohls) collected from a brown noddy tern (*Anous stolidus stolidus*) on Soldado Rock off the southwestern tip of Trinidad in 1963 (Jonkers *et al.*, 1973). The primary i.c. inoculation of the virus in suckling mice caused death and sickness in 13 days. Subsequent passages (i.c.) in suckling and adult mice caused death in 9–11 days but the virus was apathogenic by the (i.p.) inoculation route. SOL grew in Vero (where it caused persistent infections) and in XTC-2 cell lines and appeared to be sensitive to ether (Jonkers *et al.*, 1973). Further strains of SOL have been isolated from Soldado Rock from the same species of tick but associated with sooty terns

(*Sterna fuscata fuscata*). Eight strains were recovered from the sera of 8- to 12-day-old sooty tern nestlings inoculated i.c. into suckling mice (Aitken *et al.*, 1968).

There have been several isolations of SOL-like viruses since 1963. A strain of SOL was isolated from *O. (A). capensis* found in bird nests on Lake Shalla, Ethiopia, in 1975 (Hoogstraal *et al.*, 1975) and at Lambert Bay, South Africa, in 1973 (Converse *et al.*, 1981). Six strains of SOL were isolated from Bird Island (1973) and one strain from Des Noeufs Island, Amirantes group (1974), Seychelles, in the Indian Ocean. The viruses were isolated from adult *O. (A). capensis* collected on the ground in a sooty tern (*Sterna fuscata fuscata*) breeding area on Bird Island and from nests of the blue-faced booby (*Sula dactylatra melanops* Heuglin) on Des Noeufs Island. All isolates were pathogenic for suckling and adult mice by the i.c. route, with an average survival time of 5 days. Immature guinea pigs inoculated i.c. developed hindleg paralysis followed by death after 5–8 days. No antibodies to SOL were found in humans or marine birds at this site. However, persons bitten by *O. (A). capensis* in the Seychelles suffered from severe pruritus for several days. It has been shown that infected ticks can transmit SOL to 1-day-old chicks. An additional SOL isolate was recovered in the Seychelles from a pool of *Amblyomma* (Adenopleura) *loculosum* collected from a dead sooty tern (*S. fuscata fuscata*) chick on Goelette Island, Farquhar Atoll. A SOL strain was isolated from *O. (A). capensis* ticks found in gray-headed gull (*Larus cirrhocephalus*) nests on Langue-de-Barbarie in Senegal National Park (Main *et al.*, 1980). In Morocco SOL was isolated from ticks [*Ornithodoros (A). maritimus*] associated with gulls (*Larus argentatus michaellis*) at Essaouira in 1979. A person bitten by *O. (A). maritimus* in this area suffered from pyrexia (Chastel *et al.*, 1981a).

SOL has been isolated from several locations in Europe. In France, 18 strains of SOL have been isolated from Cap Frehel (Chastel *et al.*, 1979a, 1981b) and Cap Sizun, Cotes-du-Nord (Chastel *et al.*, 1981b), in 1979. The viruses were isolated from *O. (A). maritimus* ticks collected from the nests of herring gulls (*Larus argentatus*) (Chastel *et al.*, 1979a, 1981b). The strains were pathogenic to adult and suckling mice by the i.c. route but not by the i.p. route. The primary passage in suckling mice caused paralysis and death in 11–24 days but subsequent passages caused death in 6–9 days (Chastel *et al.*, 1981b). The Cap Sizun isolate was studied in detail and found to be sensitive to ether, inactivated at 60°C for 1 hr, and resistant to incubation at pH 3.0 (Chastel *et al.*, 1979b). By transmission electron microscopy the virion was found to be pleomorphic with an average diameter of 94 nm (range 70–107 nm), a membrane thickness of 8–9 nm, and projections that could occasionally be seen on the membrane surface (Chastel *et al.*, 1979b). Histopathologic changes in suckling mouse brain were most pronounced in the midbrain. There was diffuse edema and acute vasculitis with adventitial proliferation. Virions were seen in the cisternae of the Golgi apparatus and endoplasmic reticulum of neurons. These SOL strains do not hemagglutinate.

Three strains of SOL-like virus were isolated from Puffin Island, off the southeast coast of Anglesey, North Wales, in 1974 (Converse *et al.*, 1976; Johnson *et al.*, 1979). The viruses were isolated from nymphal and adult ticks *O. (A). maritimus* (Vermeil et Marguet) collected in or near nests of herring gulls (*L. argentatus Pontoppidan*) (Converse *et al.*, 1976). The isolates grew in Vero and XTC-2 cell lines (Gould *et al.*, 1983; Johnson *et al.*, 1979). Puffin Island virus has been shown to be antigenically distinct to SOL and has been proposed as a separate strain (Gould *et al.*, 1983). Another SOL-like virus was found on Great Saltee Island off the southeast coast of Eire. The virus was isolated from *O. (A). maritimus* ticks in 1976 and 1980 (Keirans *et al.*, 1976; Nuttall *et al.*, 1984). Great Saltee was placed in the Hughes group on the basis of CFT and was shown to be most closely related to SOL (Keirans *et al.*, 1976). Great Saltee virus is a proposed member of the serogroup and not a SOL strain (Nuttall *et al.*, 1986). This virus produced plaques in Vero (Nuttall *et al.*, 1986; Keirans *et al.*, 1976) and XTC-2 (Nuttall *et al.*, 1984) cell lines and was sensitive to sodium deoxycholate, ether, and chloroform (Nuttall *et al.*, 1984).

Strains of a virus closely related to the Great Saltee virus have been isolated from ticks, *Ixodes (Ceratixodes) uriae*, collected in England, the Faeroe Islands, and Iceland (Nuttall *et al.*, 1986). In England the ticks were collected from the nesting sites of common murres (guillemots, *Uria aalge*) on the Isle of May, southwest Scotland; Foula, Scotland, and Inner Farne, northeast England. The remainder of the ticks were collected from nesting sites of Atlantic puffins (*Fratercula arctica*) in Iceland (Grimsey in the north and Ellidaey in the west) and Mykines, Faeroe Islands. These isolates cross-reacted with each other and Great Saltee in NT and CF tests.

Several SOL-like viruses have been isolated in North America. SOL strains were recovered from *O. capensis* ticks collected on Manana Island, off Oahu, Hawaii (Yunker, 1975; Chastel *et al.*, 1983). Additional isolates have been found in ticks (*O. capensis*) associated with nests of western gulls (*Larus occidentalis*) on the Farallon Islands in the Pacific Ocean, west of San Francisco, in 1965 (Radovsky *et al.*, 1967). Farallon-like viruses were also recovered at Cape Lookout, Oregon, from *O. capensis* ticks associated with the common murre (*Uria aalge*) (Yunker, 1975). Farallon virus is a proposed member of the Hughes serogroup (Yunker, 1975). A SOL-like virus referred to as Sapphire 1 was isolated from a swallow tick (*Ixodes howelli*) in Montana; Sapphire 2, a similar isolate, was recovered from ticks (*Argas cooleyi*) collected from Montana [unpublished data, cited by Yunker *et al.* (1972)] and Sunday Canyon, Texas. The ticks collected in Texas were from an area inhabited by cliff swallows (*Petrochelidon pyrrhonota*). Sapphire has been proposed as a separate virus in the Hughes serogroup (Yunker *et al.*, 1972).

There are antigenic variants among the strains defined as SOL by CFT (Chastel *et al.*, 1983). A division between strains isolated from the "Old World" [associated with *O. (A.) maritimus*] and the "New World" (associated with *O. capensis* /*O. denmarki*) has been postulated.

C. Punta Salinas

PS was first isolated from a pool of 10 ticks [*Ornithodoros (A.) amblus*] collected from rocks in a guano bird colony on the Punta Salinas peninsula, Huacho, Peru, in 1967 (International Catalogue of Arboviruses, 1985). PS has also been isolated from Shinyanga, Tanzania, from a pool of 35 nymphal ticks, *Argas arboreus*, collected from resting areas of *Leptoptilos crumenifera* and Sacred Ibis, *Threskiornis ae. aethiopicus* (Converse *et al.*, 1981). The virus was isolated and passaged by i.c. inoculation of suckling mice with an average survival time of 6 days. The Tanzanian isolate was sensitive to sodium deoxycholate and did not hemagglutinate. Both strains showed complete identity in CFT.

D. Zirqa

ZIR was first isolated in 1968 from the island of Zirqa off the coast of Abu Dhabi in the Persian Gulf (Varma *et al.*, 1973). The island was investigated after complaints of illness from personnel of the British Petroleum Company Ltd. The virus was isolated from a pool of 25 nymphal ticks, *O. (A). muesebecki*, associated with ospreys, *Pandion h. haliaetus*, and socotra cormorants, *Phalacrocorax nigrogularis*. It was found to grow in primary cell cultures of *Hyalomma dromedarii* without CPE (International Catalogue of Arboviruses, 1985) and in Vero cells, where it caused persistent infections (Gould *et al.*, 1983). Humans bitten by ticks in the area developed pyrexia with allergic reactions and septic sores. ZIR was inactivated by ether and sodium deoxycholate, but there are conflicting reports on whether ZIR hemagglutinated (Chastel *et al.*, 1981a; Gould *et al.*, 1983).

The viruses have been categorized primarily on the basis of CFT (International Catalogue of Arboviruses, 1985; Converse *et al.*, 1975b, 1976; Clifford *et al.*, 1968; Main *et al.*, 1980). However, a few of the viruses have been compared using indirect immunofluorescence and neutralization tests (Gould *et al.*, 1983; Clifford *et al.*, 1968; Jonkers *et al.*, 1973). The results of tests performed on the same virus strains cannot be compared easily as there were several discrepancies, which may be attributable to different methods and reagents (e.g., sera). The following serologic comparisons are a generalized summary of data from several workers (International Catalogue of Arboviruses, 1985; Converse *et al.*, 1975b, 1976; Gould *et al.*, 1983; Clifford *et al.*, 1968; Jonkers *et al.*, 1973; Main *et al.*, 1980). HUG and RAZ are closely related by NT and show complete identity in CFT (Clifford *et al.*, 1968). HUG antiserum shows some identity with PS, FAR, SOL, and ZIR antigens by CFT (Main *et al.*, 1980; Converse *et al.*, 1976). By IFAT, HUG antiserum shared identity with PS, ZIR, SOL, and PI antigens (Gould *et al.*, 1983). In NT, HUG antiserum neutralized SOL and, to a lesser extent, PI.

SOL antiserum reacted with ZIR, PI, FAR, PS, and HUG by CFT and with PI, PS, ZIR, and HUG in IFAT (Gould *et al.*, 1983). HUG and PI antigens are not neutralized by SOL antiserum in NT. PS antiserum shared identity with HUG, FAR, ZIR, and SOL antigens but not with PI antigen in CFT. However, in IFAT, PS antiserum reacted with HUG, SOL, PI, and ZIR (Gould *et al.*, 1983). ZIR antiserum reacted with HUG, PI, PS, and FAR antigens in CFT and showed shared identity in CFT with SOL, PS, and ZIR antigens but not with FAR and HUG antigens (Converse *et al.*, 1975b, 1976). In IFAT, PI antiserum gave a limited reaction with HUG, SOL, PS, and ZIR antigens. In NT, PI antiserum neutralized HUG but not SOL. FAR antiserum showed identity with HUG, SOL, PI, and ZIR (Converse *et al.*, 1975b, 1976; Main *et al.*, 1980). SAP2 antiserum did not show any identity with SOL, HUG, PS, ZIR, or FAR by CFT (Main *et al.*, 1980).

CFT reactions in the Bunyaviridae reflect the degree of homology of the nucleoprotein, encoded by the S RNA segment (Shope, 1975). Within the Hughes group a divergent evolutionary pattern emerges, HUG and RAZ being identical, the other viruses showing a varying degree of cross-reaction consistent with such divergence, and revealing the complexity of the epitope by the number of one-way crosses that occur. NT, which is dependent on the surface glycoprotein, reflects a greater degree of differentiation. The viruses have evolved away from the type member, but the complexity of the shared epitopes on the immunogenic sites is still revealed by the one-way reactions detected. The IFAT reaction may not be specific to a single protein, and this is reflected in the broad range of cross-reactions exhibited by this test.

VII. SAKHALIN SEROGROUP

Apart from the prototype member (SAK), the Sakhalin serogroup consists of four other members: Avalon (AVA), Clo Mor (CM), Paramushir (PMR), and Taggert (TAG); AVA and PMR are closely related serologically. Fifteen strains of the type member were initially isolated from *Ixodes (Ceratixodes) putus* ticks (from nesting substrate of the guillemot *Uria aalge*) collected on Tuleniy Island, Sakhalin, U.S.S.R., between 1969 and 1971 (Lvov *et al.*, 1972). Subsequently, Thomas *et al.* (1973) isolated ten strains from the nesting sites of guillemots, cormorants, and gulls on two islands off the Oregon coast. In the study carried out by Lvov *et al.* (1972) 9% of guillemots (and no other seabirds) examined possessed CF antibodies for SAK. Furthermore, evidence for both transstadial and transovarial transmission of virus was observed in that both adult males (which do not feed) and laboratory-hatched ticks were positive for virus.

PMR virus was initially isolated from a pool of *I. signatus* ticks collected from seabird (guillemot and cormorant) nest substrates on Paramushir Island, Sea of Okhotsk, U.S.S.R., during the period 1969–1974 (Lvov *et al.*, 1976). This virus was subsequently identified as being serologically similar to, and probably a strain of, AVA. AVA virus was initially isolated between 1971 and 1972 from *I. uriae* ticks collected from seabird nesting sites on or near Great Island, Avalon Peninsula, Newfoundland, Canada (Main *et al.*,1976). AVA was also isolated from the blood of an apparently healthy herring gull chick (*Larus argentatus*) at the same site. Five other species of seabirds tested were found to be virus-free but a proportion of samples of blood from the common puffin (*Fratercula arctica*) and Leach's petrel (*Oceanodroma leucorhoa*) contained neutralizing antibodies. A further possible isolation of AVA from *I. uriae* ticks parasitizing various bird species was made at Cap Sizun, Brittany, France, during 1979 (Keirans *et al.*, 1976). The serologic identity of this isolate was unclear, but Main, at the YARU, indicated that the virus was probably AVA (cited by Keirans *et al.*, 1976). A subsequent, more extensive study of the Cap Sizun region of Brittany by Quillien *et al.* (1986) identified nine strains of AVA isolated between 1979 and 1981 (including the original isolate from this region) from *I. uriae* ticks. No CF antibodies were found in 289 seabird and landbird or 129 rodent and insectivore sera screened. However, the sera from four farmers (of 474 tested) living in the Brittany region near the reserve were found to possess CF but not neutralizing antibodies.

CM virus was first isolated in 1973 from *I. uriae* ticks collected from a guillemot colony at Cape Wrath, Scotland (Main *et al.*, 1976). A further possible isolate was made in 1979, from *I. uriae* ticks collected in puffin colonies on the Shiant Islands, Outer Hebrides (Nuttall *et al.*, 1982). However, this isolate was found in ticks that also contained an Orbivirus isolate, and the two virus types could not be separated, making it difficult to confirm the presence of this CM strain.

The fifth accepted member of the Sakhalin serogroup is TAG. Six strains of this virus were initially isolated in 1972 from *I. uriae* ticks found in tussock grass (*Poa foliosa*) and under planks near a royal penguin (*Eudyptes chrysolophus schlegeli* Finsch) rookery on the Macquarie Islands, 800 miles southeast of Tasmania (Doherty *et al.*, 1975). Of serum samples taken from a range of species (including humans) only four taken from royal penguins contained plaque-reduction neutralizing antibodies. An additional two virus isolates that were antigenically related to the Sakhalin group viruses were Tillamook virus from Three Arch Rocks, Oregon (Yunker, 1975), and Kachemak Bay virus from Gull Island, Alaska (Ritter and Feltz, 1974), although the latter has not been fully characterized.

All members of the Sakhalin serogroup were shown to have lipid envelopes by their sensitivity to both ether and detergents (Keirans *et al.*, 1976; Lvov *et al.*, 1976, 1981; Main *et al.*, 1976; Doherty *et al.*, 1975). SAK, PMR/AVA, and TAG have been shown to be both heat labile (30 min at 56°C being sufficient to inactivate the virus) and sensitive to ultraviolet irradiation (Lvov *et al.*, 1976). Only two of the viruses have been tested for stability at acid pH. The strains of AVA and CM isolated by Main *et al.* (1976) were both reported to be resistant to acid (ph 3.0), whereas the French isolate of AVA was reported to be sensitive (Quillien *et al.*, 1986).

Successful infection of suckling mice has been demonstrated with all members of this serogroup (Lvov *et al.*, 1972, 1976, 1981; Thomas *et al.*, 1973; Main *et al.*, 1976; Quillien *et al.*, 1986; Doherty *et al.*, 1975). However, in general, no demonstrable growth of virus has been observed in weanling or adult mice. Lvov *et al.* (1972) reported that SAK virus did not grow in golden hamsters, guinea pigs, rabbits, or 2-day-old chicks. However, the virus could be grown in *Culex molestus* but not *Aedes aegypti* mosquitoes by feeding them on virus-infected experimental hosts. Skvortsova *et al.* (1981) reported the successful passage of SAK in suckling rats. Both Main *et al.* (1976) and Quillien *et al.* (1986) found that AVA was pathogenic for suckling mice by the i.c. route. Occasional deaths were reported in adult mice inoculated by the i.c. route with AVA (Main *et al.*, 1976), but not when administered by other routes. No effects were observed in 1-day-old chicks. Furthermore, no viremia was observed in any of the animals inoculated with AVA, including suckling mice and *Ae. aegypti, C. pipiens,* and *Anopheles quadrimaculatus*. CM was similar to AVA with respect to most of the findings listed here, except that it was pathogenic for suckling mice by the i.p. and s.c. routes as well as the i.c. route (Main *et al.*, 1976).

The SAK viruses have all been propagated in cell culture. SAK has been reported to grow in Vero, BHK-21, and SPEV cells (Lvov *et al.*, 1981). PMR/AVA have also been grown in a number of cell lines. Lvov *et al.* (1976) found that this virus would replicate in CER, BHK-21, and duck and human embryo fibroblasts without producing a CPE. However, a CPE was induced in L, Ai, and HeLa cells. In studies on the French isolate of AVA, Quillien *et al.* (1986) could obtain neither a reproducible CPE in Vero nor growth of virus in BHK-21 cells. Estimates of the sizes of the three RNA species of cell-grown AVA have been made (Clerx *et al.*, 1981). However, they reported that it was difficult to grow AVA in cells in sufficient quantities to purify virus or immune-precipitate extracellular materials.

A polypeptide of 48kDa was identified by immune precipitation of virus-induced proteins from infected cells. Little work has been performed on TAG virus but it has been reported to plaque in PS cells (Doherty *et al.*, 1975). In contrast to TAG, the most extensively studied of the Sakhalin group is CM. Watret *et al.* (1985) reported that CM could be grown in XTC-2 cells to reasonable titers ($10^{6.6}$ pfu/ml within 30 hr). Similar results were obtained in BS-C-1 cells. In further work Watret and Elliott (1985) demonstrated the expression of virus-induced proteins in XTC-2 cells infected with CM. 45-kDa and 50-kDa proteins were observed together with a glycoprotein of 115kDa, similar to the large glycoprotein observed in QYB-infected cells by Clerx *et al.* (1981). Immune precipitation of cell extracts demonstrated the presence of gp115, the nucleoprotein, and two proteins of molecular weights 80kDa and 90kDa. It was suggested that the 80-kDa protein might correspond to the structural glycoprotein G1. Immune precipitation of culture fluids from [^{3}H]-mannose-labeled, virus-infected cells demonstrated three glycoproteins of 45, 80, and 90kDa. In subsequent pulse-chase experiments, it was found that in virus-infected cells labeled with [^{35}S]-methionine, initially only gp115 and N were observed (Watret and Elliott, 1985). However, the quantity of gp115 was found to decrease with time and the 45-, 80-, and 90-kDa proteins were expressed. It was suggested that the 45-kDa protein might have been G2 and that the 80- and 90-kDa polypeptides might be cleavage products of gp115. Two proteins of 45kDa and 50kDa molecular weight were also immune-precipitated following *in vitro* translation of mRNA extracts from CM-infected cells. The 45-kDa protein, though, was probably not gp45 as it was detected without the inclusion of microsomal membranes, necessary for glycosylation, in the translation system.

None of the members of the SAK serogroup have been found to possess HA antigens when assessed using goose erythrocytes under a variety of pH and temperature conditions (Lvov *et al.*, 1972, 1976; Main *et al.*, 1976; Doherty *et al.*, 1975). The 15 strains of SAK isolated by Lvov *et al.* (1972) were found to be antigenically related by CF test. SAK is listed as being related to AVA, CM, and TAG by CF test (International Catalogue of Arboviruses, 1985). Furthermore, Casals and Tignor (1980) reported that SAK and AVA antisera cross-reacted in HAI tests with antigens of CCHF, DUG, and HAZ, although no data were presented. The isolates made by Thomas *et al.* (1973) of SAK were found to cross-react in CF and NT tests with the original SAK isolate. However, although cross-reaction was detected, the tests failed to prove that the isolates were identical.

Antigens produced from the isolate of PMR by Lvov *et al.* (1976) did not cross-react with CCHF, HUG, SOL, NSD, QYB, or BDA. However, Lvov *et al.* (1981) subsequently demonstrated by NT, CF, and DPA tests that PMR was serologically identical to AVA. The original isolates of AVA made by Main *et al.* (1976) were demonstrated by CFT to be related to, but distinct from, CM, SAK, and TAG. Although not as clear, NT tests in suckling mice seemed to support this view. In the same study Main *et al.* (1976), using CFT, tested a range of antisera and antigens against AVA, including representatives of the following families: Herpes-, Pox-, Picorna-, Toga-, Reo-, Orbi-, Arena-, Myxo-, Paramyxo-, Rhabdo-, and Bunyaviridae. The only cross-reactions were reported to be with the SAK group. The French isolate of AVA described by Quillien *et al.* (1986) cross-reacted weakly with Upolu virus (an unassigned Bunyavirus), but was confirmed to be AVA by tests carried out at YARU [personal communication cited by Quillien *et al.* (1986)]. The serologic relationships of the CM isolate examined by Main *et al.* (1976) were similar to those of the AVA isolate (see above) with respect to other virus families investigated. However, by CFT, Main *et al.* demonstrated that CM was related to SAK, TAG, and one strain of AVA. This general trend was supported by mouse NT tests, though the data were not as clear. TAG virus has been found to be related only to SAK using the CF test (Doherty *et al.*, 1975).

VIII. DISCUSSION

The viruses that are identified as nairoviruses are included in the family Bunyaviridae on the basis of morphology and biochemistry. The genus is distinguished from the other members of the family on the basis of the greater molecular weight of the N protein and the L RNA segment, the conserved 3′ end-sequences of those genomes examined, and generalized serologic cross-reactions between the serogroups. As the gene-coding assignment of the RNA segments has not yet been performed for the nairoviruses (Bishop, 1985), the significance of segment variation cannot be assessed.

It is clear that the two most important serogroups, in public health or veterinary terms, are the CCHF and NSD groups. The CCHF serogroup, like NSD, consists of one severely pathogenic and one or two apathogenic viruses. It is not known to what degree the isolates of CCHF vary, except that they show a high degree of serologic identity, given their extremely wide distribution. It remains to be demonstrated whether the CCHF isolates are united solely in the possession of a single conserved phenotypic characteristic, or that they are, at the opposite extreme, genotypically indistinguishable. It is important, therefore, that the determination of the gene assignments of the *Nairovirus* genus is carried out, to identify the encoding regions of the markers of serologic identity, and if possible the markers of virulence (i.e., which marker identifies the *Nairovirus* genus and which distinguishes CCHF or NSD from other, less important members). This will enable the following questions to be addressed: (1) Is the marker of identity in CCHF inextricably linked with the marker of virulence? (2) Can the molecular basis for pathogenesis be determined?

The four serogroups other than NSD and CCHF in this genus consist of widely spread and serologically disparate isolates; the degree of intraserogroup cross-reaction, where detected, varies greatly. Members of these serogroups have adapted to specialized ecosystems and show a concomitant variety in serologic identity, which is apparently absent in the case of CCHF and NSD. Bishop (1985) has asserted that RNA segment reassortment can occur only between closely related viruses. It is not possible to state the reasons for the apparent lack of diversity among CCHF isolates. The absence of closely related virus strains would rule out recombination, but the apparent absence of point-mutation-related antigenic change seems to argue a functional significance even for the receptor/attachment conformation. There is no evidence for the existence of distinguishable serotypes of NSD/GAN, and in this respect the serologic cross-reactions of this group with CCHF may be significant.

REFERENCES

Abdel Wahab, K. S. E., William, R. E., and Kaiser, M. N. (1970). *Fol. Parasitol.* **17,** 355–358.

Aitken, T. H. G., Jonkers, A. H., Tikasingh, E. D., and Brooke-Worth, C. (1968). *J. Med. Entomol.* **5,** 501–503.

Antoniadis, A., and Casals, J. (1982). *Am. J. Trop. Med. Hyg.* **31,** 1066–1067.

Attia, M. A. M., and Williams, R. E. (1970). *Acta. Virol.* **14,** 145–149.

Banerjee, K., Gupta, N. P., and Goverdhan, M. K. (1979). *Ind. J. Med. Res.* **69,** 363–373.

Begum, F., Wisseman, C. L., and Casals, J. (1970a). *Am. J. Epidemiol.* **92,** 195–196.

Begum, F., Wisseman, C. L., Jr., and Casals, J. (1970b). *Am. J. Epidemiol.* **92,** 197–202.

Begum, F., Wisseman, C. L., and Casals, J. (1970c). *Am. J. Epidemiol.* **92,** 192–194.

Bishop, D. H. L. (1985). In *Virology* (B. N. Fields, ed.), pp. 1083–1110, Raven Press, New York.

Bishop, D. H. L., Calisher, C. H., Casals, J., Chumakov, M. P., Gaidamovich, S. Ya., Hannoun, C., Lvov, D. K., Marshall, J. D., Oker-Blom, N., Petterson, R. F., Porterfield, J. S., Russell, P. K., Shope, R. E., and Westaway, E. G. (1980). *Intervirology* **14,** 125–143.

Boshell, J., Desai, P. K., Dandawate, C. N., and Goverdhan, M. K. (1970). *Ind. J. Med. Res.* **58,** 561–562.

Bres, P., Cornet, M., and Robin, Y. (1967). *Ann. Inst. Pasteur* **113,** 739–747.

Buckley, S. M. (1974). *Proc. Soc. Exp. Biol. Med.* **146,** 594–600.

Casals, J. (1968). *Nature* **217,** 648–649.

Casals, J. (1969). *Proc. Soc. Exp. Biol. Med.* **131,** 233–234.

Casals, J., Henderson, B. F., Hoogstraal, H., Johnson, K. M., and Shelokov, A. (1970). *J. Infect. Dis.* **122,** 437–453.

Casals, J., and Tignor, G. H. (1980). *Intervirology* **14,** 144–147.

Cash, P. (1985). *J. Gen. Virol.* **66,** 141–148.

Causey, O. R. (1970). *Am. J. Trop. Med. Hyg.* **19,** 1123–1124.

Chastel, C., Launay, H., Rogues, G., and Beaucournu, J-C. (1979a). *C. R. Acad. Sci. Paris* **288,** 559–561.

Chastel, C., Rogues, G., and Beaucournu, J-C. (1979b). *Arch. Virol.* **60,** 153–159.

Chastel, C., Bailly-Choumara, H., and Le Lay, G. (1981a). *Bull. Soc. Path. Exot.* **74,** 499–505.

Chastel, C., Monnat, J. Y., Le Lay, G., Guigen, C., Quillien, M. C., and Beaucournu, J. C. (1981b). *Arch. Virol.* **70,** 357–366.

Chastel, C., Legoff, F., and Le Lay, G. (1983). *Acta Virol.* **27,** 51–58.

Chumakov, M. P. (1974). *Med. Virusol.* **22,** 5–18 (NAMRU-3, T90).

Clerx, J. P. M., and Bishop, D. H. L. (1981). *Virology,* **108,** 361–372.

Clerx, J. P. M., Casals, J., and Bishop, D. H. L. (1981). *J. Gen. Virol.* **55,** 165–178.

Clerx-van Haaster, C. M., Clerx, J. P. M., Ushijima, H., Akashi, H., Fuller, F., and Bishop, D. H. L. (1982). *J. Gen. Virol.* **61,** 289–292.

Clifford, C. M., Thomas, L. A., Hughes, L. E., Kohls, G. M., and Philip, C. B. (1968). *Am. J. Trop. Med. Hyg.* **17,** 881–885.

Converse, J. D., Hoogstraal, H., Moussa, M. I., Stek, M., and Kaiser, M. N. (1974). *Arch. ges. Virusforsch.* **46:** 29–35.

Converse, J. D., Hoogstraal, H., Moussa, M. I., Casals, J., and Kaiser, M. N. (1975a). *J. Med. Entomol.* **12,** 202–205.

Converse, J. D., Hoogstraal, H., Moussa, M. I., Feare, C. J., and Kaiser, M. N. (1975b). *Am. J. Trop. Med. Hyg.* **24,** 1010–1018.

Converse, J. D., Hoogstraal, H., Moussa, M. I., and Evans, D. E. (1976). *Acta Virol.* **20,** 243–246.

Converse, J. D., Moussa, M. I., Easton, E. R., and Casals, J. (1981). *Trans. Roy. Soc. Trop. Med. Hyg.* **75,** 755–756.

Dandawate, C. N. (1977). *Ind. J. Exp. Biol.* **15,** 1058–1059.

Dandawate, C. N., and Shah K. V. (1969). *Ind. J. Med. Res.* **57,** 799–804.

Dandawate, C. N., Rajagopalan, P. K., Pavri, K. M., and Work T. H. (1969a). *Ind. J. Med. Res.* **57,** 1420–1426.

Dandawate, C. N., Work, T. H., Webb, J. K. G., and Shah, K. V. (1969b). *Ind. J. Med. Res.* **57,** 975–982.

Darwish, M. A., Imam, I. Z. E., Omar, F. M., and Hoogstraal, H. (1978). *Acta Virol.* **22,** 77.

Daubney, R., and Hudson, J. R. (1934). *Parasitology* **26,** 496–509.

David-West, T. S. (1970). *Br. J. Exp. Path.* **51,** 332–339.

David-West, T. S. (1974). *Microbios* **11,** 21–23.

David-West, T. S., and Porterfield, J. S. (1974). *J. Gen. Virol.* **23,** 297–307.

David-West, T. S., Cooke, A. R., and David-West, A. S. (1975). *Trans. R. Soc. Trop. Med. Hyg.* **69,** 358.

Davies, F. G., and Mwakima, F. (1982). *J. Comp. Path.* **92,** 15–20.

Davies, F. G., Casals, J., Jesset, D. M., and Ochieng, P. (1978). *J. Comp. Path.* **88,** 519–523.

Davies, F. G., Walker, A. R., Ochieng, P., and Shaw, T. (1979). *J. Comp. Path.* **89,** 587–595.

Doherty, R. L., Carley, J. G., Murray, M. D., Main, A. J., Kay, B. H., and Domrow, R. (1975). *Am. J. Trop. Med. Hyg.* **424,** 521–526.

Doherty, R. L., Carley, J. G., Filippich, C., and Kay, B. H. (1976). *Search* **7,** 484–487.

Ellis, D. S., Southee, T., Lloyd, G., Platt, G. S., Jones, Nicola., Stamford, Susan., Bowen, E. T. W., and Simpson, D. I. H. (1981). *Arch. Virol.* **70,** 189–198.

Filipe, A. R., Calisher, C. H., and Lazuick, J. (1985). *Acta Virol.* **29,** 324–328.

Ghalsas G. R., Rodrigues, S. M., Dandawate, C. N., Gupta, N. P., Khasais, C. G., Pinto, B. D., and George, S. (1981). *Ind. J. Med. Res.* **74,** 325–331.

Gould, E. A., Chanas, A. C., Buckley, A., and Varma, M. G. R. (1983). *J. Gen. Virol.* **64,** 739–742.

Hoogstraal, H. (1979). *J. Med. Entomol.* **15,** 307–417.

Hoogstraal, H., and Feare, C. J. (1984). In *Biogeography and Ecology of the Seychelles Islands* (D. R. Stoddart, ed.), pp. 267–280, Junk, The Hague, Netherlands.

Hoogstraal, H., Kaiser, M. N., and Easton, E. R. (1975). *J. Med. Entomol.* **12,** 703–704.

Hughes, L. E., Clifford, C. M., Thomas, L. A., Denmark, H. H., and Philip, C. B. (1964). *Am. J. Trop. Med. Hyg.* **13,** 118–122.

International Catalogue of Arboviruses (1985), N. Karabatsos, ed., American Society of Tropical Medicine and Hygiene, San Antonio, TX.

Johnson, B. K., Chanas, A. C., Shockley, P., Squires, E. J., Varma, M. G. R., Leake, C. J., and Simpson, D. I. H. (1977). *Acta Virol.* **12,** 428–432.

Jonkers, A. H., Casals, J., Aitken, T. H. G., and Spence, L. (1973). *J. Med. Entomol.* **10,** 517–519.

Keirans, J. E., Yunker, C. E., Clifford, C. M., Thomas, L. A., Walton, G. A., and Kelly, T. C. (1976). *Experientia* **32,** 453–454.

Lewis, E. A. (1946). *Parasitology* **37,** 55–59.

Lvov, D. K., Timofeeva, A. A., Gromashevski, V. L., Chervonsky, V. I., Gromov, A. I., Tsyrkin, Y. M., Pogrebenko, A. G., and Kostryko, I. N. (1972). *Arch. ges. Virusforsch.* **38,** 133–138.

Lvov, D. K., Sazonov, A. A., Gromashevsky, V. L., Skvortsova, T. M., Beresina, L. K., Aristova, V. A., Timofeeva, A. A., and Zakharyan, V. A. (1976). *Arch. Virol.* **51,** 157–161.

Lvov, D. K., Leonov, G. N., Gromashevsky, V. L., Skvortsova, T. M., Shestakov, V. I., Belikova, N. P., Berezina, L. K., Gofman, Yu. P., Klimenko, S. M., Safanov, A. A., and Zakaryan, V. A. (1978). *Acta virol.* **22,** 249–252.

Lvov, D. K., Kondrashina, N. G., Berezina, L. K., Skvortsova, T. M., Gushchina, E. A., Gromashevskii, V. L., Klimenko, S. M., and Guschin, B. V. (1981). *Vop. Virusol.* **2,** 148–152.

Main, A. J., Downs, W. G., Shope, R. E., and Wallis, R. C. (1976). *J. Med. Entomol.* **13,** 309–315.

Main, A. J., Kloter, K. O., Camicas, J. L., Robin, Y., and Sarr, M. (1980). *J. Med. Entomol.* **17,** 380–382.

Matthews, R. E. F. (1982). *Intervirology* **17,** 115–118.

Miller, B. R., Loomis, R., Dejean, A., and Hoogstraal, H. (1985). *Am. J. Trop. Med. Hyg.* **34,** 180–187.

Mohan Rao, C. V. R., Dandawate, C. N., Rodrigues, J. J., Prasada, Rao, G. L. N., Mandke, V. B., Ghalsasi, G. R., and Pinto, B. D. (1981). *Ind. J. Med. Res.* **74,** 319–324.

Montgomery, E. (1917). *J. Comp. Path. Ther.* **30,** 28–57.

Moore, D. L., Causey, O. R., Carey, D. E., Reddy, S., Cooke, A. R., Akinkugbe, F. M., David-West, T. S., and Kemp, G. E. (1975). *Ann. Trop. Med. Parasitol.* **69,** 49–64.

Munz, E., Reimann, M., Fritz, T. H., and Meier, K. (1984a). *Zbl. Vet. Med. B* **31,** 537–549.

Munz, E., Reimann, M., and Jager, H. (1984b). *Zbl. Vet. Med. B* **31,** 231–239.

Nuttall, P. A., Alhaq, A., Moss, S. R., Carey, D., and Harrap, K. A. *Arch. Virol.* **74,** 259–268.

Nuttall, P. A., Kelly, T. C., Carey, D., Moss, S. R., and Harrap, K. A. (1984). *Arch. Virol.* **79,** 35–44.

Nuttall, P. A., Carey, D., Moss, S. R., Green, B. M., and Spence, R. P. (1986). *J. Med. Entomol.* **23,** 437–440.

Paul, S. D., and Dandawate, C. N. (1970). *Ind. J. Med. Res.* **58,** 556–560.

Philip, B. C. (1965). *J. Parasitol.* **51,** 252.

Quillien, M. C., Monnat, J. Y., Le Lay, G., Le Goff, F., Hardy, E., and Chastel, C. (1986). *Acta Virol.* **30,** 418–427.

Radovsky, F. J., Stiller, D., Johnson, H. N., and Clifford, C. M. (1967). *J. Parasitol.* **53,** 890–892.

Rajagopalan, P. K., Sreenivasan, M. A., and Paul S. D. (1970). *Ind. J. Med. Res.* **58,** 1195–1196.

Ritter, D. G., and Feltz, E. T. (1974). *Can. J. Microbiol.* **20,** 1359–1366.

Rodhain, F., Metselaar, D., Ardoin, P., Hannoun, C., Shope, R. E., and Casals, J. (1985). *Ann. Inst. Pasteur/Virol.* **136 E,** 243–247.

Saidi, S., Casals, J., and Faghih, M. A. (1975). *Am. J. Trop. Med. Hyg.* **24,** 353–357.

Shanmugam, J., Smirnova, S. E., and Chumakov, M. P. (1976). *Ind. J. Med. Res.* **64,** 1403–1411.

Shope, R. E. (1975). In *Virology* (B. N. Fields, ed.), pp. 1055–1088, Raven Press, New York.

Simpson, D. I. H., Knight, E. M., Courtois, Gh., Williams, M. C., Weinbren, M. P., and Kibukamusoke, J. W. (1967). *E. Afr. Med. J.* **44,** 87–92.

Singh, K. R. P., and Paul, S. D. (1968). *Ind. J. Med. Res.* **56,** 815–820.

Skvortsova, T. M., Kondrashina, N. G., Gromashevskii, V. L., and Lvov, D. K. (1981). *Vop. Virusol.* **6,** 697–698.

Smirnova, S. E. (1979). *Arch. Virol.* **62,** 137–143.

Stim, T. B. (1969). *J. Gen. Virol.* **5,** 329–338.

Swanepoel, R., Struthers, J. K., Shepherd, A. J., McGillivray, G. M., Nel, M. J., and Jupp, P. G. (1983). *Am. J. Trop. Med. Hyg.* **32,** 1407–1415.

Tantawi, H. H., Al-Moslih, M. I., Al-Janabi, N. Y., Al-Bana, A. S., Mahmud, M. I. A., Jurji, F., Yonan, M. S., Al-Ani, F., and Al-Tikriti, S. K. (1980). *Acta Virol.* **24,** 464–467.

Taylor, R. M. (1970). *Am. J. Trop. Med. Hyg.* **19,** 1115–1116.

Thomas, A. L., Clifford, C. M., Yunker, C. E., Keirans, J. E., Patzer, E. R., Monk, G. E., and Easton, E. R. (1973). *J. Med. Entomol.* **10,** 165–168.

Tignor, G. H., Smith, A. L., Casals, J., Ezeokoli, G. D., and Okoli, J. (1980). *Am. J. Trop. Med. Hyg.* **29,** 676–685.

Varma, M. G. R., Bowen, E. T. W., Simpson, D. I. H., and Casals, J. (1973). *Nature* **244,** 452.
Watret, G. E., and Elliott, R. M. (1985). *J. Gen. Virol.* **66,** 2513–2516.
Watret, G. E., Pringle, C. R., and Elliott, R. M. (1984). In *Segmented Negative Strand Viruses,* (R. W. Compans and D. H. L. Bishop, eds.), pp. 349–354, Academic Press, New York, London.
Watret, G. E., Pringle, C. R., and Elliott, R. M. (1985). *J. Gen. Virol.* **66,** 473–482.
Weinbren, M. P., Gourlay, R. N., Lumsden, W. H. K., and Weinbren, B. M. (1958). *J. Comp. Path.* **68,** 174–187.
Wood, O. L., Lee, V. H., Ash, J. S., and Casals, J. (1978). *Am. J. Trop. Med. Hyg.* **27,** 600–604.
Yu-Chen, Y., Ling-Xiong, K., Ling, L., Yu-Quin, Z., Feng, L., Bao-Jian, C., and Shou-Yi, G. (1985). *Am. J. Trop. Med. Hyg.* **34,** 1179–1182.
Yunker, C. E. (1975). *Med. Biol.* **53,** 302–311.
Yunker, C. E., Clifford, C. M., Thomas, L. A., Cory, J., and George, J. E. (1972). *Acta Virol.* **16,** 415–421.

Index